Highway and Urban Environment

ALLIANCE FOR GLOBAL SUSTAINABILITY BOOKSERIES
SCIENCE AND TECHNOLOGY: TOOLS FOR SUSTAINABLE DEVELOPMENT

VOLUME 12

Series Editor: **Dr. Joanne M. Kauffman**
6–8, rue du General Camou
75007 - Paris
France
kauffman@alum.mit.edu

Series Advisory Board:

Dr. John H. Gibbons
President, Resource Strategies, The Plains, VA, USA

Professor Atsushi Koma
Vice President, University of Tokyo, Japan

Professor Hiroshi Komiyama
University of Tokyo, Japan

Professor David H. Marks
Massachusetts Institute of Technology, USA

Professor Mario Molina
Massachusetts Institute of Technology, USA

Dr. Rajendra Pachauri
Director, Tata Energy Research Institute, India

Professor Roland Scholz
Swiss Federal Institute of Technology, Zürich, Switzerland

Dr. Ellen Stechel
Manager, Environmental Programs, Ford Motor Co., USA

Professor Dr. Peter Edwards
Department of Environmental Sciences, Geobotanical Institute, Switzerland

Dr. Julia Carabias
Instituto de Ecología, Universidad Nacional Autónoma de México, México

Aims and Scope of the Series

The aim of this series is to provide timely accounts by authoritative scholars of the results of cutting edge research into emerging barriers to sustainable development, and methodologies and tools to help governments, industry, and civil society overcome them. The work presented in the series will draw mainly on results of the research being carried out in the Alliance for Global Sustainability (AGS).
The level of presentation is for graduate students in natural, social and engineering sciences as well as policy and decision-makers around the world in government, industry and civil society.

Highway and Urban Environment

Proceedings of the 8th Highway and Urban Environment Symposium

Edited by

Gregory M. Morrison
*Water Environment Technology,
Chalmers University of Technology, Goteborg, Sweden*

and

Sébastien Rauch
*Water Environment Technology,
Chalmers University of Technology, Goteborg, Sweden*

 Springer

A C.I.P. Catalogue record for this book is available from the Library of Congress.

HE308 .H54 2007
0134111924335
Highway and Urban
  Environment Symposium
Highway and urban
  environment :
      c2007.

2009 10 16

ISBN 978-1-4020-6009-0 (HB)
ISBN 978-1-4020-6010-6 (e-book)

Published by Springer,
P.O. Box 17, 3300 AA Dordrecht, The Netherlands.

www.springeronline.com

Printed on acid-free paper

All Rights Reserved
© 2007 Springer
No part of this work may be reproduced, stored in a retrieval system, or transmitted
in any form or by any means, electronic, mechanical, photocopying, microfilming, recording
or otherwise, without written permission from the Publisher, with the exception
of any material supplied specifically for the purpose of being entered
and executed on a computer system, for exclusive use by the purchaser of the work.

The Alliance for Global Sustainability

Chairman:
    Mr. Lars G. Josefsson, President and Chief Executive Officer, Vattenfall AB
AGS University Presidents:
    Prof. Hiroshi Komiyama, President, University of Tokyo
    Dr. Susan Hockfield, President, Massachusetts Institute of Technology
    Prof. Karin Markides, President, Chalmers University of Technology
    Prof. Ernst Hafen, President, Swiss Federal Institute of Technology, Zürich
Members:
    Dr. Thomas Connelly, Chief Science and Technology Officer, DuPont
    Dr. Hiroyuki Fujimura, Chairman of the Board, Ebara Corporation
    Mr. Lars Kann-Rasmussen, Director, VKR Holding A/S
    Dr. Paul Killgoar, Director, Environmental Physical Sciences & Safety, Ford Motor Company
    Mr. Masatake Matsuda, Chairman, East Japan Railway Company
    Mr. Nobuya Minami, Advisor, Tokyo Electric Power Company, Inc.
    Prof. Jakob Nüesch, Honorary Member, International Committee of the Red Cross
    Mr. Kentaro Ogawa, Chairman of the Board & CEO, Zensho Co., Ltd.
    Mr. Kazuo Ogura, President, The Japan Foundation
    Mr. Dan Sten Olsson, CEO, Stena AB
    Mr. Motoyuki Ono, Director General, The Japan Society for the Promotion of Science
    Mr. Alexander Schärer, President of the Board, USM U. Schärer Söhne AG
    Dr. Stephan Schmidheiny, President, Avina Foundation
    Mr. Norio Wada, President, Nippon Telegraph and Telephone Corporation (NTT)
    Prof. Francis Waldvogel, President, ETH Board, Switzerland
    Ms. Margot Wallström, Member of the European Commission
    Prof. Hiroyuki Yoshikawa, President, National Institute of Advanced Industrial Science and Technology
    Dr. Hans-Rudolf Zulliger, President Stiftung Drittes Millenium, Board of Directors, Amazys Ltd.

Preface

The 8th Highway and Urban Environment Symposium (8HUES) was held on 12–14 June 2006 in Nicosia, Cyprus. 8HUES was hosted in Cyprus by the Cyprus Institute. HUES is run by Chalmers University of Technology within the Alliance for Global Sustainability (AGS).

The following facts provide a background for 8HUES:

- 150 abstracts for posters and papers were accepted
- 80 delegates (24 female) attended the symposium
- 23 countries were represented, including all continents
- 71 oral presentations at the symposium
- 20 poster presentations
- 50 written manuscripts for these proceedings

HUES was initiated by Professor Ron Hamilton at Middlesex Polytechnic (now University) in the early 1980s. The initial aim was to measure and assess challenges in highway pollution. These challenges particularly included urban photochemical smog, with an emphasis on ozone formation and particle release. The first symposium was titled "Highway Pollution" and had a clear aim to make a difference. The proceedings were published in an interdisciplinary journal.

After the first symposium, the emphasis on air pollution issues continued through to Munich in 1989 where diesel particulate issues and the relevance to health through measurements of PM10 emerged. The focus on air quality issues was also strengthened by the co-organization of the symposium by Professor Roy Harrison at the University of Birmingham from 1986 to 1998. In parallel, the symposium started to receive an increasing number of scientific contributions from the area of urban runoff, indeed, to the extent that the title of the symposium was changed to "Highway and Urban Pollution". Also, at this time the importance of science in support of policy was emerging as a cornerstone, for implementation and policy issues have continued as an aspect of the symposium to the present 8HUES.

For 8HUES, we decided to evolve the name of the symposium from "Highway and Urban Pollution" to "Highway and Urban Environment". A slight, but important, change in scope and focus. We wish to give a more positive view of our common future looking to a positive environment, rather than the more negative wording – pollution. That said, papers addressing pollution issues in the highway and urban environment remain a central part of the symposium as they help to raise awareness around issues to be solved.

The Chalmers organizers worked with the Cyprus Institute to provide both a professional organization and a scientific forum for the Eastern Mediterranean. Theodoros Zachariadis from the University of Cyrpus gave an excellent plenary lecture as did Nicolas Jarraud from the UNDP on Cyprus. One of the best poster awards was won by a joint scientific team from the Greek and Turkish communities of Cyprus (UNDP-funded project) led by Professor Guenter Baumbach from Germany.

The contributions that were presented during 8HUES and that are selected in these proceedings are well poised to provide global examples of the science required to support pathways to a positive and sustainable future in the highway and urban environment.

We will give you some example areas:

1. New data on air pollutants in emerging economies was presented, but also new research ideas on particle emissions in countries where air monitoring is more established.
2. A special session provided a series of new thoughts on the platinum group elements, and even osmium as an emerging element.
3. There were an extensive number of contributions on run-off, including quality, modelling, and treatment.
4. Ground-breaking research was presented on gender issues and exhaust emissions, on the deterioration of historic monuments, large-scale noise and Internet-based and modelled/simulated travel systems.
5. Systems analysis entered the symposium for the first time and included life-cycle assessment.
6. Finally, another new area was the study of contaminated sites, research poised to improve our urban environment.

The final presentation of 8HUES was from a Brazilian group reporting on the urban atmospheric effects of the wide-scale use of ethanol in personal vehicles and underlined the importance that HUES remains a global scientific forum.

We would like to take this opportunity to thank all who have contributed to the success of 8HUES. We would especially like to acknowledge Alexandra Priatna and Christina Lundéhn at Chalmers' AGS office whose organizational skills were essential to the success of this symposium. The staff of Water Environment Technology at Chalmers provided practical support and ensured a smooth symposium. We are indebted to Michalis Yiangou and Charita Polydorou of the Cyprus Institute for providing local assistance. Malin Kylander is acknowledged for putting together the proceedings. Finally, we would like to thank the delegates for the many valuable contributions and a highly enjoyable symposium.

The next symposium, 9HUES, is planned for 9–11 June 2008 and will be held in Madrid, Spain.

Greg Morrison
Professor
Chalmers University of Technology
Göteborg, Sweden

Sébastien Rauch
Associate Professor
Chalmers University of Technology
Göteborg, Sweden

Symposium Committees

Scientific Committee:

Graeme Batley, CSIRO Energy Technology, Australia
Peter Edwards, Swiss Federal Institute of Technology, Switzerland
Ronald S. Hamilton*, Middlesex University, UK
Noboru Harata, University of Tokyo, Japan
Roy M. Harrison*, University of Birmingham, UK
Harold Hemond, Massachusetts Institute of Technology, USA
Demetrios Lalas, Greek National Observatory, Greece
Greg Morrison, Chalmers University of Technology, Sweden
Andrés Monzon, Universidad Politécnica de Madrid, Spain
Sébastien Rauch, Chalmers University of Technology, Sweden
Matheos Santamaoris, University of Athens, Greece
*Honorary members

Executive Committee:

Greg Morrison, Chalmers University of Technology, Sweden
Alexandra Priatna, Chalmers University of Technology, Sweden
Christina Lundéhn, Chalmers University of Technology, Sweden
Michalis Yiangou, The Cyprus Institute, Cyprus
Charita Polydorou, The Cyprus Institute, Cyprus

List of Participants

Alhadi Abogrean
Environmental Sciences
University of Wales Institute Cardiff
72 Wells Str. Riverside
CF11 6DY Cardiff
UK
E.Abogrean@uwic.ac.uk

Massimo Angelone
ENEA, Italian Agency for New
Technologies, Energy, and
the Environment
Via Anguillarese, 301
00060 Rome
Italy
angelone@casaccia.enea.it

Johan Åström
Water Environment Technology
Chalmers University of Technology
412 96 Göteborg
Sweden
Johan.L.Astrom@chalmers.se

Ana Estela Barbosa
Department of Hydraulics
and Environment
Portuguese National Laboratory
for Civil Engineering
1700-066 Lisboa
Portugal
aestela@lnec.pt

Guenter Baumbach
Department of Air Quality Control
Universität Stuttgart
Pfaffenwaldring 23
70569 Stuttgart
Germany
baumbach@ivd.uni-stuttgart.de

Carolien Beckx
Integral Environmental Studies
Flemish Institute for Technological
Research
Boeretang 200
2400 Mol
Belgium
carolien.beckx@vito.be

Thomas Ruby Bentzen
Department of Civil Engineering
Aalborg University
Sohngaardsholmsvej 57
DK 9000 Aalborg
Denmark
i5trb@civil.aau.dk

Fransesco Bertolini
Milano Serravalle Milano Tangenzia
Milan
Italy
ltamburini@serravalle.it

Ersever Beyaz
Air Quality Unit
Environmental Protection
Office for TCC
Öðretmenler Cad.Hacý ali Sitesi3,
C/12-Kermiya, Ortaköy
Nicosia, Mersin10
Turkey
ersever03@yahoo.co.uk

Karin Björklund
Water Environment Technology
Chalmers University of Technology
412 96 Göteborg
Sweden
karin.bjorklund@chalmers.se

Markus Boller
Urban Water Management
EAWAG
Ueberlandstrasse 133
8600 Duebendorf
Switzerland
boller@eawag.ch

Mia Bondelind
Water Environment Technology
Chalmers University of Technology
412 96 Göteborg
Sweden
mia.bondelind@chalmers.se

Carlos Borrego
Department of Environment
and Planning
University of Aveiro
Campus Universitário
3810-193 Aveiro
Portugal
borrego@ua.pt

Susanne Charlesworth
Department of Geography
Coventry University
Prior Street
CV1 5FB Coventry
UK
s.charlesworth@coventry.ac.uk

Edward Chung
Institute of Industrial Science
The University of Tokyo
4-6-1 Komaba, Meguro-ku
153-8505 Toky
Japan
edward@iis.u-tokyo.ac.jp

Antonella De Boni
Department of Environmental Science
University of Venice
Calle Larga Santa Marta, 2137
30123 Venice
Italy
deboni@unive.it

Cristina de la Rúa
Energy System Analysis
CIEMAT
Avda Complutense 22
28040 Madrid
Spain
cristina.delarua@ciemat.es

Eduardo de Miguel
School of Mines
Universidad Politecnica de Madrid
28040 Madrid
Spain
eduardo.demiguel@upm.se

Detlef Eckhardt
Institute of Mineralogy &
Geochemistry
University of Karlsruhe
Kaiserstrasse 12
D 76161 Karlsruhe
Germany
detlef.eckhardt@img.uka.de

Mikaela Eliasson
Water Environment Technology
Chalmers University of Technology
412 96 Göteborg
Sweden
mikaela.eliasson@chalmers.se

Manuela Escarameia
Engineering Hydraulics
and Structures
HR Wallingford
Howbery Park
OX10 8BA Wallingford, Oxfordshire
UK
mme@hrwallingford.co.uk

List of Participants xv

Olalekan Fatoki
Research and Development
Directorate
University of Venda
Thohoyandou 0951
Thohoyandou
South Africa
Fatoki@univen.ac.za

Jaroslav Fisak
Department of Meteorology,
Institute of Atmospheric Physic,
Academy of Science CR
1401 Bocni II, Prague 4
141 31 Prague
Czech Republic
fisak@ufa.cas.cz

John Fitzgerald
Water Environment Technology
Chalmers University of Technology
412 96 Göteborg
Sweden
john.fitzgerald@chalmers.se

Adalgiza Fornaro
Department of Atmospheric Sciences
Astronomy, Geophysics and
Atmospheric Sciences Institute
Rua do Matão, 1226, Cidade
Universitária
05508-090, São Paulo
Brazil
fornaro@model.iag.usp.br

Johan Fredriksson
Water Environment Technology
Chalmers University of Technology
412 96 Göteborg
Sweden
johfr514@student.liu.se

Reinaldo Fujii
Department of Epidemiology
University of São Paulo

Avenida Dr. Arnaldo 715, Cerqueira
Cesar sala 217
CEP: 01246-904, São Paulo
Brazil
rfujii@metrosp.com.br

Kensuke Fukushi
Integrated Research System
for Sustainability Science
The University of Tokyo
7-3-1 Hongo, 113-8654 Tokyo
Japan
fukushi@esc.u-tokyo.ac.jp

Hülya Genç-Fuhrman
Institute of Environment & Resources
Technical University of Denmark
Byg. 115
2800, Kongens-Lyngby
Denmark
hyg@er.dtu.dk

Ana Maria Graciano Figueiredo
Laboratório de análise por ativaçao
neutrônica
Instituto de Pesquisas Energéticas e
Nucleares – IPEN-CNEN/SP
Av. Prof. Lineu Prestes 2242, Cidade
Universitária
05508-000, São Paulo
Brazil
anamaria@ipen.br

Magnus Hallberg
Land & Water Resources Engineering
Royal Institute of Technology
Brinellvägen 28
100 44 Stockholm
Sweden
mhal@kth.se

Noboru Harata
Department of Urban Engineering
The University of Tokyo

113-8656 Tokyo
Japan
nhara@ut.t.u-tokyo.ac.jp

Robert Hares
Capita Symonds, UK
robert.hares@capita.co.uk

Paul Höglund
Transforsk – Transresearch
Jungfrudansen 38
17156 Solna
Sweden
pghoglund@nordplanab.se

Thorkild Hvitved-Jacobsen
Department of Life Sciences
Aalborg University
Sohngaardsholmsvej 57
9000 Aalborg
Denmark
thj@bio.aau.dk

Luc Int Panis
IMS
VITO
Boeretang 200
2400 Mol
Belgium
luc.intpanis@vito.be

Neemat Jaafarzadeh
Environmental Health
Ahwaz Joundishapour Medical Sciences University
Ahwaz University PO Box 61355-29
Ahwaz
Iran
N_jaafarzadeh@yahoo.com

Ahmand Jamrah
Sultan Qaboos University
Muscat
Sultanate of Oman
jamrah@squ.edu.om

Nicolas Jarraud
United Nation Development Programme
Cyprus
nicolas.jarraud@undp.org

Linda Katzenellenbogen
Water Environment Technology
Chalmers University of Technology
412 96 Göteborg
Sweden
linda.katzenellenbogen@chalmers.se

Michaela Kendall
Uludag University/Health Effect Institute
16059 Bursa
Turkey
michaela_kendall@yahoo.co.uk

Savvas Kleanthous
Department of Labour Inspection
Cyprus
skleanthous@dli.mlsi.gov.cy

Christina Lundéhn
Water Environment Technology
Chalmers University of Technology
412 96 Göteborg
Sweden
christina.lundehn@chalmers.se

Heidi Ina Madsen
Department of Life Sciences
Aalborg University
Sohngaardsholmsvej 57
9000 Aalborg
Denmark
him@bio.aau.dk

Claudia Martins
Departamento de ciencias e engenharia do ambiente
Faculdade de ciencias tecnologia/UNL
P-2829-516 Caparica

Portugal
claudia.martins@iambiente.pt

Andrés Monzón
TRANSyT
Universidad Politécnica de Madrid
c/ Prof Aranguren s/n
28040 Madrid
Spain
ampardeiro@caminos.upm.es

Greg Morrison
Water Environment Technology
Chalmers University of Technology
412 96 Göteborg
Sweden
greg.morrison@chalmers.se

Marina Neophytou
Department of Civil and
Environmental Engineering
University of Cyprus
75 Kallipoleos Ave, PO Box 20537
1678, Nicosia
Cyprus
neophytou@ucy.ac.cy

Asbjørn Haaning Nielsen
Department of Life Sciences
Aalborg University
Sohngaardsholmsvej 57
9000 Aalborg
Denmark
ahn@bio.aau.dk

Malin Norin
NCC Construction Sverige AB
405 14 Göteborg
Sweden
malin.norin@ncc.se

Stefan Norra
Institute of Mineralogy &
Geochemistry
Universität Karlsruhe
Kaiserstrasse 12
76128 Karlsruhe
Germany
stefan.norra@img.uka.de

George Papageorgiou
Department of Mathematics
and Statistics
University of Cyprus
PO Box 20537
1678 Nicosia
Cyprus
gnp@ucy.ac.cy

Pedro José Pérez-Martinez
TRANSyT
Universidad Politécnica de Madrid
c/ Prof Aranguren s/n
28040 Madrid
Spain
pjperez@caminos.upm.es

Gustavo Perrusquia
Water Environment Technology
Chalmers University of Technology
412 96 Göteborg
Sweden
gustavo.perrusquia@wet.chalmers.se

Thomas Pettersson
Water Environment Technology
Chalmers
Sven Hultins gata 8
41296 Göteborg
Sweden
thomas.pettersson@chalmers.se

Vincent Pettigrove
Research and Technology
Melbourne Water
PO Box 4342
3001 Melbourne
Australia
vin.pettigrove@melbournewater.com.au

Magnus Quarshie
Centre for Cycling Expertise
Ghana Institution of Engineers,
Institute for transportation engineers
PO Box BT 446 Tema
Ghana
delin@idngh.com

Paulo J Ramisio
Departamento de Engenharia
Civil-Campus de Gualtar
Universidade do Minho
4704-533 Braga
Portugal
pramisio@civil.uminho.pt

Sébastien Rauch
Water Environment Technology
Chalmers University of Technology
Sven Hultins gata 8
41296 Göteborg
Sweden
sebastien.rauch@wet.chalmers.se

Gunno Renman
Land & Water Resources Engineering
Royal Institute of Technology
Brinellvägen 28
SE-100 44 Stockholm
Sweden
gunno@kth.se

Stephanie Roulier
Institute of Terrestrial Ecology
ETHZ
Universitätstrasse 16
8092 Zürich
Switzerland
Stephanie.Roulier@env.ethz.ch

Lian Scholes
School of Health and Social Sciences
Urban Pollution Research Centre

Middlesex University, Queensway
EN3 4SA Enfield
UK
L.Scholes@mdx.ac.uk

Lars-Ove Sörman
Water Environment Technology
Chalmers University of Technology
412 96 Göteborg
Sweden
lars-ove.sorman@chalmers.se

Anderson Spohr Nedel
Department of Atmospheric Sciences
University of São Paulo
Avenida Dr. Arnaldo 715, Cerqueira
Cesar sala 216
CEP: 01246-904 São Paulo
Brazil
asnedel@model.iag.usp.br

Michele Steiner
Urban Water Management
EAWAG
Ueberlandstrasse 133
8600 Duebendorf
Switzerland
michele.steiner@eawag.ch

Ann-Margret Strömvall
Water Environment Technology
Chalmers University of Technology
412 96 Göteborg
Sweden
ann-margret.stromvall@chalmers.se

Doris Stüben
Inst. Für mineralogie und geochemie
Universität Karlsruhe
Kaiserstrasse 12
76187 Karlsruhe
Germany
doris.stueben@img.uni-karlsruhe.de

Peter Suppan
Institute for Meteorology
Atmospheric Environmental Research
(IMK_IFU)
Forschungszentrum Karlsruhe GmbH
Kreuzeckbahnstr. 19
82467 Garmisch-Partenkirchen
Germany
peter.suppan@imk.fzk.de

Fabio Luiz Teixeira Gonçalves
Department of Atmospheric Sciences
University of São Paulo
Rua do Matão 1226
05508-090 São Paulo
Brazil
fgoncalv@model.iag.usp.br

Marek Vach
Laboratory of Environmental
Geochemistry
Institute Of Geology, Academy
of Science, Czech Republic
Rozvojová 135, Praha 6
CZ-165 02 Prague
Czech Republic
vach@gli.cas.cz

Erik Verhaeven
Vehicle Technology, Flemish Institute
for Technological Research
Boeretang 200
2400 Mol
Belgium
erik.verhaeven@vito.be

Sophie Veschambre
LCABIE, UMR CNRS 5034 UPPA
The University of Pau
Hélioparc, 2 avenue Pierre Angot
64053 Pau cedex 9

France
sophie.veschambre@univ-pau.fr

Jes Vollertsen
Dept of Life Sciences
Aalborg University
Sohngaardsholmsvej 57
DK 9000 Aalborg
Denmark
jv@bio.aau.dk

Neil Ward
Department of Chemistry
University of Surrey
Staghill
GU2 7XH Guildford
UK
n.ward@surrey.ac.uk

John Watt
Centre for Decision Analysis and Risk
Management
School of Health and Social Sciences
Middlesex University
Queensway, Enfield
UK
j.watt@mdx.ac.uk

Britt-Marie Wilén
Water Environment Technology
Chalmers University of Technology
412 96 Göteborg
Sweden
britt-marie.wilen@chalmers.se

Ronald Wyzga
Department of environment
EPRI
3412 Hillview Ave. PO Box 10412
CA 94303-0813, Palo Alto
USA
rwyzga@epri.com

Yvonne Young
Water Environment Technology
Chalmers University of Technology
412 96 Göteborg
Sweden
yvonne.young@chalmers.se

Theodoros Zachariadis
Economics Research Centre
University of Cyprus
PO Box 20537
1678 Nicosia
Cyprus
t.zachariadis@ucy.ac.cy

Contents

I. Vehicle Consumption and Emissions and Traffic Management 1

Engineering–economic simulations of sustainable transport policies
T Zachariadis 3

Life cycle environmental benefits of biodiesel production and use in Spain
C de la Rúa, Y Lechón, H Cabal, C Lago, L Izquierdo, R Sáez 13

Engine management for Flex Fuel plus compressed natural gas vehicles
O Volpato, F Theunissen, R Mazara, E Verhaeven 23

Mobility and environment in Spain
PJ Pérez-Martínez 35

Influence of gear-changing behaviour on fuel use and vehicular exhaust emissions
C Beckx, L Int Panis, I De Vlieger, G Wets 45

Effect of speed reduction on emissions of heavy duty lorries
L Int Panis, I De Vlieger, L Pelkmans, L Schrooten 53

Gender-linked disparity in vehicle exhaust emissions? Results from an activity-based survey
C Beckx, L Int Panis, M Vanhulsel, G Wets, R Torfs 63

Floating Automotive Data Collection
A Schaerf, S Kumra, R Mazara, L Pelkmans, E Verhaeven 71

Reducing car trip and pollutant emissions through strategic transport planning in Madrid, Spain
A Monzón, AM Pardeiro, LA Vega 81

Evaluation of car control measures based on an Internet-based travel survey system
N Harata, S Aono 91

Integrating cycling in Bus Rapid Transit system in Accra
ML Quarshie 103

II. Air Pollution and Air Quality 117

Evaluation of hydrogen peroxide in rainwater in downtown São Paulo
MA Santos, JJ Pedrotti, A Fornaro 119

The comparison of pollutant concentrations in liquid falling and deposited precipitation, and throughfall
J Fisak, P Chaloupecky, D Rezacova, M Vach, P Skrivan, J Spickova 129

Wet deposition at Llandaff station in Cardiff
E Abogrean, G Karani, J Collins, R Cook 143

Monitoring the atmospheric deposition of particulate-associated urban contaminants, Coventry, UK
SM Charlesworth, C Booty, J Beasant 155

Size, morphological and chemical characterization of aerosols polluting the Beijing atmosphere in January/February 2005
S Norra, B Hundt, D Stüben, K Cen, C Liu, V Dietze, E Schultz 167

Air pollution levels in two São Paulo subway stations
RK Fujii, P Oyola, JCR Pereira, AS Nedel, RC Cacavallo 181

Air quality nearby different typologies of motorways: Intercomparison and correlation
CS Martins, F Ferreira 191

Assessment of air pollution in the vicinity of major alpine routes
P Suppan, K Schäfer, J Vergeiner, S Emeis, F Obleitner, E Griesser 203

The relative impact of automobile catalysts and Russian smelters on PGE deposition in Greenland
S Rauch, J Knutsson 215

Cultural heritage stock at risk from air pollution
J Watt, E Andrews, N Machin, R Hamilton, S Beevers, D Dajnak, X Guinart, P de la Viesca Cosgrove. 223

III. Contaminated Environments and Remediation 233

Organic contaminants in urban sediments and vertical leaching in road ditches
A-M Strömvall, M Norin, TJR Pettersson 235

The use of an epiphyte (*Tillandsia usneoides* L.) as bioindicator of heavy metal pollution in São Paulo, Brazil
AMG Figueiredo, CA Nogueira, B Markert, H Heidenreich, S Fränzle, G Liepelt, M Saiki, M Domingos, FM Milian, U Herpin 249

On-line matrix separation for the determination of PGEs in sediments by ICP-MS
A De Boni, W Cairns, G Capodaglio, P Cescon, G Cozzi, S Rauch, HF Hemond, C Boutron, C Barbante. 259

Determination of PGE and REE in urban matrices and fingerprinting of traffic emission contamination
M Angelone, F Spaziani, C Cremisini, A Salluzzo 271

Sorption behaviour of Pt, Pd, and Rh on different soil components: results of an experimental study
J Dikikh, J-D Eckhardt, Z Berner, D Stüben 283

Reactive soil barriers for removal of chromium(VI) from contaminated soil
A-M Strömvall, M Norin, H Inanta 295

Cleaning of highway runoff using a reactive filter treatment plant – a pilot-scale column study
G Renman, M Hallberg, J Kocyba 309

Heavy metal removal efficiency in a kaolinite–sand media filtration pilot-scale installation
PJ Ramísio, JMP Vieira 319

IV. Storm Water 331

Site assessment of road-edge grassed channels for highway drainage
M Escarameia, AJ Todd 333

Evaluation of the runoff water quality from a tunnel wash
AE Barbosa, J Saraiva, T Leitão 345

An investigation of urban water and sediment ecotoxicity in relation to metal concentrations
L Scholes, R Mensah, DM Revitt, RH Jones 359

Establishing a procedure to predict highway runoff quality
in Portugal
AE Barbosa 371

A field microcosm method to determine the impact
of sediments and soils contaminated by road runoff
on indigenous aquatic macroinvertebrates
V Pettigrove, S Marshall, B Ryan, A Hoffmann 385

Assessment of storm water ecotoxicity using a battery
of biotests
L Scholes, A Baun, M Seidl, E Eriksson, M Revitt, J-M Mouchel 399

Is catchment imperviousness a good indicator of ecosystem health?
V Pettigrove 411

V. Storm Water Treatment 427

Evolution on pollutant removal efficiency in storm water
ponds due to changes in pond morphology
TJR Pettersson, D Lavieille 429

Characterization of road runoff and innovative treatment
technologies
M Boller, S Langbein, M Steiner 441

Development and full-scale implementation of a new
treatment scheme for road runoff
M Steiner, S Langbein, M Boller 453

Reactive filters for removal of dissolved metals
in highway runoff
M Hallberg, G Renman 465

Designing filters for copper removal for the secondary
treatment of storm water
H Genç-Fuhrman, P Steen Mikkelsen, A Ledin 475

Modelling the oxygen mass balance of wet detention
ponds receiving highway runoff
HI Madsen, J Vollertsen, T Hvitved-Jacobsen 487

Monitoring and modelling the performance of a wet pond for
treatment of highway runoff in cold climates
*J Vollertsen, SO Åstebøl, JE Coward, T Fageraas, HI Madsen,
AH Nielsen, T Hvitved-Jacobsen* 499

Can we close the long-term mass balance equation for
pollutants in highway ponds?
TR Bentzen, T Larsen, MR Rasmussen 511

VI. Environmental Assessment and Effects 521

Cardiovascular and respiratory variability related to air
pollution and meteorological variables in Oporto,
Portugal – preliminary study
JM Azevedo, F Gonçalves, AR Leal 523

Component analysis on respiratory disease variability
at São Paulo
FLT Gonçalves, MS Coelho, MRD Latorre 535

Management and optimization of environmental information
using integrated technology: a case study in the city of
São Paulo, Brazil
AS Nedel, P Oyola, RK Fujii, RC Cacavallo 543

Using Bayesian inference to manage uncertainty in
probabilistic risk assessments in urban environments
E Chacón, E De Miguel, I Iribarren 551

An assessment framework for urban water systems – a
new approach combining environmental systems with
service supply and consumer perspectives
C Lundéhn, GM Morrison 559

Large area noise evaluation
E Chung, A Bhaskar, M Kuwahara 579

I. Vehicle Consumption and Emissions and Traffic Management

Engineering–economic simulations of sustainable transport policies

T Zachariadis

Economics Research Centre, University of Cyprus

Abstract

The paper describes an engineering–economic model of the transport sector in the European Union (EU). The model starts with a simulation of the economic behaviour of consumers and producers in order to calculate the modal split and allocate the vehicle stock into vintages. Then a technology-oriented algorithm calculates air pollutant and greenhouse gas (GHG) emissions. The paper describes briefly the methodological approach, but its main focus is the presentation of results from scenarios that attempt to simulate sustainable transport policies. A major conclusion is that individual measures cannot address sufficiently the sustainability concerns. Therefore, policies combining technical and fiscal measures would be most appropriate.

Introduction

Sustainable development is a concept that for a long time had been accepted as an important ingredient in formulating long-term strategies. Apart from atmospheric pollution concerns in urban areas as well as climate change, the emergence of a great number of additional sustainability concerns in recent years (concerning bio-diversity, transport congestion, social exclusion, regional imbalances with their attendant political risks, etc.) has posed particular challenges to analysts with respect to integration and quantification of these problems. One of the major issues in this agenda is transport, which is worldwide accepted as a priority area in sustainability discussions [1,2]. Work on sustainable transport is well in progress, both in the research field and in policy-oriented studies, concentrating primarily on emissions of air pollutants (causing health problems) and greenhouse gases (GHGs) (affecting climate change), and expanding to other sustainability concerns such as congestion, noise, and accidents.

In order to respond to the requirements of a European research study that was wide-ranging in scope and addressed other sustainable development concerns as well (MINIMA–SUD project under the European Commission's 5th Framework Programme), a transport simulation and policy assessment tool had to be used. Based on the experience collected so far, a new model was developed, which is described in more detail elsewhere [3]. The second section of this paper provides a brief overview of the methodological approach. The third section, which is the paper's main focus, describes the assumptions and results of several scenario runs that attempt to simulate policies aiming to promote sustainable transport. Items of future work are outlined in the concluding section.

Methodological overview

The model covers the whole transport sector (road and rail transport, inland shipping and aviation) in the 15 countries that were European Union (EU) member states in the beginning of 2004. It covers the major sustainability issues associated with the transport sector and is designed so as to enable integrated policy exploration jointly with models addressing different aspects of sustainability. Its underlying database is consistent with those of EU institutions such as the EU's statistical service (Eurostat), the European Commission, and the European Environment Agency (EEA). It links the choice of transport mode and technologies with economic variables such as income and generalized transport costs, but at the same time is technology-rich so as to simulate sustainability impacts in a sufficiently precise manner. Depending on the values of exogenously set policy variables such as taxes, regulations, or infrastructure investments, the demand and supply modules interact in order to reach equilibrium, where generalized prices satisfy demand and supply for transport services. The outcome of this iterative process is the determination of the overall demand for passenger and freight transport, travelling speeds, and associated costs. These variables are then fed in the next modules, where the vehicle stock is calculated and allocated into vintages, technology shares are assessed, and the sustainability-related impacts are determined.

A major assumption made in the model is that the choice of transport mode, both for consumers (as regards passenger transport) and for producers (as regards freight transport), is based on economic considerations. More specifically, a microeconomic optimization framework is assumed: consumers will allocate their income expenditures for passenger mobility to different transport modes so as to maximize their overall welfare (or utility). Similarly, producers will allocate their expenditures for the transport of goods to different freight transport modes so as to minimize their total costs. Consumer and producer choices are described as a series of separable choices, which create a nesting structure (decision tree). All utility/cost functions of the decision tree, which describe the behaviour of economic agents, were assumed to have constant elasticities of substitution (CES) functions.

Having determined the transport activity evolution by transport mode, vehicle size, and road type and the associated costs, the model becomes more detailed for road transport. It calculates the evolution of the road vehicle stock and the distance travelled annually by vehicle type and size and by road type. This stock is further decomposed into age cohorts, according to an initial exogenous age distribution in the base year and assumptions on the evolution of scrapping rates. Following the assessment of the age distribution for a given year, technology shares are calculated. These depend both on European emissions legislation and technology-related parameters such as total vehicle travel costs, availability of a given alternative technology/fuel, and policy measures assessed in a specific scenario. The model includes the 113 technology classes of the COPERT III methodology [4], which is the state-of-the-art methodology for vehicle emissions calculation in Europe, as well as several alternative road vehicle technologies/fuels. Air pollutants covered are CO, non-methane volatile organic compounds (NMVOC), oxides of nitrogen (NO_x), particulate matter (PM), SO_2, and lead. GHGs addressed are CO_2, CH_4, and N_2O. Simpler approaches are applied for non-road transport modes. Sustainability indicators other than air emissions are assessed with the aid of simple methodologies, as large-scale transport policy assessment studies often do not require more detailed coverage of these issues.

The model requires a comprehensive set of technical and economic data for its application. Moreover, in order to be acceptable for policy-making purposes across Europe, it uses international data sources and established forecasts as a basis for both base year data and baseline transport activity and emissions outlook. The two major statistical sources for base year calibration are energy and fuel price data from Eurostat and European Commission's statistical pocketbook data on passenger kilometre/ton kilometre (pkm tkm^{-1}) and vehicle stock for the year 2000. Numerous additional data sets have been evaluated and are used where necessary, with information coming from the European Conference of the Ministers of Transport, the United Nations Economic Commission for Europe, the International Energy Agency, the International Public Transport Union, the Association of European Airlines, and many research projects. Zachariadis [3] provides an extensive list of data sources that were used. In order to prepare the data for their meaningful use within the model, an extensive calibration procedure is conducted.

Simulation of sustainable transport policies

In order to assess the impact of policies on transport-related sustainability indicators, a number of policy instruments were selected, shocks (i.e., large "doses" of each instrument) were applied, and their impact on these indicators was evaluated. Ten scenarios were selected so that: (1) several types of policy instruments are examined and (2) the applied instruments sometimes yield conflicting impacts on sustainability indicators.

In *Policy 1*, vehicles powered with compressed natural gas (CNG) and fuel cells are subsidized by 50% of their pre-tax purchase cost. This applies to cars, light, and heavy trucks alike. At the same time, fuel supply is assumed to progress considerably in order to respond to the increasing demand for CNG and methanol. In *Policy 2*, the tax imposed on diesel fuel used by private vehicles (i.e., cars and trucks) is doubled, a measure directed primarily towards curbing PM emissions. According to *Policy 3*, from 2006 onwards, "Euro V" emission standards are implemented in cars and light trucks instead of "Euro IV" ones, but their purchase and maintenance costs are 40% higher because of their "premature" introduction in the market. "Euro V" vehicles are assumed to be 10% more fuel efficient and to emit 24–50% less NO_x, NMVOC, and PM than the corresponding "Euro IV" technologies. The justification for assuming the corresponding fuel economy improvements and increases in car purchase prices is provided in [3]. *Policy 4* assumes that investment expenditure for road infrastructure doubles throughout the outlook period; this applies to both the urban and non-urban road network. In *Policy 5*, fares of all public transport modes (buses, trams, metro, and trains) are subsidized by 50%. According to *Policy 6*, all automobile trips in urban areas are charged on average with additional 3 euros (in 2000 prices); the latter measure would correspond, e.g., to an increase of 5–10 euros of car-parking prices per day (depending on the fraction of cars that have to pay for their parking space in each city), or to a charge of 3–8 euros per trip in a road pricing scheme (depending on the share of residents of urban areas who may not be affected by such a charge). *Policy 7* assumes that, in an attempt to accelerate scrapping of old cars and renew the vehicle fleet, the pre-tax purchase cost of all new passenger cars that replace old ones is subsidized by 50%.

Besides these exercises, three additional scenarios were applied with combinations of some of the above instruments. *Policy 8* is a combination of road pricing and advanced emission standards (i.e., of scenarios 3 and 6). *Policy 9* assumes, on top of *Policy 8*, subsidies on CNG and fuel cell vehicles, thus combining scenarios 1, 3, and 9 by assuming that revenues raised from road user charges are used in order to boost the use of alternative propulsion systems and fuels. Finally, *Policy 10* combines road pricing (scenario 6), advanced emission standards (scenario 3), and subsidies on bus and rail fares (scenario 5), thereby assuming that revenues collected by discouraging private vehicle use are directly utilized to support public transport modes.

Some impacts of these policies on sustainability indicators, expressed as percentage changes from the baseline run, are shown in Table 1 (results of the baseline run were presented in [3]). The following paragraphs provide some comments on these results.

Policy 1 causes negligible changes in generalized prices, aggregate transport activity, and speeds. The major change is observed in the fuel mix after 2010, with CNG and fuel cell vehicles emerging dynamically. By 2020, alternative fuel vehicles will account for about 3%, 18%, 11%, and 29% of the total fleet of cars, buses, light trucks, and heavy trucks, respectively. This results in reduced consumption of conventional fuels and increases in demand for CNG and methanol. The decline of CO_2 emissions is projected to be limited. NO_x emissions are expected to fall slightly; emissions of other pollutants will decline much more as neither CNG nor fuel cell vehicles emit NMVOC, PM, SO_2, or lead.

Despite the considerable rise in the costs of diesel vehicles, *Policy 2* causes an overall increase of less than 2% in user costs of passenger cars and 2–10% in those of trucks. Passenger kilometres of diesel cars decrease by 3–14% (mainly under urban peak driving conditions), but gasoline cars benefit most from this decline, so that total pkm of cars fall negligibly. As the use of trucks is largely inelastic to cost increases, road freight tkm decreases by only 2–7% to the benefit of rail. Energy demand and CO_2 emissions decline slightly due to the higher transport price. Because of the increase in

Table 1. Results of policy scenarios, expressed as relative change in sustainable development indicators compared to the baseline scenario in the year 2020

Impact of Policy Exercises on Sustainable Development Indicators in the Year 2020 (%)

Indicator	Policy 1	Policy 2	Policy 3	Policy 4	Policy 5	Policy 6	Policy 7	Policy 8	Policy 9	Policy 10
Passenger Transport Intensity (pkm GDP^{-1})	0.1	−0.5	−3.6	1.2	2.1	−2.6	0.0	−6.1	−9.2	−3.6
Freight Transport Intensity (tkm GDP^{-1})	0.0	−2.5	−0.2	2.2	0.1	−0.1	0.1	−0.4	−0.3	−0.4
Energy Intensity of Transport (toe GDP^{-1})	−1.8	−1.2	−8.2	1.9	0.8	−3.8	−0.1	−11.7	−14.6	−10.7
CO2 Emissions of Transport (Mt)	−1.5	−1.4	−8.3	1.9	0.4	−3.8	−0.1	−11.7	−14.2	−11.1
NOx Emissions, Total (kt)	−2.0	−3.6	−7.8	0.6	1.0	−1.9	−0.6	−9.8	−11.9	−11.3
NOx Emissions, Urban (kt)	−3.6	−4.6	−12.6	−1.4	1.8	−6.4	−0.9	−18.0	−23.2	−15.4
NMVOC Emissions, Total (kt)	−4.6	1.3	0.8	0.4	0.1	−4.6	−0.7	−4.4	−11.7	−4.2
NMVOC Emissions, Urban (kt)	−3.3	1.6	1.6	0.8	0.1	−6.1	−0.8	−5.1	−11.5	−5.0
PM Emissions, Total (kt)	−5.7	−7.8	−12.6	1.0	0.2	−5.0	−0.7	−17.5	−26.5	−17.0
PM Emissions, Urban (kt)	−6.0	−8.9	−13.6	−2.3	0.2	−9.7	−0.7	−22.1	−32.2	−21.5
SO2 Emissions, Total (kt)	−0.7	−4.4	0.3	−2.8	0.3	−0.4	−0.1	0.1	−0.2	0.6
SO2 Emissions, Urban (kt)	−7.4	−1.2	−10.2	−0.8	0.4	−9.1	−0.1	−18.2	−28.0	−17.4
Lead Emissions, Total (kt)	−2.2	3.3	−12.6	1.9	−0.8	−8.2	−0.1	−20.0	−26.9	−20.7
Lead Emissions, Urban (kt)	−2.6	4.0	−11.6	−0.5	−1.2	−15.0	−0.1	−24.9	−31.0	−25.9
Urban Road Congestion (travel time in hours)	−0.1	−0.6	−1.3	−5.7	−0.1	−3.5	0.0	−4.7	−5.9	−4.7
Noise Emissions in db(A)	0.0	−0.1	−0.5	1.6	0.0	0.0	0.0	−0.5	−0.9	−0.5
Fatalities from Road Accidents (000)	−0.1	1.1	3.5	−2.0	0.9	15.5	0.0	19.6	24.0	20.8

The accelerated introduction of "Euro V" technology (*Policy 3*) at higher purchase costs is projected to make car travel more expensive, particularly in non-urban areas, where capital costs account for the major part of total costs. This will slightly improve congestion levels, which is expected to affect particularly freight transport where time costs dominate. Buses and rail in urban areas are projected to gain about 4–5% after 2020. In non-urban transport, high-speed rail and aviation are expected to benefit most. Energy consumption will fall substantially, with some switch to alternative fuels. Air pollutant emissions (particularly NO_x and PM) will also fall significantly due to lower "Euro V" emission levels and the shift towards public transport.

In *Policy 4*, driving becomes somewhat cheaper and time spent in urban driving declines by about 6% throughout the outlook period. The effects are not as pronounced as might be expected because of a "rebound effect": improved congestion conditions reduce time and fuel costs, make car driving more attractive, and lead to even higher passenger and freight transport intensity and energy demand. Pollutant emissions change very little. Despite higher driving speeds, which would normally lead to more road accidents, better infrastructure improves road safety, so that the number of deaths from road accidents remains about the same as in the baseline.

Heavy subsidies of public transport fares (*Policy 5*) render these transport modes much more attractive, but this induces an overall increase in the use of trains and buses without affecting the use of cars significantly. Although pkm of buses and rail increase in total by 11–16% and 27–29%, respectively over the outlook period, pkm of cars decline only by approximately 1%. Here again, the "rebound effect" is evident: improved congestion conditions encourage private car travel, so that the overall impact on car use is very small. Emissions remain essentially unchanged.

Imposing urban road user charges (*Policy 6*) produces more remarkable effects in the first years of implementation. Urban travel costs rise considerably, thus reducing congestion by over 4% and increasing transport activity of urban buses and tram/metro by more than 25% and 15%, respectively in the 2010–2020 period. A marked improvement is forecast for peak-hour driving speeds, which leads to fuel savings and to a reduction of 4.4% in energy demand and CO_2 emissions by 2010. In the absence of measures to reduce emissions from diesel bus engines, pollutant emissions are projected to fall moderately.

Policy 7 does not affect travel costs, car ownership, and car use, as the subsidies address only those cars that enter the market in replacement of old ones. However, it causes a significant acceleration in the renewal of the car park, so that scrapping rates increase by 15–20%. As a result of the faster penetration of "Euro IV" cars, energy use and pollutant emissions decrease, particularly in the 2006–2015 period. The overall effect is limited though, as the subsidies apply to passenger cars only; emission levels of trucks and aviation remain unchanged.

The impacts of scenarios 3 and 6 are effectively added up in the case of *Policy 8*. More substantial improvements in energy intensity and CO_2 emissions (–12% from 2020 onwards) and pollutant emissions (up to –25% for urban PM and lead emissions) are achieved in this way. However, as average speeds rise, the impact on accident rates is also added up: road fatalities are 20% more than in the baseline.

Policy 9 is an additional step further: all sustainability indicators except road fatalities improve more than in scenario 8. Passenger transport intensity becomes 9% lower after 2015. After 2020, penetration of alternative fuels will result in over 20% lower energy demand and CO_2 emissions. The improvement is more pronounced in emissions of PM, SO_2, and lead. On the other hand, lower congestion levels are expected to give rise to more accidents and fatalities than in the baseline. Figure 1 illustrates the cumulative impact of the three distinct policy instruments that have been simulated in this scenario on transportation energy demand and NO_x emissions. It is evident that the introduction of "Euro V" standards accounts for the largest part of the improvements compared to the baseline, whereas road pricing and alternative fuel subsidies yield together about as much improvement as "Euro V" standards on their own.

Finally, the effectiveness of *Policy 10* is somewhat lower than that of Policy 9, because the use of passenger cars falls less remarkably due to the "rebound effect" mentioned in the case of scenario 5.

Conclusions and outlook

The policy scenarios presented here reconfirm the widely expressed assertion that individual policy measures cannot respond sufficiently to the diverse sustainability concerns associated with transportation [1,2]. In order to achieve improvements in energy intensity, CO_2 emissions, congestion, and air pollutant emissions, packages of measures are necessary. Strategies that promote advanced technologies can mainly affect air pollution and to a lesser extent energy demand, whereas traffic-related measures can primarily improve congestion and thus energy intensity and emissions as long as appropriate clean technologies are in place. Therefore, a suite of policies

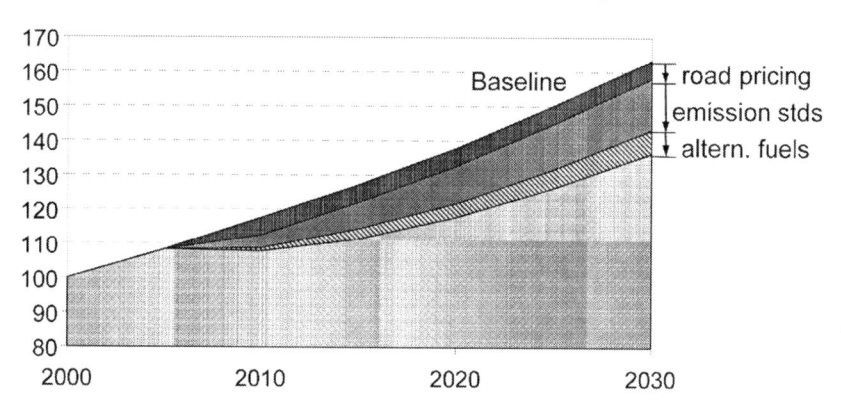

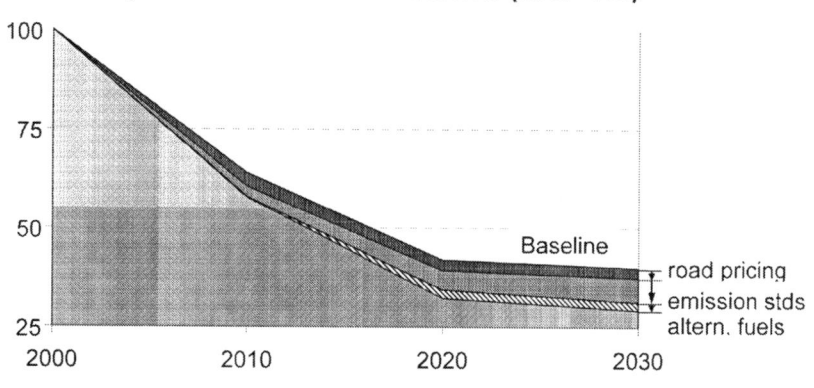

Fig. 1. Cumulative impact of road pricing, emission standards, and alternative fuel subsidies on the evolution of energy demand and urban NO$_x$ emissions, 2000–2030

combining promotion of advanced "conventional" technologies and alternative fuels with interventions to reduce demand for transport would be most suitable to address the variety of sustainability issues.

For a comprehensive analysis of policy options, the social cost of measures has to be assessed with the aid of appropriate economic methods (such as the welfare losses or gains because of changes in consumer/producer surplus), and this is a future direction for research.

Acknowledgements

The major part of the analysis and model development presented here was funded by the European Commission within the MINIMA–SUD (Methodologies to Integrate Impact Assessment in the Field of Sustainable Development) project of the 5th Framework Programme on Research and Technological Development, although model assumptions and results reported in this paper are basically different from those produced in the project. The author wishes to thank Nikos Kouvaritakis and Nikos Stroblos for their substantial guidance and suggestions in model development.

References

1. European Commission (2001) White Paper on European transport policy up to 2010: time to decide. Brussels. Also at http://europa.eu.int/comm/energy_transport/en/lb_en.html
2. World Business Council for Sustainable Development (2004) Mobility 2030: meeting the challenges to sustainability. Geneva. Also at http://www.wbcsd.org
3. Zachariadis T (2005) Assessing policies towards sustainable transport in Europe: an integrated model. Energ Policy 33:1509–1525
4. Ntziachristos L, Samaras Z (2000) COPERT III – Computer programme to calculate emissions from road transport – Methodology and emission factors. Technical Report No. 49. Copenhagen: European Environment Agency. Also at http://reports.eea.eu.int/Technical_report_No_49/en

Life cycle environmental benefits of biodiesel production and use in Spain

C de la Rúa, Y Lechón, H Cabal, C Lago, L Izquierdo, R Sáez

CIEMAT. Energy Department, Avda. Complutense 22, 28040, Madrid Spain. Fax: 00 34 913466005

Abstract

The Spanish Ministry of Environment has launched forth into a project related to the evaluation of the environmental impacts of biodiesel production in Spain, in order to support its biofuels promotion policies. The objective of this project is to evaluate the environmental impacts of four different fuels composed by biodiesel from several crude vegetable oils and waste vegetable oils in comparison to diesel EN-590 along their whole life cycle, and to identify the opportunities to reduce the environmental impacts. Biodiesel is produced from sunflower oil, soybean oil, rapeseed oil, and palm oil, and also from waste vegetable oils. The transformation technology is that currently used in the Spanish biodiesel plants.

Introduction

The Spanish Renewable Energy Plan 2005–2010 and the 2003/30/EC Directive of the European Parliament in relation to the promotion of the use of biofuels and other renewable fuels have generated an important public discussion on the real environmental benefits of these biofuels compared to the conventional fuels used in the transport activities.

The necessary increase in the use of biofuels in order to reach the objectives described in the directive and to achieve the Kyoto Protocol commitments, should be accompanied by detailed analysis of the produced environmental impacts in order to know which fuels should be promoted and the best way to do it.

The environmental impacts of the production of biofuels depend, to a large extent, on the particular conditions in which they are produced. These conditions vary in each country due to the use of different raw materials, transformation processes, and the end-use technologies and blends. The Spanish Ministry of Environment has launched forth into a project related to the evaluation of the environmental impacts of biofuels production in Spain. Bioethanol production in Spain was analysed in a previous stage. In this project, environmental impacts of biodiesel production in Spain have been studied. The study presented in this paper has been carried out and financed within the framework of this initiative.

Methodology

Life cycle assessment

Life cycle assessment has been chosen as the methodology to evaluate the environmental loads of the studied fuels. This methodology is described in the international standards series ISO 14040. The main stages in the methodology are: (1) goal definition and scoping, (2) inventory analysis, (3) impact assessment, and (4) interpretation.

Goal definition and scoping
The objectives of the study are to:

- Evaluate the environmental impacts of different blends of biodiesel from crude vegetable oils with diesel EN-590 and biodiesel of waste vegetable oil also with diesel EN-590, in comparison to those of diesel EN-590 along their whole life cycle.
- Identify and evaluate the opportunities to reduce the environmental impacts along the life cycle of the studied fuels.

The studied systems are defined as:

- System BD10A1: 10% biodiesel from crude vegetable oils (10% sunflower oil, 40% soybean oil, 25% rapeseed oil, and 25% palm oil) and 90% diesel EN-590
- System BD100A1: 100% biodiesel from crude vegetable oils (10% sunflower oil, 40% soybean oil, 25% rapeseed oil, and 25% palm oil)
- System BD10A2: 10% biodiesel from waste vegetable oils and 90% diesel EN-590

- System BD100A2: 100% biodiesel from waste vegetable oils
- System Diesel EN-590: 100% diesel EN-590

Diesel EN-590 has been considered to have a 5% of biodiesel from crude vegetable oils, also in the blends $BD10_{A1}$ and $BD10_{A2}$.

The functional unit (FU) has been defined as the amount of each fuel, expressed in megajoules (MJ), needed to drive 1 km in a diesel passenger vehicle in the driven cycle defined by the 98/69/EC Directive, and is used to compare the studied systems.

In order to define the FU, Ford [1] has provided data of diesel (without biodiesel addition) consumption. Fuel consumptions for diesel EN-590 with 5% biodiesel from crude vegetable oils, and for $BD10_{A1}$, $BD100_{A1}$, $BD10_{A2}$, and $BD100_{A2}$ in the same vehicle have been estimated based on the lower heating value (LHV) of each fuel and are shown in Table 1.

Limits of the systems

- Geographical limits: This study takes into account the use of the studied fuels in Spain. Original national data have been used, whenever it has been possible. However, some stages of the life cycle are located outside the Spanish boundaries. In these cases, average industrial data or primary data reported in scientific papers have been used.
- Temporal limits: The time horizon considered for the production systems and the different technologies is the year 2005.
- Stages excluded from the study are the construction and maintenance of capital goods as infrastructure, machinery and equipment, transport vehicles, and diesel passenger vehicles.
- Soil as part of the productive system: The soil down to the depth of the water table is included within the system limits in the agricultural stages and is considered as a productive system and not as part of the environment.

Table 1. Fuel consumption of the studied fuels

	LHV (MJ L^{-1})	Fuel consumption (L km^{-1})
Diesel EN-590	34.950	0.05400
Diesel EN-590 (+5% BD)	34.840	0.05416
$BD10_{A1}$	34.642	0.05448
$BD100_{A1}$	32.840	0.05747
$BD10_{A2}$	34.702	0.05438
$BD100_{A2}$	33.440	0.05673

Description of the systems

Figure 1 shows all the systems included in the life cycle analysis.

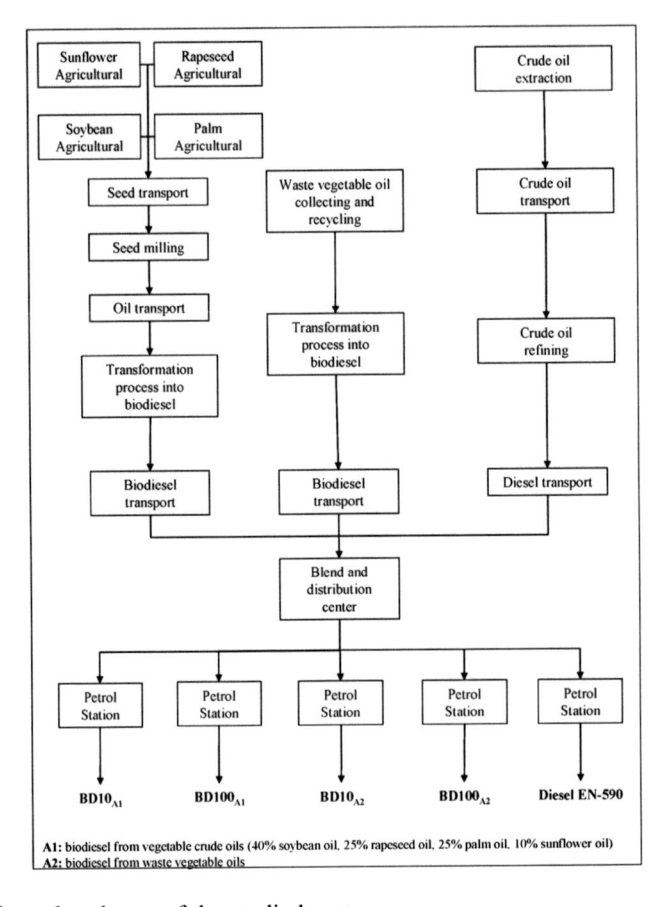

Fig. 1. Life cycle scheme of the studied systems

A1: biodiesel from vegetable crude oils (40% soybean oil, 25% rapeseed oil, 25% palm oil, 10% sunflower oil)
A2: biodiesel from waste vegetable oils

Hypothesis used in the inventory of biodiesel from crude vegetable oils

The methodology developed in the EU Concerted Action AIR3-CT94-2028 "Harmonisation of Environmental Life Cycle Assessment for Agriculture" [2] has been the base of the methods used in the agricultural stages of the biodiesel from crude vegetable oil inventory.

- Biodiesel is produced from several seeds in a mixture of 40% soybean oil, 25% rapeseed oil, 25% palm oil, and 10% sunflower oil. Sunflower seeds are all produced in Spain. Of the rapeseeds 5% are produced in Spain, whereas 95% are cultivated in France. Soybeans are imported from the USA. Palm seeds are produced and imported from Thailand as palm oil. The Laboratory of Agro Energy of the Botanic and Vegetal Production, Department of the Higher Technical School of Agricultural Engineers of the Polytechnical University of Madrid, has provided national seed production data. Soybean production data have been taken from Sheehan et al. [3]. Data from rapeseed cultivation have been taken from Ceuterick et al. [4]. Data used for the palm seed cultivation and palm oil production have been taken from Gheewala et al. [5].
- Growing biomass is an important sink of atmospheric CO_2. This CO_2 is used in the photosynthesis process in the crop in order to form its organic carbon compounds: the seed, the aerial part, and the roots.
- The CO_2 contained in the seed will be transformed into oil and flour in a first step. The flour will be mainly used as animal food and finally this organic carbon will evolve into CO_2 again. This emission will not be considered since the carbon has been previously fixed during the plant growth. In the case of the palm oil production, palm fibres are consumed as fuel in oil mills and taking into account the same argument, no CO_2 emissions from the boilers are considered. In the palm oil production, other by-products are obtained, such as the kernel, that is sold to other mills to be used as raw material, and the shells, sold as fuel. Although the organic carbon will become CO_2, it is not taking into account either.
- The oil is transformed into biodiesel, yielding glycerine as by-product. This glycerine will also eventually become CO_2, but, following the same reasoning as with the other by-products, these emissions are not taken into account.
- The roots of the plant and the aerial biomass that is not collected and left in the field, will be, with time degraded, emitting in the process CO_2 that was fixed during the growth. These CO_2 emissions have not been taken into account. No additional net fixation of C in the soil due to rhizodeposits has been taken into account.
- During the combustion of biodiesel in the vehicle engine, CO_2 will be emitted. But, as it has been mentioned before, these emissions have not been considered in the inventory.
- N_2O emissions from the field produced after the application of nitrogen fertilizers contribute to the greenhouse gas (GHG) balance in the agricultural systems. Data of these emissions can be found in the literature

but with a very variable range based on the conditions under which the experiments and measures are made. Studies performed in Spain have reported N$_2$O emissions rates of around 1% of the N applied in the fertilization [6]. This is the emission factor that has been used in this study in the domestically grown seeds since it is within the range of values reported by the Intergovernmental Panel on Climate Change (IPCC) methodology.

- Allocation rules have been applied in the processes where several by-products are originated, following the methodology described in the ISO-14040 series. These processes are the agricultural production processes, the extraction of oil from the different seeds, and the transformation process of the vegetable oil into biodiesel. The extension of the system limits has been carried out in order to avoid the allocation of the environmental loads to the different by-products.

Hypothesis used in the inventory of biodiesel from waste vegetable oils

- Data used for the collection and recycling of the waste vegetable oils have been provided by companies involved in these activities in Spain.
- Waste vegetable oil is a non-hazardous residue that must be treated properly. In this study, waste vegetable oil has been considered as the initial raw material and the stages and processes in which this oil has participated before have not been taken into account.
- In order to estimate the distances driven to collect the waste vegetable oils, the ArcGIS 9.1 tool has been used. The municipality of Alsasua has been considered as the central collection point. Taking into account the towns where the waste oil is collected from and the road network, average distances have been estimated.
- Allocation rules had to be considered in the transesterification process. To avoid the allocation procedures, the extension of the system limits has been made.

Hypothesis used in the diesel EN-590 inventory

- Data belong to a Spanish refinery with fluid catalytic cracking and have been obtained from Annual Operating Plan (AOP). AOP has also provided data on the origin of the crude oil. Other data such as the extraction and transportation of the crude oil have been taken from literature.

- Allocation rules have been applied in the crude oil extraction process, where natural gas is a by-product, and in the refinery process, where other several refinery products are produced.
- In this inventory, allocation of the environmental loads due to the by-products has been done based on the energy content of the different products.

Results

Results presented in this work are related to two relevant aspects: (1) energy use in the production and distribution of the studied fuels and (2) GHG emissions along the whole life cycle of each considered fuel.

Regarding the energy use, it has been estimated the fossil energy used in the production and distribution of the fuels as well as the energy efficiency of the processes.

Life cycle energy use and efficiency

The energy efficiency calculated in this study has been taken as the fossil energy ratio defined as the ratio of the fuel product energy to the amount of fossil energy required to produce the fuel. Results of the fossil energy consumption in the production and distribution of each studied fuel are shown in Table 2.

The larger the amount of biodiesel in the blend, the lower is the fossil energy consumption, with a reduction of 6% in the case of $BD10_{A1}$, and 7% in the case of $BD10_{A2}$. The production of biodiesel from crude vegetable oils saves 1.42 MJ of fossil energy per kilometre driven compared to the production of diesel EN-590, which amounts a saving of 77% fossil energy. Besides, biodiesel production from waste vegetable oil saves 1.64 MJ km^{-1} driven, that means a saving of 87% fossil energy. Figure 2 shows which

Table 2. Fossil energy used in the production and distribution of the studied fuels

	Fossil energy (MJ km^{-1})
Diesel EN-590	1.881
$BD10_{A1}$	1.767
$BD10_{A2}$	1.742
$BD100_{A1}$	0.433
$BD100_{A2}$	0.237

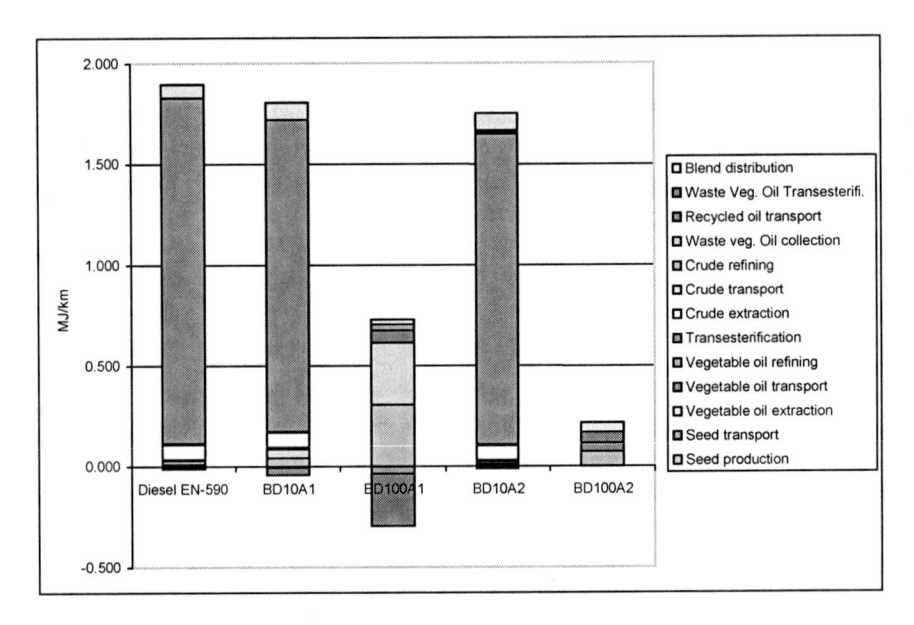

Fig. 2. Fossil energy consumption by life cycle stage

stages of the production and distribution of the studied fuels are the main consumers of fossil energy.

Crude oil refining is the fuel life cycle stage that represents the highest fossil energy consumption in the Diesel EN-590, BD10$_{A1}$, and BD10$_{A2}$, while for BD100$_{A1}$, the highest consumption corresponds to the agricultural stages. Oil extraction is also an important consumer of fossil energy.

The production of biodiesel from waste vegetable oil is the less fossil energy consumer. One of the reasons for this behaviour is that the waste oil has been considered as the raw material and it has not been taken into account the environmental loads of its production as food oil. Fossil energy ratio is shown in Table 3. In all cases, fossil energy contained in the fuel is higher than the fossil energy needed to produce and distribute the fuel with the only exception of Diesel EN-590.

Table 3. Fossil energy ratio

	Fossil energy ratio $MJ_{fuel}/MJ_{fossil\ energy}$
Diesel EN-590	1.00
BD10$_{A1}$	1.07
BD10$_{A2}$	1.08
BD100$_{A1}$	4.36
BD100$_{A2}$	7.96

Table 4. GHG emissions in the life cycle of the studied fuels (g CO_2 eq km^{-1})

	CO_2	N_2O	CH_4	GHG total
Diesel EN-590	151.8	5.6	1.0	158.4
$BD10_{A1}$	141.2	7.9	1.3	150.4
$BD10_{A2}$	139.7	5.5	1.0	145.3
$BD100_{A1}$	38.0	29.1	3.6	70.7
$BD100_{A2}$	24.5	4.5	0.9	29.9

Life cycle greenhouse gas emissions

CO_2, N_2O, and CH_4 emissions along the whole life cycle of the studied fuels have been calculated and are shown in Table 4.

In terms of CO_2 eq, production and use of pure biodiesel from crude vegetable oils avoid the emission of 88 g CO_2 eq km^{-1} driven compared to production and use of diesel EN-590. That means a saving of 55%. In the case of pure biodiesel from waste vegetable oils, production and use avoid the emission of 128 g CO_2 eq km^{-1} driven, 81% emission reduction. Blends $BD10_{A1}$ and $BD10_{A2}$ can avoid the emission of 8 and 12 g CO_2eq km^{-1} driven, respectively.

Paying attention to the different stages of the life cycle of each fuel, it can be noticed (Fig. 3) that the most important emissions in the life cycle of biodiesel from crude vegetable oils ($BD100_{A1}$) are produced in the seed production and in the oil extraction stages. In the life cycle of biodiesel

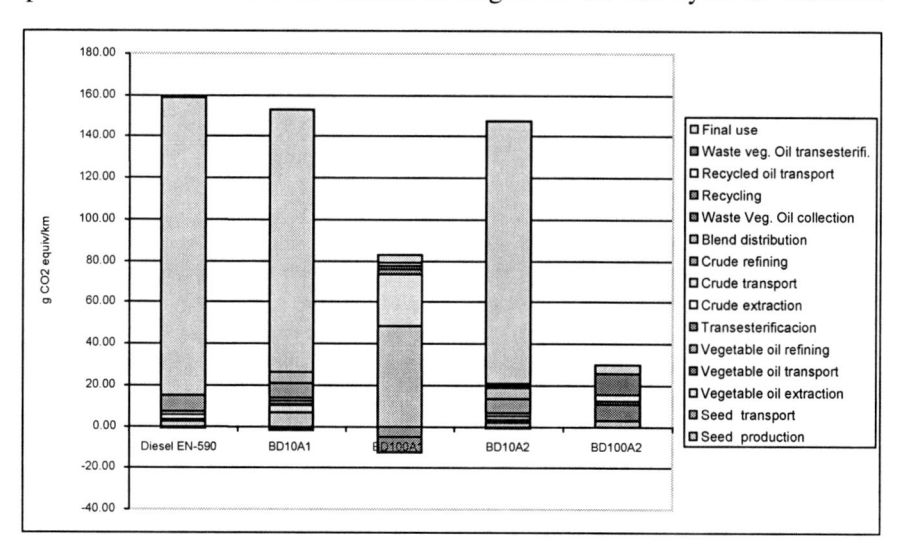

Fig. 3. Greenhouse gas emissions distribution by life cycle stage

from waste vegetable oils (BD100$_{A2}$), the transesterification process contributes the most to the GHG emissions followed by the waste oil collection stage.

Conclusions

Results obtained from the life cycle analysis conducted show a clear benefit of the use of both biodiesel from crude vegetable oil as well as biodiesel from waste vegetable oil in comparison with Diesel EN-590. The production and distribution of the blends with biodiesel mean an important saving of fossil energy that amounts 0.114 MJ km^{-1} for BD10$_{A1}$ and 0.139 MJ km^{-1} in the case of BD10$_{A2}$. The larger the amount of biodiesel in the blend, the more important the energy saving is. Production and distribution of BD100$_{A1}$ save 1.45 MJ km^{-1}, for BD100$_{A2}$ save 1.64 MJ km^{-1}.

In terms of CO_2 emissions, BD100$_{A1}$ avoids 88 g CO_2 eq, while BD100$_{A2}$ avoids 128 g CO_2 eq km^{-1} driven that means a saving in GHG emissions of 55% and 81%, respectively.

As a consequence of this study, some improvements in the biodiesel production process have been proposed. As far as the energy use is related, cogeneration plants in the oil extraction processes could optimize the energy consumption. Besides, the use of residual biomass, as in the case of the palm oil extraction, could reduce the fossil energy consumption. Regarding the agricultural procedures, it is proposed to reduce the consumption of fertilizers and cultural labours without affecting the seed production efficiency, and to find new seeds with higher production. The waste vegetable oil collection process could be improved in order to reduce the number of trips.

References

1. Ford (2003) Personal communication
2. Audsley E et al. (1997) Harmonization of environmental life cycle assessment for agriculture. Final report. Concerted Action AIR2-CT94-2028
3. Sheehan J et al. (1998) Life cycle inventory of biodiesel and petroleum diesel for use in an urban bus. US Department of Agriculture and US Department of Energy
4. Ceuterick D, Spirinckx C (1997) Comparative LCA of biodiesel and fossil diesel fuel. Vlaamse Instelling voor Technologisch Onderzoek, Mol, Belgium
5. Gheewala SH, Pleanjai S, Garivait S (2004) Environmental evaluation of biodiesel production from palm oil in a life cycle perspective. The Joint International Conference on "Sustainable Energy and Environmental (SEE)". Thailand
6. Vallejo A (2003) Personal communication

Engine management for Flex Fuel plus compressed natural gas vehicles

O Volpato,[1] F Theunissen,[1] R Mazara,[1] E Verhaeven[2]

[1]Delphi, South America Technical Center, Brazil
[2]Flemish Institute for Technological Research (VITO), Mol, Belgium

Abstract

Ethanol has been used in Brazil as a passenger vehicle fuel since 1979. Until the year 2000, vehicles were made to run exclusively with either gasoline or ethanol. The MultiFuel engine control module (ECM) was developed allowing vehicles to use fuels with any ethanol percentage, relying only on the existing oxygen sensor as opposed to an add-on ethanol sensor for the percent ethanol evaluation. The use of tank fuel level information allowed for far more robust ethanol percent detection and improved drivability. Four years later, compressed natural gas (CNG) capability was integrated into the MultiFuel technology. Prior to that, vehicles using CNG normally required a second ECM. The MultiFuel integrated with CNG capability is known as the Tri-Fuel system. It uses only one ECM, seamlessly controlling both liquid (ethanol blends and gasoline) and CNG fuels with little power loss, and excellent drivability and fuel consumption. This paper will present both the MultiFuel and Tri-Fuel technologies.

Introduction

In the quest for diminishing its dependence on foreign petrol along with its vast agricultural capabilities, Brazil launched the "ProAlcool" program giving tax incentives to alcohol vehicles. The first ethanol (E100) vehicles appeared in 1979. In order to solve engine starting difficulties in cold weather, an auxiliary gasoline tank, fuel pump, and injector solenoid were used. E100 vehicles became a success and after 6 years, they accounted for nearly 80% of the Brazilian market [1]. Later, due to an ethanol shortage, the market for E100 vehicles diminished, and by 1990 they accounted for only 1% of the total production [1].

The fuel mixtures are named in function of their ethanol content. The name E85 refers to a mixture of 15% gasoline and 85% ethanol. With the rising costs of petrol and anticipated future shortages, in the year 2000 the Brazilian government again fostered the use of ethanol in vehicles. In response, the MultiFuel engine management system was developed where any fuel mixture between E20 and E100 could be used. The production of flex fuel vehicles has reached the 40% mark and is increasing, as shown in Figure 1. The accumulated production for 2005 shown is an estimate.

In addition to the flex fuel advent, Brazil has been experiencing an abundance of compressed natural gas (CNG), which is expected to last for decades. In response, a system using only one ECM was developed. The system integrates CNG capability into the MultiFuel algorithms. This technology is known as Tri-Fuel. In the following sections, the MultiFuel and Tri-Fuel systems are discussed in detail, highlighting their uniqueness and advantages over existing technologies. In addition, some comments about the benefits of flex fuel and CNG vehicles are made. The term flex fuel is used in reference to any vehicle that can use fuels from E20 to E100; the details given below are about the MultiFuel technology.

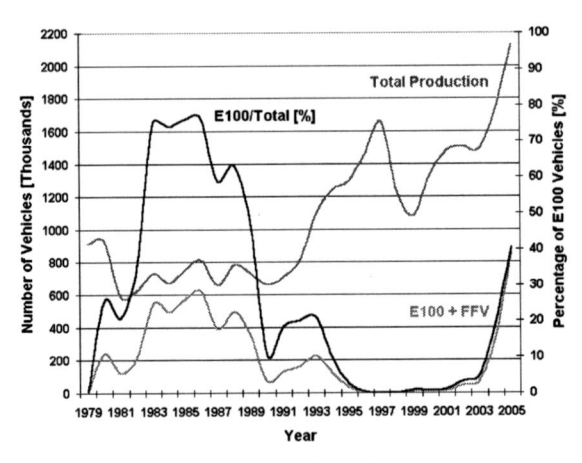

Fig. 1. Brazilian production of ethanol vehicles [1]

MultiFuel system (E20 to E100)

An engine control module (ECM) with algorithms that evaluate the ethanol percentage in fuel by software using the oxygen sensor feedback was developed. The ECM uses this information to adjust the amount of fuel and spark advance in order to meet all the requirements of performance, fuel economy, and emissions.

In Brazil, the fuel stations sell two types of fuel: E20 and E100. E100 is 100% hydrated ethanol (7% water, 93% ethanol). E20 is 80% gasoline (either regular or premium) and 20% ethanol. Since 1992, the use of lead-based octane booster has been forbidden in Brazil and the amount of ethanol in gasoline is regulated by the government, varying from 20% to 26%. The ethanol added to pure gasoline to make the E20 gasoline consists of ethyl alcohol dehydrated to remove the 5% of eutectic water remaining after normal distillation. See Figure 2 for ethanol and E20 blends in Brazil.

The percent ethanol evaluation algorithm is only launched after a refuelling is detected, so as to avoid a change in engine and component characteristics being interpreted as an ethanol percentage change in fuel. Because of differences in octane rating, stoichiometric air/fuel ratio (AFR) and vapour curve, the base engine design and engine management system should be different between various ethanol blends and thus result in different vehicles. Until the year 2000, vehicles were made to run exclusively with gasoline or ethanol. Flex fuel vehicles should have a means to evaluate the current ethanol content in the fuel in order to adjust fuelling and spark ignition parameters.

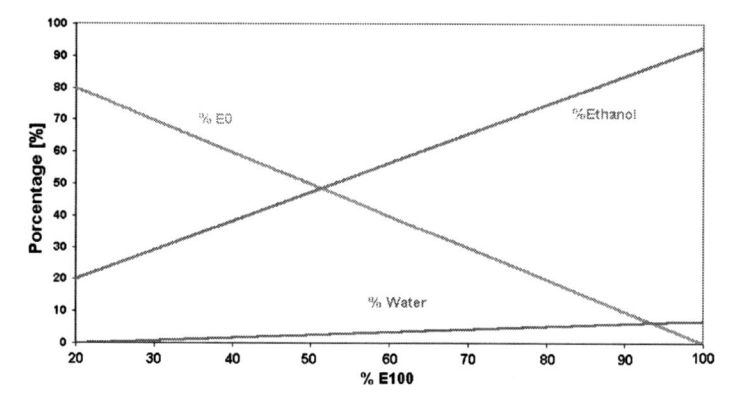

Fig. 2. Ethanol and gasoline blends

Flex fuel vehicles that use an ethanol sensor are already commercially available in some markets, but have a high add-on cost to the vehicle. This makes them economically unfeasible in Brazil. Another disadvantage is that E100 fuel has 7% water and the ethanol sensor cannot distinguish between water and alcohol.

Sensorless percent ethanol estimation

The use of the existing oxygen sensor for percent ethanol estimation is possible because the stoichiometric AFR changes in function with the percent of ethanol in the fuel. Fuels with high percentages of ethanol burn lean when low percentage ethanol fuel injection parameters are in operation and vice versa. Figure 3 illustrates this fact.

The ECM monitors the fuel tank level to determine when a refuelling occurs, setting the MultiFuel algorithm in one of the several learning modes and minimizing driveability side effects due to different fuels. The algorithm then adjusts the fuelling to address the updated alcohol content in the fuel, taking into account that the ECM's closed loop control keeps the AFR at stoichiometric value.

In order to distinguish the closed loop fuel control change due to alcohol content change from changes due to long-term component characteristic changes, the alcohol percentage learn algorithm is active only for a limited time window.

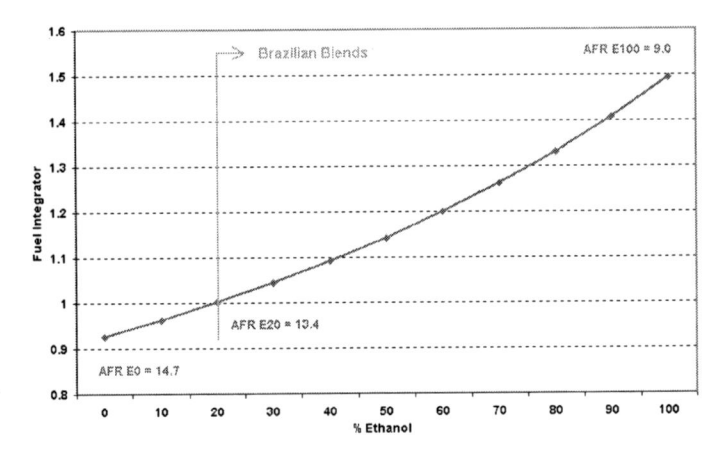

Fig. 3. ECM's fuel integrator versus air/fuel ratio

The quantity of "old" fuel inside the fuel line, fuel rail, and inside the injector body is accounted for by the triggering of the ethanol percentage learning mode. These features are unique to the MultiFuel flex fuel technology and represent a major advantage.

Based on this alcohol content, the spark timing is advanced or retarded to either take advantage of the engine's power or to reduce the likelihood of knock. The AFR is also modified based on the alcohol content. As the alcohol content increases, the AFR decreases.

Advantages of a sensorless system are clear: It is a low-cost system as it has no add-on components, it does not require a new engine management system layout and it does percent ethanol estimation right after combustion without sensor-to-injector delays.

Tri-Fuel systems (E20 to E100 + CNG)

After the construction of a CNG gas duct between Bolivia and the main Brazilian cities, Brazil began to experience an abundance of CNG. On top of that, recently huge CNG fields were discovered in Brazil. Using the available CNG could free Brazil from importing petrol. As a result of this CNG availability, vehicles using CNG have proliferated. Due to its lower cost, many taxi fleets are converting their vehicles to run with CNG as a dual-fuel system. Add-on systems, normally a second ECM installed in "parallel" with the original one, have been used for this purpose. However, the resulting engine performance and fuel economy are less than the ideal. These systems also interfere with existing sensors and actuators, making installation cumbersome and requiring changes designed to "cheat" on the original ECM diagnostics.

The majority of those systems consisted of a CNG tank, safety valve, refuelling valve, shut-off valve, CNG pressure regulator, a "venturi type" single point injector, a CNG switch, and electronic control units used to avoid the logging injector and oxygen sensor malfunctions by the original vehicle's ECM. Optionally, a spark advance control module was added. Later, more sophisticated CNG systems appeared but they continue to be installed "in parallel" with the original ECM with many splices in the wiring harness. Their purchase and mounting costs are usually high. By the end of 2004, a cost-effective MultiFuel–CNG system was developed using only one ECM, seamlessly controlling both fuel systems with very little power loss in aspirated engines and excellent drivability and fuel consumption.

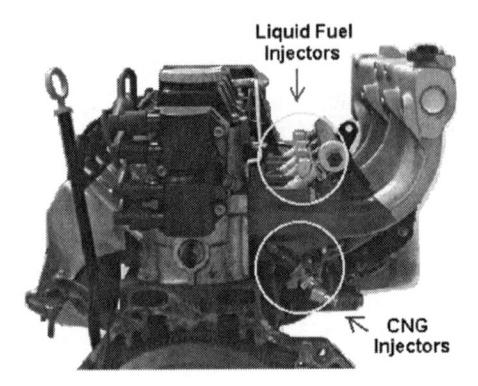

Fig. 4. Corsa's engine with liquid fuel injectors and CNG injectors

A Corsa 1.8 L, 4-cylinder engine was the first production Brazilian vehicle with a single ECM flex fuel–CNG system. The same engine used for the MultiFuel vehicle was used without modifications. The engine fitted with CNG injectors can be seen in Figure 4

System performance

One requirement for Tri-Fuel vehicles was to use the same engine as MultiFuel vehicles and get the best performance with all three fuels. CNG exhibits a narrow flammability range that goes from 5% to 15% in volume of air [4,5]. In order to meet emission regulations, the system runs at stoichiometric AFR using a current production oxygen sensor. The performance for CNG is lower than the theoretical optimum but if the engine is changed to achieve this, it could not be used with either E100 or E20. The achieved performance is very close to the optimum, considering the characteristics of the engine used. With CNG, the maximum torque is lower because the fuel volume accounts for 7% of engine displacement. With liquid fuels, the maximum torque point is achieved at rich conditions and it is approximately 5% higher than the stoichiometric torque [2]. When the E100 effect takes place, its high vaporization heat cools down the intake air, increasing the volumetric efficiency and resulting in higher torques than E20 [2]. A table with basic characteristics of the three fuels is shown in Table 1 [4,5].

Performance comparison

Performance data for Tri-Fuel vehicles using E100, E20, and CNG were taken and are shown in Table 2. Note that only a 13% performance loss

Table 1. Fuel properties table

	Boiling temperature (°C)	Stoichiometric AFR	Vaporization heat (KJ kg^{-1})	Fuel volume stoich mix (%)
E100	78	9.0	850	6.5
E20	25–175	13.4	418	2.7
CNG	–	17.2	–	6.9

	Calorific value (MJ kg^{-1})	Calorific value stoich mix (MJ kg^{-1})	Calorific value stoich mix (MJ m^3)	Octane RON
E100	16.9	2.69	3.89	107
E20	39.0	2.71	3.58	92
CNG	49.5	2.72	3.72	120

resulted with CNG, a much better figure than the ones for most of the add-on CNG systems. The Tri-Fuel vehicle passes the current emissions requirements for all tree fuels and any blend of E20 and E100. A catalyst with a slightly larger noble metal charge and better high temperature handling was developed for CNG usage. The vehicle performance with liquid fuel blends between E20 and E100 will exhibit an intermediate performance.

Table 2. Vehicle performance with E100, E20, and CNG

Corsa MultiPower			
Fuels	E22	E100	CNG
Engine type	\multicolumn{3}{c}{C18XE 4 Cylinder SOHC}		
Displacement (L)	\multicolumn{3}{c}{1.8}		
Compression ratio	\multicolumn{3}{c}{10.5}		
MaxPower (kW)	77.2	80.2	67.2
@rpm	5400.0	5400.0	5400.0
Torque (Nm)	170.0	178.0	148.0
@rpm	3000.0	3000.0	2800.0
Transmission type	\multicolumn{3}{c}{5 gear manual}		
Top speed (km h^{-1})	182.0	184.0	175.0
Time from 0–100 km h^{-1} (s)	10.6	10.4	11.2
Fuel economy – per NBR7024			
City (km h^{-1})	11.0	7.6	14.7
Highway (km L^{-1})	16.3	11.0	18.7
Relative fuel cost per km (%)	100.0	82.0	43.0

Benefits of flex fuel and CNG vehicles

"In 2000, over 40% of automobile fuel consumption and 20% of total motor vehicle fuel consumption in Brazil was ethanol. According to one estimate, around US$140 billion would have been added to Brazil's foreign debt if ethanol had not been used as a fuel over the past 25 years" [7]. "Since its launch in March 2003, the flex fuel car, which can run on any mixture of ethanol and petrol, has captured a large slice of Brazil's vehicle market. In 2004, nearly 330,000 units were sold and analysts predict further growth this year. At about R$1.20 (46 US cents, 36 euro cents, 25 pence) per litre, ethanol costs nearly half the price of regular gasoline. Because of a shortage of ethanol in the late 1980s, many motorists were hesi tant to buy cars running only on the fuel. The new flex fuel car gives them the security of being able to switch to petrol partially or entirely."[8]

Lower dependence on foreign fuel

The consumption of diesel accounts for about 25% of the petrol based energy matrix in Brazil, considering that the transportation of goods is mainly done by diesel-powered trucks over a continent-sized country. In addition, the Brazilian petrol is not the best for diesel production; for that reason, petrol and diesel is imported and Brazilian petrol and gasoline is exported [9] (see Fig. 5).

Figure 6 shows an estimated evolution of the passenger vehicle fleet up to 2010 [9]. This data was used to make an estimate of the savings on petrol imports. The use of ethanol and CNG visibly contributes to reducing Brazil's foreign debt. Figure 7 illustrates the amount of petrol saved when considering that the average passenger vehicle is driven 25,000 km year^{-1} with the following consumption rates: gasoline: 12.7 m L^{-1}; ethanol: 9.9 km L^{-1} and CNG: 11.3 km L^{-1}. The values for ethanol and CNG are expressed in equivalent gasoline mileage. Also 1 m^3 of gasoline is equivalent to 5.636 barrels of petrol [9].

Lower emission of greenhouse gases

Vehicles fuelled with ethanol and CNG are a good environmental choice when considering air quality and global warming issues [9]. Carbon monoxide and non-methane hydrocarbon emissions are reduced with flex fuel and CNG vehicles. These elements form ground-level ozone, the main component of smog [9]. The ethanol has negative greenhouse gas (GHG)

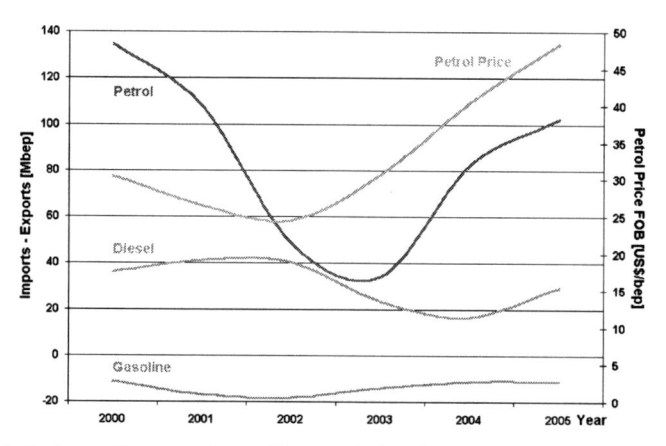

Fig. 5. Trade balance for petrol, gasoline, and diesel

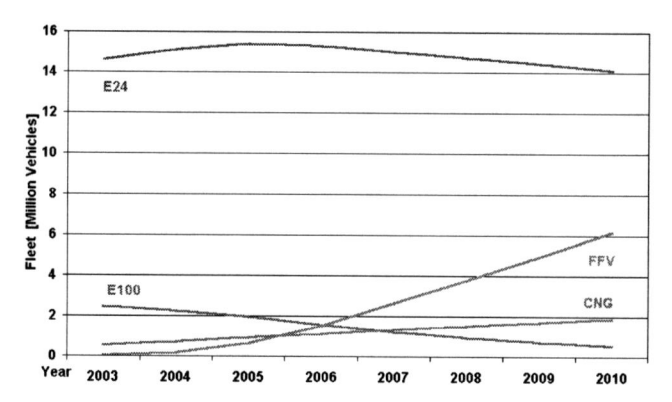

Fig. 6. Passenger vehicle fleet evolution

emissions because of carbon uptake sequestration during growth of sugar cane. All the carbon sequestered during biomass growth is released back to the air during combustion of ethanol in vehicles [10]. The amount of GHG, on an energy basis, released during production/distribution (well to tank) and released by vehicles (tank to wheel); the overall efficiency of vehicles' engines and the emitted GHG on g km^{-1} can be seen on Table 3. Data from [10] and [11] were used to construct Table 3. In Brazil, the production of ethanol is efficient because all energy used in the process comes from sugar cane, e.g., electricity comes from the burn of sugar cane, trucks use either ethanol or methane, made from sugar cane moist. The distribution, however, still uses diesel.

According to one study, the ethanol produced in Brazil from sugar cane, considering the complete CO_2 life cycle, emits 461 g L^{-1} (best scenario) to

Table 3. Green house gases life cycle for different fuels

Fuel	GHG Well–tank (g mmBTU^{-1})	GHG Well–wheel (g mmBTU^{-1})	Overall engine efficiency (%)	GHG emitted (g km^{-1})
E0	21,000	76,477	16.7	344
E20	–	–	–	289
E100	57,000	76,218	16.7	68
CNG	16,000	60,185	16.9	312

1 mmBTU = 1,000,000 BTU, factor (mmBTU km^{-1}) = 4, 9e – 4

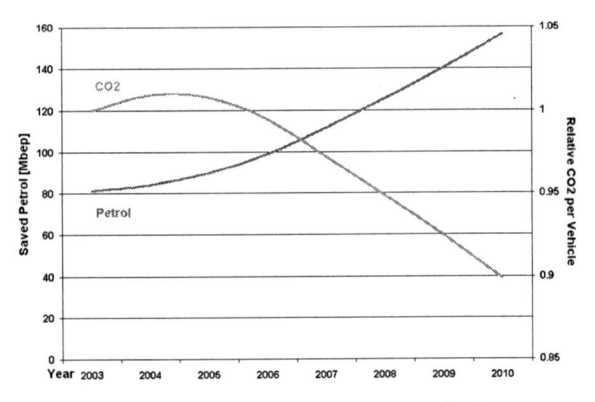

Fig. 7. Estimated savings on petrol imports and CO_2 emissions (considering complete CO_2 life cycle)

572 g L^{-1} (worst scenario) while ethanol produced in the USA from corn emits 1392 g L^{-1} to 1459 g L^{-1} [11]. Considering the complete CO_2 life cycle, compared to gasoline (E0), the production of CO_2 is 80% lower for ethanol and 9% lower for natural gas [9,10]. This information was used to calculate the amount of CO_2 emitted, on a per vehicle basis, shown on Figure 7.

The considered time frame does not show a significant improvement in the decrease of CO_2 emissions due to the fact that most vehicles are not enabled to use alternative fuels; however, as older vehicles are replaced by flex fuel vehicles, the improvement will be significant in a decade or two.

Conclusions

The reduction of fossil oil dependence and emissions of toxic and GHGs will gradually be realized by an increase in the use of alternative fuels.

Drivers can choose to use E20, E100, or CNG depending on cost, availability, or preference. Automakers and system solution providers work together to use alternative fuels without increasing system costs.

An efficient and cost-effective solution has been developed for the use of gasoline, ethanol, and CNG on the same engine using only one ECM, seamlessly controlling both fuel systems with very little power loss in aspirated engines, excellent drivability, and fuel consumption. Such technology clearly contributes to a healthier environment and higher alternative fuels demand.

References

1. Brazilian Automotive Industry Yearbook, 2004
2. Richard S (1999) Introduction to internal combustion engines, 3rd edn. SAE and Macmillan, Washington, DC
3. Theunissen FMM (2003) Percent ethanol estimation on sensorless multi fuel systems: advantages and limitations. SAE Technical Paper no. 2003–01–3562
4. Bosch R (1996) Automotive handbook, 4th edn. Wiley, New York
5. FuelTable, US Department of Energy. Also at www.eere.energy.gov
6. Combustíveis para Motores Endotérmicos, Enfoque no Gás Natural Veicular. Diocles Dalavia. Petrobrás
7. Renewable energy in developing countries (2003) Renewable energy world, July–August
8. Brazil taking the lead in biofuels (2005) Petrol World August
9. Gás Natural na Matriz de Combustíveis Veiculares (2005) Luciana Bastos de Freitas Rachid, Petrobras
10. Well-to-Tank Energy Use and Greenhouse Gas Emissions of Transportation Fuels (2001) North American analysis, vol. 2, Argonne, IL
11. Dias de Oloveira ME, Vaughan BE, Rykiel J Jr (2005) Ethanol as fuel: energy, carbon dioxide balances, and ecological footprint. BioScience 55:593–602

Mobility and environment in Spain

PJ Pérez-Martínez

Universidad Politécnica de Madrid, Centro de Investigación del Transporte (TRANSyT). C/Profesor Aranguren s/n 28040 Madrid, Spain

Abstract

This study summarizes most recent findings on the contribution of transport, especially road transport, to air pollution and emissions of greenhouse gases (GHG) in Spain. The evaluation of the national emissions inventory and the calculation of vehicle emissions during the period 1990–2003, using the database developed by the Direction of Air Quality, Spanish Ministry of Environment, reflects a significant reduction of ozone precursors (–21.8%), acidifying substances (–7.4%), and particles (–2.2%). The economical growth (31%) and the transport demand growth, passengers (76%) and freight (49.9%), are coupled.

This study compares vehicle emissions with current pollution concentrations and analyses air quality relating it with the European relevant legislation. The concentrations are lower than the permissible values of pollutants fixed by the directives, except the particles.

The emissions of GHG have grown 47.1%, at a much higher rate than the 15% increment warned by the Kyoto Protocol and one of the most urgent present priorities is to reduce them. Even though the transport, which a passenger uses like road, rail, and air has improved its efficiency by 15%, 41%, and 50% respectively, additional measures are needed to reduce emissions and energy consumption. These measures require time to be effective.

Transport problems and environment

In the context of a constantly increasing transport demand, in terms of passengers and freight, transport is the main anthropogenic source of certain atmospheric pollutants [1]. The emissions of these polluting substances make the air concentrations of these substances exceeding the limits, in many sites, the levels considered to have harmful effects on human health and the environment. Therefore, transport is a growing source of carbon dioxide (CO_2) emissions, and CO_2 is the most important GHG [2].

The national emissions inventory, database developed by the Spanish Ministry of Environment, gives time series of the most important atmospheric pollutants and GHG. The transport sector contributes to 20.1% of total national emissions of acidifying substances (AS) in 2003. Similarly, transport is the main source of ozone precursors (OP) emissions and contributes to 28.2% in 2003. Finally, transport is the second main source of PM_{10}, and contributes to 32.2% of total national emissions. The transport sector is the second main source of GHG and accounts 28% of total emissions [3].

This study reviews transport and environment trends in Spain and discusses if the current transport system is sustainable from the energy and environment point of view. These discussions are supported by a national database of transport, environment, economy, and demography, which was also used for developing indicators of transport sustainability and for statistical analysis [4]. The paper concludes with recommendations for an energy efficiency strategy, suggesting approaches, policies, and measures that can make transportation relevant in the attainment of higher energy efficiency.

Growing mobility

In Spain, transport demand grew considerably during the period 1990–2003, both passengers (76%) and freight (49.9%). Transport demand increments are higher than economic activity growth, as gross domestic product (GDP) per capita grew 31% during that period. Similarly, transport GHG emissions, increased by 47.1%. Transport AS emissions, OP, and PM, decreased by 7.4%, 21.8%, and 2.2%, respectively (Table 1). Figure 1 shows emissions, GDP, and transport demand trends during the period 1990–2003. There is a correlation between transport demand and GHG emissions growth, closer in freight than in passengers transport; similarly GDP growth correlates transport demand, as is showed in all international trends. It is significant

Table 1. Transport and environment indicators in 1990, 2003, relative growth rates in 1990 basis, and annual growth

Indicator	1990	2003	Mean	Change 1990–2003 (%)	Mean change (% year^{-1})
Green house gases (CO_2 kg eq inhabitant^{-1} year^{-1})	1.760	2.589	2.174	47	3.0
Acidifying substances (AS kg eq inhabitant^{-1} year^{-1})	0.510	0.470	0.51	−7	−0.6
Ozone precursors (NMVOC[a] kg eq inhabitant^{-1} year^{-1})	42.000	33.000	40.000	−22	−1.9
Particulate matter (PM kg eq inhabitant^{-1} year^{-1})	20.000	20.000	21.000	−2	−0.2
Passenger-km (pkm inhabitant^{-1} year^{-1})	5.832	10.262	8.194	76	4.4
Tons-km (tkm inhabitant^{-1} year^{-1})	6.261	9.387	7.620	50	3.2
Gross domestic product[b] (€ inhabitant^{-1} year^{-1})	10.183	13.304	11.729	31	2.1

[a] Non-methane volatile organic compounds
[b] 1995 constant prizes

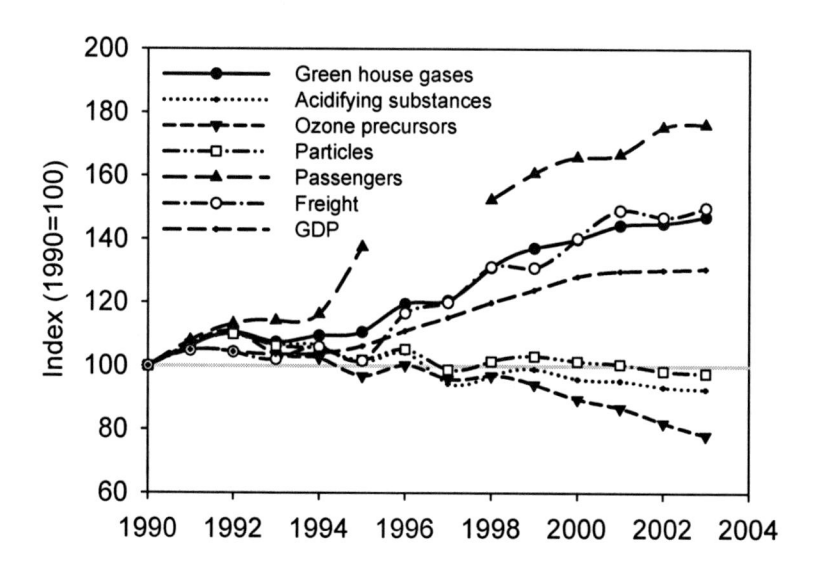

Fig. 1. Trends in green house gases, acidifying substances, ozone precursors, particles, passengers and freight transport, and gross domestic product

Table 2. Pearson coefficients for sustainability indicators. Period adjustment 1990–2003

Indicator:	Correlated with:	P-values
GHG (CO_2 kg eq inhabitant^{-1} year^{-1})	pkm inhabitant^{-1} year^{-1}	0.976[a]
	tkm inhabitant^{-1} year^{-1}	0.986[a]
	GDP (€ inhabitant^{-1} year^{-1})	0.994[a]
AS (AS kg eq inhabitant^{-1} year^{-1})	OP (NMVOC kg eq inhabitant^{-1} year^{-1})	0.903[a]
GDP (€ inhabitant^{-1} year^{-1})	Passengers (km inhabitant^{-1} year^{-1})	0.972[a]
	tkm inhabitant^{-1} year^{-1}	0.989[a]

[a]Significant at 99% confidence level

that passengers' mobility growth is higher, for what, lifestyle factors must be considered, which influence this trend.

The analysis of the data consisted of simple bivariate regressions to show whether there are certain basic relationships between the former indicators or not and whether these relationships are consistent across time or not. Table 2 shows the correlations and all values are statistically significant at the 95% confidence interval. Transport demand, passengers and freight, and GDP are positively correlated with GHG emissions, as are shown by Pearson product moment coefficients (P-values) 0.97, 0.98, and 0.99, respectively. AS emissions are positively correlated with OP emissions (0.9). GDP is positively correlated with transport demand, passengers (0.97), and freight (0.99).

Air quality

In this study emission levels of several pollutants are analysed. The average annual concentrations of different pollutants during last years (from 1995 to 2003): NO_2, PM_{10}, SO_2, CO, decreased by 12%, 16.3%, 52.4%, and 53.3%, respectively. Only ozone (O_3) increased in its concentration by 5.3%. Figure 2 shows average annual concentrations in Spain. These results are derived from urban roadside stations where pollution mainly comes from vehicle emissions: 73, 105, 93, 87, and 82 stations for NO_2, PM_{10}, SO_2, O_3, and CO, respectively.

The European directives determine and regulate the maximum permissible air concentration values for different pollutants. These air concentration values could not exceed a number of days/hours per year. Therefore, the average annual concentrations of different pollutants could not exceed certain values. Directive 1999/30/EC establishes, among others, NO_2, PM_{10},

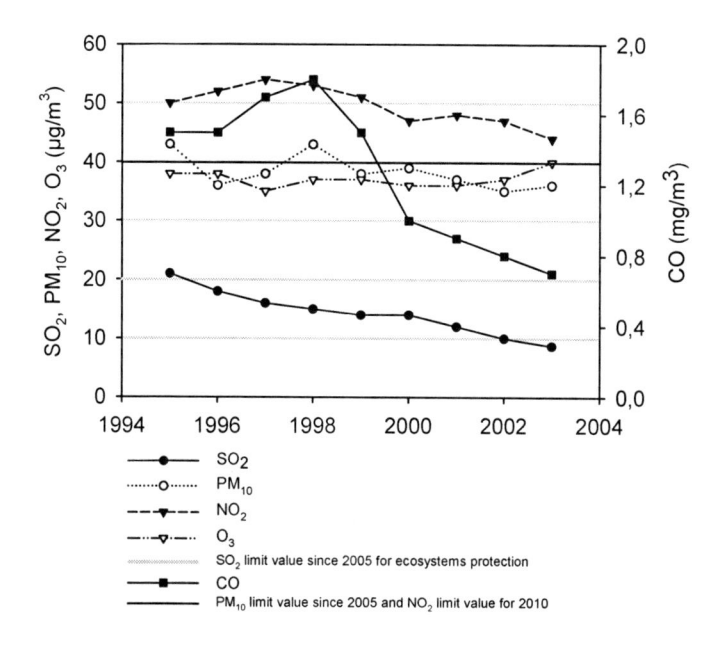

Fig. 2. Trends in average pollution concentrations in Spain of SO_2, PM_{10}, NO_2, O_3, and CO (1995–2003)

and SO_2 concentration limit values. Similarly, Directives 2000/69/EC and 2002/3/EC establish concentration limit values of CO and O_3.

Figure 3 shows air pollution exceeds, expressed in number of hours/days, for NO_2, PM_{10}, SO_2, and O_3 by population range of the Spanish municipalities and for the period 1995–2003. For NO_2 and PM_{10}, the maximum permissible values fixed by the European directives are exceeded certain number of times per year in many Spanish municipalities. PM_{10} exceeded allowable values, though the number by which it exceeds follows a negative trend for municipalities between 250,000 and 500,000 inhabitants. The number by which NO_2 exceeds follows a negative trend for all population ranges and in 2002 only the value for municipalities bigger than 500,000 inhabitants was critical. For the other pollutants, SO_2 and O_3, the average number of exceedances do not exceed the limit values according with the directives. These results are derived from urban roadside stations with real time monitoring provided: 96, 127, 115, and 110 stations for NO_2, PM_{10}, SO_2, and O_3, respectively. Therefore, Figure 3 presents the limit values of the exceedances (18, 35, 24, and 25) together with the years when the directives are effective.

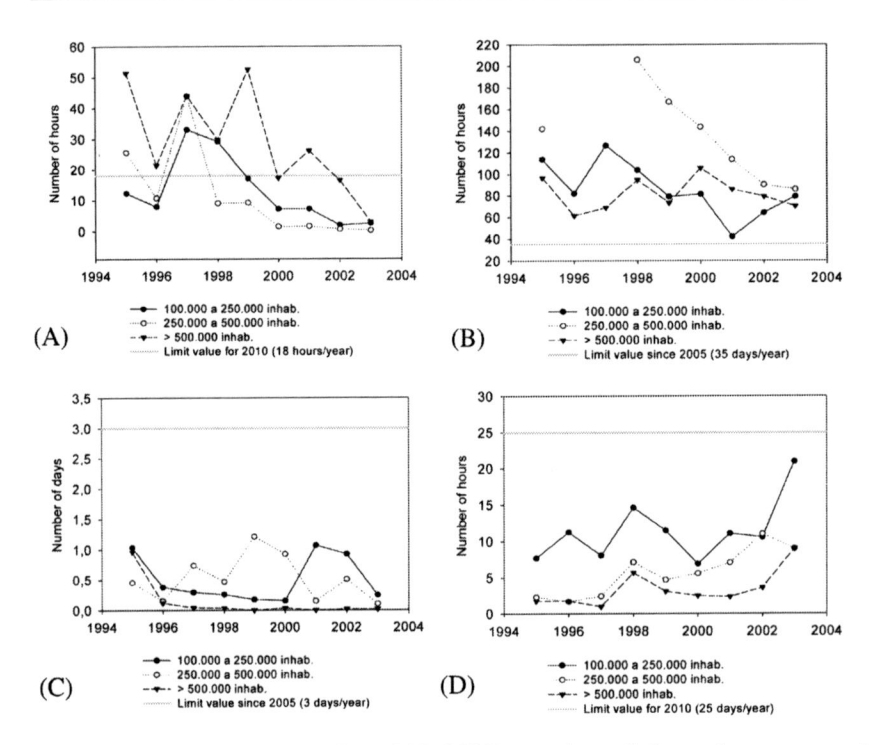

Fig. 3. Air pollution in Spain (1995–2003), number of times the concentration levels exceed emission permissible limits for: NO_2 h 200 µg m^{-3} (A), PM_{10} 24 h 50 µg m^{-3} (B), SO_2 24 h 125 µg m^{-3} (C), and O_3 24 h 120 µg m^{-3} (D)

Consumption and energy efficiency

Transport energy consumption during the period 1990–2003 grew by 57.5%, and in 2003 it reached 38 million tons of oil equivalent (Mtoe). Figure 4 shows energy consumption forecast, horizon 2012, considering the trend curve. The graph includes the consumption levels corresponding to the Kyoto Protocol, which limits total emission rates in Spain, hence consumption, to 15%, compared to 1990 levels. Similarly, Figure 4 includes the levels integrated in the E4 strategy of the National Plan of Emission Rights 2005–2007, which pretends to reach 48 Mtoe in the year 2012, 4.7 Mtoe less than the 53 corresponding to the trend scenario. To accomplish the objective, the E4 includes measures oriented to decrease GHG and improve energy efficiency [5]. E4 distinguishes between short and mean/long-term measures. Measures of short term are traffic management- (development of Urban Transport Plans in cities bigger than 100,000

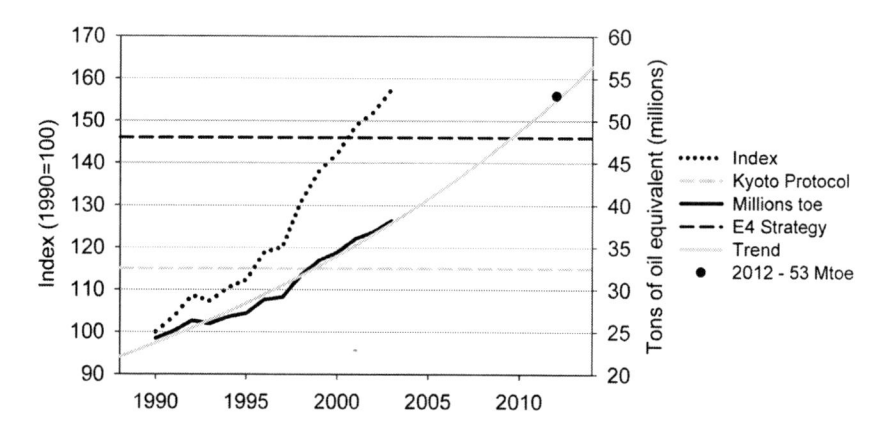

Fig. 4. Estimation of transport energy consumption in Spain, future scenarios, and normative framework

inhabitants), congestion charges, parking constraints, traffic calming (commute transport plans, improvement, and development of public transport), promotion of non-motorized transport systems, and improvement of freight transport logistics in the city. Measures of mean/long term are efficient use of transport (road fleet management, driving, and best practises) and energy efficiency improvement of vehicles (alternative fuels and modal shifts).

The efficiency during the period 1990–2003, expressed in kilograms of oil equivalent per passenger kilometre (koe pkm^{-1}) or ton kilometre (koe tkm^{-1}), improves by 15% in road passenger transport, and worsens by 7.1% in road freight transport. Rail and air passenger transport, improve their efficiencies 40.8% and 50%, respectively. Figure 5 shows changes in consumption, demand per inhabitant, and energy efficiency. Air passenger transport has experienced the highest demand increments, coupled with small increments in GHG and energy consumption. This is due to big technological improvements that air passenger transport is a high cost transport mode, which could assume additional costs of these improvements. Therefore, oil cost is an important percentage of total costs and energy savings ensure the profitability of air services. Railway, which was electrified in those lines with high demand, has experienced GHG reductions coupled with new rolling stock and, energy efficiency improvements. Road has improved its efficiency and experienced GHG emissions increments smaller than demand increments, even though energy consumption reduction was scarce due to congestion increase and higher share of metropolitan trips.

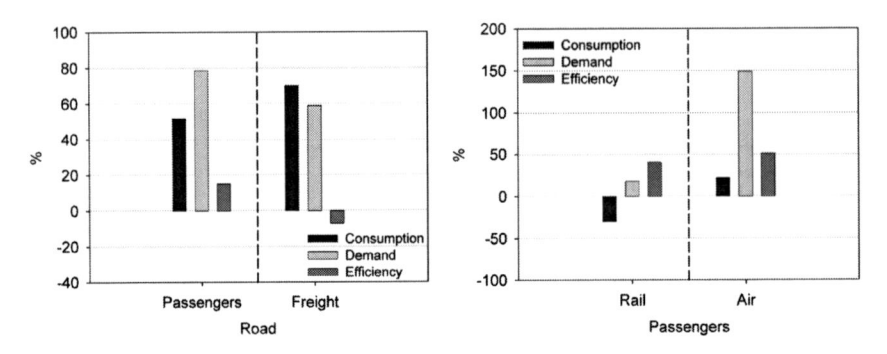

Fig. 5. Energy consumption change, demand and road transport energy efficiency (passengers and freight), and rail transport and air transport (only passengers), 1990–2003

Conclusion and discussions

Transport is a growing source of GHG emissions in Spain. GHG emissions growth is related to transport demand and socio-economic circumstances [6]. In the same manner, transport is an important source, although not growing, of atmospheric pollutants emissions. Pollutant emissions data are correlated with emission data and reflect the importance of PM_{10} and NO_2 air concentrations. The air concentrations of PM_{10} exceed the thresholds of the European directive. The concentrations of other pollutants do not exceed these thresholds though NO_2 concentrations are close. All the concentrations of pollutants, except O_3, are decreasing.

Transport energy use will grow exponentially in Spain in the next years at least that transport activity and income could be decoupled [7]. It is necessary to stimulate measures and policies which help to revert to energy consumption current trends [8]. In this line, the implementation of efficient policy measures, effective transport systems and technology developments, might offer possibilities to control emission and consumption in a short and long term [9]. However, it is doubtful whether the proposed measures are sufficient to bring about this energy consumption and emission reductions and meet the Kyoto Protocol alone. These measures demand time to be effective and need to be supported by changes on lifestyles to influence transport growth in the following decades [10].

The success in the improvement of passenger transport efficiency is partially due to new cars and improvement of the traffic management. The technological developments, partially due to a compromise of cars manufacturers to fabricate less pollute vehicles but also by a shift to diesel vehicles, have contribute to these improvements [11]. However, continued efforts and additional measures are needed to reduce energy use per transport activity unit. Despite of new fleet (improvement in energy efficiency of vehicles and engines) and improvements in operational factors (speed, network characteristics, and dynamics), freight transport has not improved its efficiency. For this reason, shifts to relatively low energy intensity forms of transport, such as rail, are advisable [12].

References

1. Lenz HP, Prüller S, Gruden D (2003) Means of transportation and their effect on the environment. In: Gruden D (ed.) The handbook of environmental chemistry, vol. 3, Part T, Traffic and environment. Springer, Berlin, pp 107–173
2. Sperling D (2004) Environmental impacts due to urban transport. In: Nakamura H, Hayashi Y, May AD (eds.) Urban transport and the environment: an international perspective. Elsevier, Oxford, pp 99–189
3. Ministerio de Medio Ambiente (2005) Perfil Ambiental de España 2004. Informe basado en indicadores. Centro de Publicaciones de la Secretaría General Técnica, 239 pp
4. Pérez-Martínez PJ, Monzón de Cáceres A (2005) Informe sobre transporte y medio ambiente. Trama 2005. Elorrieta I, San Miguel M (eds.) Centro de Publicaciones Secretaría General Técnica Ministerio de Medio Ambiente, Madrid, 108 pp
5. Ministerio de la Presidencia (2004) Real Decreto 1866/2004, de 6 de septiembre, por el que se aprueba el Plan nacional de asignación de derechos de emisión, 2005–2007. Boletín Oficial del Estado 216:30616–30642
6. Wohlgemuth N (1998) World transport energy demand modelling. Energ Policy 25:1109–1119
7. OECD Environment Directorate (2003) Analysis of the links between transport and economic growth. Project on decoupling transport impacts and economic growth. Paris, France, 94 pp
8. Bose RK, Srinivasachary V (1997) Policies to reduce energy use and environmental emissions in the transport sector. Energ Policy 25:1137–1150
9. DeCicco J, Mark J (1998) Meeting the energy and climate challenge for transportation in the United States. Energ Policy 26:395–412
10. Rodenburg CA, Ubbels B, Nijkamp P (2002) Policy scenarios for achieving sustainable transportation in Europe. Transport Rev 22:449–472
11. Johansson B (1995) Strategies for reducing emissions of air pollutants from the Swedish transportation sector. Transport Res A 29:371–385
12. Vanek FM, Morlok EK (2000) Improving the energy efficiency of freight in the United States through commodity-based analysis: justification and implementation. Transport Res D 5:11–29

Influence of gear-changing behaviour on fuel use and vehicular exhaust emissions

C Beckx,[1,2] L Int Panis,[1] I De Vlieger,[1] G Wets[2]

[1]Flemish Institute for Technological Research, Boeretang 200, 2400 Mol, Belgium
[2]Transportation Research Institute, Hasselt University, Wetenschapspark 5 Bus 6, 3590 Diepenbeek, Belgium

Abstract

This study explores the influence of gear-changing behaviour on vehicular exhaust emissions and fuel consumption using real drive cycles as an input. As many as 235 different drive cycles, recorded from people participating in a survey, were imported in an emission simulation tool called Vehicle Transient Emissions Simulation Software (VeTESS). Emissions and fuel consumption were calculated with VeTESS using two different gear change assumptions (normal and aggressive). This paper reports on the differences in vehicle exhaust emissions between trips made with those two different settings.

Introduction

The largest potential to improve fuel use and reduce pollutant emissions in road transport probably lies in enhancing vehicle technology. However, such an approach involves a relatively large implementation time and considerable costs. An effective way to improve fuel economy in the short term is to aim at a change in driver behaviour and promote an environment-friendly driving style.

Environment-friendly driving includes different behavioural aspects to obtain a more fuel-efficient driving, one of them implying a selective use of gears. By shifting gear early one can avoid high engine speeds and therefore achieve a reduction of emissions and fuel consumption. When applying this kind of soft measures it is therefore important to assess the potential benefits of these actions.

Methodology

This section first describes how data about driving behaviour were collected and then presents how this information can be used to estimate the vehicle exhaust emissions caused by every trip.

Data collection

Data were obtained in a small-scale travel survey, collecting trip information from 32 participating respondents. During a period, varying from two days to 1 week, these respondents were asked to fill in a travel diary and to activate a personal digital assistant (PDA) with built-in Global Positioning System (GPS) receiver when making a trip. The use of this device allows acquiring accurate information about the travel behaviour of the individual respondents. It is able to collect second-by-second trip information (speed, location, etc) for every vehicle trip during the survey period. In total 303 vehicle trips were reported by the respondents of which 235 trips were recorded completely by the GPS receiver. After data processing, 235 speed profiles were developed for the calculation of emission estimates and fuel consumption.

The VeTESS emission model

Within the European Union (EU) 5th framework project DECADE (2001–2003), a vehicle level simulation tool was developed for the simulation of fuel consumption and emissions of vehicles in real traffic transient operation conditions. The final simulation tool, which is called Vehicle Transient Emissions Simulation Software (VeTESS), calculates emissions and fuel consumption made by a single vehicle during a defined "drive-cycle" [1]. The following description of the model can be found more detailed in [2].

In the project DECADE a new method for characterizing engine behaviour was developed, including the description of transient effects. Within the new measuring procedure, the effect on emissions and fuel consumption of sudden torque changes (in a step of about 0.2 s) at constant speed are recorded on an engine test bed. Based on three independent variables from the experimental procedure, namely engine speed, engine torque, and change in torque, four parameters are defined for each pollutant: steady state emission rate, jump fraction, time constant, and transient emissions. The steady state emission rate is the rate at which the pollutant is produced as the engine runs under steady state, i.e., at constant speed and torque. The jump fraction characterizes the fraction by which the emission rate increases or decreases after a change in torque not taking into account the dynamic behaviour. The time constant is a measure for the time required to approach the steady state emission value after a torque change. The transient emission is a discreet amount of additional pollutant generated after the change of torque. The overall emissions of the trip are obtained by adding up the emissions produced under the different load conditions during the drive cycle.

VeTESS calculates the emissions per second for CO_2, CO, NO_x, HC, and particulate matter (PM), but, for the moment, detailed engine maps are only available for three types of passenger cars: a Euro II LGV, a Euro III diesel car, and a Euro IV petrol car. Since all the participants in the survey drove a diesel car, most vehicle kilometres in Belgium are covered by diesel vehicles, and VeTESS produces the best results for diesel cars, this study was limited to the EURO III diesel car, described in Table 1.

Gear-changing behaviour

VeTESS uses specific gear change rules to determine the gear change points for the vehicle [4]. Three options are available to simulate aggressive, normal, or gentle style driving. A custom option is also available allowing

Table 1. The EURO III diesel car in the VeTESS emissions model [3]

Make of car	Skoda Octavia 1.9 Tdi
Engine size	1896 cm^3 diesel engine
Fuel system	Direct injection
Euro class	EURO III certified
Max. power	66 kW at 4000 rpm
Max. torque	210 Nm at 1900 rpm
Engine aspiration	Turbo + intercooler
Exhaust gas recirculation	Yes
Emissions control device	Oxidation catalyst

the user to alter the values to suit a particular driving style. In this study a comparison was made between the "normal" and the "aggressive" gear-changing behaviour and therefore only those two options will be discussed.

When selecting the "normal" gear-changing assumptions VeTESS will simulate average engine speeds and an average number of gear changes over a given route. "Normal" gear-changing settings will assume a gear shift to a higher gear when the engine speed exceeds 55% of the maximum engine speed. In case of the EURO III diesel car in the model, this maximum engine speed amounts 4800 rpm. The "aggressive" gear-changing assumptions on the other hand will allow higher engine speeds and less engine torque than values used during normal driving. This will result in a larger number of gear changes. When using this "aggressive" setting, gear shifting will occur at 80% of the maximum engine speed.

Analysis

The analysis aimed at answering the following question: What can be the influence of gear-changing behaviour on vehicle exhaust emissions? By using the recorded driving cycles as an input in the emission model VeTESS, emissions and fuel consumption were therefore calculated according to two different gear-changing assumptions. The relative difference of the emission and fuel-use estimates for those two assumptions was discussed.

Results

This section presents the results of the calculations where 235 real driving cycles where converted into emission estimates using the VeTESS emission tool. The results from the emission estimates are presented for the trips made by two different gear-changing assumptions: normal and aggressive.

Tables 2 and 3 present the calculated values for the total emissions and the emission factors, respectively. In Tables 2 and 3 the results indicate that an aggressive gear-changing behaviour will result in higher emissions of CO_2, NO_x, PM, and HC and in an increased fuel consumption. This conclusion accounts for the average total values as well as for the emission factors of those pollutants. The emissions of CO seem to be influenced differently since an aggressive gear shifting apparently implies a decrease of the average CO emissions per trip. A paired two-sided t-test was performed on

Table 2. Average total emissions and fuel consumptions per trip using two different gear-changing assumptions

	Fuel (L)	CO_2 (g)	CO (g)	NO_x (g)	PM (g)	HC (g)
Normal	0.76	1996.98	0.43	9.31	0.09	0.99
Aggressive	0.95	2475.43	0.28	11.91	0.14	1.24
t-test (p-value)	<0.05	<0.05	<0.05	<0.05	<0.05	<0.05

Table 3. Average emission factors and fuel consumptions per trip using two different gear-changing assumptions

	Fuel (100 L km^{-1})	CO_2 (g km^{-1})	CO (g km^{-1})	NO_x (g km^{-1})	PM (g km^{-1})	HC (g km^{-1})
Normal	7.15	186.59	0.05	0.97	0.01	0.09
Aggressive	9.19	240.21	0.03	1.1	0.01	0.12
t-test (p-value)	<0.05	<0.05	<0.05	<0.05	<0.05	<0.05

the results to check the differences between the values of different gear-changing settings. The statistical test demonstrated that the differences were all significant ($p < 0.05$).

Tables 4 and 5 present the relative difference of the "aggressive" estimates in comparison with the "normal" estimates to demonstrate the extra emissions one can cause by using an aggressive gear-changing behaviour. In these tables, the results indicate that one will increase the emissions of CO_2 and the average fuel consumption per trip by 30% when applying the aggressive gear-changing settings in the VeTESS model instead of the normal settings. This means that one can save an average of 30% of the fuel consumption per trip by avoiding an aggressive gear change.

The results for the NO_x emissions also indicate an increase of the emissions when the aggressive gear settings were applied. NO_x emissions will increase by 15% when changing gear at higher engine speeds. Concerning the emissions of PM and HC, the results show an average increase of the pollutant emissions of 41% and 38%, respectively. The impact on CO emissions on the other hand shows an average decrease of the emissions by 30%. Apparently the emissions of this pollutant are influenced differently than the pollutant mentioned before.

Table 4. Average total emissions and fuel consumptions using two different gear-changing assumptions. Relative difference of aggressive to normal settings (%)

	Fuel (L)	CO_2 (g)	CO (g)	NO_x (g)	PM (g)	HC (g)
Average %	29.45	29.47	−29.67	15.79	41.59	38.80
SD	8.41	7.88	37.90	17.02	29.44	14.11

Table 5. Average emission factors and fuel consumptions using two different gear-changing assumptions. Relative difference of aggressive to normal settings (%)

	Fuel (100 L km^{-1})	CO_2 (g km^{-1})	CO (g km^{-1})	NO_x (g km^{-1})	PM (g km^{-1})	HC (g km^{-1})
Average %	29.27	29.35	−29.38	15.74	22.84	38.58
SD	7.79	7.86	43.97	17.08	40.54	14.81

Discussion

The emission model VeTESS calculated the emission values based on second-by-second speed measurements of 235 vehicle trips. Both, the total amount of emissions as well as the emission factors were calculated according to two different gear-changing settings. The results indicate that, on average, the use of an aggressive gear-changing behaviour results in an increase of the emissions of CO_2, NO_x, PM, and HC. Fuel consumption will increase by 30% when using an aggressive gear use. The impact on CO emissions appears to be opposite to the other impacts and needs to be studied in future research. Apparently the emission of this pollutant is influenced differently when changing gears at higher engine speeds.

Another aspect that needs to be studied more into detail is the impact on the emissions of PM. The impact of an aggressive gear shifting on the emissions of this pollutant seems to display large variations between the different values. Possibly there is a difference between the different "kinds" of trips, but this assumption needs more research.

Further research should also include other vehicle types to study the impact of gear changing on vehicle exhaust emissions. This study was performed by using the engine details of only one EURO III diesel car, but the same methodology should be applied at other vehicles to gain more insight into the consequences of this kind of measures and obtain some general results. At the moment the VeTESS model is less accurate for gasoline vehicles. This is due to the important role the catalyst plays, an issue which should be investigated in more depth in the future.

Conclusions

This paper demonstrates how a selective use of gears can contribute to a more environment-friendly driving behaviour. By using the driving cycles of real trips as an input in the emission model, VeTESS the benefits of using a more gentle driving behaviour were estimated. When changing gears early, one can reduce fuel consumption by 30% on average and prevent the extra emissions due to running up the engine. Future research will include more vehicle types in the study to generate more results and the differences in impact between the different pollutants will be studied thoroughly.

References

1. Pelkmans L, Debal P, Hood T, Hauser G, Delgado MR (2004) Development of a simulation tool to calculate fuel consumption and emissions of vehicles operating in dynamic conditions. In: Proceedings of the society of automotive engineers international conference, June 2004, Toulouse, France, 2004-01-1873, ISBN 0-7680-1480-8
2. VITO (2003) Scientific report 2003. Energy, environment and materials. Responsible publisher: Dirk Fransaer
3. Beevers SD, Carslaw DC (2005) The impact of congestion charging on vehicle speed and its implications for assessing vehicle emissions. Atmos Environ 39:6875-6884
4. VeTESS User Manual. Software version V1.18B

Effect of speed reduction on emissions of heavy duty lorries

L Int Panis, I De Vlieger, L Pelkmans, L Schrooten

Flemish Institute for Technological Research (VITO), Boeretang 200, B-2400 Belgium

Abstract

In many European countries the speed limit for trucks is under discussion or review. The speed limit for heavy trucks is 80 km h^{-1} in most countries, but 90 km h^{-1} in Belgium. We investigated the effect of reducing the speed limit on fuel consumption and emissions of CO_2, NO_x, and particulate matter (PM).

To ensure robust conclusions under a strict deadline, our evaluation used two existing, complementary approaches: the macroscopic emission model TEMAT and the microscopic emission simulation model VeTESS. Both models show a CO_2 reduction between 5% and 15%. The results for NO_x and PM were ambiguous.

Introduction

In June 2005 the Flemish Transport Minister proposed to lower the maximum speed for trucks on highways from 90 to 80 km h^{-1}. This resulted in an enormous wave of critique from various stakeholders. Reference was made to time losses, economic losses, and serious doubts were cast over the assumed environmental and safety benefits. Unfortunately most of the discussion was on the basis of ideology and prejudice. Scientific analysis was either ignored or was unavailable for use in the discussion at that time.

Given the strict deadlines for the debate in parliament, the Flemish Institute for Technological Research (VITO) conducted a fast screening to assess the possible effects on emissions of the proposed policy. We defined the scope of the study to include only two scenarios:

1. A reduction of the maximum speed from 90 km h^{-1} to 80 km h^{-1} for trucks (12–40 t; equipped with a speed limiter) on motorways.
2. A reduction of the maximum speed of trucks having a gross weight between 3.5 and 12 t (those currently not equipped with a speed limiting device) from 100 km h^{-1} to 90 or 80 km h^{-1}.

Only the direct effect of the speed reduction on the emissions of trucks was taken into account. Second and third order effects on the speed distribution of other vehicles and their emissions were not included in this screening. For this reason the third question of the minister, about the effect of a complete ban on overtaking by trucks, was not answered because it would require the use of sophisticated traffic simulation models, which was far outside the scope of the allocated budget.

Methodology

Scope, general approach, and models used

Given the context it was decided to perform a screening based on existing tools to ensure that existing scientific knowledge would be used in the political discussions. In this respect it reflects a common situation for scientists involved in policy advice. Because there was broad agreement that the emissions to air would be the dominant impact, it was decided to estimate the emissions for the three most important pollutants in the transport sector: CO_2, NO_x, and particulate matter (PM) [1,2]. To ensure robust conclusions the problem was tackled from three angles simultaneously. We discuss the results of two complementary models and compare them with results from an inquiry among truck manufacturers.

The fleet-based emission model TEMAT

Fleet emission factors (in grams per vehicle-km – vkm) were estimated with the macroscopic model TEMAT [3]. The model essentially uses well established COPERT III emission functions but was updated with information from our own emission measurements [4] and results from the Emission Factor Handbook [5,6], the COST 346 report, as well as the European

ARTEMIS project (mainly for NO_x and PM) because earlier versions tended to underestimate the "real" on the road emission for some of the more advanced engines (Euro II and later standards). The model was fed with both theoretical maximum speeds as well as generalized average speeds on highways (which are typically between 3% and 6% lower).

There are two disadvantages to using the TEMAT model in this study:

1. A break at 12 t, important from a policy viewpoint, was unavailable from the TEMAT model (developed from a technical perspective for optimal emission modelling cfr. COPERT III).
2. COPERT functions for heavy duty vehicles are inherently unreliable at speeds above 100 km h^{-1}.

The vehicle-based VeTESS model

The same scenarios were also evaluated with a completely different model. VeTESS is a microscopic model that estimates fuel consumption and emissions for a single vehicle on the basis of specific (second-by-second) speed profiles, gear choice and the efficiency of all elements of the power train, and other vehicle characteristics. VeTESS calculates total engine power demand and uses 3D engine maps to estimate emissions.

The main disadvantage of this complex model is that detailed dynamical engine maps are available for only three trucks. All three comply with the Euro II standard. Nevertheless, differences found between these trucks may hint at the reliability of the results. The reader is referred to [7] for a detailed description of the VeTESS model.

For the current situation the model was run with a compilation of speed profiles measured on Flemish highways in normal traffic. The maximum speed is legally limited to 90 km h^{-1} and the average real speed is approximately 86–87 km h^{-1}. Small variations that occur between 85 and 90 km h^{-1} can be attributed to the presence of other vehicles. The measured speed profiles were then converted to lower speeds to reflect a change in the legal speed limit. The speed variation was left unchanged whereas the average speed in scenario 1 was 77–78 km h^{-1}.

Integrated analysis and uncertainty

To arrive at an integrated conclusion and test its robustness, emission reductions for the entire fleet were converted to external environmental costs using the methodology described in [8]. Uncertainty was assessed by Monte Carlo (MC) techniques in which single value assumptions were replaced

Results

General context

Heavy trucks (32–40 t) are the most important emitters for each of the pollutants studied. Their emissions equal at least two-thirds of the total emissions (of trucks) because of their large share in the total mileage driven on highways. Smaller trucks are driven less far, less frequently, and relatively more on other road types.

Results for policy scenario 1

In Table 1 the relative emissions under scenario 1 are given. Summarizing we can say that the total CO_2 emission would decrease by 5–10%. This trend is consistent for all weight classes and years. Results from the detailed vehicle-based modelling (VeTESS) confirm the results for CO_2

Table 1. Emissions at 80 km h^{-1} relative to 90 km h^{-1} (From TEMAT)

Emission (t)	Max. speeds (%)			Real speeds (%)		
	2005	2010	2020	2005	2010	2020
CO_2 3.5–7.5	86	86	86	86	86	86
7.5–16	86	86	86	89	89	89
16–32	92	92	92	95	95	95
32–40	91	91	91	94	94	94
Fleet average	91	91	91	94	94	94
NO_x 3.5–7.5	84	84	84	84	84	84
7.5–16	89	89	89	93	93	93
16–32	94	94	94	97	97	97
32–40	105	105	105	105	105	105
Fleet average	102	102	103	103	103	103
PM 3.5–7.5	97	97	97	98	98	97
7.5–16	102	102	102	103	103	102
16–32	97	97	98	98	98	98
32–40	106	106	106	105	105	106
Fleet average	103	104	104	103	104	104

(Table 2), which makes the results more credible. In absolute numbers the CO_2 emission factors would on average drop by approximately 100 g km^{-1} if the policy resulted in a decrease from 90 to 80 km h^{-1}. Using more realistic estimates of the impact of the policy on real traffic speeds yields a reduction of only 50–70 g km^{-1}.

For the other pollutants, TEMAT predicts an increase in the fleet average emissions of NO_x and PM with 2–3% and 3–4%, respectively (Table 1). Emission factors of NO_x derived from both models show a decrease for most types of trucks (3.5–32 t). In sharp contrast an increased emission is simulated for the heaviest trucks (+0.2–0.5 g km^{-1}) (Fig. 1). As a result the fleet averaged emission factor is also higher.

The results for PM are even more confusing (Fig. 2). PM emission factors decrease for the 3.5–7.5 and 16–32 t weight classes and increase for the 7.5–16 and 32–40 t weight classes. All changes (increases and decreases) become smaller in the future. Because of the dominance of the largest trucks the fleet average emission factor also increases.

A high R^2 was reported in MEET for this emission function, indicating it was based on a small sample. The lack of consistency between the effects for the different classes indicates a large amount of uncertainty.

Detailed modelling of the engine and vehicle characteristics with VeTESS provides us with an additional set of results (Table 2). Emissions of NO_x are expected to decrease for the selected vehicles, while only one vehicle showed a very small increase in PM emissions.

Faced by these partially conflicting and uncertain findings we looked for tools to convert the result into one integrated conclusion. The observed emission differences where converted to (avoided) external costs [8]. In this way we could balance the positive effect for CO_2 against the impacts for NO_x and PM by expressing them in similar monetary units. All the assumptions were converted to probability distributions and used in MC

Table 2. Relative emissions for different speed reductions (From VeTESS)

	CO_2 (%)	NO_x (%)	PM (%)
Scenario 1: 90 km h^{-1} · 80 km h^{-1}			
IVECO Eurocargo 7,500 kg	84	71	84
IVECO Eurocargo 12,000 kg	86	72	100
MAN 30,000 kg	91	89	103
Scania 30,000 kg	90	85	n.a.
Scenario 2: 100 km h^{-1} · 90 km h^{-1}			
IVECO Eurocargo 7,500 kg	73	85	71
IVECO Eurocargo 12,000 kg	80	88	67

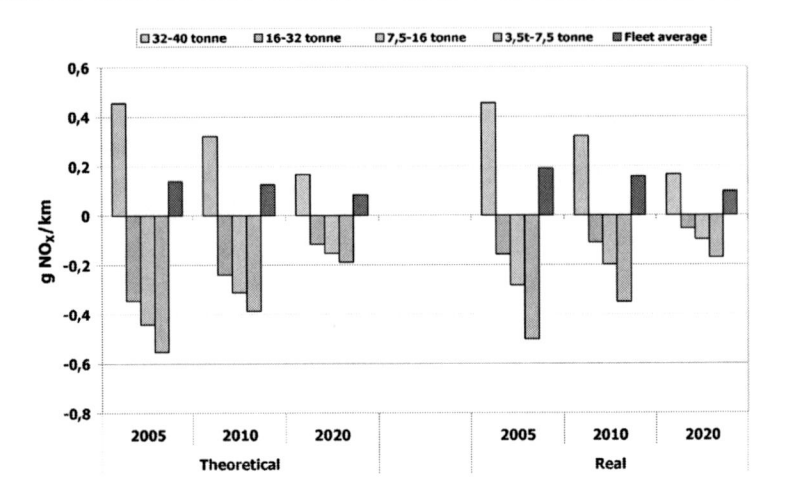

Fig. 1. Scenario1, difference in the NO_x fleet emission factors (From TEMAT)

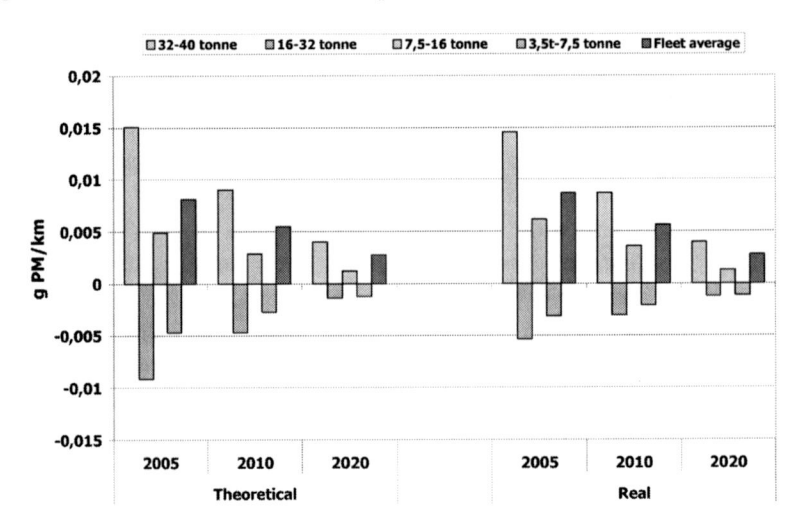

Fig. 2. Absolute difference in the PM fleet emission factors for the 3.5–32 t trucks, 90 km h^{-1} compared to 80 km h^{-1} (From TEMAT)

Table 3. Avoided external costs under scenario 1 (median, 95% CI)

2005	6 million euros (48; −34)	62% chance for positive Outcome
2020	10 million euros (38; −8)	85% chance for positive Outcome

Table 4. Relative emissions for lighter truck categories (From TEMAT)

$100 \geq 80$ km h^{-1}	Emissions (t)	Max. speed (%) 2005	2010	2020	Real speed (%) 2005	2010	2020
CO_2	3.5–7.5	72	72	72	73	73	73
	7.5–16	73	73	73	77	77	77
NO_x	3.5–7.5	69	69	69	69	69	69
	7.5–16	76	76	76	82	82	82
PM	3.5–7.5	91	91	91	92	93	93
	7.5–16	103	103	102	104	104	105

analysis. The total benefit of the proposed policy was found to be positive and likely to increase. Nevertheless, the range of results was wide and the possibility that a speed reduction could have negative environmental impacts could not be ruled out (Table 3).

Results for policy scenario 2

This scenario only looks at effects on trucks below 12 t because they are the only class that can drive faster than 90 km h^{-1}. Policy options for these vehicles include the installation of mandatory speed limiters like those used on all heavier trucks at this moment (90 km h^{-1}) or reducing their speed even further to 80 km h^{-1} similar to the proposed policy under scenario 1 for heavier trucks.

In the results presented in Table 4 we see that there is a consistent and significant decrease in the emissions of CO_2 and NO_x (−13% to 28% and −12% to 31%, respectively). For PM there is a difference between both weight classes. The PM emissions for the smallest category (3.5–7.5 t) would decrease by 3–9% but trucks in the 7.5–16 t class could see their PM emissions increase by up to 5%.

Discussion

The finding that emissions of CO_2 decrease but emissions of NO_x and PM could increase is consistent with the results from other studies. Using other models and different fleets [5,11] similar results were obtained. The COST 346 [12] working group decided that a further speed reduction (below 80 km h^{-1}) would not affect fuel consumption but would increase emissions of NO_x and PM. The large uncertainties are blamed on the fleet composition [5,12] and on difficulties to derive a typical driving pattern [13].

The choice of gear features among the most prominent changes to the driving pattern and is likely to be influenced by changes to the speed limit. This may be more important in urban locations than on highways. The VeTESS model was therefore used to study the effect of different gear-shifting strategies in connection with different speed limits (see [2] for details). Our conclusions were confirmed for any gear-shifting strategy for any speed reduction down to 80 km h^{-1} although we found some variation in the magnitude of the effect. Further speed reductions below 80 km h^{-1} however, resulted in much higher emissions for some (but not all) trucks although the fuel consumption remained fairly stable. These results were presented to and discussed with both individual manufacturers and the Advisory Committee on Environmental Aspects (ACEA) expert group. It is clear that they design and build long distance haulage trucks to minimize fuel consumption at the most prevailing speed limits in Europe (80 km h^{-1} on highways). The optimum is between 80 and 85 km h^{-1}, which confirms our findings. They would, however, not elaborate on effects on the other emissions because only the standardized driving cycles are used in the engine certification.

Under scenario 2 we found that all changes in emissions are larger when the speed reduction is more important. On the other hand the effects are expected to decrease significantly in the future with more advanced fleets. The absolute difference then becomes very small and given the small share of light trucks on highways, the overall effect is likely to be negligible. Manufacturers confirm that fuel consumption is not perceived as an issue in this case. Manoeuvrability and other qualities prevail.

Conclusions

1. All results consistently indicate that lower maximum speeds for trucks on motorways result in lower emissions of CO_2.
2. Results for NO_x and PM are not consistent and uncertain but probably too small to offset the clear benefits of the CO_2 reduction.
3. The chances that this policy has environmental benefits increases over time (with future fleets that are technologically more advanced).
4. The magnitude of the overall environmental benefits is very uncertain but averages between 6 and 10 million euros per year.
5. Scenario 2 has clear environmental benefits, but the magnitude of the effects is much smaller than those of scenario 1.

This study was heavily debated in parliament and on national radio and TV. Although it answered some questions and raised others, people found it hard to ignore the conclusions. Paradoxically none of the parties referred to the large uncertainties. The discussion faded away and later reports on safety and economic effects drew less attention. The proposal was later discussed by the Belgian government that recently decided not to change the speed limits but to impose a ban on overtaking on all two-lane motorways.

References

1. MIRA (2004) Flemish environment. Report at www.milieurapport.be (in Dutch)
2. De Vlieger I, Schrooten L, Pelkmans L, Int Panis L (2005) 80 km/h for trucks. Report to the Flemish ministry of transport, 49 pp
3. De Vlieger I, Pelkmans L, Verbeiren S, Cornelis E, Schrooten L, Int Panis L (2005) Sustainability assessment of technologies and modes in the transport sector in Belgium, Belgian science policy, Brussels
4. Lenaers G, Pelkmans L, Debal P (2003) The Realisation of an on-board emission measuring system serving as a R&D tool for ultra low emitting vehicles. Int J Veh Design 31:253–268
5. HBEFA (2004) Infras handbook emission factors for road transport. Version 2.1, Vienna 2004
6. R. Pischinger et al. (2002) Update of the emission functions for heavy duty vehicles in the handbook emission factors for road traffic. TU Graz, December 2002
7. Pelkmans L, Debal P, Hood T, Hauser G, Delgado MR (2004) Development of a simulation tool to calculate fuel consumption and emissions of vehicles operating in dynamic conditions. SAE 2004 Spring Fuels & Lubricants, 2004-01(1873), SAE Int., Warrendale, PA
8. Int Panis L, De Nocker L (2001) Belgium. In: Friedrich R and Bickel P (eds.) Environmental external costs of transport. Springer Berlin
9. Rabl A, Spadaro J (1999) Damages and costs of air pollution: an analysis of uncertainties. Environ Int 25:29–46
10. Int Panis L, De Nocker L, Cornelis E, Torfs R (2004) An uncertainty analysis of air pollution externalities from road transport in Belgium in 2010. Sci Tot Environ 334–335:287–298
11. IEA (2005) Saving oil in a hurry, OECD/IEA, ISBN 92-64-109414
12. COST 346 (2005) Emission and fuel consumption from heavy duty vehicles. Draft final report, August 2005
13. Cornelis E, Broekx S, Cosemans G, Pelkmans L, Lenaers G (2005) Impact of traffic flow description and vehicle emission factor selection on the uncertainty of heavy-duty vehicle emission calculation, Vol. 85. VKM-THD, Mitteilungen

Gender-linked disparity in vehicle exhaust emissions? Results from an activity-based survey

C Beckx,[1,2] L Int Panis,[1] M Vanhulsel,[2] G Wets,[2] R Torfs[1]

[1]Flemish Institute for Technological Research (VITO), Boeretang 200, 2400 Mol – Belgium
[2]Transportation Research Institute, Hasselt University, Wetenschapspark 5 Bus 6, 3590 Diepenbeek, Belgium

Abstract

This study explores the relationship between the vehicle exhaust emissions caused by a trip and the characteristics of the driver involved. The hypothesis formulated is that certain "groups" of individuals produce more emissions (per kilometre) than others and therefore should be treated differently when aiming vehicle emission reduction. To support this hypothesis an activity-based (AB) survey collected speed profiles and driver characteristics of different car drivers. The speed profiles of the individual trips served as input for the emission model Vehicle Transient Emissions Simulation Software (VeTESS), to calculate the instantaneous emissions made by a single vehicle. This paper reports on the differences in vehicle exhaust emissions between trips made by men and women.

Introduction

Policy measures to reduce vehicle exhaust emissions often include campaigns to induce efficient trip chaining, environment-friendly driving behaviour, or reduced car use. When applying these soft measures it is not just important to assess the potential benefits but also to determine the proper target group for each of these actions to obtain the most efficient results. People displaying different travel behaviour will, for example, respond to different policy measures.

To gain an insight into the travel behaviour of people travel surveys can provide useful information. A travel survey generally starts with standard questions about the person and then questions his/her travel behaviour. As a result the travel behaviour can often be classified according to certain personal characteristics like age, education, and household structure.

A recently developed travel survey is the activity-based (AB) survey, asking people to fill in an activity diary during several days. The data collected through this survey can then be used to develop an AB transportation model, which aims at predicting the trips and the activities conducted by people [1]. To acquire this kind of information in more detail but still taking into account the respondent's burden, new data collecting technologies are being developed [e.g., 2]. Accurate information on facets like activity, location, and route choice, for example, are nowadays often obtained through the use of a personal digital assistant (PDA) with built-in Gobal Positioning System (GPS) receiver.

Methodology

This section describes how the database with details on emissions and driving behaviour was obtained and how the data were processed.

Data collection

Data were obtained in a small-scale AB survey collecting activity diary data using self-reporting of activities and trips by respondents in a paper activity diary. A PDA with built-in GPS receiver was used to acquire information about the exact location of activities and to provide more accurate information on the reported trips (route choice, trip distance, driving speed, etc.)

This study was actually intended as a pilot survey preceding a large-scale AB survey (2500 households) for the development of an AB transportation model. Therefore the original data set from this small-scale survey contained only information of 32 respondents, 15 men and 17 women, varying in age, education, income, etc. They all filled in the activity diary for a period varying from 2 days to 1 week. In total 1014 trips were reported in the paper activity diaries of which 303 trips were made as a car driver.

Data processing

After data collection, data were organized and converted into usable formats. Trips were classified according to different variables like trip purpose, gender of the driver, age of the driver, number of accompanying persons, etc. The GPS logs, consisting of second-by-second information on location, time, speed, and date were downloaded from the GPS receiver. Next, these NMEA, GPRMC sentences were converted into formats usable for further analysis. The following step in the data processing was linking the activity diary data and the GPS logs based on the trip departure and ending times. A manual check was finally performed to ensure that all the GPS records associated with vehicle trips were included in the analysis and, if necessary, trip timings were adjusted. After this processing procedure 235 vehicle trips remained.

Emission modelling

Within the European Union (EU) 5th framework project DECADE (2001–2003), a vehicle level simulation tool was developed for the simulation of fuel consumption and emissions of vehicles in real traffic transient operation conditions [3]. A specific task in the project was to develop and validate a method for calculating very accurately dynamic emissions, and thereby reaching higher accuracy than traditional emission simulation modelling. The final simulation tool, which is called Vehicle Transient Emissions simulation Software (VeTESS), calculates emissions and fuel consumption made by a single vehicle during a defined "drive-cycle" [3]. The VeTESS emissions model uses new methods based on experimental characterization of engines and aims to provide a more realistic simulation by incorporating transient engine behaviour [4]. Together with the associated speed profiles, the actual power demands allow a detailed calculation of emissions.

VeTESS calculates the emissions per second for CO_2, CO, NO_x, HC, and particulate matter (PM), but for the moment detailed engine maps are only available for three types of passenger cars: a Euro II LGV, a Euro III diesel car, and a Euro IV petrol car. Since all the participants in the survey drove a diesel car and, moreover, most vehicle kilometres in Belgium are covered by diesel vehicles, this study was limited to the EURO III diesel car, described in Table 1. The other assumptions used in the model include flat terrain, "normal" driving, and gear change assumptions, and no air conditioning or additional payload carried by the vehicle.

Table 1. The EURO III diesel car in the VeTESS emissions model [4]

Make of car	Skoda Octavia 1.9 Tdi
Engine size	1896 cm^3 diesel engine
Fuel system	Direct injection
Euro class	EURO III certified
Max. power	66 kW at 4000 rpm
Max. torque	210 Nm at 1900 rpm
Engine aspiration	Turbo + intercooler
Exhaust gas recirculation	Yes
Emissions control device	Oxidation catalyst

Speed profiles, based on the instantaneous speed data from the GPS receiver, were composed for every detected vehicle trip and used as input for the VeTESS tool. The model output consisted of second-by-second emission data, total emission data, and emission factors for every trip.

Driving behaviour

A set of driving parameters was calculated based on the speed and acceleration profiles from the GPS receiver. The driving parameters applied in this study include average speed, average positive acceleration, relative positive acceleration (RPA), and the percentage of stop time (PST) (see Table 2). RPA is calculated from the power that is needed for all vehicle accelerations in the cycle, divided by the distance driven. It gets high when the driving pattern includes a lot of high power-demand accelerations and is found to increase exhaust emissions and fuel consumption [5]. RPA is calculated as:

$$\frac{1}{x}\int_0^T va^+ dt,$$ where TM = total cycle time (s), v = speed (m s^{-1}), a^+ = positive acceleration (m s^{-2}) and x = total distance (m).

Table 2. Driving pattern parameters for the study

Driving pattern parameter	Denotation	Unit
Average speed	v_{avg}	km h^{-1}
Average positive acceleration	a^+_{avg}	m s^{-1}
Relative positive acceleration	RPA	m s^{-1}
Percentage of stop time	PST	%

Analysis

Vehicle trips were classified according to the characteristics of their driver. The analysis aimed at answering the following question: Is there a relationship between the "type" of the car driver and its driving behaviour and/or amount of vehicle emissions caused per trip? To answer this question, the calculated emission factors and driving parameters had to be linked with the trip driver information. Since the trip number was attached to every calculated value, each driving pattern and emission estimate could be coded with information concerning the driver of the vehicle trip. In this study, due to the small scale, trips were classified either as made by a man, or as driven by a woman. An analysis was then performed to find out if there was a gender-linked disparity in vehicle exhaust emissions.

Results

This section presents the results from the small-scale survey where 32 respondents participated in an AB survey with GPS tracking technology. Both the results from the emission estimates and the driving parameters are presented for the trips made by men and the trips made by women. In total 235 trips were analysed.

Emission estimates

The emission model VeTESS calculated the emission values based on second-by-second speed measurements. Both the total amount of emissions as well as the emission factors were calculated. Tables 3 and 4 present, respectively, the average total emission values and the average emission factors per trip as calculated by the emission model. An unpaired two-sided t-test was performed to check the differences between the values of different trip purposes ($p < 0.05$). The results in Table 3 show clearly that there is a difference between the total emission values of trips made by men and women. In the first place, the average distance of a trip driven by a man is almost twice as high comparing to the average distance of trips made by woman ($p < 0.05$). Mainly due to this, the total fuel consumption and the total amount of emissions per trip are also higher for trips made by men. When taking into account the distance values and calculating the emission factors for every trip, Table 4 shows that the differences between men and women remain significant. The mean emission factors for trips made by women are always higher than the values for trips made by men ($p < 0.05$).

Table 3. Distance, total emissions, and fuel consumption. Averages per trip and per gender

	Distance (km)	Fuel (L)	CO_2 (g)	CO (g)	NO_x (g)	PM (g)	HC (g)
Male	16.34	1.05	2733.99	0.56	12.86	1.36	0.11
Female	9.9	0.63	1651.51	0.36	7.65	0.82	0.08
t-test	<0.05	<0.05	<0.05	<0.05	<0.05	<0.05	<0.05

Table 4. Emission factors and fuel consumption. Averages per trip and per gender

	Fuel (100 L km^{-1})	CO_2 (g km^{-1})	CO (g km^{-1})	NO_x (g km^{-1})	HC (g km^{-1})	PM (g km^{-1})
Male	6.77	176.78	0.04	0.89	0.01	0.08
Female	7.32	191.18	0.06	1.00	0.01	0.09
t-test	<0.05	<0.05	<0.05	<0.05	<0.05	<0.05

Driving behaviour

Driving parameters used in this study include average speed (vavg), average positive acceleration (a^+_{avg}), RPA and PST. Table 5 presents the calculated driving parameter values for the trips in this pilot study.

The average driving speed for trips made by men is significantly higher than the average speed for trips made by women ($p < 0.05$). Trips with a male driver have an average speed of 40.45 km h^{-1} whilst the trips with female chauffeurs are driven at an average speed of 33.30 km h^{-1}. The parameters concerning the positive acceleration, a^+_{avg} and RPA, and the driving parameter PST are all higher for trips made by women ($p < 0.05$).

Table 5. Driving parameters. Averages per trip and per gender

	v_{avg} (km h^{-1})	a^+_{avg} (m s^{-2})	RPA (m s^{-2})	PST (%)
Male	40.45	0.56	0.22	20.13
Female	33.30	0.64	0.25	26.57
t-test	<0.05	<0.05	<0.05	<0.05

v_{avg} average driving speed; a^+_{avg} average positive acceleration; RPA relative positive acceleration; PST: percentage of stop time

Discussion

The methodology and results presented in this paper demonstrate that useful information can be obtained by enlarging an AB survey with GPS technology and linking information of the car driver to other parameters, like emission estimates and driving parameters. This kind of travel survey does not only provide the necessary information for the calculation of all these parameters, but also offers useful information on the travel behaviour of people. Information is provided on different facets like the reason of the trip, the trip distance, and the driving behaviour. Individuals displaying the same travel behaviour can be classified into groups, e.g., according to age, income, gender, etc. When aiming policy measures to reduce the traffic air pollution, this kind of information can then be used to work with "target groups" for every action. Since every "group" will respond differently to policy measures, these actions can then be tuned to the proper target group. Based on the results of the pilot study one could, for example, suggest focusing on the target group "men," when considering policy measures like car pooling or the use of public transport, since they seem to make only a few trips a day, mainly for work. An analysis of the travel behaviour of women, on the other hand, reveals that women seem to combine several short trips (going to work, shopping, bringing/getting children, etc.), limiting their possibilities to leave the car at home. But since the driving behaviour of women apparently is the main cause of the higher emission estimates per kilometre, this target group would benefit more from other policy measures, like environment-friendly driving tips.

Still there are some aspects that need to be considered when applying this methodology. Firstly, the travel survey procedure needs to be updated with more recent technology reducing the respondent's burden and increasing the detail of information. Trip detection needs to be done automatically and the paper work needs to be reduced. Secondly, the emissions should be calculated more realistically. At present, the vehicle emissions are calculated assuming only one vehicle type for all the recorded speed profiles. This assumption needs to be validated since changes in vehicle type have an impact on emission simulations. On the other hand, the use of only one vehicle type offers advantages for the analysis excluding the influence of the vehicle type. Another aspect that needs our attention deals with gear-changing behaviour. When calculating the emissions, the default gear-changing values were used as provided within the VeTESS model. Since gear-shifting behaviour can have a great influence on the emission exhaust, this needs to be taken into consideration. If possible, future research could include an in-vehicle tool recording this information.

The first results from this pilot study demonstrate the application of the developed methodology and indicate the field of application. Real explications and conclusions based on the results will require more data. Future research will therefore also include a large scale survey.

Conclusions

This paper demonstrates the methodology to link information on the driver of a trip to driving parameters and emission estimates. This kind of approach provides useful information when aiming target groups for policy measures. This method includes the completion of an AB survey extended with GPS technology and the calculation of parameters like vehicle exhaust emissions and driving parameters. In this pilot study, differences were found between these parameters depending on the gender of the trip driver. Women seem to emit more emissions per kilometre than men, but more data are needed to acquire meaningful results and explanations. Future research will therefore include the application of the developed methodology on a large-scale survey.

References

1. Ettema K, Timmermans H (1997) Theories and models of activity patterns. In: Ettema, K., Timmermans H. (eds.) Activity-based approaches to travel analysis. Pergamon, Oxford
2. Kochan B, Janssens D, Bellemans T, Wets G (2005) Collecting activity-travel diary data by means of a hand-held computer-assisted data collection tool. Proceedings of the 10th EWGT meeting/16th mini EURO conference, September 2005, Poznan, Poland
3. Pelkmans L, Debal P, Hood T, Hauser G, Delgado MR (2004) Development of a simulation tool to calculate fuel consumption and emissions of vehicles operating in dynamic conditions. In: Proceedings of the society of automotive engineers international conference, June 2004, Toulouse, France, 2004–01–1873, ISBN 0–7680–1480–8
4. Beevers SD, Carslaw DC (2005) The impact of congestion charging on vehicle speed and its implications for assessing vehicle emissions. Atmos Environ 39:6875–6884
5. Ericsson E (2005) Variability in driving patterns over street environments in three European cities. Presented at the 14th international conference "transport and air pollution", June 2005. Graz, Austria. Published in proceedings of the same conference

Floating Automotive Data Collection

A Schaerf,[1] S Kumra,[2] R Mazara,[3] L Pelkmans,[4] E Verhaeven[4]

[1]University of Udine, Italy
[2]NetPEM, New Delhi, India
[3]Delphi, South America Technical Centre, Brazil
[4]Flemish Institute for Technological Research (VITO), Mol, Belgium

Abstract

In its early stage, tracking devices were primarily used to locate stolen vehicles. However functionalities can be expanded very far. Built-in functions can be added and extra sensor signals can be added to the standard registrations. Especially, fleet management solutions are envisaged currently. With the use of intelligent transport systems, service companies can provide live traffic data, picked up from sensors that monitor vehicle speed, combined with incident reports. The tracking device can be in direct contact with a central server or can send regular reports and data.

Within the Floating Automotive Data Collection (FADC) project (carried out with the support of the European Union (EU)–Asia Information Technology and Communications (IT&C) Programme) experiments were done on the combination of a tracking device with automated traffic jam detection. On specific routes in Belgium and in India, speeds and acceleration profiles were monitored as a function of location, in order to automatically detect deviations from normal fluent traffic. The detection can be done on-board the vehicle and in case of traffic jam detection, the device will send a message to a central server, which maps different traffic jam detection massages. The system will of course operate optimal if a sufficient number of vehicles in the traffic are equipped with such device.

Introduction

FADC project

The project Floating Automotive Data Collection (FADC) is carried out with the support of the EU–Asia Information Technology and Communications (IT&C) Programme [1]. The project involves three partners: Flemish Institute for Technological Research (VITO) in Belgium, the University of Udine in Italy, and Network for Preventive Environmental Management (NETPEM) in India [2].

The project aims at providing information and tools for helping Asian small and medium enterprises (SMEs) to optimize transport management from a starting point to a destination point and hence to minimize fuel consumption and environmental pollution. Direct input would be provided for optimized transport routing techniques and automatic detection of traffic jam situations.

The main activities in the project envisage the calibration of the fleet management functionalities of a mobile and in-built vehicle device to local requirements in India. Activities include dynamic floating data collection, quantification of fuel consumption, and pollutant emissions related to traffic conditions, organization of dedicated workshops, and development of a reference framework for pilot applications in specific types of roads in Delhi, India.

Vehicle tracking and traffic jam detection

In its early stage, tracking devices were mostly used to locate stolen vehicles. However functionalities can be expanded very far. Built-in functions can be added and extra sensor signals can be added to the standard registrations. Especially fleet management solutions are envisaged currently. With the use of intelligent transport systems, service companies can provide live traffic data, picked up from sensors that monitor vehicle speed, combined with incident reports. The tracking device can be in direct contact with a central server or can send regular reports and data.

Within the FADC project the idea of combining a tracking device with automated traffic jam detection was looked into. The idea was that an in-built tracking device monitors speeds and acceleration profiles as a function of location, and automatically detects deviations from normal fluent traffic. The detection is done on-board the vehicle and in case of traffic jam detection, the device sends a message to a central server, which maps different

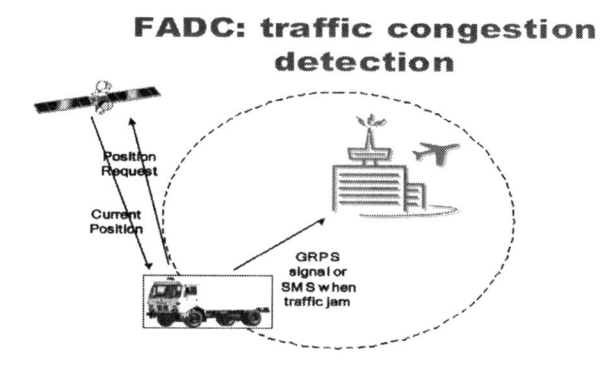

Fig. 1. System for automated traffic jam detection [2]

traffic jam detection massages. The system will of course operate optimal if a sufficient number of vehicles in the traffic are equipped with such device.

Working principle

There are two main parts in the entire system. On the one hand, reference data needs to be collected and typical speed profiles need to be determined for various roads (off-line). This database can then be used in the on-board tracking system to detect if there is a traffic jam.

The off-line component, called "instructor" runs on a server, and performs two main tasks:

- Makes a statistical analysis of data collected along a set of training runs and manages a set of *road profiles*. Each road profile contains a set of statistical information about a single road segment. This module can handle information collected also from non-systematic runs and is able to update the road profile accordingly.
- Allows the user to build a custom *route profile* (i.e., a collection of road profiles) by extracting it from the road profiles database. A route profile is used as configuration data for the detector module.

The second part is the "detector", which actually runs on the on-board device and relies on the *route profile* provided by the instructor. This software module analyses the real-time data coming from the sensors of the on-board device and checks them against the profile in order to detect abnormal traffic conditions. The traffic jam detection algorithm is running in the vehicle and is comparing the actual speed with the speed stored in the profile of that road previously calculated based on test drives.

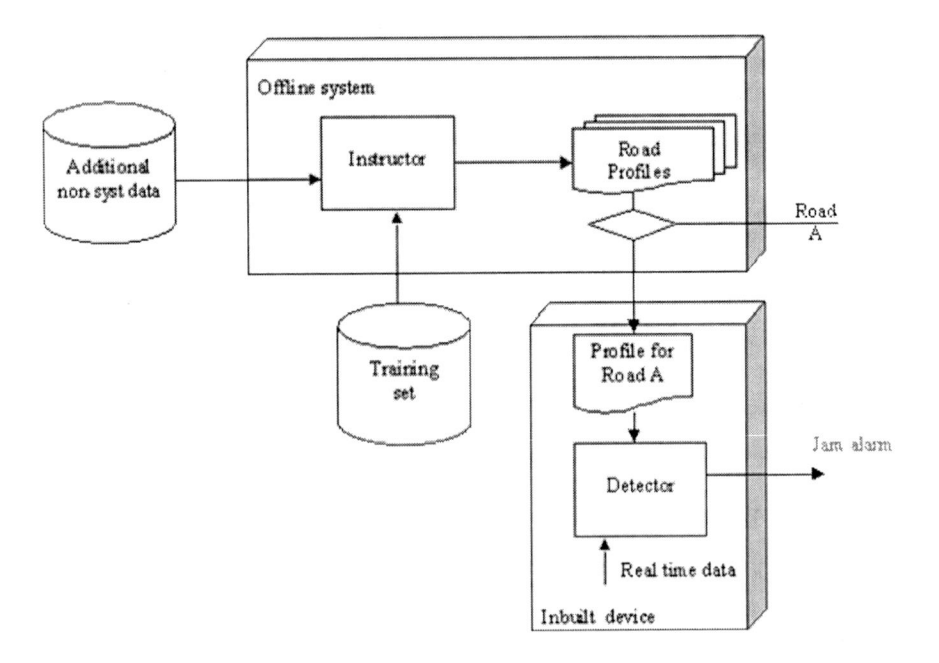

Fig. 2. General architecture of the system

Instructor

The instructor is provided with a *training set* database, made up of systematic runs on given routes and (possibly) with *non-systematic data* collected during additional runs.

The training set data are collected in a set of *test measurements* performed on a set of roads selected in an a priori design phase. The design phase consists in identifying the set of road of interest (each of them is called a *segment*) by means of two *way-points*, i.e., the geographic position of the beginning and of the end of the road. Furthermore, each segment has associated some information about road type (urban, rural, or highway) and the maximum speed limit on that road. In the data collection for the training set, the driver signals the presence of a traffic jam, so that the sensors data are augmented with a field that records this event. This allows to recognize and to discard those data from the training set and permits to perform the statistical analyses on "normal conditions" only.

Detector

The detector is a control module that processes the real-time data coming from the on-board sensors and the Global Positioning System (GPS) device in order to identify abnormal traffic situations. The configuration of the detector is given by a route profile, generated by the instructor as the output of a user request.

The data is sampled by the microcontroller on the on-board device at fixed intervals. Each sample includes, among other data, the position of the vehicle (latitude, longitude expressed in degrees, primes, and decimals of primes), the speed, the current time, and a validity flag.

The detector processes each sample in real time and maintains a record called *traffic status* that summarizes the history of the run. For each sample, based on its values and on the record, the detector issues a traffic alert level (the possible values are *normal, low alert, high alert,* and *jam*). Figure 3 shows the flow diagram of the different steps in the detector algorithm.

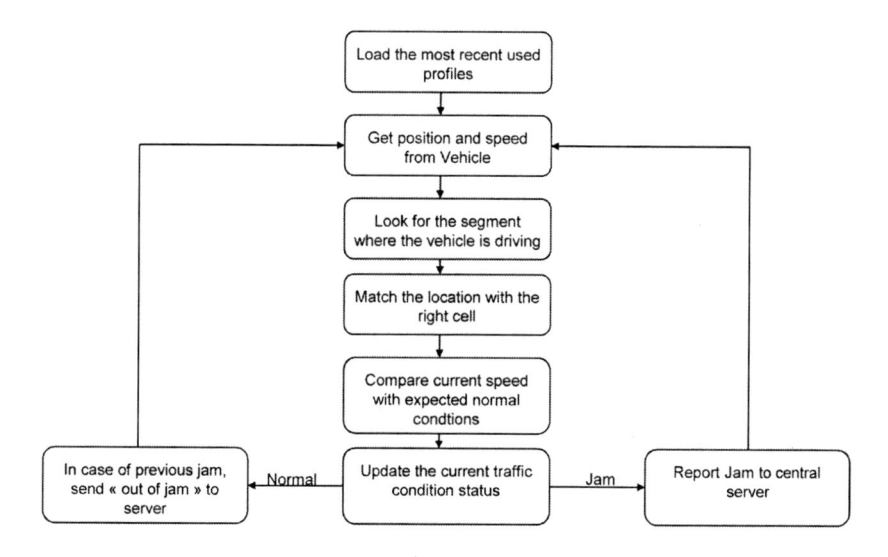

Fig. 3. Flow diagram of detector algorithm

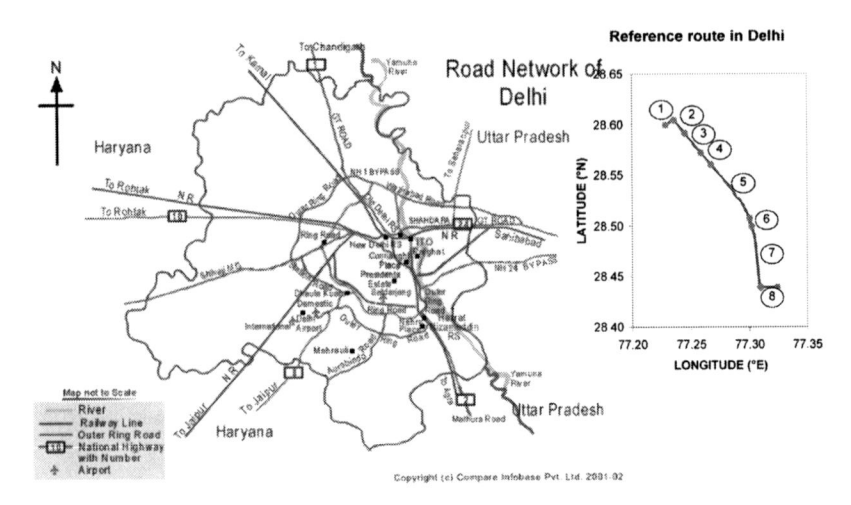

Fig. 4. Test area and reference route in Delhi

Application in Delhi, India

Reference measurements

As a basis for feeding the database with typical speed profiles as a function of specific roads and road types, reference measurements were performed on a fixed route in Belgium and in Delhi, India. The tests in Delhi were most on and around Mathura Road (Fig. 4).

The following eight segments were defined:
1. Subramaniam Bharti Marg: city traffic (1.0 km)
2. Dr. Zakir Hussain Marg: city traffic (1.7 km)
3. Mathura Road: suburb traffic (2.6 km)
4. Mathura Road: suburb traffic (1.5 km)
5. Mathura Road/NH 2 motorway (with regular stops) (6.8 km)
6. Border Zone Delhi: always slow traffic (0.9 km)
7. Mathura Road/NH 2 motorway continue until turn left (6.4 km)
8. Sector 31 – rural traffic (1.5 km)

All segments are also driven back. The tests were driven with two diesel Toyota Qualis 2.4D taxis of Lucky Taxi Company in Delhi. Figure 5 shows a typical speed profile on the Delhi reference route as a function of the distance. It gives a good indication of where stops are situated (usually related to crossings). Segments are also indicated.

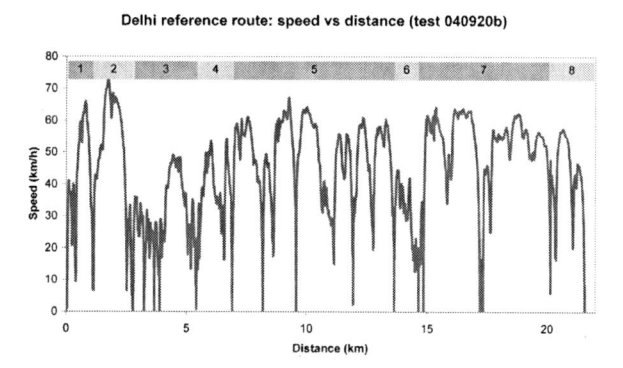

Fig. 5. Typical speed profile in Delhi track

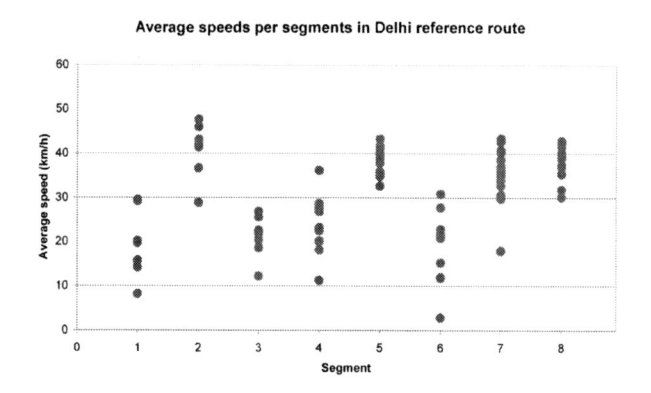

Fig. 6. Average speeds recorded on every segment

Figure 6 shows the average speeds recorded on every segment.

- Segment 1 (city traffic): variation of average speed between 8 and 30 km h^{-1}. Variations are due to standstill at crossroads, other traffic and traffic jams.
- Segment 2 (city traffic): variation of average speed between 30 and 50 km h^{-1}. Variations are due to standstill at crossroads or some other traffic.
- Segment 3 (suburb traffic): variation of average speed between 10 and 30 km h^{-1}. Variations are due to standstill at crossroads, other traffic and traffic jams.
- Segment 4 (suburb traffic): variation of average speed between 10 and 40 km h^{-1}. High variations, related to the traffic circumstances.

- Segment 5 (motorway traffic): variation of average speed between 30 and 45 km h^{-1}. Motorway circumstances are different from the European situations. No traffic jams were recorded in this part.
- Segment 6 (border area): variation of average speed between 3 and 30 km h^{-1}. High amount of queuing for the border control.
- Segment 7 (motorway traffic): variation of average speed between 20 and 45 km h^{-1}. Generally traffic goes quite fluently in this part.
- Segment 8 (rural traffic): variation of average speed between 30 and 45 km h^{-1}. Generally quite calm road.

Speed limit within the city area is 60 km h^{-1}, outside the city area there is no speed limit, but generally speeds do not exceed 100 km h^{-1} (linked to other traffic).

Road profiles

The detector contains the profile of the road, calculated based on the many test drives in the reference road in Delhi. The detector will on the other hand capture live position and speed information from the GPS receiver mounted on the roof of the vehicle. Figure 7 shows a typical example of the profiles in Delhi. The idea of the profiles is clear: based on the reference measurements three values have been defined for each part of the segment (every 100 m): an average speed, maximum speed, and minimum speed. These values will be used in the traffic jam detection algorithm. On some spots a minimum speed of 0 km h^{-1} is taken. These are generally stops for crossings and traffic lights. Between traffic lights the minimum speed is generally above zero.

The algorithm will first try to detect whether the vehicle is on the reference road or not. After that, once on the reference road, the detector will, based on live position and speed data, decide whether the vehicle is entering, driving, or leaving a traffic jam. A signal will be given when the vehicle is in a traffic jam. This signal can then be sent to the traffic monitoring room for traffic jam management.

Validation

In December 2004 the methodology was first tested in Belgium and in January 2005, in Delhi. Traffic jam was actually detected when the recorded speed was outside the calculated profiles. Figure 8 shows an example where three times a traffic jam was detected (marked by the stars).

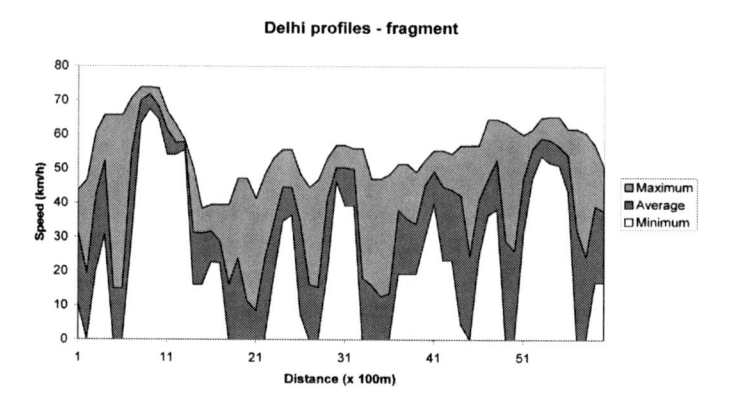

Fig. 7. Typical road profile in Delhi track

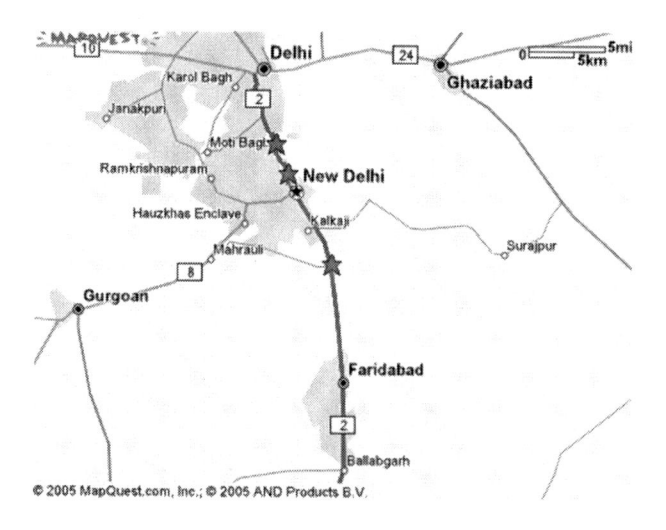

Fig. 8. Traffic jam detections on Mathura Road, Delhi

Conclusions

The paper showed the approach used to test a system of traffic jam detection, integrated in a tracking device. The scope of the FADC project ends at the detection and the signalling of traffic jam to a central server. Traffic jam management and the feedback to the drivers will be covered by a new project that will be the follow-up of FADC. The idea is to finalize a precommercial product that will help SMEs to reduce their fleet management costs and hence improve environmental protection due to less time spent in traffic jam, better information on coming traffic jam hence reducing speed on time, hence reducing gas emission and pollution. It is strongly believed that SMEs and other local institutions involved in transport as well as fleet companies, insurance companies, vehicle leasing companies, etc. can all take the benefits of the project results and ensure the financial sustainability. The project results and its further developments have indeed applications in the fleet market and can adequately be exploited according to different opportunities. One of the most reliable possibilities is based on a major involvement of local SMEs, which provides the funding needed as donors or customers of an independent service provider.

References

1. Asia IT&C website:http://europa.eu.int/comm/europeaid/projects/asia-itc/cf/index.cfm
2. FADC website: www.vito.be/FADC

Reducing car trip and pollutant emissions through strategic transport planning in Madrid, Spain

A Monzón,[1] AM Pardeiro,[1] LA Vega[2]

[1]TRANSyT-Transport Research Centre, Univ. Politécnica de Madrid
[2]Universidad Pedagógica y Tecnológica de Colombia

Abstract

Madrid is suffering a rapid suburbanization process where population and jobs are moving out of the central city. This process produces an imbalanced mobility pattern and more car dependency. Car pressure is increasing as well as its negative environmental impacts. Pollutant emissions and noise exposure form one of the key problems in central areas.

To reverse the present situation, a number of actions were envisaged constituting a whole strategy to maintain current standards of mobility and to avoid its negative effects. This strategy endeavours to maintain the social and economic vitality of the urban centre, in a sustainable framework, in order to reduce externalities and to enhance citizen quality of life. It includes transport policy measures such as pedestrianization, parking pricing, bus priorities, and new metro lines. Additionally, plans exist to improve the inner orbital to bypass out traffic, a measure that would seem to be contrary to the rest of the strategy. On the air quality control side, a low emissions zone has been designed in central districts.

The paper presents an evaluation of the effects of the new orbital on both mobility patterns and air quality levels. Paradoxically, the new motorway could reduce pollutant emissions and improve mobility in the entire city.

Mobility patterns and environmental problems

The Madrid Region, with its 5.4 million inhabitants, is moving towards a polinuclei metropolitan area (Fig. 1), where the denser core is offset by a metropolitan ring. Mobility rates by crowns indicate that this serves as dormitory and business/mix areas. However, its high population density and the quality of the public transport provided place Madrid at the top of public transport patronage within Europe [1].

Mobility patterns are not uniform. Their complexity is related to many factors such as population densities, land uses, and the provision of different means of transport [2]. Bus and metro services are in plentiful supply in the City of Madrid, whereas interurban bus and rail services have their higher share in the metropolitan ring. All these performance patterns induce very different transport costs, both by trip and by passenger-km (pkm), for each individual transport mode involved and for its geographical environment. They also produce different environmental and social effects associated with different trip lengths between the denser central districts (Madrid City) and its suburbs. But it is clear that there is a steady growth in trips in outlying areas, which are heavily car dependent.

Figure 2 shows that car trips are competitive with public transport both in central and in suburban areas. This is surprising because the central area is fairly congested and average speed is about 14 km h^{-1}. The consequence

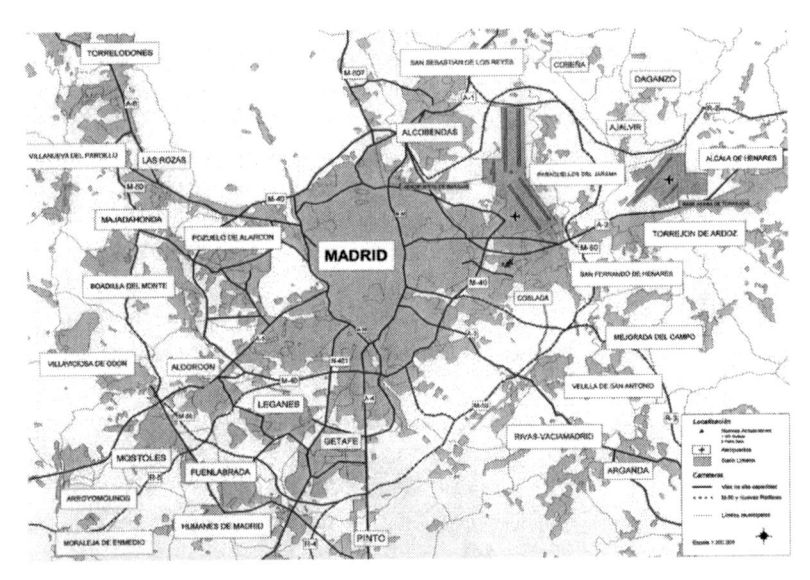

Fig. 1. Madrid metropolitan area

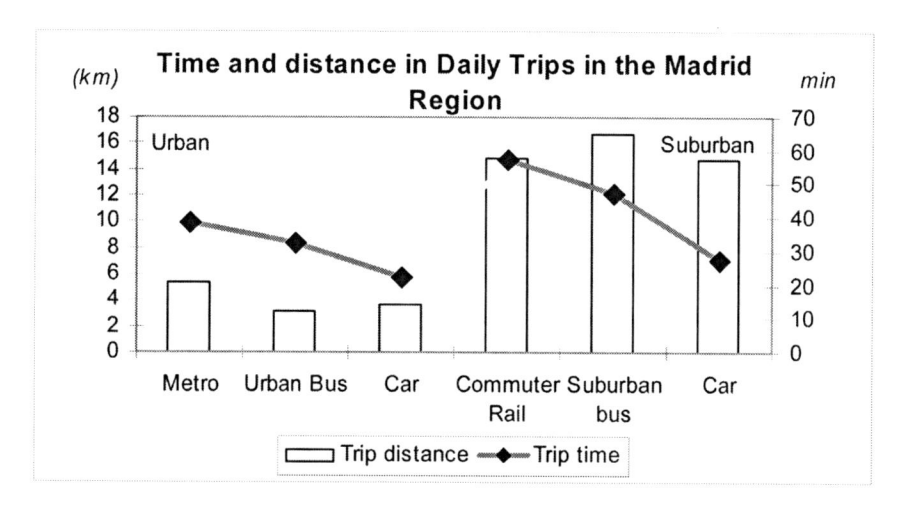

Fig. 2. Time and distance in urban/suburban trips in the Madrid region

is that, although Madrid has a good supply of public transport, car is still an attractive option even in central areas. Following the Zahavi principle [3–5], users choose their transport option according to time spent.

On the other hand, economic activities in Madrid City have their biggest competitor in the towns within the Madrid metropolitan belt. Many companies and commercial activities are moving out towards the periphery in the search for cheaper and greater space for their businesses. Consequently, Madrid City is losing part of its competitiveness owing to congestion and other agglomeration-based disutility [6]. Figure 3 shows the comparison in mechanized trips between zones in the Madrid region. It is clear that the Madrid Central Business District and periphery are losing trips in favour of outer areas in the metropolitan ring.

Cairncross [7] predicts the death of distance of cities like Madrid where new activity location patterns could change their functionality and competitiveness. The new concept of meta-town planning [8,9] indicates a way forward in the form of a collaborative approach among means of transport and development policies to achieve better mobility patterns, and to improve citizen quality of life in residential areas. A systematic approach to this vision is formulated by the Transportation Research Board (TRB) [10]. Monzón [11] analyses the relationships among local and arterial roads in the metropolitan planning process. In fact, Madrid City is already experiencing disadvantages against alternative locations in its metropolitan ring. It is necessary to design a comprehensive policy to reorganize mobility in order to preserve liveable streets in central areas and, at the same time,

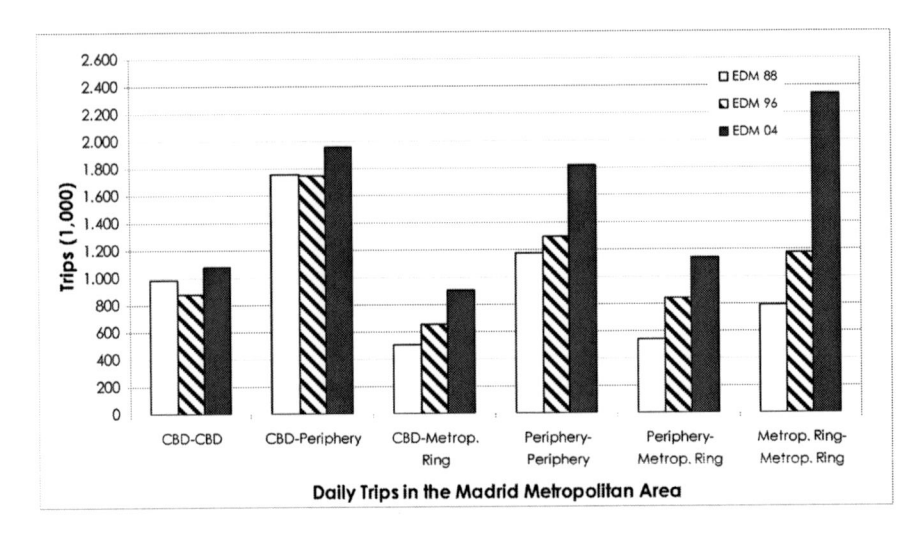

Fig. 3. Evolution in mechanized mobility in the Madrid region

to attract new businesses and keep the old ones alive if its attractiveness and economic vitality is to be increased [12].

Sustainable development, however, has to cope with three aspects, namely economic, environmental, and mobility-related [13]. This whole process is producing substantial environmental problems. Longer trips and greater car dependency are steadily increasing the emission of pollutants and global warming gases [14].

Tables 1 and 2 show the level of emissions in Madrid City in 2002 caused by traffic. These values are too high and Madrid City Council has launched an environmental strategy to reduce emissions over the 2006-2010 period [15]. It includes a reduction of NO_x emissions by as much as 16,500 t, implying 27% less than current values. There is no specific target for particulate matter emissions (PM_{10}) but there are many measures designed to

Table 1. Annual traffic-related emissions

	NO_x (t)	PM_{10} (t)	CO_2 (kt)	COVNM (t)
Year 2002	22,585	1591	4268	18,908

Table 2. Contribution of traffic emissions to the total

	NO_x	PM_{10}	CO_2	COVNM
% of total	77.03%	74.79%	51.15%	33.60%

reduce the emission of these particles in line with European legislation. Spain has to reduce its CO_2 emissions by 14,500 t in the near future. Madrid has taken on board the challenge of reducing a quantity equivalent to 4.8% of Spain's total emissions. The environmental strategy includes several measures to reduce traffic rates, comprising one of the key problems, particularly in the Convention of Biological Diversity (CBD).

A new strategy

A number of activities are in place to recover city vitality and environmental quality through a series of initiatives in various sectors, including transport. The main principles for achieving a better mobility distribution are to reduce car pressure and increase pedestrian trips (PT) in the CBD, to foster PT in radial trips travelling in and out of the CBD and to alleviate car pressure on local streets by conveying traffic through the orbital road surrounding the CBD.

To achieve these goals a number of policy actions have been started:

- Pedestrianization of historical zones: pedestrian priority, improved accessibility, greener streets, and squares.
- Parking restrictions: a parking pricing scheme, called SER, has been implemented in a wide area since 2001. More than 110,000 parking spaces are now under control, covering most of the Madrid CBD.
- Improving PT provision and the seamless mobility targets: reserved bus lanes, extensions to the metro network, intermodal interchanges, etc.
- Redesign of the M-30 inner orbital motorway: this road is undergoing total redesign to attract car traffic out of the CBD. The main goal is to improve the level of service and to reduce accidents. The number of lanes will remain approximately the same, but the more uniform design and better integration with urban arteries is intended to reduce congestion. Three tunnels will also be built to avoid traffic nuisance and to improve environmental quality in some stretches such as the river banks and two parks.

The first three actions are "traditional" ones in any transport policy package designed to foster PT patronage and improve walking and cycling [16,17]. The fourth one, however, is quite controversial. The concept of improving an urban motorway seems to be an action going in the opposite direction. Many planners and media have protested over this scheme that will eventually attract more cars towards the city centre and will, in principle, produce bigger environmental impact.

Mobility impacts

The rationale behind the approach is to model the impacts of the alternatives – to do nothing or to rebuild the M-30, thereby reducing traffic pressure on the urban grid and improving its safety conditions (the scheme proposed).

We developed a model to test the medium-term impacts on mobility, energy consumption, and pollutant emissions. It forecasts the different impacts of the project modelled against the "do nothing" scenario. Two different time horizons were tested – immediately post-completion of the works (2007) and the medium-term evolution (2012). Although absolute figures vary, the tendency and location of effects are about the same. Results are given in Table 3.

The scheme to rebuild the M-30 orbital will reduce car pressure on urban streets both in the CBD and in the city periphery. On the contrary, it will attract many more cars along the renovated motorway. The outer M-40 ring road will see its traffic load reduced. In short, the M-30 will concentrate many of the trips that are currently attempting to obtain their shortest route across the city or bypassing it via the outer ring road. The change in the behaviour of the first group will reduce car pressure on the CBD and its environmental impacts. Current M-40 users moving to the M-30 will produce less CO_2 and pollutant emissions, but in more central areas, also meaning less dispersion effect.

Driving conditions in the CBD will be better, even in the areas surrounding the M-30, owing to less congestion. This implies lower pollutant emissions [18,19].

Table 3. Impacts of the M-30 Orbital Improvement Scheme

	2007		2012	
	Traffic flow variation (veh-km)	Trip time variation (veh-h)	Traffic flow variation (veh-km)	Trip time variation (veh-h)
CBD	–91,581	–6,057	–118,843	–8,699
M-30	909,465	2,559	1,053,843	1,478
City periphery	–129,065	–4,228	–154,861	–5,492
M-40	–577,650	–9,884	–613,402	–11,716
Metropolitan ring	–292,912	–1,000	–303,871	–3,364

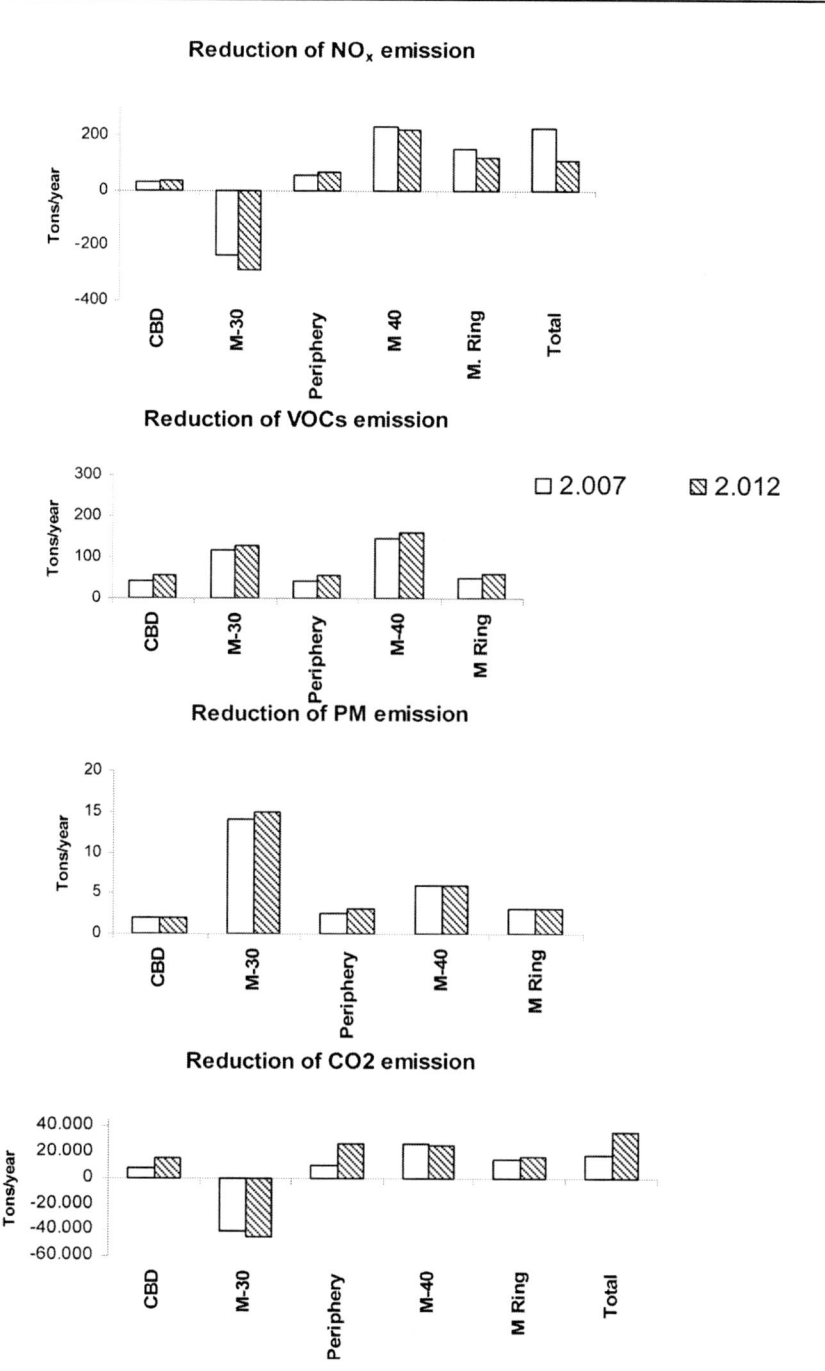

Fig. 4. Reduction of emissions associated with the Madrid strategy

Environmental effects

A further noticeable advantage of improving this "urban" motorway will be a reduction of certain environmental impacts and, above all, in the number of accidents. This is also paradoxical as although more traffic will be attracted to central areas, the overall nuisance will be less significant. The explanation is related to the results shown in the previous section and the effects of the three tunnels that will avoid noise and direct pollutant emissions in some densely populated central districts.

Pollutant emissions are mainly due to traffic in the city. The characteristics of the construction lower the impacts on the environment and the number of traffic accidents. The new M-30 route will be more uniform and this will alleviate congestion. Additionally, almost 8 km of tunnels among the total 32.6 km will reduce environmental problems precisely where the population exposed is greatest.

The results of the assignment model feed the emissions model based in COPERT [20]. The calculations take into account the road type, the distance and the vehicle type. Overall, emissions will be reduced as illustrated in Figure 4. Higher CO_2 and NO_x emissions will, however, appear in the M-30 orbital, but far less important than the reductions in other zones. Emissions inside the tunnels will be filtered and treated to eliminate PM and other pollutants. We could therefore conclude that the new scheme will produce better quality standards and environmental impact in the streets as a result of the concentration of heavy traffic on only one artery – the M-30.

Conclusions

As a result of the modelling exercise and the evaluation process, we could forecast that the improvement of the M-30 inner orbital is consistent with other sustainable transport policies in the City of Madrid. The concentration of traffic flow on this orbital could alleviate traffic pressure on the streets and reduce environmental impact.

The environmental benefits are important as the scheme will reduce CO_2 and pollutant emissions throughout the metropolitan area by concentrating them on the M-30, but with a clearly positive overall balance.

Madrid will also be recuperating economic vitality with mixed developments of housing and commercial areas that benefit from the better quality of the central districts.

The key elements of the positive strategy are the combination of push and pull measures in the design of complementary policy measures to control traffic in Madrid. The three central points of the strategy are reducing traffic in the CBD, attracting more passengers to PT, and improving traffic conditions in the M-30 urban orbital as an alternative to crossing the city and avoiding longer trips in the periphery.

The M-30 urban orbital will have more capacity as a result of the lower traffic congestion. In addition, traffic will be reduced in the CBD because the M-30 takes it up. In both cases, the M-30 and the CBD, traffic conditions will improve and the emission of pollutants drop off.

Further actions are envisaged, such as reducing the average driving speed along the M-30 to obtain further reduction of emissions and accidents.

References

1. EMTA (2004) Eurobarometer 2002. European Metropolitan Transport Authorities, Paris
2. Newman P, Kenworthy J (1999) Sustainability and cities. Overcoming automobile dependence. Island Press, Washington, DC
3. Zahavi Y, Talvitie A (1980) Regularities in travel time and money expenditure. Transport Res Rec 750:13–19
4. Schafer A (2000) Regularities in travel demand. An international perspective. BTS, J Transport Stat 3(3)
5. Schafer A, Victor D (2000) The future mobility of the world population. Transport Res A 34:171–205
6. EC (2001) White paper European transport policy for 2010: Time to decide. European Comission, Luxembourg
7. Cairncross F (1997) The death of distance: how the communications revolution will change our lives. Orion, London
8. Hall P (1998) Cities in civilization: culture, technology and urban order. Weidenfeld & Nicolson, London
9. Schreyer C (2004) External cost of transport. INFRAS-IWW. UIC, Zurich/Karlsruhe
10. Transportation Research Board (2002) A guide to best practices for achieving context sensitive solutions. NCHRP Report 480. Washington, DC
11. Monzón A, Vega Báez LA, Pardeiro AM (2005) Assessment of the effects of improving the Madrid inner ring road to a more balanced and sustainable metropolitan mobility. European Transport Conference, Strasbourg, France
12. Boarnet M, Crane R (2001) Travel by design: the influence of urban form on travel. Oxford University Press, Oxford
13. ECMT (2001). Assessing the benefits of transport. European Conference of Ministers of Transport. OECD, Paris
14. Gruden D et al. (2003) Means of transportation and their effects on the environment. Traffic and environment. The handbook of environmental chemistry. Springer, Berlin

15. Ayuntamiento de Madrid (2003) Libro Blanco de la calidad del aire en el municipio de Madrid. Madrid
16. Minken H et al. (2003) A methodological guidebook. Developing sustainable urban land use and transport strategies. Prospects D14. European Comission, Oslo
17. WCTRS (2004). Urban transport and the environment. An international perspective. Elsevier, Tokyo
18. André M, Joumard R, Hickman AJ, Hasse D (1994) Actual car use and operating conditions as emission parameters: derived urban driving cycles. Sci Tot Environ 146–147:225–233
19. Joumard R (1999) Methods for estimation of atmospheric emissions from transport: European scientist network and scientific state of the art. COST Action 319. Final report
20. COPERT III (2000) Computer program to calculate emissions from road transport. Methodology and emission factors. EEA, Copenhagen

Evaluation of car control measures based on an Internet-based travel survey system

N Harata, S Aono

Department of Urban Engineering, School of Engineering, University of Tokyo

Abstract

This paper presents the development of an Internet-based travel survey system for designing car control measures for a new campus site. This system was developed for collecting stated preference (SP) data on commuter trips after relocation to the new campus. Using the personal attributes of the respondents and transportation network data, the system produced customized alternatives for the SP questionnaire. It was found that clear differences exist between stay patterns and modes of travel of faculty, clerks, and students. Finally, the effective control measures of car use were designed and it was found that the Internet-based travel survey system is useful for finding effective package measures.

Introduction

The relocation of large-scale facilities such as workplaces, schools, universities, and commercial facilities has a considerable impact on the surrounding transport infrastructure; the relocation affects traffic congestion, parking problems, road safety, and environmental impact. Sometimes these projects themselves include restructuring of the existing transportation network. Transportation planners should facilitate the use of multiple modes of transport to reduce the impact. Therefore, multimodal transportation planning and the assessment of its impact are very important for development projects.

The relocation of a facility causes behavioural changes in people with regard to mode choice, destination choice, and induced travel. In the long run, residential choice, and land-use pattern will also change. In order to forecast the changes and assess their impact, the SP data should be collected from potential commuters or visitors to the facility. However, because it is likely that they do not have sufficient knowledge on the level-of-service (LOS) variables of choice alternatives with regard to personal factors (social status, geographical location, car availability, etc.), it is difficult to collect reliable data using traditional travel survey methods. In such a situation, computer-based interactive survey methods integrating the Geographic Information System (GIS) function can be effective [1–4].

This paper presents the development of an Internet-based travel survey system for designing car control measures for a new campus site. The system was developed for collecting stated preference (SP) data on commuter trips after relocation to the new campus [3]. By using this collected data, the combined effects of the control measures of car use are examined.

Development of the survey system

The new campus of the University of Tokyo in Kashiwa City was completed in April 2006. About 2000 faculty, clerks, and students commute to the campus. It was expected that the transportation infrastructure around the campus would be heavily affected by the newly established commuter trips. There are three railway stations in the proximity of the campus; one of them was under construction at the time of this study (July 2002). Therefore, multimodal transportation planning and impact assessments were needed.

This Internet-based travel survey system was developed for collecting commuter modes and route choice preferences of faculty, clerks, and students of the University of Tokyo who had already moved or were planning to move to the new campus at Kashiwa. The survey system had a client-server type architecture. On the server side, the Web server software was installed (Microsoft Corp. Internet Information Server 4.0) on the server PC to handle requests from clients and transmit the questionnaire; the Web GIS software (MapInfo Corp. MapXtreme 2.02) was installed to handle and display spatial data, and the recoding database (Microsoft Corp. Access format) was used to record answers from the respondents. On the

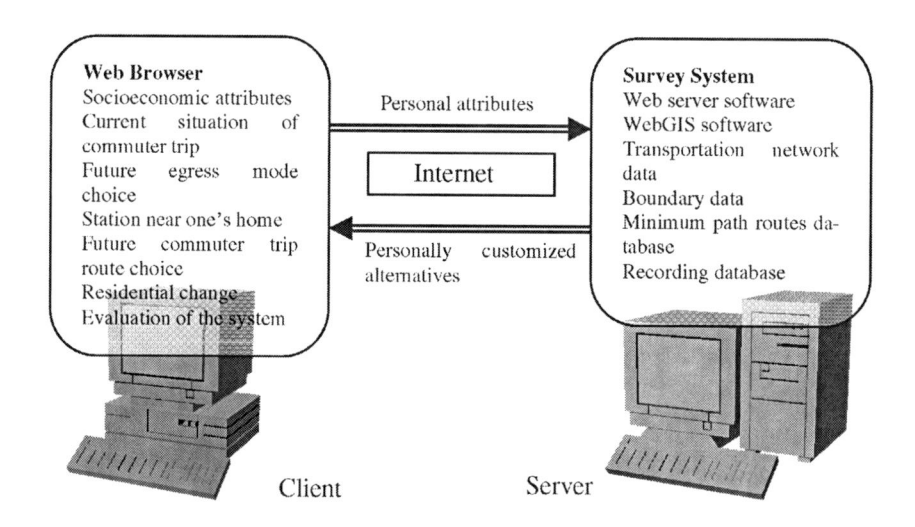

Fig. 1. Outline of the survey system

client side, a respondent can access the website from a PC by using a Web browser and answer the displayed questionnaire. Figure 1 shows the conceptual outline of the system.

The respondents of the survey were asked about their personal attributes, current commuter trip situations, and stations near their homes. Then, they were required to answer the SP questionnaire about the egress mode choice from the three stations near the campus. The alternatives were walking, bicycles, motorbikes, and buses. The time required and out-of-pocket costs per month were used as the LOS variables.

Figure 2 shows an example of the SP questionnaire with regard to commuter route choice. In the questionnaire, a hypothetical new railway line that would directly connect to the inner city of Tokyo was supposed to be established. The alternatives included a maximum of six routes – three railway routes that correspond to three egress stations, the express bus from Tokyo station, and two car routes with/without toll. If respondents did not own a car and had no future intention of using a car, the two car routes mentioned above were excluded from the alternatives.

In order to customize the questionnaire to correspond to the geographical location of a respondent, the system used Web GIS software. The respondents had inputted the station near their homes using the map displayed. Then, the system searched for minimum path routes to the campus from the minimum path routes database files for each alternative route.

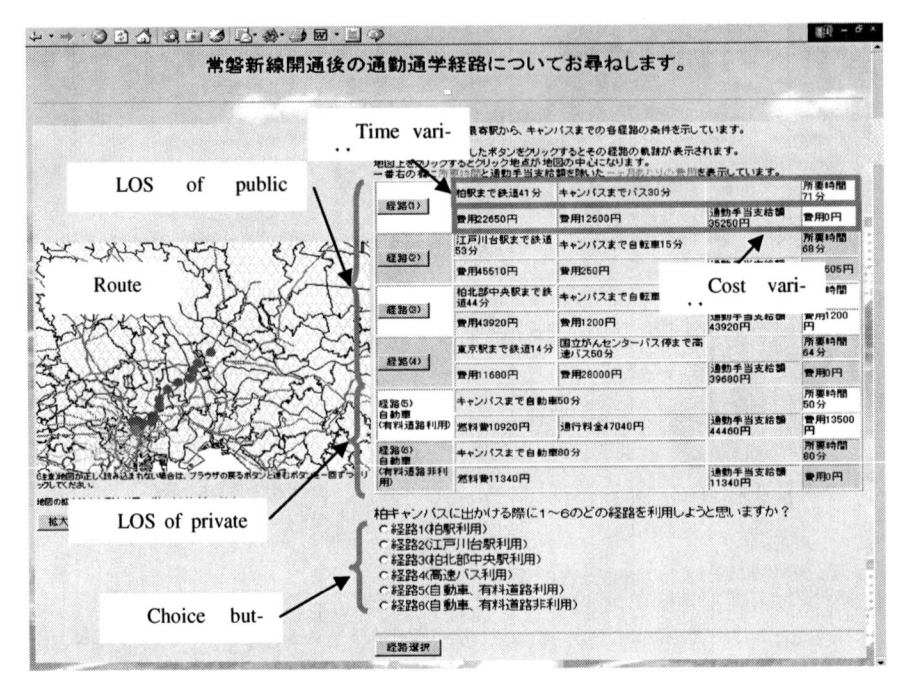

Fig. 2. Commuter route choice questionnaire

The LOS variables for the main mode included the time required and the out-of-pocket costs per month. For faculty and clerks, commuter allowance was considered. These values were computed from future railway and road network data. The LOS variables for the egress modes included time and costs of each mode selected by the respondent in the questionnaire. The respondent was then asked to select the most preferable route from the alternatives.

The survey

The respondents of the survey were the faculty, clerks, and students of the University of Tokyo. Most of them used PCs routinely, could access the Internet, and had their own email addresses. Therefore, email could be used to recruit the respondents. The population was relatively very homogeneous in terms of accessibility to the Internet; therefore, the problem of representativeness of the population could be avoided; this is a serious problem in an Internet survey.

Table 1. Attributes of valid respondents

Gender	Male	79.8%	Female	
Position	Faculty and clerks	41.4%	Students	58.6%
Driver's license	Yes	85.9%	No	14.1%
Commute to the new campus	Commutes presently	34.4%	Does not commute	65.6%

The recruiting mails had been dispatched in July 2002. Two months later, there had been 657 accesses to the website. Of these, 326 of them provided valid responses (Table 1).

The respondents who had already commuted to the campus at Kashiwa were required to provide their usual arrival and departure times at the campus. Figure 3 shows these results. There are clear differences in stay patterns according to the position of the respondents.

The egress bus lines from the nearer two of the three alternative stations (Stations 2 and 3) are not operated frequently: in particular in the evening and night. Therefore, when estimating the modes and route share, the LOS variables in the return trips should be considered as well as in the trips from home to the campus.

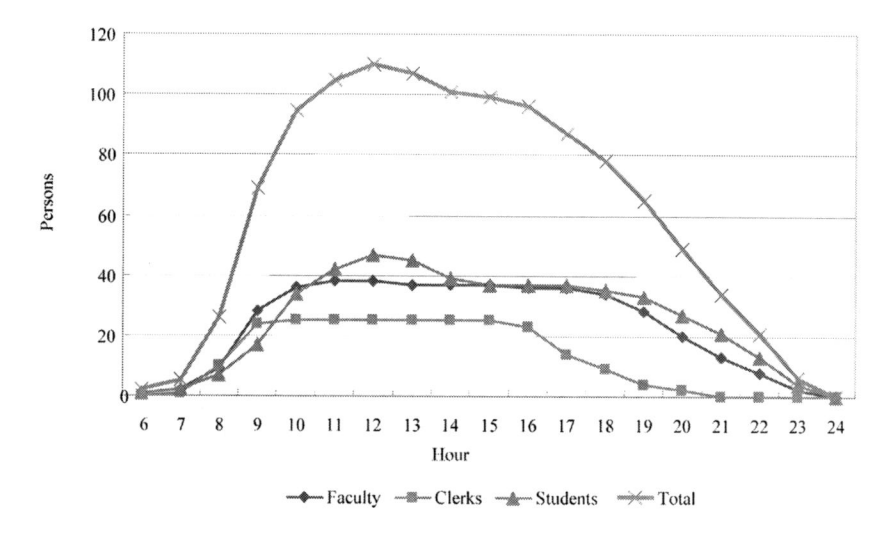

Fig. 3. Number of respondents in the campus

Results of stated preference questionnaire

Table 2 shows the shares in the SP data with regard to the egress modes.

Table 2. Shares in the SP data with regard to the egress modes (%)

	Station 1 F&C	Station 1 S	Station 2 F&C	Station 2 S	Station 3 (new) F&C	Station 3 (new) S
Walking	–	–	4.2	3.6	7.6	4.2
Bicycle	1.7	35.9	26.3	75.4	33.1	77.2
Motorbike	4.2	9.6	3.4	3.6	3.4	2.4
Bus	94.1	54.5	66.1	17.4	55.9	16.2

F&C: faculty and clerks; S: students

Station 1 is about 5 km away from the campus. Stations 2 and 3 (new station) are about 3 km away. It could be consistent result that many students choose the bicycle as the egress mode.

Table 3 shows shares in the SP data with regard to the commuter routes. For most cases, travel route using station 1 was longer and cheaper as compared to station 3. Many of the faculty and clerks could compensate for all their costs by using their commuter allowance up to 50,000 (JPY/month). On the other hand, students had to incur their costs by themselves. This factor affected the choice results between the two groups.

Estimation of commuter mode and route choice models

Considering the abovementioned results, the collected data was divided into two groups – group F&C (faculty and clerks) and group S (students). For each group, the nested logit type commuter mode and route choice models were estimated. Figure 4 shows the assumed choice structure of

Table 3. Shares in the SP data with regard to the commuter routes (%)

	F&C	S	Total
Station 1	15.8	49.7	36.0
Station 2	8.1	10.3	9.4
Station 3 (new)	47.6	26.5	35.1
Express bus	9.8	0.6	4.4
Car (using toll road)	6.2	1.3	3.3
Car (not using toll road)	12.5	11.6	12.0

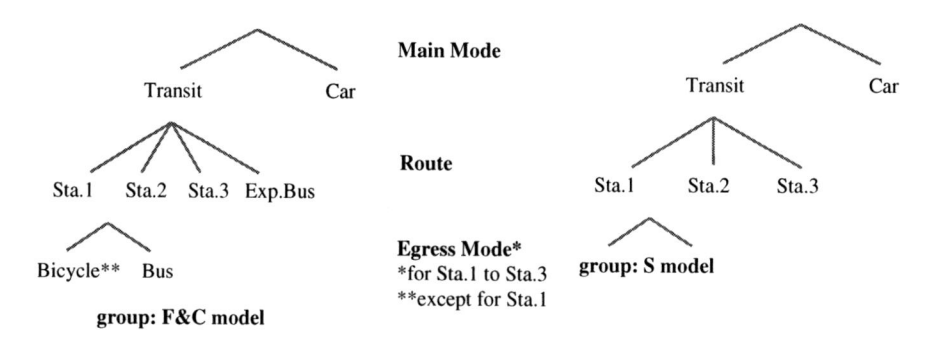

Fig. 4. Choice structures of the mode and route choice models

both the models. Also, parameters of time variable are assumed to be the same for different level of the nested structure

Tables 4 and 5 show that it was possible to estimate statistically significant nested logit models for both the group F&C and the group S. Because of the upper limit of the commuter allowances, some respondents could not completely compensate their monthly cost when buses were selected as the egress mode. Therefore, the cost variable was adopted on the level of the egress mode choice in the group F&C model.

Table 4. Estimation results of the group F&C model

Levels	Variables	Parameter estimates	t(0)	t(1)
Main mode choice	Car specific	0.352	0.518	
	Scale parameter 2	0.646	2.09*	1.150
Route choice	Time	−0.121	−4.72**	
	Exp. bus specific	−2.330	−2.69**	
	Scale parameter 1	0.557	4.78**	3.80**
Egress mode choice	Time	−0.121	−4.72**	
	Out-of-pocket cost/month	−2.05E-04	−2.72**	
	Bicycle-specific	−0.213	−0.783	
L(0)	−			−199.010
L(θ)	−			−161.010
Adjusted ρ^2	−			0.182
Number of samples	−			101.000

*95% level of significance; **99% level of significance

Table 5. Estimation results of the group S model

Levels	Variables	Parameter estimates	t(0)	t(1)
Main mode choice	Car–specific	0.824	0.953	
	Scale parameter 2	0.467	2.35*	2.69**
Route choice	Time	−0.133	−4.37**	
	Cost/month	−1.82E-04	−3.42**	
	Scale parameter 1	0.786	2.77**	0.7560
Egress mode choice	Time	−0.133	−4.37**	
	Cost/month	−1.82E-04	−3.42**	
	Bicycle-specific	1.750	3.02**	
	Bus-specific station1	3.010	5.00**	
L(0)	−			−266.71**
L(θ)	−			−193.58**
Adjusted ρ^2	−			0.2690
Number of samples	−			139.0000

*95% level of significance; **99% level of significance

Effects of control measures of car use

Using the models above, the effects of car control measures were estimated. The total number of commuters comprised 400 faculty, 200 clerks, and 1400 students. Because of the limitation of data, it was assumed that the geographical distribution of the commuters was similar to the SP samples. The arrival/departure time pattern in actual commuting samples was assigned to the SP samples that had not reported their arrival/departure times. The express bus was excluded from the alternatives because it had been abandoned after the opening of the new railway line. The alternatives in the egress mode choices were bicycles and buses. Because bicycles were very popular as an egress mode, large number of bicycles were used without restriction. However, there should be a limitation with regard to bicycle availability, parking site, or bicycle ownership. Initially, the number of person who could use bicycles was restricted to one in four people.

Table 6 shows the estimated mode and route share considering trips from home to campus only. In this case, the LOS variables of the morning time were used.

Table 6. Mode and route share: considering home to the campus trips only

	Station 1		Station 2		Station 3		Car	Total
	Bicycle	Bus	Bicycle	Bus	Bicycle	Bus		
Faculty	0	104	4	49	22	151	70	400
Clerks	0	49	2	25	9	67	48	200
Students	69	902	31	53	66	69	210	1400
Total	69	1055	37	127	97	287	328	2000

On the other hand, Table 7 shows the results considering both home to the campus trips and return trips. In this case, the time variables of the egress buses are the averages of the home to the campus trips and the return trips. In the return trips, the waiting time at the campus bus stop (headway/2) was considered. Considering the low LOS in the return trips, the number of cars used increased by 14.3%. This was the base case.

In measure 1, it was assumed that availability of bicycles improved due to a bicycle rental system or provision of sufficient number of parking sites. Then, the number of persons who could use bicycles was one in two people. In this case, the number of cars used was 353 (–5.9% from the base case). In measure 2, the egress bus service was assumed to improve and waiting time at the bus stop was reduced to 5 min for the entire day. A shuttle bus service in the evening and night, and a bus location system by which responders can determine the optimum time to leave a lab/office to board a bus will be instrumental in reducing the waiting time. The number of cars used was 352 (–6.1%). In measure 3, a parking fee of JPY 3000/month was collected from faculty and clerks and JPY 1500/month from students. The number of cars used was 343 (–8.5%).

Finally, Table 8 shows the combined effect of these three measures. The number of cars used was 304 (–18.9%). Further, Figure 5 shows the temporal change in the number of cars parked in the campus between the base case and the package case.

Table 7. Mode and route share in the base case

	Station 1		Station 2		Station 3		Car	Total
	Bicycle	Bus	Bicycle	Bus	Bicycle	Bus		
Faculty	0	113	7	26	32	138	84	400
Clerks	0	53	4	15	12	63	53	200
Students	93	868	37	28	79	57	238	1400
Total	93	1034	48	69	123	258	375	2000

Table 8. Mode and route share: the package of three measures

	Station 1		Station 2		Station 3		Car	Total
	Bicycle	Bus	Bicycle	Bus	Bicycle	Bus		
Faculty	0	107	10	36	57	126	64	400
Clerks	0	47	7	22	22	59	43	200
Students	166	743	69	34	145	46	197	1400
Total	166	897	86	92	224	231	304	2000

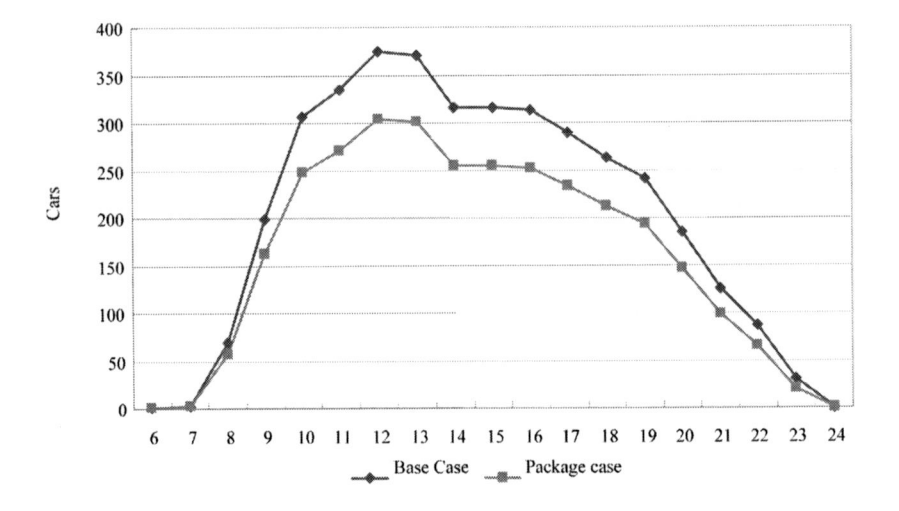

Fig. 5. Number of cars parked in the campus: the base case and the package case

Conclusions

This paper presents the design of car control measures for a new campus site based on an Internet-based travel survey system.

The survey system using the Web GIS function can customize the SP questionnaire corresponding to the position of the respondent in the university, his/her geographical location, and car availability. There were clear differences in the stay patterns and modes of travel of faculty, clerks, and students.

Based on the Internet-based travel survey, it was possible to estimate the statistically significant nested-logit type mode and route choice models for both the F&C group and the S group. Using the estimated models, effective control measures of car use were designed. At first, the importance of low LOS for egress buses in the evening and night were examined by considering both home to campus trips and return trips. Then, the effects of improvement in bicycle availability, bus accessibility, and parking fee control were estimated. Finally, it was found that car use can be reduced by about 19% by the package of all three measures.

In conclusion, it is shown that the proposed Internet-based travel survey system is useful for designing effective package of car control measures in the new campus based on the detailed and reliable response data.

Because the campus has been completely relocated, another survey is needed to ensure a precise and updated evaluation of the car control measures. In the future survey, a more customized questionnaire should be developed by considering the variation of arrival/departure time pattern depending on the day of the week and the railway time table.

References

1. Polak J (2000) The Use of computer-based methods in stated preference research. In: PTRC perspectives 4. Stated preference modelling techniques, pp 73–90
2. Nossum A (2005) Stated preference survey on Internet. *Transportøkonomiks Institute summary report*
3. Aono S, Ohmori N, Harata N (2001) A development of a computerized travel behavior data collection method. Infrastructure Planning Rev 18:123–128
4. Aono S, Ohmori N, Harata N (2004) Development of an Internet-based travel survey system. Proceedings of the international symposium on city planning, pp 41–50
5. Aono S, Ohmori N, Harata N (2006) Development of a Web-GIS based scheduling simulator on holiday non-work activities. Proceedings of the international conference on traffic and transportation studies

Integrating cycling in Bus Rapid Transit system in Accra

ML Quarshie

Centre for Cycling Expertise (CCE), Accra, Ghana

Abstract

Since the introduction of motor vehicles, practicing engineers and city planner have underestimated the potential of non-motorized transport. As traffic congestion has become severe in many large cities, especially in the developing world, other transport alternatives need to be considered and efficient public transport systems should be given utter priority as they have enormous environmental, social and commercial benefits. There are several options of public transport system. However, in recent times the Bus Rapid Transit (BRT) system option is fast catching up with developing countries. Accra has in recent years been grappling to put in place a public transport system. The BRT concept has been introduced and the World Bank has pledged to support it having ascertained its feasibility [1,2]. Inevitably, there has arisen the need to integrate cycling into the BRT system owing to the increasing use of bicycles in Accra. The integration will afford commuters the opportunity to combined different modes in the most efficient, time and cost-effective manner. This would therefore require strategic planning to link cycle routes to terminals on major route.

Introduction

Transport as simply defined is the movement of people and goods. Obviously the common forms of movement since creation had been non-motorized transport. In the last 80 years when the motorcar emerged, decision makers, practicing engineers, and planners, have grossly underestimated these common and prominent forms. Whilst several developed cities face the challenge of turning the wheels back such is not the case of third world cities. Traffic congestion has become phenomenal with several of cities. Instead of addressing transport several of these cities continue to mitigate congestion by just improving the road infrastructure. The question is what if there is no more space to widen roads to accommodate the ever-increasing dependence of the motorcar? There is therefore the need for decision makers, practicing engineers, and planners in Third World cities to appreciate what transport really is, i.e., especially from road users point of view. Obviously there is no doubt that efficient public transport systems should be given priority as they have enormous environmental, social, and commercial benefits. In most developing cities such as Accra the public transport system is characterized by what is termed the individual public transport. This is a scenario where individual owners of small vans in the case of Accra called the "tro-tro" either form unions or operate routes. This obviously is not and has not been reliable. With the growing of Accra's population at a nearly 4% growth rate and the attendant traffic congestion there is no doubt that the city requires an efficient public transport system. The growing cyclist population also calls for concern and the need to integrate cycle facilities into the existing road network and necessarily integrating cycling into any effective public transport system in order to facilitate movement from areas or zones that may not have motorable roads notwithstanding the fact that the cyclists ply on these routes anyway. A clear understanding of public transport dispensation over time is necessary for discussion.

Accra's public transport: a serious look

A trail of public transport in Ghana

Since independence, public transport has been characterized by what is termed individual public transport. That is to say individuals own mini vans and over time come together to dispense this service. In 1969, the government passed the Omnibus Services Authority Decree, which nationalized all City, Municipal, Urban, and Local Council bus undertakings within one unitary body responsible both for the planning and the provision of public transport services. The Omnibus Services Decree of 1972 then split these two functions by creating a separate Licensing Authority to regulate the omnibus sector. The Omnibus Services Authority continued in existence but with the sole objective of bus service provision in its specified areas. Later legislation concerning the commercialization of service delivery bodies within the public sector resulted in its restructuring as OSA Transport Ltd.

In 1996, the government decided to privatize its passenger transport undertakings, which by then also included the State Transport Corporation and City Express Services Ltd. However, it was unable to find any buyers for these businesses at the time, which resulted in their continuing decline in the absence of public investment in rolling stock and other assets. Urban public transport services were largely replaced by private sector provision of Para-transit, known locally as tro-tro. Individual owners came together to dispense public transport service under a union (i.e., The Ghana Private Road Transport Union – GPRTU), which was affiliated to the Trades Union Congress), though smaller bodies such as the Progressive Transport Owners Association have since emerged.

The current administration, first elected in 2000, came to power on a platform that included a strong commitment to improving the transport sector in general and urban transport services in particular. Despite these commitments, however, it has been reluctant to relinquish total control over public transport fares as required by the legislation. It has also reversed the policy of the previous administration of public procurement of rolling stock for leasing to private operators through GPRTU [3].

Technical conditions

Typical traditional transportation organization can be grouped in three main levels: the individual owners or family owners of usually one or two vehicles, the drivers, and the unions, which usually gather most of the owners and represent them before the authorities.

This system of organization exists in Accra. Currently, the minibus operators in Accra are mostly small private owners with only one or two vehicles, but they are organized into the syndicates (i.e., GPRTU with approximately 17,000 members, the Ghana Transport Cooperative, and Progressive Transport Association – PROTOA). All of these groups are coordinated by the Ghana Road Transport Coordinating Council (GRTCC).

The GRTCC regulates services, sets fares. Over the years, most of the stations have been controlled by GPRTU. Efforts are being made to accommodate the other groups. These para-transit operators run without subsidies and have operated for years, even under financially stressful times. However, the vehicles are typically older, polluting, and are run for longer than is safe. They follow no predictable schedule – routes and times.

The traditional transportation system has reached its most difficult scenario as an industry. The prime motivation of the transportation business is to transport the greatest number of people at the smallest possible cost. This should translate to the least number of capital goods (i.e., buses) without neglecting minimum levels of quality service for the users. Nevertheless, the transportation industry has managed to break this basic business principle, in detriment to itself.

The traditional scheme between the unions and the owners is as follows: the transport authority authorizes permits to service the public routes of the city. However, due to low levels of capacity and the proliferation of the competition, the industry took a wrong turn. Instead of having companies that own a large fleet of buses and have paid employees like any formal, organized, and productive industry, it has been fragmented into thousands of individual owners, competitors, and mini-entrepreneurs. This has caused an indiscriminate and rapid increase of the fleet and has begun the Penny War for the owner and/or driver. The Penny War is the fight to find the resources to pay for the affiliation to the union, for day-hires, for small repairs, to pay for the use of terminals, and then for the fuel, for food, and education for the family. This Penny War is being fought in the streets as drivers compete for passengers and the passengers end up bearing the costs of the Penny War. The resources are never enough to do preventive maintenance or for depreciation of the vehicle, and barely for the strictly necessary repairs. Obviously, the conditions for safety and health and the quality of service for the citizens are being compromised by the reality that currently persists in the streets.

The government has made the proactive decision to help resolve these problems that face the majority of the population that uses this service. However, the road system has been designed and continues to be developed for the extreme minority of the higher income population that has access to a private vehicle. But for the city to be competitive, this subject of mobility must be one of the critical factors for its development.

Accra opts for BUS Rapid Transit system

The origins of the BRT can be traced back to Latin American planners and officials seeking a cost effective solution to the dilemma of urban transport. Facing a high population growth from the citizenry dependent upon public transport and having limited financial resources to develop car-based infrastructure. The planners were challenged to create a new paradigm. The result is the BRT, a surface metro system that utilizes exclusive right of way bus lanes. The developers of the Latin American BRT concept astutely observed that the ultimate objective was to swiftly, efficiently, and cost-effectively move people rather than cars.

In general, BRT is a high-quality customer-oriented transit that delivers fast, comfortable, and cost effective urban mobility. The main characteristics of BRT systems include:

- Segregated bus ways
- Rapid boarding and alighting
- Clean, secure, and comfortable stations and terminals
- Efficient pre-board fare collection
- Effective licensing and regulatory regimes for bus operators
- Clear and prominent signage and real time information displays
- Transit prioritization at intersections
- Modal integration at stations and terminals
- Clean bus technologies
- Sophisticated marketing identity
- Excellence in customer service

One main goal of BRTs is to improve quality of life and city competitiveness. BRTs seek to be more efficient systems that yield significant reductions in travel time and reduce traffic fatalities, as well as harmful air and noise pollution. They are designed to provide full accessibility to disabled, elderly, and children. Service must be high quality and low cost. High quality standards for cleaning, security, and maintenance of buses and stations, and respect of schedules must be maintained at a reasonable price for users and low capital costs for taxpayers. However, it still needs to have the possibility for the government to afford infrastructure costs, for the private sector to recover bus acquisition, and operation costs from fares without public subsidies, and for the users to be able to afford such fares.

Components of the BRT system

Five basic organizational components that can be found in the BRTs are:

1. A government that regulates and controls this public service, represented by the independent transit authority
2. The companies that render the transportation service, that is to say, the companies that own and operate the buses
3. The companies in charge of the fare collection system
4. The entity in charge of the administration and payments of the resources earned from the bus fares. (For the foregoing four components, it is necessary to develop the BRT Operational Model and the Financial Model)

5. The infrastructure specially built for the system, which should not only include the main routes, feeder routes, stations, and maintenance yards, but also parallel routes, platforms, and bicycle routes and bicycle parking close to the system

Accra BRT study background

Accra, the capital city and main administrative and commercial centre of Ghana, had a population of 1,659,000 in 2000, while the populations of the other urban centres were: Tema 506,000 and Ga 550,000. The combined effect of growth and migration will increase the population of Greater Accra Metropolitan Area (GAMA) to just less than 4 million by 2013, and by 2023 the combined population of the three assemblies will exceed 5 million.

Population growth and increasing rates of car ownership are expected to increase the number of cars in GAMA from 181,000 in 2004 to over 1 million in 2023. Under the prevailing conditions less than 5% of the population of Ghana own private motor vehicles.

The highest traffic volumes are found in the Winneba Road and Liberation Road corridors, which have volumes over 50,000 vehicles per day at certain points. Approximately 10,000 vehicles enter the central area of Accra within the Ring Road in the morning peak hour and on a typical weekday 270,000 vehicle trips are made into, or out of, the Accra central area. Higher volumes, of approximately inbound 16,000 vehicle trips in the morning peak hour and 300,000 daily vehicles trips in both directions, cross into the area inside the motorway extension.

These vehicle trips correspond to 50,000 inbound passenger trips into the Accra central area and 85,000 into the area inside the motorway extension in the morning peak hour. Approximately 1.3 million passenger trips per day are estimated to enter or leave the area within the Accra Ring Road and 1.6 million passenger trips into or out of the area within the motorway extension. Almost 84% of these passenger trips are made by public transport. Over half (56%) of daily passengers are carried by tro-tros, and a further 15% by taxi. Approximately 1 million passenger trips are made each day into and out of the central area of Accra using trotros and taxis. In Accra tro-tros and taxis carry an average number of passengers per trip of 13 and 2.3, respectively. These vehicles are inefficient in terms of the amount of road space used, and congestion caused, to transport each passenger.

Congestion is a major problem on arterial routes with 70% of major roads operating at unacceptable level of service at some time during the day (<20 km h^{-1}). Considerable scope therefore exists to improve the efficiency of people movement through a shift from low capacity public transport vehicles to large and double-decker or articulated buses with the potential to carry over 100 passengers. Government control over the operation of public transport is very limited; however, the private operations are strictly controlled by trade unions of which the most powerful is the GPRTU. The overall quality of public transport is poor, most vehicles are old, and maintenance standards are extremely low. High vehicle maintenance costs arising from poor road surfaces and limitations imposed on earnings by the acute congestion on the urban roads constrain the operators to invest in new vehicles. The limited number of vehicles and their low capacities result in long waiting times during the morning and evening rush hours.

The study considered seven major arterials in the city namely:

1. BRT 1 Guggisberg Avenue from Mpoase to Convention of Biological Diversity (CBD)
2. BRT 2 Winneba Road from Mallam to CBD
3. BRT 3 Nsawam Road from Apenkwa to CBD
4. BRT 4 Liberation Road and Independence Avenue from Tetteh Quarshie to CBD
5. BRT 5 Ring Road from Korle Bu to Labadi
6. BRT 6 Motorway Extension from Mallam to Tetteh Quarshie
7. BRT 7 Labadi Road from the Accra CBD to Tema

After a thorough evaluation and assessment of key factors, the BRT2 Winneba Road from Mallam to CBD was proposed for the first BRT pilot route. The BRT 2 is a six-lane east-west corridor. At present, in some sections hawkers have occupied a lane whilst public transport operators have also occupied a lane or two.

The main corridor would have six feeder routes. Several communities, which abound the present unofficial feeder route stations, do not have proper local access roads. The opportunity for integrating cycling facilities into the overall system is strategic and imminent. Commuters would therefore have the option of parking and riding from feeder station or even further parking and riding from the main corridor stations or major terminals.

The need for integrating cycling into BRT system

Non-motorized Transport Service (NMT) feeder services do add value to the BRT. This provides commuters who do not have direct access to the service on the trunk route to ride or walk to terminal and or stations, park and continue journey on the bus. This would improve safety for cyclists and contribute drastically to reducing air pollution from the transport sector. This integration is a means to liveable city where people from all walks of life rich or poor, no matter the status in society, and income levels are equitably provided for in terms of transport, feel safe and secure, have a sense of belongingness, a sense of respect. Transport is not just a means but a strong tool for integrating society when the right choices are made for a city. Bogotá's depicts a strong case where this integration of cycling or NMT with BRT is a success. Schoolchildren especially who walk or cycle have then a friendly and safe environment to commute.

Some planners perceive the difficulty of this integration but a good knowledge of cycling planning proves the possibility of integrating cycle facilities in existing road network. The bottleneck usually exists at intersection but this could be easily integrated.

The Winneba Road case

Provision for regular bike commuters on this route should be made especially at at-grade intersections to facilitate movement in the perpendicular direction to the BRT route. The NMT feeder requires all-weather, safe, and continuous pedestrian walkways and cycle ways to the interchange points and/or designated stations. This service facilitates the movements of commuters in low incomes areas and brings generally some dignity to commuters. Besides the cycle routes, parking facilities for cyclist does improve the system when commuters are confident of not losing their bikes. In cities where this facility exists, the integration of cycling into the BRT has been enhanced. Graphic road section of the BRT 2 route already have all-weather walkways and cycle ways.; except that the cycle ways begin from the Obetsebi–Lamptey circle and ends at the signalized intersection in front the Toyota Ghana showroom. The current provision though discontinuous does have some user unfriendliness, especially at intersections that must be improved. Cyclists are therefore forced to use the motor traffic way. Under this project the cycle route would have to be continuous between the trunk terminal at Mallam and the CBD. This will be on either side of the main road. Intersections shall be design preferably at-grade to incorporate NMT. Depending on the function of the bus feeder service routes the appropriate bike facilities should be provided. Obviously these are distributor collectors and bikes lanes of 1.5 m wide should be provided. The following routes should be considered under the project for bike facilities:

1. Mallam to CBD through Graphic Road 11 km
2. Santa Maria to Odorkor station 5.9 km
3. Anyaa to Odorkor station 9.4 km
4. Gbawe to Mallam Trunk terminal 7.5 km

Full-scale bicycle parking facilities shall be provided at the trunk terminals and a medium-scale parking facilities provided at selected trunk station. This will comprise bike-parking racks under aesthetically designed sheds where commuters can store bikes safely. The free parking facilities shall have full security for which payment is factored in the bus service.

The following modal interchanges should be considered for full-scale bicycle parking facility. The stations shall have medium-scale bicycle parking facilities as shown below:

- Mallam Trunk Terminal
- Kaneshie Trunk Terminal
- CMB Trunk Terminal
- Odorkor Station
- Darkuman Junction
- Liberty House

Cycling in Accra: a recent study

The Centre for Cycling expertise's study of Accra's cycling situation [4] does give a clear picture of the city taking to cycling. Despite the difficulties and lack of traffic safety cyclists are increasing by day. This may be attributed to high transport fares, which are a ripple effect from increases in petroleum products. The following are some of the key issues of the study.

Bike ownership and ability to cycle

Bike ownership and ability to cycle is highest among age group below 20 years and the 20–30 years bracket. As the population ages, bikes ownership reduces. Nevertheless, it is envisaged that a greater percentage of the population would have the ability to cycle and own bikes as the population ages.

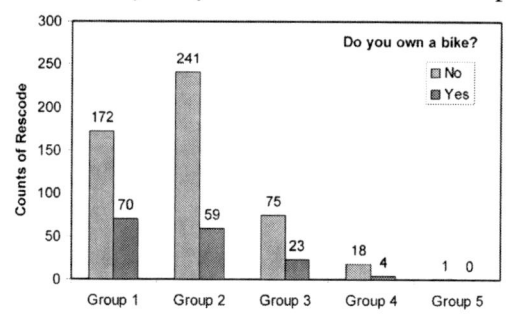

Fig. 1. Bike ownership per age group

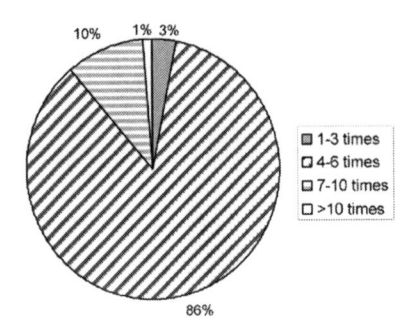

Fig. 2. Bike weekly usage in Accra

Frequency of bike usage

Among those who cycle the study did establish the frequency of use of bikes (Fig. 2). The high usage of 4–6 times a week represents students and artisans.

Obstacles to riding

A point of interest therefore, was to discover hindrances to bike use within this group (Fig. 3). Largely it was due to lack of traffic safety that most people would not cycle and this represented 50% of the views. This was followed by no safe parking places for bikes (21%) then fear of lack of social safety (9%). This had to do with unfriendliness of desired routes and dark areas in the evening.

The study discovered, from people who did not own bikes, what use they would put a bike to should they own one in future (Fig. 4). Whilst 31% would use it over weekends for exercise, 25% indicated that they would like to park and ride on a bus with a bike.

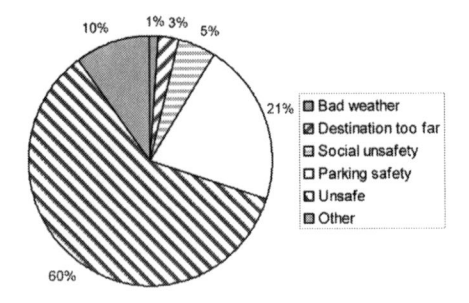

Fig. 3. Reasons for not using a bike

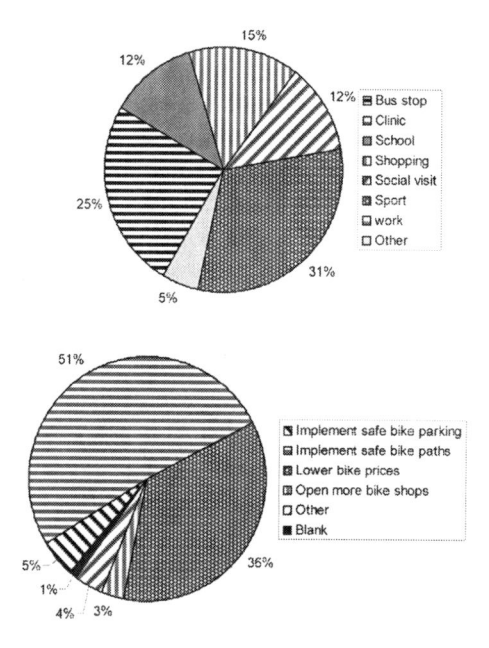

Fig. 4. Predicted use for persons who do not own a bike *(left)* and recommended measures for increase bike use *(right)*

The study also found out what responders propose government should do regarding cyclists (Fig. 4). Out of the total of 700 responders, 51% indicated that government should implement a safe network of bicycle paths, whilst 36% advocated that the price of bicycle should be made affordable.

Attitudes towards cycling

Generally in many developing cities such as Accra attitude towards cyclists is quite worrisome. It is obvious that most decision makers drive motor vehicles and for that matter almost always prepare and implement plans that favour motorists. People make cities happen and for that matter city planners and engineers should appreciate how people commute in cites and/or built-up areas in order to plan equitably. Transport as indicated is the movement of people and goods. This, of course, includes walking and cycling. It is quite ambivalent to find implementation of road networks in cities, which clearly leave out these modes of travel. Some have argued, for example, that cycling is not our culture. The question is Ghana has no car manufacturing industry neither have we discovered fossils fuel; is driving motor vehicle then our culture? Understandably we have adopted the motor vehicle to move from place to place. Why then is cycling not looked at as a mode of transport but rather as a cultural disposition? Motor vehicles are on the verge of choking our cities with attendant pollution and rapid increases in accidents, unfortunately many decision makers see motor vehicles as an epitome of development. What we ought to understand is that cities and transport is about people. People must therefore be the centre of the provision of transport infrastructure.

Obviously, one major issue is the lack of technical know-how of integrating NMT into road network. The general lack of knowledge in this area provides a platform and interface for Civil Society Organizations (CSOs) such as community and continuing education (CCE) to stand in the gap. The Centre recently organized a workshop for city planners and engineers to discuss issues on NMT including planning and design. The Centre has also supported some staff in government agencies to attend international conferences on NMT to give them a better appreciation of the facts.

Owing to the fact that there is no existing plan that characterizes NMT, the Centre has been working assiduously in collaboration with the Department of Urban Roads to develop the first ever Bicycle Masterplan (BMP) for the Capital City and eventually for the regional capitals. As part of this the Centre has conducted mobility survey studies to provide information on cyclists especially as this is non-existent. The BMP document, which will be a strategic policy document, will at the same time provide information for a comprehensive network of the National Capital territory and provide some details such as what type of bike facility is required for a particular route. Already, the Centre has also supported international consultants working for the Department of Urban Roads to provide for the relevant bicycle infrastructure in preliminary designs in recent contract.

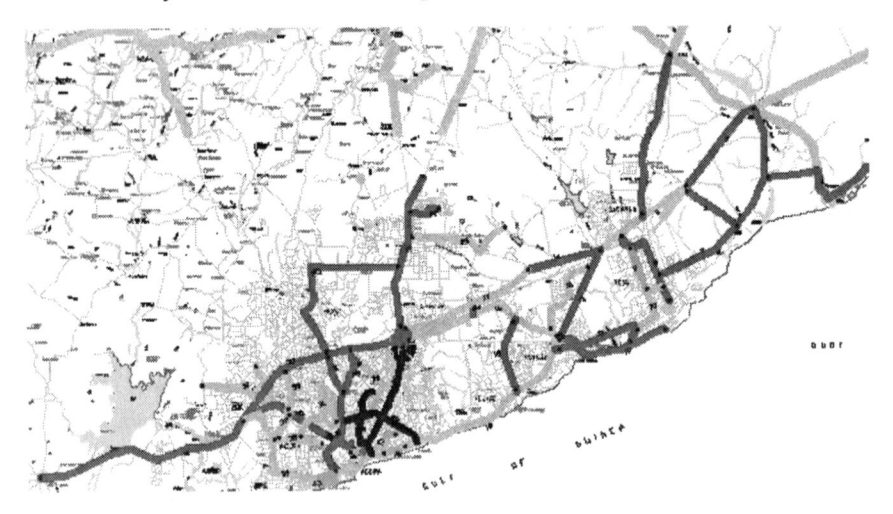

Fig. 5. Accra bicycle master plan

Portions of the networked developed have been adopted. Figure 5 shows the proposed plan for the city's bicycle network.

Economic significance of the integration

An efficient and effective public transport system such as the proposed BRT system for Accra brings with it enormous economic significance. Of course, with the integration with cycling, the city is yet to experience savings in time and money. Ghana's urban economy contributes 46% of the GDP.

Analysis on the corridor shows that estimated annual trips for 2004 the year of study were 69,400,000. With the proposed system annual time savings is estimated to be 10 million man-hours, which is expressed in money terms as US$ 4.2 million.

The health impacts of the vehicular emissions cannot be underestimated. The Ghana Health Services believes that many respiratory and heart problems is contributed by vehicular emissions. A move such as the integration of cycling with BRT will certainly reduce emissions and impact positively on the health of people.

Conclusions

In conclusion, a BRT system integrated with cycling is no less a system to pursue for Accra. The following benefits shall be derived:

- Significant time savings
- Reduced travel costs indirectly putting money in people's pockets
- Reduction in vehicular emissions
- Dignity of people improved with improved urban transport services.
- Significant travel safety for commuters especially schoolchildren and the physically challenged
- High sense of personal and mutual respect and a sense of belongingness
- A healthy population
- Enhanced business environment for the private sector
- Positive impacts to growth in all sectors of the economy

To facilitate the process the following is inevitably required:

- Support for the policy to introduce a high capacity bus system for the major urban settlements

- Support for the creation of the legal and institutional framework for efficient delivery of urban transport services
- Support for the comprehensive integration of land use and transport
- Support for the policy to introduce strong environmental quality standards for procuring motor vehicles and monitoring of same

References

1. Institute for Transportation and Development Policy (2004) A prefeasibility study for a bus rapid transit and cycle ways for Accra (A UNDP/GEF funded study)
2. DHV Consultants (2005) Bus rapid transit options and identifications and prefeasibility study: draft final report. A study for the Department of Urban Roads of the Ministry of Transportation, Accra
3. Smith A (2004) A World Bank study on involving private participation in improved public transport services
4. Quarshie ML (2004) Cycling in Accra: an in-depth study. Centre for Cycling Expertise, Accra

II. Air Pollution and Air Quality

Evaluation of hydrogen peroxide in rainwater in downtown São Paulo

MA Santos,[1] JJ Pedrotti,[1] A Fornaro[2]

[1]Departamento de Química, Universidade Presbiteriana Mackenzie, Rua da Consolação, 896, 01302-907, São Paulo, SP, Brazil
[2]Departamento de Ciências Atmosféricas, Instituto de Astronomia, Geofísica e Ciências Atmosféricas (IAG/USP), Rua do Matão, 1226, 05508-090, São Paulo, SP, Brazil

Abstract

The concentration of hydrogen peroxide (H_2O_2) in rainwater samples collected between April 2003 and April 2004 in downtown São Paulo were determined. The concentrations of H_2O_2 ranged from 2.29 to 48.6 μmol L^{-1}, with an average value ($n = 70$) of 13.1 (±11.3) μmol L^{-1}. The higher concentrations were observed in spring and summer. The volume-weighted mean (VWM) of the free H^+ was 6.5 μmol L^{-1}, corresponding to a pH of 5.2. The analysis of SO_4^{2-}, NO_3^-, and $HCOO^-$, in the same set of samples, showed an average concentration of 11.7 (±10.9), 20.0 (±18.0), and 6.18 (±12.4) μmol L^{-1}, respectively. The $HCOO^-$ concentrations showed correlation ($r = 0.43$) with H_2O_2, which can indicate reactions in aqueous phase where these two species are formed.

Introduction

The acidification of wet precipitation is a well-known environmental problem in large urban areas and high industrialized regions. The effects of the acid precipitation on the different ecosystems have been known since the

19th century, when the main cause of the acid rain was attributed to sulphur dioxide (SO_2) and nitrogen oxides (NO_x) emitted to the atmosphere from the coal burn used for domestic heating and as power source in industrial plants [1]. Acid rain can cause several environmental damages including lake acidification, adverse effect on the vegetation, corrosion of buildings and monuments, and modification on chemical composition of soils [2].

The increasing acidity of the liquid phase in the atmosphere has been associated to the presence of inorganic acids like nitric (HNO_3) and sulphuric (H_2SO_4) and some organic acids like acetic (CH_3COOH) and formic ($HCOOH$), produced by oxidation reactions of sulphur, nitrogen, and carbon compounds, respectively. Acidifying compounds in the atmosphere have been produced in liquid and gas phase. In the aqueous phase of the troposphere, when the pH is lower than 5.0, the hydrogen peroxide (H_2O_2) is considered the most effective oxidant agent for the conversion of dissolved SO_2 to sulphuric acid, the main contributor of the rainwater acidity [3].

Among the various oxidant compounds in the atmosphere, H_2O_2 and organic hydroperoxides have received great attention from several groups in last 25 years mainly for acting as reservoirs of OH and HO_2 radicals, producing and consuming this chemical species by cyclic reactions in the atmosphere [4], and also for presenting adverse effects on plants and human health [5,6].

The role of H_2O_2 on atmosphere chemistry has been studied under different aspects including the formation of sulphuric acid [7], the ozone decomposition, and oxidation of organic compounds. The presence of H_2O_2 in the liquid phase (cloud, rain, and snow) of the atmosphere can be attributed to two sources: dissolution of H_2O_2 gas phase ($K_H = 1.2 \times 10^5$ mol L^{-1} atm^{-1} at 25°C) and direct production of H_2O_2 by chemical reactions in the liquid phase [8].

H_2O_2 determination in gas and liquid phase over continents and oceans has been investigated extensively in the last decades. The results of these studies had shown that the ambient concentration of H_2O_2 ranges typically from 0.1 to 2.0 ppb in the gas phase and from $<10^{-7}$ to 10^{-4} mol L^{-1} in fog, cloud, and rainwater, with higher concentrations observed in the summer and lower concentrations during the winter [8,9]. The first results of the H_2O_2 evaluation in 20 rainwater samples collected between September 2001 and March 2002 in São Paulo downtown were described previously [10].

The atmospheric emissions in the metropolitan area of São Paulo (MASP) are due to pollutants released by industries and, principally, by the large fleet of light and heavy vehicles, reaching approximately 8 million, with an estimate of 2.5 million running every day. Diesel, hydrated ethanol, and gasohol (gas + 25% of ethanol) are the most common fuels used by vehicles. According to estimates of the São Paulo State Environmental

Agency (CETESB) the fluxes of the main pollutants emitted into the atmosphere were 1.7 million ton year^{-1} of carbon monoxide (CO), 404,000 t year^{-1} of hydrocarbons (HC), 371,000 t year^{-1} of nitrogen oxides (NO$_x$), 63,000 t year^{-1} of inhalable particles (PM$_{10}$), and 38,000 t year^{-1} of sulphur oxides (SO$_x$).

The vehicular fleet is mainly responsible for the emissions of these pollutants (98% CO, 97% HC, 96% NO$_x$, 55% SO$_2$, and 40% PM$_{10}$) [11]. Since the 1970s, CETESB has been monitoring these pollutants, the ozone and also some meteorological parameters. In 1988, CETESB and National Commission of the Environment (CONAMA) initiated a progressive emissions control program for automotive vehicles (PROCONVE), addressing CO, SO$_2$, hydrocarbons, NO$_x$, aldehydes, and particles.

After that, a decrease in SO$_2$ concentrations was observed, with annual average of 65 μg m^{-3} in 1983 and 15 μg m^{-3} in 2003. This indicates that PROCONVE's legal actions, such as the increase of the fuel quality (desulphurization, substitution of tetraethyl lead by ethanol) and the introduction of electronic injection and catalytic converters in the cars, were effective to reduce some pollutant emissions. However, this decrease was not observed in NO$_x$ and hydrocarbons, which produce ozone.

The quality of the air in MASP has been harmed a great deal because of this tropospheric ozone. The overrun of the O$_3$ standard (160 μg m^{-3}) was stronger in October, during 1997 and 2003 [11]. Also the same photochemical process which produces O$_3$, also produces H$_2$O$_2$ in the atmosphere. It is important to highlight that there is no official net in Brazil for monitoring the rainwater chemical composition.

This study presents the magnitude of wet precipitation of H$_2$O$_2$ in the same sampling site for 34 events (70 samples) occurring between April 2003 and April 2004 and a correlation analysis between H$_2$O$_2$, hydrogen ion, nitrate, sulphate, and formate. The effects of rainfall and of the air temperature on the H$_2$O$_2$ concentration were also analysed.

Experimental

Site sampling

The city of São Paulo (Fig. 1) is located in the south-eastern region of the São Paulo State, Brazil, around 45 km from the Atlantic Ocean, at an altitude of 780 m. The MASP is one of the most populated regions in the world, with more than 17 million inhabitants distributed in 1747 km^2 of unplanned urban area.

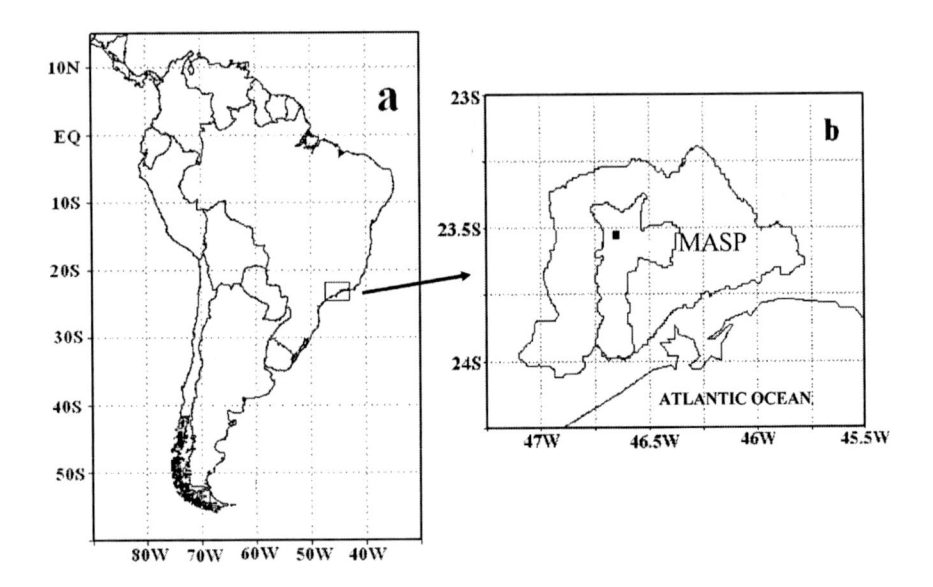

Fig. 1. Localization of MASP in the south-east of Brazil: (a) South America and Brazil (b) MASP and the São Paulo city. The symbol ■ shows the central region of São Paulo city.

Methodology

Wet-only (WO) samples were collected with an automatic rainwater collector, model G.K. Walter, within the campus of Mackenzie University, in downtown São Paulo. On arrival at the laboratory, each sample was weighed for volume determination. After this step, an aliquot of sample was filtered with 0.45 µm porous Teflon membrane (Millipore) to remove solid particles and, subsequently, the spectrophotometric determination of H_2O_2 was carried out. Afterwards, the sample was separated into aliquots in the high-density polyethylene (PE) flasks for additional analytical determinations, which were carried out 1 week after the rain event. For pH an unfiltered aliquot was preserved at 4°C in a refrigerator, while for chromatographic determinations a filtered aliquot was stored in a freezer at −18°C.

All reagents used were of analytical grade. The solutions were prepared with ultra pure water (Nanopure, Barnstead, USA). The enzyme peroxidase (E.C.1.11.1.7 − 115 U mg^{-1}) was purchased from Sigma (St. Louis, Missouri, USA). All the reference solutions were prepared just before their use.

The pH values were measured with a Digimed DM-20 pH meter provided with a glass electrode combined with an Ag/AgCl (KCl sat.) reference electrode.

The chromatographic analyses were made by ion chromatography Metrohm model 761 provided with an electrical conductivity detector. The mobile phase was a solution of Na_2CO_3 4.0 mmol L^{-1}/$NaHCO_3$ 1.2 mmol L^{-1} and column Metrohm suppressor was regenerated with 50 mmol L^{-1} H_2SO_4.

The H_2O_2 determinations in rainwater samples were carried out by using a flow injection analysis system combined with a spectrophotometer Femto, model 600, provided with a micro flow-through cell of 10 mm light path. The spectrophotometric detection is based on the reaction of H_2O_2 with phenol and 4-aminoantipyrine in the presence of peroxidase enzyme that yields a quinoneimine dye (red compound) with a maximum absorption at 505 nm.

Results and discussion

The chemical composition data of the 70 samples correspond to 34 events of rain. Pluviometry was 420 mm, 30% of the total 1400 mm rainfall between April 2003 and April 2004. The concentrations of H_2O_2 in rainwater samples varied between 2.29 and 48.6 µmol L^{-1} with an average value and standard deviation of (13.1 ±11.3) µmol L^{-1}. However, most samples have concentrations below 10 µmol L^{-1}. On the other hand, for the same rainfall amount, there was a great variability of H_2O_2 concentration (Fig. 2).

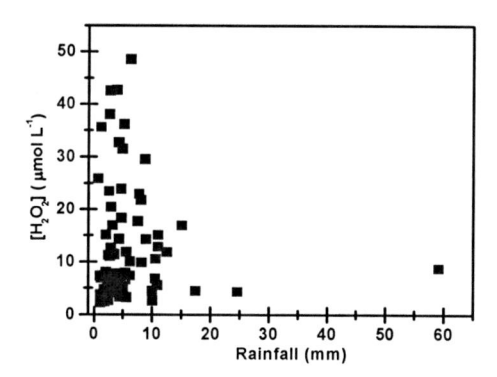

Fig. 2. Rainfall influence in the hydrogen peroxide concentration

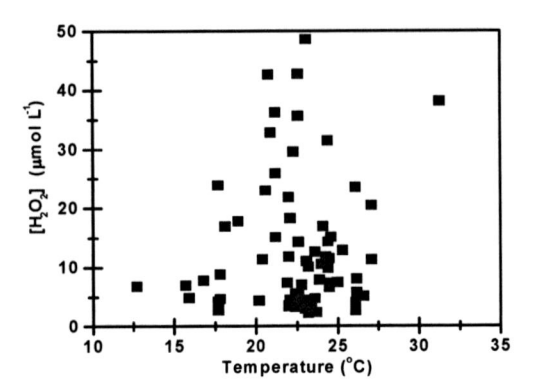

Fig. 3. Air temperature influence in the rainwater H_2O_2 concentrations

The effect of the temperature indicated that there was a range of optimum temperature for the presence of peroxide in rainwater (Fig. 3). This temperature range was between 20 and 25°C. Higher temperatures may cause H_2O_2 decomposition, while lower temperatures do not provide conditions for photochemical reactions.

Table 1 shows great variation in the H_2O_2 concentrations in the different seasons, being the maximum media value observed in the spring, which corresponds to the period between dry and rainy seasons. In the spring the first rainfalls start and remove the particles which had been accumulated in the atmosphere during the dry winter. Therefore, sunnier and hot days become more frequent, favouring the photochemical reactions and the H_2O_2 formation. Moreover, it is important to consider that the ozone concentration is higher during springtime, mostly in October [11].

Table 1. Seasonal variation of the hydrogen peroxide concentrations

	\multicolumn{4}{c}{H_2O_2 (µmol L^{-1})}	Samples number
	Min.	Max.	Mean	(±SD)	VWM	
Spring	2.74	42.6	18.4	(13.2)	14.6	21
Summer	2.46	48.6	11.0	(10.6)	11.7	37
Autumn	2.29	14.3	7.46	(4.06)	7.71	8
Winter	6.81	23.9	15.9	(4.10)	14.4	4
Annual	2.29	48.6	13.1	(11.3)	12.2	70

VWM: volume-weighted mean

The analysis of SO_4^{2-}, NO_3^-, and HCOO–, in the same set of samples, showed an average concentration of 11.7 (±10.9), 20.0 (±18.0), and 6.18 (±12.4) µmol L^{-1}, respectively. The volume-weighted mean (VWM) of the free H$^+$ was 6.5 µmol L^{-1}, corresponding to a pH of 5.2. Figure 4 shows the sulphate, nitrate, formate, and hydrogen ions concentrations versus H_2O_2 concentration. In these rainwater samples, H_2O_2 concentration had the same magnitude order as that in these ions. A little correlation ($r = 0.43$, $p = 0.0002$) among H_2O_2 and formate ion was observed, while no significant correlation was observed among the other ions, except for sulphate and nitrate ($r = 0.85$, $p < 0.0001$).

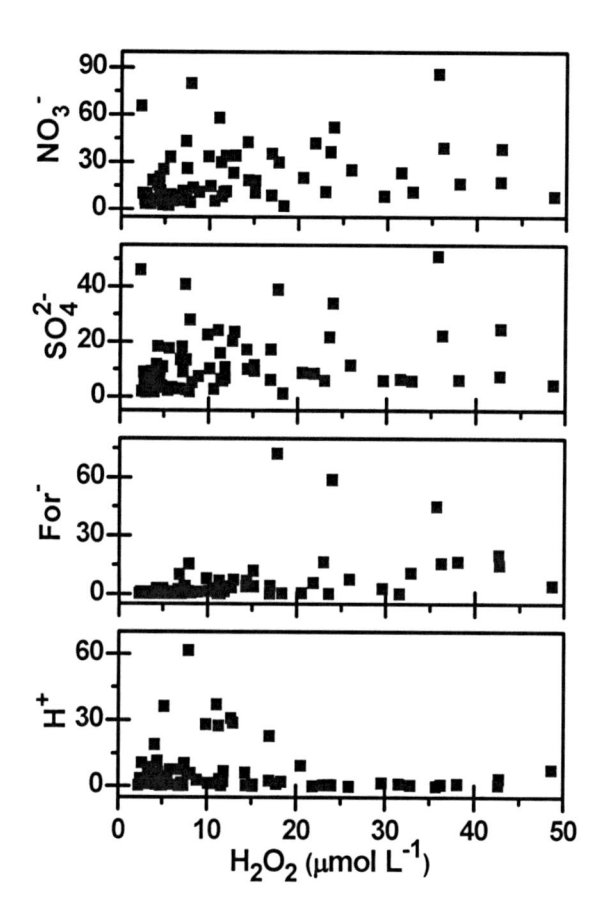

Fig. 4. Sulfate, nitrate, formate, and hydrogen ion concentrations (µmol L^{-1}) versus hydrogen peroxide concentrations

These results can be interpreted as incorporation by different processes of these compounds in rainwater. Sulphate and nitrate were dissolved from particles incorporated by processes in- and below-cloud, while formate and H_2O_2 were produced by reactions in liquid phase [3]. The HCOOH and H_2O_2 formation happens through oxidation of formaldehyde by OH radicals:

$$HCHO(g) + H_2O \rightleftharpoons CH_2(OH)_2 \quad (1)$$

$$CH_2(OH)_2 + OH\cdot \rightarrow CH(OH)_2\cdot + H_2O \quad (2)$$

$$CH(OH)_2\cdot + O_2 \rightarrow HCOOH + HO_2\cdot \quad (3)$$

$$HO_2\cdot + HO_2\cdot \rightarrow H_2O_2 + O_2 \quad (4)$$

Table 2 shows that the H_2O_2 concentrations found in the rainwater downtown São Paulo were similar to those in other sampling sites described in the literature. However, among studies presented in Table 2, the north-west of Spain is highlighted, since lower peroxide concentrations are due to high SO_2 concentrations from the thermal power plant in the region [5]. In this region, the mean and maximum values of the non-sea salt sulphate concentration were 40.8 and 236 µmol L^{-1}. On the other hand, the mean and maximum values of the sulphate concentration were 11.7 and 51.0 µmol L^{-1}, respectively, in rainwater samples from São Paulo, therefore showed to be much lower than the values observed in Spain.

Table 2. Hydrogen peroxide concentrations in rainwater from different regions worldwide

Sampling site	Min.	Max.	Mean (±SD)	Sampling period	Samples number
		(µmol L^{-1})			
São Paulo #	2.29	48.6	13.1 (11.3)	4/2003–4/2004	70
São Paulo [10]	0.50	78.1	20.3 (17.0)	9/2001–3/2002	20
Santiago do Chile [12]	0.10	32.0	6.60*	1997	6
Miami [7]	0.30	38.6	6.90*	4/1995–10/1996	80
Kyoto (daytime) [13]	1.20	65.0	12.9	1999–2000	29
North Carolina [4]	0.13	48.4	12.0 (11.0)	10/1992–10/2004	61
South and central Atlantic Ocean [15]	3.50	71.0	26.0 (22.0)	5–6/1996	25
North-west of Spain (thermal power plant region) [5]	<0.002	2.70	0.36 (0.42)	1933–1995	138

Min: Minimum, Max: Maximum, SD: standard deviation, #: present study
*Volume-weighted mean

Conclusions

The concentrations of H_2O_2 and some major ions were measured in rainwater samples collected between April 2003 and April 2004 in the central region of São Paulo city. The results ranged from 2.29 to 48.6 µmol L^{-1} and the average value was 13.1 µmol L^{-1}, showing seasonal variability and dependence on the air temperature. These concentrations with values above 20 µmol L^{-1} were obtained in rainwater samples collected between September and January, during spring and summer. Higher concentrations were observed in air temperature between 20 and 25°C. The positive relation between H_2O_2 and formate suggested that both compounds were produced in the liquid phase, through oxidation of formaldehyde by OH radicals. The magnitude of H_2O_2 concentration in rainwater is also an indicative of the strongly oxidant capacity of the atmosphere in São Paulo city.

Acknowledgements

The authors are grateful to Fundação de Amparo a Pesquisa do Estado de São Paulo – FAPESP (proc. 01/09838–0) and Mackpesquisa for financial support.

References

1. Cowling EB (1982) Acid precipitation in historical perspective. Environ Sci Technol 16:110A–123A
2. Satake E, Kojima S, Takamatsu T, Shindo J, Nakano T, Aoki S, Furukuyama F, Hatakeyama S, Ikuta K, Kawashima M, Kohno Y, Murano K, Okita T, Taoda H, Tsunoda K (2001) Acid Rain 2000 – conference summary statement – looking back to the past and thinking of the future. Water Air Soil Pollut 130:1–16
3. Fornaro A, Gutz IGR (2003) Wet deposition and related atmospheric chemistry in the São Paulo metropolis, Brazil: part 2 – contribution of formic and acetic acids. Atmos Environ 37:117–128
4. Reeves CE, Penkett SA (2003) Measurements of peroxides and what they tell us. Chem Rev 103:5199–5218
5. Peña RM, García S, Herrero C, Lucas T (2001) Measurements and analysis of hydrogen peroxide rainwater levels in a northwest region of Spain. Atmos Environ 35:209–219

6. Hasson AS, Paulson SE (2003) An investigation of the relationship between gas-phase and aerosol-borne hydroperoxides in urban air. Aerosol Sci 34:459–468
7. Deng Y, Zuo Y (1999) Factors affecting the levels of hydrogen peroxide in rainwater, Atmos Environ 33:1469–1478
8. Gunz DE, Hoffmann MR (1990) Atmospheric chemistry of peroxides: a review. Atmos Environ 24A:1601–1633
9. Sakugawa H, Kaplan IR, Cohen Y (1990) Atmospheric hydrogen peroxide – does is share a role with ozone in degrading air quality? Environ Sci Technol 24:1452–1462
10. Gonçalves C, Matos RC, Fornaro A, Pedrotti JJ (2002) Avaliação preliminar de peróxido de hidrogênio em águas de chuva na região central de São Paulo, Anais do XII Congresso Brasileiro de Meteorologia
11. CETESB (2004) Relatório de Qualidade do Ar no Estado de São Paulo –2003. Secretaria do Meio Ambiente, Série Relatórios – ISSN 0103–4103, São Paulo (www.cetesb.sp.gov.br)
12. Ortiz V, Rubio MA, Lissi EA (2000) Hydrogen peroxide deposition and decomposition in rain and dew water. Atmos Environ 34:1139–1146
13. Yamada E, Tomozawa K, Nakanishi Y, Fuse Y (2002) Behavior of hydrogen peroxide in the atmosphere and rainwater in Kyoto, and its effect on the oxidation of SO_2 in rainwater. Bull Chem Soc Jpn 75:1385–1391
14. Willey JD, Kieber RJ, Lancaster RD (1996) Coastal rainwater hydrogen peroxide concentration and deposition. J Atmos Chem 25:149–165
15. Yuan J, Shiller AM (2000) The variation of hydrogen peroxide in rainwater over the south and central Atlantic Ocean. Atmos Environ 34:3973–3980

The comparison of pollutant concentrations in liquid falling and deposited precipitation, and throughfall

J Fisak,[1] P Chaloupecky,[1] D Rezacova,[1] M Vach,[2] P Skrivan,[2] J Spickova[2]

[1]Institute of Atmospheric Physics, Academy of Sciences of the Czech Republic, 1401 Bocni II, 141 31 Prague 4, Czech Republic
[2]Institute of Geology, Academy of Sciences of the Czech Republic, Rozvojova, 14 131 Prague 6, Czech Republic

Abstract

The paper presents mutual comparison of samples of individual kinds of precipitation – deposited from fog, precipitation below the tree canopies (the throughfall), and bulk – and wet-only (WO) precipitation on an open place. The throughfall samples reach the highest conductivity values and maximum mean concentrations of individual components with the exception of Cu, Pb, and Cd. The throughfall exhibits also the lowest mean pH values. It was also found that the mean concentrations of several components in fog water are significantly higher than those in the WO samples, but not by one order of magnitude, as it is presented in several papers.

The majority of components in throughfall exhibit considerably higher concentrations throughout the autumn season. On the other hand, concentrations of only several distinct components in fog water exhibit an increase in autumn (K, Mn, Fe, Be, Cd, Al, Ba, and Rb), in effect without any changes remain values and concentrations of pH, Na, F⁻, Cl⁻, Cu, Pb, and As. Concentrations or values of other components are higher in spring than in autumn.

Introduction

The atmosphere is purified from contaminants, as well as from dry fallout, through the wet precipitation. Each type of precipitation is entrapping the atmospheric pollutants in different ways and intensity. This is caused also by different residence time of precipitation and its activity in contaminated earth-bound atmospheric layers. The falling precipitation (rain, drizzle) interacts with the atmosphere for the shortest time, while the deposited precipitation indicates the longest reaction time, as it originates from fog or low cloudiness (in higher altitudes). The highest concentration of pollutants, however, shows the below-tree-canopy precipitation (the throughfall) that contains both the washed out solid aerosol and the leached metabolites of the vegetation.

At the Milesovka Observatory (837 m a.s.l.), the fog water has been sampled and analysed since 1999. Since the second half of 2004, there have been also collected samples of blue spruce (*Picea pungens* Engelm.), and beech (*Fagus sylvatica* L.) throughfall. Since 2005, samples of falling precipitation (both the monthly bulk samples and the WO samples) have been collected as well. Consequently, there is an excellent opportunity to compare the pollutant concentration in various types of the precipitation samples collected at a single place with same local contamination and emission load from the local industry. Another advantage comes from the location of site, which is situated in a heavily polluted region. In spite of considerable improvements in the mentioned region (complete desulphurization of local large power plants in the 1990s and attenuation of mining activities in last 10 years), territory around the Milesovka Observatory belongs to the most polluted regions of the Czech Republic. In our contribution, attention will be focused on the comparison of concentrations of selected ions and trace elements in the individual precipitation types.

So far, concentrations of pollutants were compared in particular kinds of precipitation individually. The samples of falling precipitation and water from fog and rime were sampled on various sampling sites, often were compared results presented by various authors. In this contribution, various kinds of samples collected at single locality are compared. All instruments determined to collect precipitation (WO, bulk, fog water, and throughfall) are distributed on an area of 50×50 m. Even here, however, emerged certain disproportions caused especially by the seasonal occurrence of fog.

Experimental

Site description

The meteorological Milesovka Observatory (50°33'17"N, 13°55'57"'E, 837 m a.s.l.), which belongs to the Institute of Atmospheric Physics, Academy of Science of the Czech Republic (ASCR), is situated in the northern part of the Czech Republic at the top of the conical isolated hill Milesovka (Fig. 1). The region around the Milesovka Mt. is one of the most

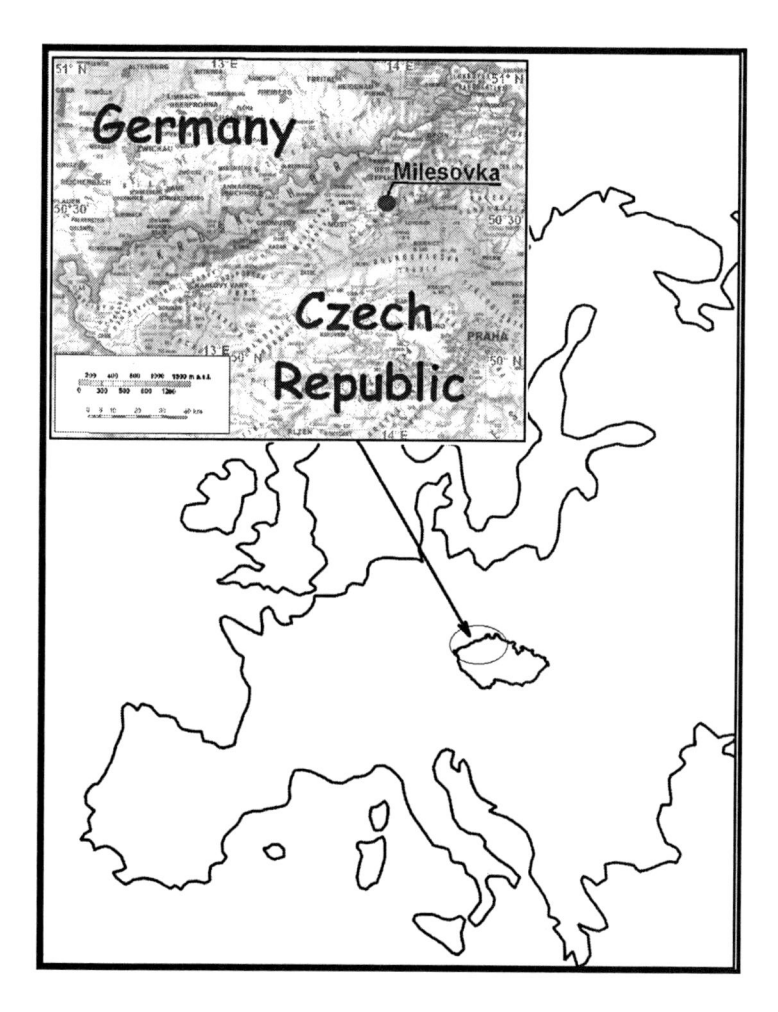

Fig. 1. Location of the Milesovka Observatory

polluted areas in the Czech Republic, with numerous coal-burning power plants, opencast lignite mines, and chemical facilities. Most of the power plants have been equipped with the desulphurization technology or ceased operation in the second half of the 1990s. Power plant desulphurization in this region was completed at the end of 1997. There is a road with a heavy traffic of both freight and public transport near the Milesovka Mt.

The collectors and their cleansing

The samples of fog water were collected using the active fog water collector (AFWC) (Fig. 2), described in Tesar et al. [1], Daube et al. [2], or Fisak and Rezacova [3]. Sampling runs during time periods of above-zero air temperature. At the Milesovka Mt. such periods usually cover the time intervals from March to the end of October, depending on actual meteorological conditions. The AFWC measurements were switched on (off) automatically by using the present weather detector (PWD21) device when visibility dropped (exceeded) the threshold of 1 km. Sampling goes on during the whole fog event and the sampling bottles are replaced when the sample volume exceeds approximately 100 ml. The collector efficiency and the stored fraction of fog droplet size spectrum could not be exactly determined. A fog event was defined by a time period with the visibility lower than the threshold value of 1 km. The visibility was not allowed to exceed the threshold for more than 60 min during the fog event. Just after sampling termination the AFWC cartridge (position 3 in Fig. 2) is taken away,

Fig. 2. Active fog water collector (1: sampling box; 2: fan; 3: cartridge with Teflon fibres; 4: tripod; 5: collection bottle)

Fig. 3. Bulk precipitation collector

cleansed by hot water vapour, and rinsed by warm demineralized water. The cleansing procedure was applied also to the sampling box (position 1 in Fig. 2) and it was repeated routinely each week even when no sampling took place.

The bulk precipitation (BP) collectors (Fig. 3) consist of a glass funnel protected by polyethylene (PE) casing whose upper rim is arranged in a sawtooth pattern to protect samples from contamination by birds. The mouth of glass funnel is leading to 1 L PE collection bottles. The collection bottles were pre-dosed with 2.5 mL of diluted (22% v/v) HNO_3 (Merck, Suprapur) to prevent the adsorption of dissolved elements onto the walls of the sampling bottle. The resulting aqueous sample therefore represents the sum of dissolved solutes in precipitation together with the forms of the solutes weakly bound to the solid particles of deposition. Aliquots of BP used for determination of anions, as well as for laboratory pH and conductivity measurements, were collected separately.

The throughfall collector (Fig. 4) is additionally equipped with a glass conical bulb placed onto the funnel so that the throughfall flows down into the collecting bottle without contamination by needles or other organic debris. Further details of the sampling techniques were described in Skrivan et al. [4,5]. The throughfall sampling site is equipped by four collectors used for spruce throughfall (STH) sampling and five collectors used for sampling of beech throughfall (BTH).

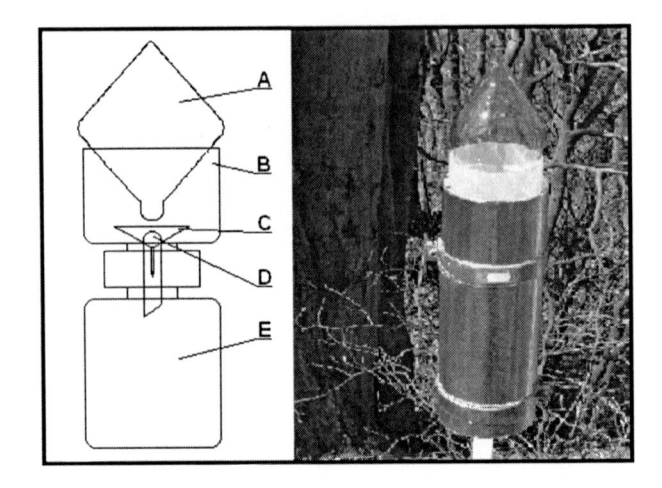

Fig. 4. Throughfall collector. (A: glass conical bulb; B: polyethylene holder; C: glass funnel; D: glass bubble; C: collection bottle)

Fig. 5. Wet-only collector

The WO collector (Fig. 5) is equipped with PE sampling funnel. It is opened (closed) automatically at the beginning (end) of the precipitation episode. The WO collector opening (closing) is driven with a rainfall detector. The rainwater pours into the 1 L PE bottle. The sample is taken away from the collector after the finished precipitation episode and stored until its elaboration, similarly to the samples of other kinds of precipitation, in darkness at temperature up to +5°C.

All samples are filtered using the 0.45 μm membrane filter. The samples determined for the study of trace elements are acidified (samples of BP in advance, samples of throughfall additionally after the filtration) by diluted

HNO$_3$ (Merck, Suprapur) to prevent losses by adsorption. The filtrates were stored at +4°C in a cooler until the analysis.

The samples of BP determined for the study of major cations and anions were also employed to the conductivity and pH measurement. The bottles were washed in hot distilled water and then rinsed out until the conductivity of the rinsing water decreased below 4 µS.

Analytical methods

The major cations and anions, represented by Na, K, Mg, Ca, NH$_4^+$, F$^-$, Cl$^-$, NO$_3^-$, and SO$_4^{2-}$ were analysed in the laboratories of Czech Geological Survey, Prague. Analyses of trace metals (Cu, Mn, Fe, Zn, Pb, Be, As, Sr, Cd, Al, Ni, Ba, and Rb) have been carried out in the Geological Institute, ASCR. Applied methods for the chemical analyses and their determination limits are summarized in Table 1.

Table 1. Analysed elements and ions, analytical methods, and determination limits

Elemention	Type of analysis	Determination limit
Fe, Mn, Zn	Flame analysis (acetylene-air) (FAAS) Atomic absorption spectrometer VARIAN, model SpectrAA 300	30 µg L^{-1} (Fe); 20 µg L^{-1} (Mn); 10 µg L^{-1} (Zn)
Al, Be, Cd, Cu, Mn, Pb, Sr, Ni, Ba, Rb	Electrothermic atomization in graphite tube atomizer Varian GTA 96 (GFAAS) Atomic absorption spectrometer VARIAN, model SpectrAA 300	1 µg L^{-1} (Al); 0.1 µg L^{-1} (Be, Cd); 0.5 µg L^{-1} (Cu, Pb, Sr, Mn)
As	Hydride-generation technique Atomic absorption spectrometer VARIAN, model SpectrAA 300 + vapour generation accessory (VGA-76)	0.5 µg L^{-1}
Ca, K, Mg, Na	AAS-flame analysis; Perkin Elmer 3100	0.1 µg L^{-1}
Cl$^-$, NO$_3^-$, SO$_4^{2-}$	High-pressure liquid chromatography (HPLC); Schimadzu SPD-6A	0.1 mg L^{-1} (Cl$^-$); 0.3 mg L^{-1} (NO$_3^-$); 1 mg L^{-1} (SO$_4^{2-}$)
F$^-$	Ion-selective electrode (ISE) method	0.02 mg L^{-1}

Results: sample elaboration

Mean concentration values of selected elements were calculated for the rainwater collected both on an open place and below-tree canopies (the throughfall) and for water collected from fog (FW). The precipitation water was collected both as the WO and BP. The data set was elaborated

Table 2. Processed sampling periods of various types of precipitation at Milesovka Observatory

Precipitation type	Processed period	
	From	To
STH	VI/2004	IX/2005
BTH	VI/2004	IX/2005
FW	III/2004	IX/2005
WO	III/2005	IX/2005
BP	IV/2005	IX/2005

as a whole and with respect to the seasonal character of fogs the data were also separately processed for the spring and autumn seasons. With respect to the limited size of the data set, its extensive statistical elaboration was not realized. Table 2 presents the sampling periods for the individual precipitation types.

The analysed components of WO samples are not so extensive as those of the STH and BTH, or as of the FW and BP samples. The sampling of BP at the Milesovka site has been initiated later and the number of samples insufficiently covers the so-called transient seasons. The data elaboration involved calculations of the concentration ratios of individual sampled components among STH, BTH, and BP, accordingly the bulk samples, and FW and WO, i.e., samples collected at fog and precipitation episodes (Table 2). The results are presented in Tables 3 and 4.

Table 3 shows that the lowest mean pH (4.1), the highest mean conductivity (304.6 µS), and concentrations of selected components (except for Cu, Pb, and Cd) are reached in STH samples. On the contrary, the lowest mean conductivity value (21.8 µS) and concentrations of selected components (except for Cu, Pb, As, and Al) exhibits the open place BP. The maximal values of mean concentrations of Pb and Cd were found in WO samples (Table 3).

In spite of the fact that the chemical analyses of WO samples do not involve all components as the FW samples, we compared at least the components common to both sample types. Both sample types were collected on the open place at precipitation (fog) episodes.

Table 3 indicates that the FW samples exhibit lower pH values than the WO ones. Similarly, the conductivity and concentrations of several components (Al, SO_4^{2-}, Fe, Zn, and Cu) are significantly higher in FW than in the WO samples. On the other hand, considerably lower (than in WO samples) are the concentrations of Cd, Pb, and K in the FW. Table 4 shows the sequence of concentrations in the individual kinds of precipitation for all studied components.

Table 3. Mean component concentrations or values in various types of precipitation

Precipitation type	n.s.	Cond. (µS)	pH	Na	Mg	K	Ca	NH$_4^+$ (µg L^{-1})	NO$_3^-$	F$^-$	SO$_4^{2-}$	Cl$^-$	Cu	Mn
STH	16	304.64	4.07	7951	2415	13874	9202	19220	56963	534	65879	13705	15.47	206.80
BTH	16	144.33	4.84	2354	934	8076	4024	3770	16399	197	17445	4380	2.59	185.19
WO	44	27.03	4.95	395	131	546	960				4075		7.92	14.59
BP	5	21.80	4.61	133	57	280	243	386	2112	15	3157	423	3.01	13.24
FW	39	126.25	4.38	530	156	420	886	9914	16702	194	17944	1122	18.22	24.07
FW/WO		4.67	0.89	1.34	1.20	0.77	0.92				4.40		2.30	1.65
THS/BP		13.97	0.88	59.63	42.61	49.55	37.81	49.77	26.98	34.95	20.87	32.43	5.15	15.62
THB/BP		6.62	1.05	17.66	16.47	28.84	16.54	9.76	7.77	12.90	5.53	10.36	0.86	13.98
FW/BP		5.79	0.95	3.98	2.76	1.50	3.64	25.67	7.91	12.74	5.68	2.65	6.06	1.82

		Fe	Zn	Pb	Be	As µg L^{-1}	Cd	Sr	Al	Ni	Ba	Rb
STH	16	531.33	334.80	2.31	0.20	2.13	0.83	49.04	541.53	3.43	19.42	15.07
BTH	16	83.63	27.25	0.64	0.03	0.85	0.18	12.08	91.08	0.60	9.15	11.40
WO	44	23.15	48.03	10.03			6.90		17.65			
BP	5	74.64	26.95	1.05	0.01	0.26	0.10	1.42	40.65	0.44	1.93	0.54
FW	39	85.58	168.14	6.10	0.03	1.84	3.74	2.97	95.23	1.28	8.82	2.88
FW/WO		3.70	3.50	0.61			0.54		5.40			
STH/BP		7.12	12.42	2.20	29.98	8.33	8.07	34.43	13.32	7.77	10.06	28.14
BTH/BP		1.12	1.01	0.61	3.94	3.31	1.72	8.48	2.24	1.35	4.74	21.29
FW/BP		1.15	6.24	5.82	4.30	7.17	36.51	2.08	2.34	2.89	4.57	5.37

Table 4. Size of component concentrations in individual precipitation types

Components	Concentrations/values order	Comments
Cond., Na, Mg, Mn	STH > BTH >> FW > WO > BP	
K, Ca	STH > BTH >> WO > FW > BP	
Fe, Al	STH >> FW > BTH > BP > WO	
SO_4^{2-}	STH > FW > BTH >> WO > BP	
Cu	FW > STH > WO > BP > BTH	Whole precipitation types
Zn	STH > FW > WO > BTH > BP	
Pb	WO > FW > STH > BP > BTH	
Cd	WO > FW > STH > BTH > BP	
pH	WO > BTH > BP > FW > STH	
F^-, Cl^-, Sr, Ba, Rb	STH > BTH > FW > BP	
NH_4^+, NO_3^-, Be, As, Ni	STH > FW > BTH > BP	Without WO

The results of chemical analyses from autumn and spring seasons were processed separately. Tables 5 and 6 were eked out with the ratios of concentrations or values of individual components from autumn and spring seasons. Tables 5 and 6 show that concentrations of several selected components embody a seasonal character. The concentrations of selected components in STH and BTH (except for Fe in STH and Fe and Al in BTH) are approximately double-fold higher in autumn than in spring (Tables 5 and 6).

The other kinds of atmospheric precipitation (FW and WO) exhibit seasonal differences only in selected components. Maximum mean values of conductivity, pH, and concentrations of Na, K, and SO_4^{2-} were found in WO samples during the autumn season. Maximum mean concentrations of Mg, Ca, Mn, Fe, and Al were observed in WO samples throughout the spring season (Tables 5 and 6). In case of FW, the maximum values of conductivity, pH, and concentrations of Mg, Ca, and SO_4^{2-} were found in spring, and mean concentrations of Na, K, Mn, Fe, and Al were found in autumn.

Table 5. Mean concentrations (values) of selected components in various types of precipitation in transitional year period

Precipitation type	n.s.	cond.? (μS)	pH	Na	Mg	K	Ca	NH$_4^+$ (μg L^{-1})	NO$_{3-}$	F$^-$	SO$_4^{2-}$	Cl$^-$	Cu	Mn
Autumn														
STH	4	338.33	4.66	11,187	3743	28,037	13,913	30,117	77,633	726	10,5310	23,730	24.93	320.00
BTH	4	201.00	4.71	2817	1467	13,240	6373	3897	17,097	301	27,677	6583	2.53	327.50
WO	9	29.59	5.26	843	108	395	759				4589			5.03
FW	23	94.40	4.54	524	163	429	876	8434	13119	183	17752	1196	13.08	18.46
Spring														
STH	3	350.00	3.67	5120	1865	8300	7200	16934	59828	590	56103	6434	14.95	160.00
BTH	3	122.33	4.70	1887	613	4030	2767	6287	24115	207	15511	2904	4.17	98.00
WO	16	22.93	4.83	295	130	360	1297				3801			10.74
FW	9	113.61	5.17	503	391	331	1317	12559	17570	165	21272	1098	11.79	12.10
Autumn/Spring														
STH		1.0	1.3	2.2	2.0	3.4	1.9	1.8	1.3	1.2	1.9	3.7	1.7	2.0
BTH		1.6	1.0	1.5	2.4	3.3	2.3	0.6	0.7	1.5	1.8	2.3	0.6	3.3
WO		1.3	1.1	2.9	0.8	1.1	0.6				1.2			0.5
FW		0.8	0.9	1.0	0.4	1.3	0.7	0.7	0.7	1.1	0.8	1.1	1.1	1.5

Table 6. Mean concentrations (values) of selected components in various types of precipitation in transitional year period

Precipitation type	n.s.	Cond (µS)	pH	Fe	Zn	Pb	Be	As (µg L^{-1})	Cd	Sr	Al	Ni	Ba	Rb
Autumn														
STH	4	338.3	4.7	607.5	425.0	1.6	0.2	2.6	1.1	73.8	680.8	6.2	27.3	30.8
BTH	4	201.0	4.7	64.5	23.3	0.5	0.0	0.7	0.2	20.0	93.6	0.7	15.2	14.6
WO	9	29.6	5.3	11.0							8.0			
FW	23	94.4	4.5	68.1	131.7	3.7	0.0	1.4	5.2	2.0	97.2	0.9	10.4	19.8
Spring														
STH	3	350.0	3.7	640.0	305.0	2.7	0.2	2.6	0.7	32.8	628.0	2.6	17.0	8.0
BTH	3	122.3	4.7	105.3	27.3	0.8	0.0	1.7	0.2	8.1	120.3	0.8	5.7	5.6
WO	16	22.9	4.8	25.1							17.3			
FW	9	113.6	5.3	39.2	169.3	4.2	0.0	1.5	0.6	4.0	34.8	1.3	3.4	7.3
Autumn/Spring														
STH		1.0	1.3	0.9	1.4	0.6	0.9	1.0	1.6	2.2	1.1	2.4	1.6	3.9
BTH		1.6	1.0	0.6	0.9	0.7	0.7	0.4	1.0	2.5	0.8	1.0	2.7	2.6
WO		1.3	1.1	0.4							0.5			
FW		0.8	0.9	1.7	0.8	0.9	1.6	0.9	8.6	0.5	2.8	0.7	3.1	2.7

Conclusions

According to the expectation, the coniferous growth most effectively purifies the near-surface layer of atmosphere. As it is shown in Tables 3 and 4, the STH reaches maximum values in conductivity and in mean concentrations of individual chemical components except for Cu, Pb, and Cd. Likewise, the lowest mean pH value is recorded in STH. Further in the sequence are the samples of BTH and FW. The BTH samples exhibit the highest mean conductivity value and higher concentrations of Na, Mg, K, Cl⁻, Sr, and Rb than in FW samples. Comparable mean concentrations in both sample types were observed in F⁻, Fe, Be, Al, and Ba. The remaining components (NH_4^+, NO_3^-, SO_4^{2-}, Zn, Pb, As, Cd, and Ni) exhibit higher mean concentrations in FW than in BTH samples. High mean concentrations of the individual components in STH and BTH are not surprising, as the samples contain also the washed out solid atmospheric aerosol (the fallout) and the metabolic products of trees leached from their assimilatory organs. Pronounced seasonal differences are caused both by the varying state of development of foliage in spring and autumn, and chiefly by differences in the metabolic activity of the forest growth in these two annual seasons.

It is, however, necessary not to underestimate high concentrations of particular components in FW samples, as the influence of deposited precipitation on the tree growth is more complex and as it acts for a longer time (hours and even days) compared to the falling (vertical) precipitation (WO, BP). In contrast to the unambiguous increase of concentrations of majority of components in throughfall, concentrations of several components in FW show increase in autumn (K, Mn, Fe, Be, Cd, Al, Ba, and Rb), practically without any change remain concentrations and values of pH, Na, F⁻, Cl⁻, Cu, Pb, and As. Concentrations or values of remaining components are higher in spring than in autumn.

References [e.g., 6] denote the concentrations of individual components in FW even by one order higher than in WO. These conclusions followed mainly from analyses of samples that were not collected on the same place. As was already mentioned, the sampling of FW and WO on the top of Milesovka Mt. occurred on the same place (distance of the FW and WO collectors is 10 m only). Unfortunately, the analyses of WO samples lack any information concerning the concentrations of NH_4^+, NO_3^-, and Cl⁻. Table 3 nevertheless confirms that the maximal value of FW to WO ratio was detected in SO_4^{2-} (4.4). Concentrations of Mg, K, and Ca seem to be almost identical in both types of samples. It is therefore possible to state that the mean concentrations of several components in FW samples are considerably higher than in WO. It cannot be however unambiguously stated that the mean concentrations of components in FW are by one order higher than those in WO. As the concentrations of pollutants in FW depend mainly on local conditions, it is very difficult to extrapolate the results from one sampling site to another. Likewise, the relatively low distances (on order of several tenths of kilometres) between the sampling localities of FW and WO samples can play an important role in the determination of FW to WO ratios. These conclusions will have to be verified on a larger set of analytical results covering longer time span of monitoring and the results will have to be extended by the analyses of as yet missing ions (NH_4^+, NO_3^-, and Cl^-).

Acknowledgements

The results described in this paper were obtained under the support of the Grant Agency of the Czech Republic (Project No. 205/04/0060) and of the Academy of Sciences of the Czech Republic (Project No. 1QS200420562).

References

1. Tesar M, Elias V, Sir M (1995) Preliminary results of characterisation of cloud and fog water in the mountains of southern and northern Bohemia. J Hydrol Hydromech 43:412–426
2. Daube B, Kimball KD, Lamar PA, Weathers KC (1987) Two new ground-level cloud water sampler designs which reduce rain contamination. Atmos Environ 21:893–900
3. Fisak J, Rezacova D (1999) Sampling of fog and low cloud water at the meteorological observatories Milesovka and Kopisty. Meteorol Bull 52:144–148 (In Czech)
4. Skrivan P, Navratil T, Burian M (2000) Ten years of monitoring the atmospheric inputs at the Cernokostelecko region, Central Bohemia. Scientia Agriculturae Bohemica 31:139–154
5. Skrivan P, Minarik L, Burian M, Martinek J, Zigova A, Dobesova I, Kvidova O, Navratil T, Fottova D (2000) Biogeochemistry of beryllium in an experimental forested landscape of the Lesní potok catchment in Central Bohemia, Czech Republic. GeoLines (Ocassional Papers in Earth Sciences of the GLÚ AV ČR) 12:41–62
6. Kirkaite A, Sopauskas K (1976) On some parameters of scavenging of the atmosphere by the ground level condensation products. In: Styra B (ed.), Physical aspects atmospheric pollution. Atmospheric Physics 2. Mokslas, Vilnius, pp 82–90 (In Russian)

Wet deposition at Llandaff station in Cardiff

E Abogrean,[1] G Karani,[1] J Collins,[2] R Cook[1]

[1] School of Health Sciences, University of Wales Institute of Cardiff (UWIC), Western Avenue, CF5 2YB, Cardiff, UK
[2] Business School, University of Wales Institute of Cardiff (UWIC), Western Avenue, CF5 2YB, Cardiff, UK

Abstract

Wales receives rainfall for about 210 days in a year. Chemical characteristics of rainfall and its seasonal variation at the Llandaff campus station located in Cardiff, Wales were studied for the period from June 2003 to July 2004. The station is located in an urban area about 3 km north of Cardiff centre. The rainwater samples were collected weekly using bulk precipitation techniques and analysed for pH and major ion concentration by atomic absorption spectrophotometer (AAS). The samples were collected from two sample collectors at different heights and separated by about 200 m, the ground level of house sample being designated (Hs) and upper level of roof sample (Rs).

Over the period of investigation, the volume of rainfall measured was almost the same for Hs and Rs. The pH of Hs and Rs was 5.6 and 5.3, respectively. Average wet deposition of ions in parts per million (ppm) of Hs and Rs were: Mg^{2+} (0.56, 0.73), Na^+ (7.3, 8.0), Ca^{2+} (3.17, 3.10), K^+ (3.03, 1.76), and Cl^- (10.6, 12.0). The ratio of the total average concentration of chloride to that of sodium for Hs and Rs are 1.45 and 1.5, respectively, which is close to the ratio of seawater, 1.8. Seasonal variations for some major ions appear to be pronounced. Generally, the maximum pH occurred in the autumn season and the minimum pH in winter season for both samples. Dust, insects, and tree debris found in both samples may have been the reason for the increased pH, and was more pronounced in the Hss. This study provides the influence of wet deposition in Cardiff.

Introduction

Acid species and wet deposition into terrestrial environments can potentially affect the health of the ecosystem. In order to make an assessment of the input of materials into terrestrial system sources, the composition of atmospheric deposition needs to be determined [1,2]. The pollutants can originate in the atmosphere from both natural and anthropogenic sources. The main sources of the anthropogenic pollutants are vehicles, industrial, and power-generating plants [3,4].

Wet deposition is defined as the process by which atmospheric compounds are attached to and dissolved in cloud and precipitation droplets and delivered to the earth's surface by rain, hail, or snow [5,6]. During the last 3 decades, the chemistry of precipitation has been widely investigated in many industrialized areas, such as Western Europe, the USA, southern China, and Canada, which is significantly affected by acid rain [7,8]. In 1995, bulk precipitation was sampled at weekly intervals from 25 collectors located across Wales in rural stations. The precipitation chemistry was dilute, acidic (overall mean pH 4.9), and dominated by sea salts with sodium to chloride ratio close to that of seawater [9].

In order to gain an insight into rainwater composition and acidity in Cardiff, research was carried out from June 2003 to July 2004 at the atmospheric monitoring station of the University of Wales Institute, Cardiff (UWIC). The objective of this study was to determine the variation in chemical composition of the rainwater samples collected at an urban station in Cardiff City (the Llandaff station) using bulk collectors, and explain the possible reasons for the seasonal variation of the concentration of major ions for the period between June 2003 and July 2004.

Location, materials, and method

The rainfall data used in this study was collected at UWIC (Llandaff campus). The Llandaff station is located at 53°08′N; 3°11′W. It is in the north of Cardiff in an urban area about 3 km from the centre. There are no major industrial activities in the area. Figure 1 shows the location of the station and the surrounding areas. The site of the station has been chosen in accordance with

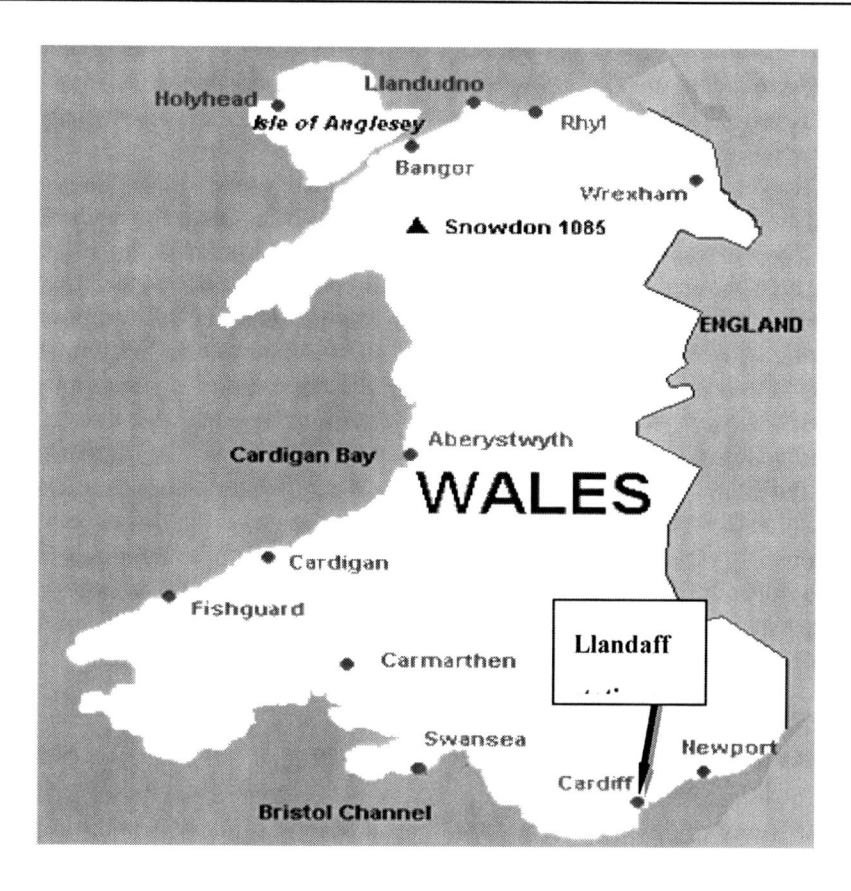

Fig. 1. Map showing the location of Llandaff rainfall collection sites in Cardiff (www.wikipedia.org)

general guidelines so that background precipitation in the region can be obtained.

Two bulk collectors were used in sample collection. Each collector comprised a 200 mm diameter polypropylene funnel mounted above a 3 L polythene collecting bottle. The funnel and bottle assembly were located within a stainless steel canister with an air gap of 25 mm. A three-legged stainless steel wire frame extended above the metal sleeve to just below the funnel opening to provide support for the funnel and for the bird guard protruded above the funnel. The sampling head was mounted on a plastic-coated aluminium stand with the funnel rim 1.75 m above the ground surface. The funnel was rinsed using deionized water and shaken dry when each sample was collected. The house collector was fixed on a steel table 1.0 m above ground level (2.75 m above ground level for Hs). The roof collector was fixed on the roof of a three-floor building, which is about 12 m above the ground level (13.75 m above ground level for Rs). The distance between the two collectors is about 200 m.

The precipitation samples were collected weekly every Monday at 9 a.m. local time and returned to the laboratory at UWIC, where the volume of the sample was recorded using a graduated cylinder and the pH was measured immediately using a Fisherbrand Hydrus 300 pH meter. The pH meter was calibrated before and after each set of precipitation samples at a minimum of two points in the expected pH 4.0 and 7.0 buffer solutions. After pH measurement, the samples were filtered through 11 μm pore size Whatman No. 1 membrane filter paper. The samples were transferred into high-density polyethylene storage bottles and stored in a refrigerator at 4°C until they were analysed. Although low-temperature storage could be as long as 6 months without any appreciable change on major ion concentration [10], the samples were analysed within 2 weeks after collection.

Samples were analysed for sodium, calcium, magnesium, and potassium ions, using atomic emission/absorption techniques (Solaar 969A Spectrophotometer) and analysed chloride using an ion-selective electrode.

Results

Ionic compositions of rainwater samples from the Llandaff station over a 14-month period, from June 2003 to July 2004, were utilized in this study. The total number of the samples was 61. A statistical summary of volume-weighted mean (VWM) concentration of major ions in the rain samples and their pH values along with the mean precipitation amounts for the period between June 2003 and July 2004 are presented in Table 1.

The mean measured volume was almost the same for Hs (555 mL) and Rs (558 mL). This indicates that the flux of rainfall is the same at different levels, and also the effect of wind is minimal. The five ions involved were four cations and one anion: namely Na^+, Mg^{+2}, K^+, and Ca^{+2} as cations and Cl^- as anion. The average pH of house sample (Hs) and roof (Rs) sample rainwater in the Llandaff stations for the period of study was 5.6 ± 0.5 and 5.3 ± 0.4, respectively. These values are higher than the reported results (pH 4.9) across Wales [9]. The cause of higher values of pH in this research might be due to dust, files, and insects, which contaminated the bulk

Table 1. Selection of statistical parameters in precipitation chemistry data

	n	Mean	SD	Max.	Min.
V of Hs (mL)	61	555	604	2450	0.0
V of Rs (mL)	61	558	630	2640	0.0
pH (Hs)	51	5.6	0.5	7.3	4.2
pH (Rs)	51	5.3	0.4	6.9	4.0
Cl$^-$ of Hs (ppm)	51	10.6	4.8	36.3	2.8
Cl$^-$ of Rs (ppm)	51	12.0	6.7	61.2	1.8
Na$^+$ of Hs (ppm)	51	7.3	3.2	29.3	2.8
Na$^+$ of Rs (ppm)	51	8.0	4.1	35.7	1.3
Mg^{2+} of Hs (ppm)	51	0.56	0.4	3.9	0.05
Mg^{2+} of Rs (ppm)	51	0.73	0.5	4.2	0.06
K$^+$ of Hs (ppm)	51	3.03	4.8	37.5	0.1
K$^+$ of Rs (ppm)	51	1.76	1.6	16.4	0.1
Ca^{2+} of Hs (ppm)	51	3.17	3.5	29.4	0.2
Ca^{2+} of Hr (ppm)	51	3.10	3.5	29.9	0.3

samples especially in the Hss. Usually, the pH of the rainwater is around 5.6, due to a carbonate buffer as a result of CO_2 dissolved in rain droplets [11]. The pH of the rainwater is determined by the relative composition of acids and bases, i.e., relative composition of anions and cations present in the rainwater.

Mg^{2+}, Na^+, Ca^{2+}, K^+, and Cl^- ion concentrations in rain samples were analysed in this study as shown in Table 1. The mean concentration of ions in Hs and Rs were: Mg^{2+} (0.56, 0.73), Na^+ (7.3, 8.0), Ca^{2+} (3.17, 3.10), K^+ (3.03, 1.76), and Cl^- (10.6, 12.0), all units being ppm. The effect of sea-spray may have been more evident on the Rss, since Mg^{2+}, Na^+, and Cl^- ions in the Rss were higher than in Hss which was located about 14 m above the ground level, but the soil alkalinity may have had an effect on the Hss, since Ca^{2+} and K^+ ions of Hss were higher than in Rss. However, this concentration is significantly lower than the concentration in the urban area when compared with the results of [12] because of lower dust particle concentration in the urban atmosphere originating from the soil.

The ratio of the total average concentration of chloride ratio to that of sodium for Hs and Rs is 1.45 and 1.50, respectively, that is closed to the ratio of seawater at 1.8. These values support the possible effects of sea-spray.

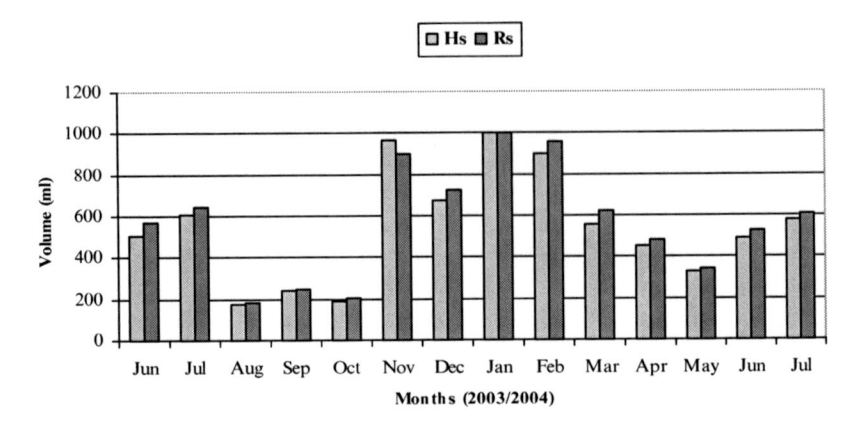

Fig. 2. Monthly mean of rainfall in Llandaff, Cardiff

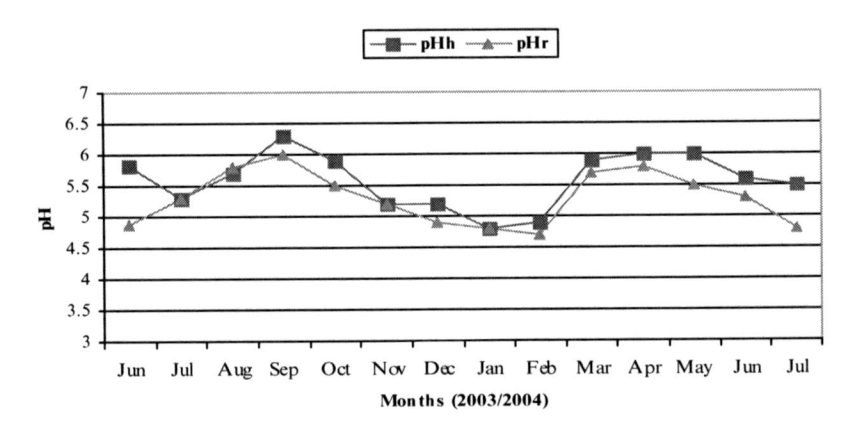

Fig. 3. Monthly mean of pH in Llandaff, Cardiff

Figure 2 shows monthly mean rainfall taken from the Llandaff stations between June 2003 and July 2004. The maximum rainfall has been measured in January (1000 mL) and the minimum was in August (180 mL) for both collectors (Hs and Rs).

Figure 3 shows the variation of pH over the period from June 2003 to July 2004 for both collectors. The maximum pH for Hs observed in September was 6.2 and the minimum in January was 4.8. Also, the maximum pH of the Rs observed in September was 5.9 and the minimum value in January was 4.7. In general, variable weather patterns cause the seasonal difference in precipitation chemistry.

Seasonal variation

The climate of Cardiff and surrounding areas is wet and cold for most of the year [13]. The temperature in Cardiff ranges from 2.0 to 21.0°C with a maximum temperature of 21.0°C in July and with a minimum temperature of 2.0°C in January [12]. In order to look into the seasonal variations of ion concentrations, the rain data were separated into four groups: summer samples (June to August), autumn samples (September to November), winter samples (December to February), and spring samples (March to May) and the results presented in Table 2.

The seasonal mean value for Rss of pH ranged between 4.7 and 5.5, with a maximum value of pH 5.5 and a minimum value of pH 4.7 for autumn and winter, respectively. On the other hand, mean pH values for

Table 2. Seasonal volume-weighted mean concentration (ppm) of major ions of rainwater, samples at the Llandaff stations in Cardiff

Season	Summer	Autumn	Winter	
Total precipitation (Hs) (mL)	462	465	859	445
Total precipitation (Rs) (mL)	484	451	895	481
pH of Hs	5.6	5.8	5.00	5.5
pH of Rs	5.2	5.7	4.7	5.3
Mg^{+2} (Hs)	0.75	0.46	0.47	0.5
Mg^{+2} (Rs)	0.82	0.68	0.78	0.6
Na^+ (Hs)	6.9	7.7	6.6	8.2
Na^+ (Rs)	8.9	9.8	8.0	6.8
Ca^{2+} (Hs)	4.9	2.5	1.5	2.7
Ca^{2+} (Rs)	4.8	2.6	1.4	2.4
K^+ (Hs)	5.5	2.6	0.9	1.5
K^+ (Rs)	2.7	1.7	1.2	1.0
Cl^- (Hs)	8.3	13.5	13.3	10.9
Cl^- (Rs)	9.9	14.8	14.1	11.8

spring and summer were similar (spring and summer values are 5.2 and 5.3, respectively), but the value of pH for the Hs is about 5% more than the Rss in all seasons which might be due to the contamination of Hss by dust, insects, and leaf.

It is clearly seen in Table 2 that Cl^- and Na^+ concentrations are much higher than other ion concentrations measured in this study. Cl^- concentration was found to be at maximum in autumn for both samples, (Hs = 13.5 ppm and Rs = 14.8 ppm) and at a minimum in the summer season (Hs = 8.3 ppm and Rs = 9.9 ppm). Na concentrations are at maximum for Hss in spring, i.e., 8.2 ppm and Rss in autumn, i.e., 9.8 ppm, but the minimum value of Hss was 6.6 ppm in winter and for Rs recorded 6.8 ppm in spring.

The minimum concentration of Ca^{2+} and K^+ occurred in winter with 1.5 and 0.9 ppm for Hs, and 1.4 and 1.2 ppm for Rs, respectively. The maximum concentration recorded in the summer with 4.9 and 5.5 ppm for Hs, and 4.8 and 2.7 ppm for Rs, respectively. Furthermore, the minimum concentration of Mg^{2+} for Hs occurred in the autumn, 0.46 ppm but for Rs was in the spring season, 0.6 ppm and maximum concentrations were in the summer season, for Hs 0.75 ppm and for Rs 0.82 ppm, respectively.

Statistical analysis

Relationships between measured parameters were examined through analysis of linear correlation [14]. The correlation matrix for all samples is given in Table 3. We consider the correlation is good if $0.7 \leq r \leq 0.8$ and strong if $r > 0.9$ [1].

Table 3 shows a good correlation between Mg^{2+} and Na^+ in both samples ($r = 0.7$) and strong correlation between Mg^{2+} and Na^+ in Rss ($r = 0.9$), which might be due to effect of sea-spry especially in Rs. In addition, strong correlation is found between Mg^{2+} in Hs and K^+ in Rs ($r = 0.9$) and good correlation between Mg^{2+} and K^+ in Rs ($r = 0.8$). The reason might possible to the interaction between dusts and sea-spry. Furthermore, there is a good correlation between Na^+ in both samples and Cl^- in Hs ($r = 0.7$), which maybe due to the effect of sodium chloride. Although the relationship between Ca^{2+} in both samples and K^+ in Hs is strong correlation ($r = 0.9$), there is a good correlation between Ca^{2+} in both samples and K^+ in Rs ($r = 0.7$). This might have caused by dust-rich local and surrounding limestone environment around Cardiff.

Table 3. Correlation coefficients of precipitation chemistry data at Llandaff stations in Cardiff

	Vol. (Hs)	Vol. (Rs)	pH (Hs)	pH (Rs)	Mg^{2+} (Hs)	Mg^{2+} (Rs)	Na$^+$ (Hs)	Na$^+$ (Rs)	Ca^{2+} (Hs)	Ca^{2+} (Rs)	K$^+$ (Hs)	K$^+$ (Rs)	Cl$^-$ (Hs)
Vol. (Hs)													
Vol. (Rs)	1.0												
pH (Hs)	−0.9	−0.9											
pH (Rs)	−0.7	−0.7	0.7										
Mg^{2+} (Hs)	−0.1	−0.1	−0.0	0.2									
Mg^{2+} (Rs)	−0.0	−0.0	−0.2	0.2	0.9								
Na$^+$ (Hs)	0.2	0.2	−0.1	0.2	0.7	0.7							
Na$^+$ (Rs)	0.0	−0.0	−0.2	0.3	0.7	0.9	0.6						
Ca^{2+} (Hs)	−0.5	−0.5	0.3	0.5	0.6	0.6	0.2	0.6					
Ca^{2+} (Rs)	−0.5	−0.5	0.2	0.5	0.6	0.6	0.1	0.6	1.0				
K$^+$ (Hs)	−0.5	−0.6	0.2	0.5	0.6	0.6	0.1	0.6	0.9	0.9			
K$^+$ (Rs)	−0.4	−0.4	0.1	0.4	0.9	0.8	0.4	0.6	0.7	0.7	0.7		
Cl$^-$ (Hs)	0.3	0.3	−0.2	0.2	0.3	0.5	0.7	0.7	0.3	0.3	0.2	0.1	
Cl$^-$ (Rs)	0.6	0.6	−0.5	−0.1	−0.1	0.2	0.4	0.5	−0.2	−0.2	−0.3	−0.3	0.7

Conclusions

An investigation of chemical composition of rainwater was carried out at the Llandaff stations in Cardiff, Wales during the period of June 2003–July 2004. Precipitation data of 14 months were analysed to find out the variations in the chemical composition of the rainwater samples collected from an urban stations reflecting the background concentration of various ions present in rainwater. Seasonal variations of the concentration of major ions in rainwater samples within the period considered, namely June 2003 and July 2004, were examined.

1. Observation that the pH of the ground level is higher than the pH of elevated for all of the seasons might be due to contamination of the ground level by dust, (alkalinity), insects, and leaves which are found in Hss, and the reason for contamination might be because the house collector is fixed 2.75 m above the ground level, but the roof collector was 13.75 m above the ground level.
2. The minimum average pH for both samples was in autumn and winter but the maximum average was in spring and summer seasons.
3. The seasonal variation for some major ions is pronounced: the maximum concentration of Na^+ and Cl^- appeared in winter and autumn, but maximum concentration of Mg^{2+}, Ca^{2+}, and K^+ appeared in summer and spring.
4. Weak correlations are found between pH and Cl^- ion for all season because of the neutralization with alkaline particles, consequently strong correlation between Ca^{2+} ion and K^+ ion.
5. The dust-rich local and surrounding limestone environment might have caused the high concentration of Ca^{2+} and K^+ in Cardiff area. The relatively high concentration of Na^+, Mg^{2+}, and Cl^- may due to the effect of seawater.

References

1. Olias M, Nieto JM, Sarmiento AM, Ceron JC, Canovas CR (2004) Seasonal water quality variations in a river affected by acid mine drainage: the Odiel River (South West Spain). Sci Tot Environ 333:267–281
2. Zeng GM, Zhang G, Huang GH, Jiang YM, Liu HL (2005) Exchange of Ca^{2+}, Mg^{2+}, and K^+, and uptake of H^+, NH_4^+ of the subtropical forest canopies influenced by acid rain in Shaoshan forest located in Central South China. Plant Sci 168:259–266
3. Topcu S, Incecik S, Atimtay AT (2002) Chemical composition of rainwater at EMEP station in Ankara, Turkey. Atmos Res 65:77–92
4. Krusche AV, de Camargo PB, Cerri CE, Ballester MV, Lara LBLS, Victoria RL, Martinell LA (2003) Acid rain and nitrogen deposition in a sub-tropical watershed (Piracicaba): ecosystem consequences. J Environ Pollut 121: 389–399
5. Krupa SV (2002) Sampling and physico-chemical analysis of precipitation: a review. Environ Pollut 120:565–594
6. Staelens J, De Schrijver A, Van Avermaet P, Genous G, Verhoest N (2005) A comparison of bulk and wet-only deposition at two adjacent sites in Melle (Belgium). Atmos Environ 39:7–15

7. Yu S, Gao C, Cheng Z, Cheng X, Cheng S, Xiao J, Ye W (1998) An analysis of chemical composition of different rain types in 'Minnan Golden Triangle' region in the southeastern coast of China. Atmos Res 47–48:245–269
8. Page T, Beven KJ, Whyatt D (2004) Predictive capability in estimating changes in water quality, long-term responses to atmospheric deposition. Water Soil Air Pollut 151:215–244
9. Reynolds B, Lowe JAH, Smith RI, Norris DA, Fowler D, Bell SA, Stevens PA, Ormerod SJ (1999) Acid deposition in Wales, the results of the 1995 Walsh acid water survey. Environ Pollut 105:251–266
10. Galloway JN, Likens GE (1978) The collection of precipitation for chemical analysis. Tellus 30:71–82
11. Sopauskiene D, Jasineviciene D (2006) Changes in precipitation chemistry in Lithuania for 1981–2004. J Environ Monit 8:347–352
12. Bayraktar H, Turalioglu FS (2005) Composition of wet and bulk deposition in Erzurum, Turkey. Chemosphere 59:1537–1546
13. PGAR (1987) Acid deposition in the United Kingdom 1981–1985. A second report of the United Kingdom Review Group on Acid Rain, Warren Spring Laboratory, Stevenage
14. Singh AK, Mondal GC, Kumar, S, Singh KK, Kamal, KP, Sinha (2006) Precipitation chemistry and occurrence of acid rain over Dhanbad, Coal City of India. Environ Monit Assess. 18 August

Monitoring the atmospheric deposition of particulate-associated urban contaminants, Coventry, UK

SM Charlesworth, C Booty, J Beasant

Department of Geography, Environment & Disaster Management, Coventry University, UK

Abstract

Based on the determination of Zn, Cu, Pb, Ni, and Cd in soils, street dusts, and sediments from the urban area of Coventry, UK, two sets of bioaccumulators were located around Coventry City Centre. The first set of 12 peat-containing seed trays were located on a busy road traffic island with a second set placed for comparison with a rural area. One tray was removed each month from each site for analysis. Whilst the concentrations of most metals were higher in the urban area, the temporal trends were similar, with the concentrations of Pb indicating that urban levels had declined. The second set of biomonitors of three moss bags were located at 16 sites in the City Centre; one bag was removed every 2 months, making a 6-month study in all. The temporal trends for the moss bags were generally increasing concentrations with time, although two sites declined in the last sampling month. Concentrations of Zn and Cu in the street dusts, sediments, and soils were high, as was Zn in both biomonitors and Ni in just the moss bags.

Introduction

The potential hazard of road and pavement dusts and gully pot sediments to the urban environment has been studied fairly widely [1,2]. From these studies it has emerged that, whilst the residence time of dusts on the street is fairly short, they still retain the signature of past industry and as a consequence, levels of metals associated with them can be significantly above both background and statutory guideline values.

Sources of these contaminants to urban deposits include the naturally weathered underlying geology, soil, as well as anthropogenic emission sources, which in the UK include motor vehicles (23.7%) as well as general industrial combustion plants (7.8%) including large power plants (17.4%), uncontrolled domestic coal burning and major mining, quarrying, and construction activities (27.4%) [3].

However, whilst studies such as these give a snap shot of conditions on the pavement surface or in the gully pot, it can give no indication of the changing concentrations of metals with time which monitoring airborne particulate-associated metals can.

This study therefore seeks to:

1. Establish the spatial distribution of particulate-associated heavy metals in urban soils, street deposits, and gully pot sediments
2. Establish the temporal distribution of atmosphere-derived dust-associated metals.

Experimental – Method

Coventry in the West Midlands, UK (see Fig. 1a,b) was chosen since several studies of metals in street dusts had already been carried out there [e.g., 2]. Coventry has a population of 0.3 million and the road network is structured around a central ring road with nine major roads radiating from it (Fig. 1c). Much of the area within the ring road is pedestrianized or with restricted vehicular access. A detailed description of the sampling and analytical methodology for the soils, street dusts, and gully pot sediments are given in [2] and so this is not repeated here; however, the locations of the biomonitors were based on the street dust and soil sample sites, so this will be described.

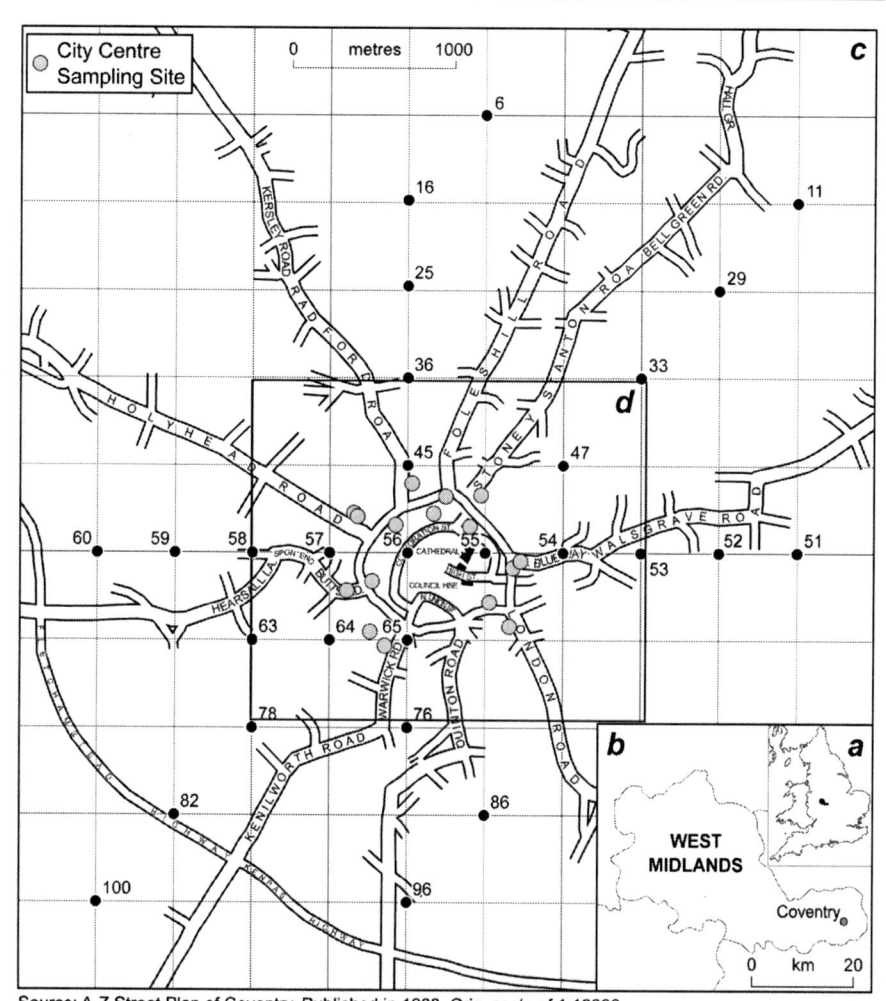

Fig. 1. Location of a. the West Midlands, b. Coventry in the West Midlands, c. sampling sites in Coventry, d. City center (black circles indicate soils, street dusts, and sediment locations, and grey circles indicate where the moss bags were located)

Soils, street dusts, and sediments

Sampling sites for the street dusts and gully pot sediments were based on a street map of Coventry which divided the city into 0.4 km² grid squares. Soil and material from the pavement, gully pot, and road gutter was collected as close as possible to the intersection of the grid squares (some samples are located in Fig. 1c). After analysis of the spatial distribution of heavy metals in these samples, it was decided to concentrate on the centre of the city for the soils, making 25 samples in all, as is shown in Fig. 1c.

Biomonitors

The results of the soil and dust analysis (detailed later) indicated that the major repository of all the heavy metals was associated with the nine main roads leading off the ring road. It was decided, therefore, to locate the biomonitors in association with these major road junctions, as shown by the grey circles in Fig. 1c. Two approaches were used: a busy road traffic island was chosen ("Sky Blue Way" on Fig. 1c) and 12 seed trays containing peat placed in the middle in January 2003. As a control, 12 similar trays were deployed at a rural site in Warwick, 20 km to the south-east of Coventry. Coarse gravel was placed in the bottom of the trays to ensure sufficient drainage and the peat was screened to <2 mm to remove any large lumps. One tray from each site was retrieved each month and the surface material removed using a plastic scoop to a depth of approximately 0.5 cm. The sample was dried, gently disaggregated, and stored in a plastic bag prior to digestion. Unfortunately, it was only possible to collect trays for 7 months of the trial since the remaining five at the urban site were destroyed when the grass on the island was mown.

The second set of biomonitors used were moss bags of the type developed initially by [4], and used widely ever since due to their high surface to volume ratio, simple anatomy, and lack of cuticle [5], as well as their substantial cation exchange capacity [6]. They are also cheap, and large numbers can be deployed in the same trial [7].

The moss bags were made up of 30 g of moss and 20 × 20 cm nylon pond netting enclosed using plastic-coated garden wire. Three bags were suspended from trees and fences 2 m from the ground [8], at 16 sites on 15 April 2005. One bag per site was to be retrieved from each site on the same date in June, August, and October. However, by June only 11 sites had moss bags left, by August there were only 10, and only 4 sites were eventually monitored for the full 6 months. Once collected, the moss was removed from the nylon netting, dried at 50°C and gently disaggregated. All of the biomonitor material was wet-digested using a microwave prior to analysis by inductively coupled plasma optical emission spectroscopy (ICP-OES).

For both biomonitors, samples of the original material was analysed for a background value. Digestion of certified reference materials using this procedure suggested that extraction efficiencies lay between 93% and 98% and results reported in this paper have not been corrected for recovery efficiency.

Results and discussion

Street dusts, sediments, and soils

Many of the urban deposits collected from Coventry exceeded background [2], but on comparison with published background values [9], contaminated land exposure assessment (CLEA) soil guideline values (SGVs) [10] where available and ICRCL trigger concentrations [11] where not, it was discovered that Cu in particular and to some extent Zn, were particularly high. Similar to the street dust results, soil Cu and Zn had the highest concentrations (Table 1), so their spatial distributions are provided in Figure 2. These show that the highest concentrations of both elements appear to be associated with main roads. However, when the results are sorted by land use, it was found that residential gardens were an important repository for all of the heavy metals apart from Cu (Table 2), as well as roads and industrial soil, whereas parks, scrub, and agricultural land were the least important. Many of these gardens are close to the roads and it would be expected

Table 1. Percentages of heavy metals exceeding background, ICRCL trigger concentration, and CLEA SGV in soils from Coventry City Centre, $n = 24$

	Ni	Cd	Zn	Cu	
Percentage above background	29	38	100	79	21
Percentage above CLEA/trigger	0	0	21	2	0
Percentage above ICRCL action	0	0	0	0	0

Table 2. Relative loadings in Coventry soils for each metal in each land use

Zn	Roadside> gardens> industrial> parks> scrub> agricultural
Pb	Gardens > industrial> scrub > agricultural > roadside > parks
Cu	Industrial > roadside> gardens> scrub > parks> agricultural
Ni	Industrial > gardens>roadside> agricultural > scrub> parks
Cd	Industrial>gardens> parks> scrub> roadside>agricultural

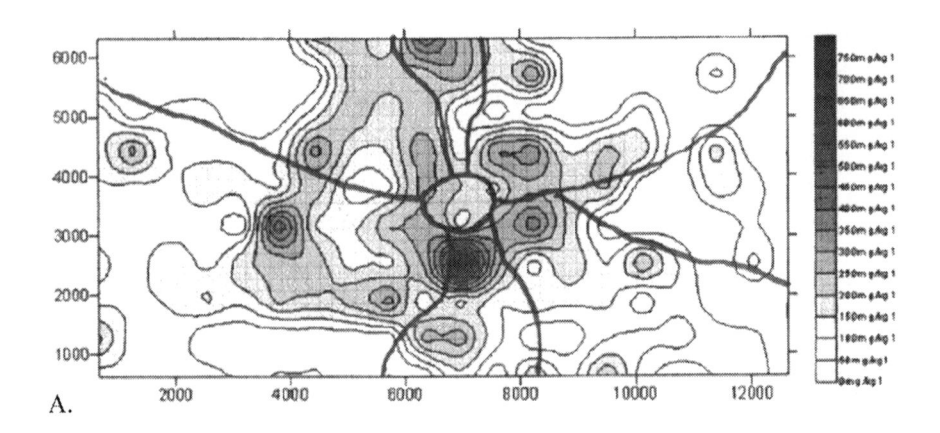

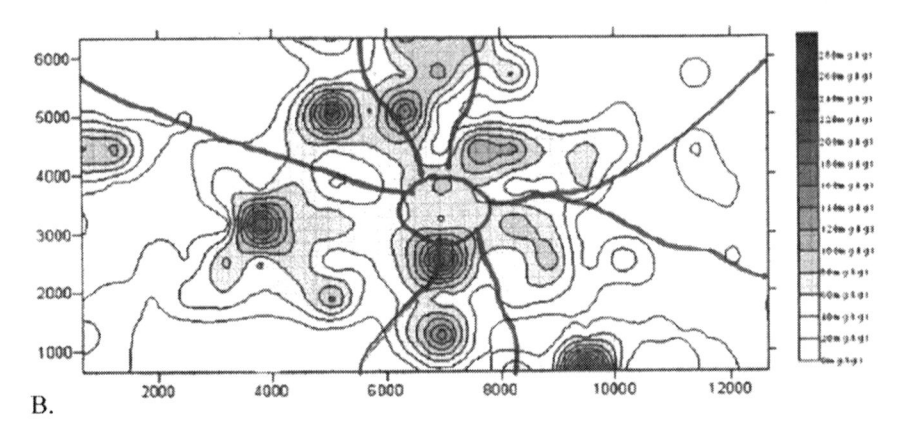

Fig. 2. Spatial distribution of Zn (A) and Cu (B) concentrations in soils in Coventry (mg kg^{-1})

that they had the potential to become contaminated. The lowest concentrations for all metals were found in the soils from parks and agricultural use.

A comparison of metal concentration in soils, dusts, and gully pot sediment using Kolmogorov–Smirnov and given in Table 3 indicates that there is no significant difference between Cd in soils and dusts, a significant difference between the Ni in soils, pavements, and gully pots at the 5% level, but no significant difference with the road gutters and a significant difference between Cu, Zn, and Pb in the soils and all the dusts and sediments.

Table 3. Kolmogorov–Smirnov test of significant difference between the soils, dusts, and gully pot sediments. Where NS = no significant difference; $P = 0.05$ or 0.01 where the significant difference is at the 5% or 1% level, respectively

Soil and pavement	Cd	Zn	Ni	Pb	Cu
significance	0.08	2.05E^{-05}	0.02	0	0
P	NS	0.01	0.05	0.01	0.01
Soil and gutters	Cd	Zn	Ni	Pb	Cu
significance	0.46	6.63E^{-05}	0.08	4.8E^{-05}	0
P	NS	0.01	NS	0.01	0.01
Soil and gully pots	Cd	Zn	Ni	Pb	Cu
significance	0.33	2.66E^{-07}	0.01	1.1E^{-08}	0
P	NS	0.01	0.05	0.01	0.01

Biomonitors

As outlined above, because all of the moss in the bag and the top 0.5 cm of the peat were removed for analysis from the biomonitors, the results presented here will have been diluted with uncontaminated material and will therefore be an underestimate. It was found that metal concentrations declined over the period of monitoring at two sites; Junction 4 at the ring road end of the Foleshill Road and particularly the Belgrade (Fig. 3). It is difficult to compare these results with statutory guidelines, since absolute concentrations are generally not used, and it is these which were needed for comparison against the street dusts and soils. Published soil background [9], CLEA SGV, or ICRCL trigger were therefore used, and it was found that all biomonitored Zn exceeded background; both peats (rural and urban), but not the moss exceeded Cu and Pb background and all the moss was higher than Ni background. None of the Zn, Cu, or Pb results in the biomonitored material exceeded statutory levels; however, the Ni concentrations in all of the moss were above guidelines, but nowhere near ICRCL action levels. Figure 4 shows the results of 7 months monitoring at the urban and rural sites. In general, the trends are similar with the possible exception of Zn. The Pb levels at both sites are very similar suggesting that whilst they do vary at around 10 times background, concentrations in the urban airborne dusts have declined to those found in rural areas.

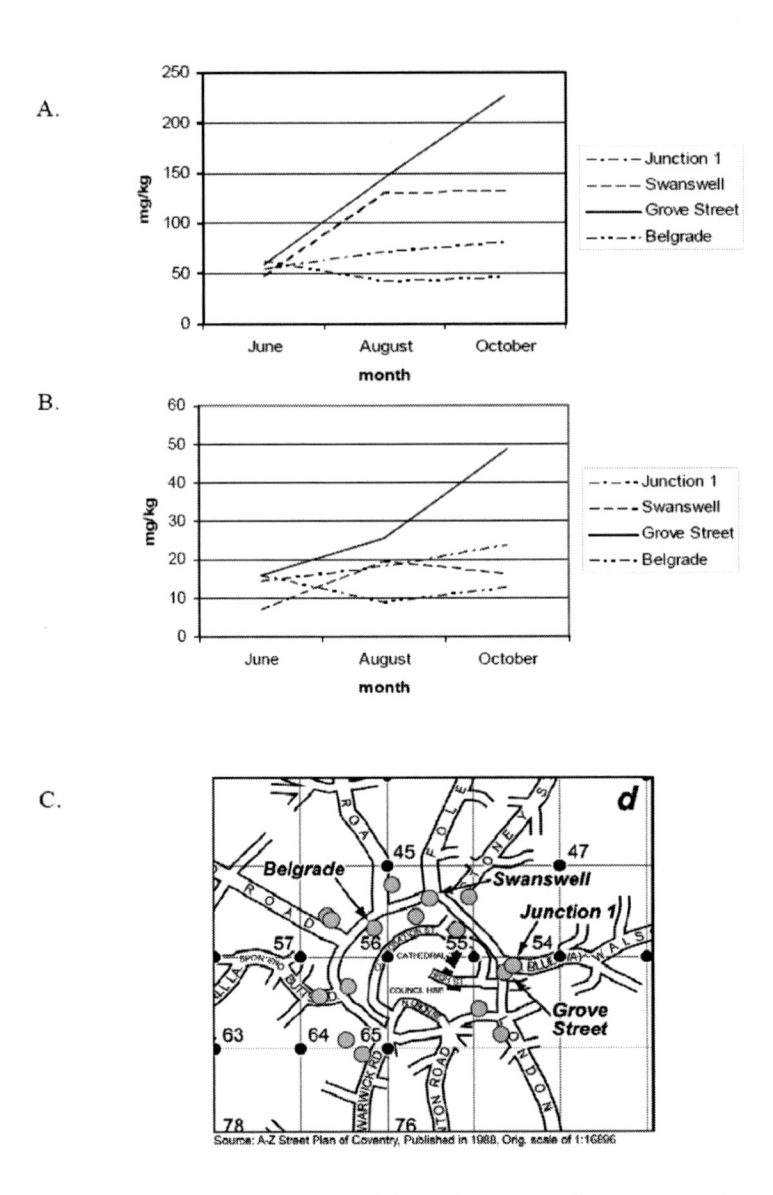

Fig. 3. Temporal trends in Zn (A) and Cu (B) concentrations in moss bags for 6 months at four sites in Coventry, shown in C

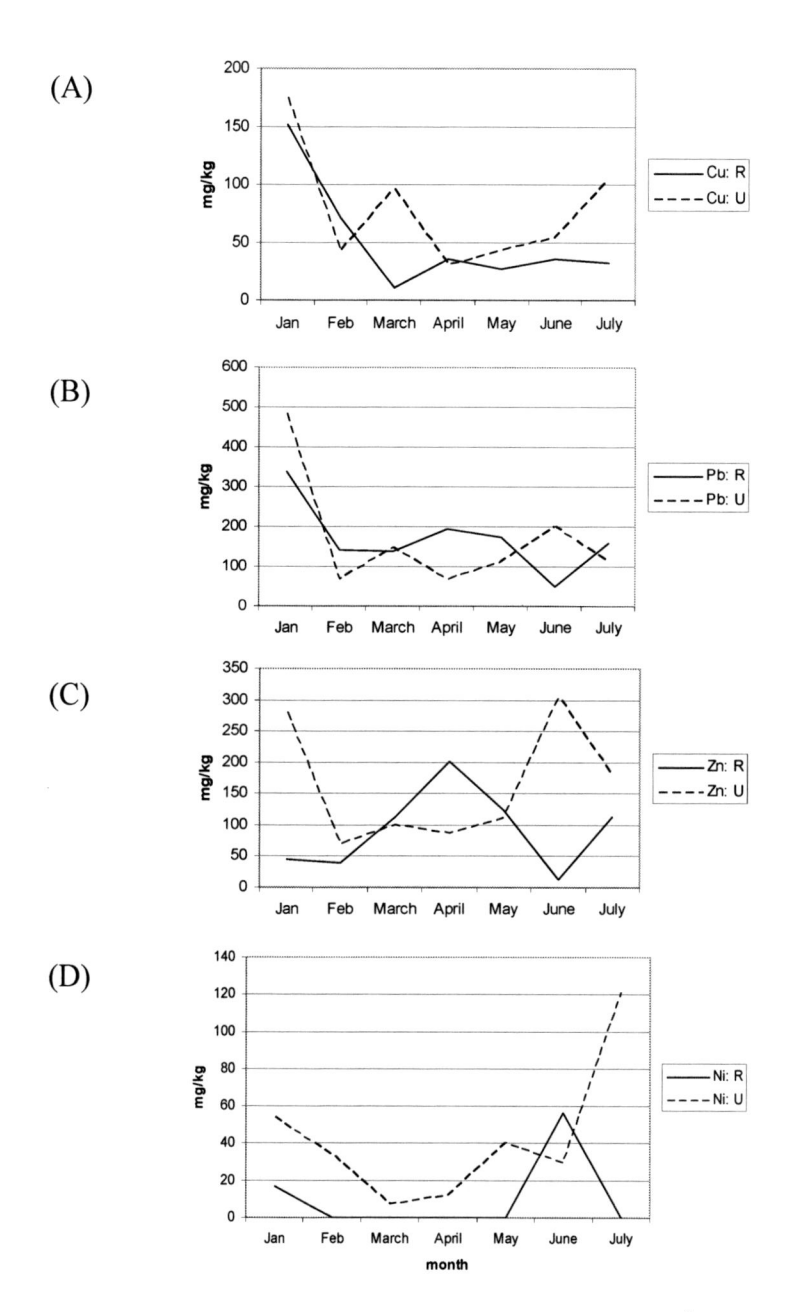

Fig. 4. Temporal trends in heavy metal concentrations (mg kg^{-1}) Cu (A), Pb (B), Zn (C), and Ni (D) in peat biomonitors

Conclusions

Findings from earlier studies of Coventry street dusts [1,2] flagged up concerns regarding activities of radionuclides and concentrations of metals, particularly Cu and Zn in pavement and road gutter dust. However, concentrations of Pb appear to be declining commensurate with the introduction of unleaded petrol. The extension to these studies using biomonitors has confirmed the decline of Pb to rural levels, although neither rural nor urban concentrations are as low as soil background. Street dusts retain the pollution history of the area in which they are found [1], since one of their main sources is surrounding soils [12]. Soils are "the ultimate sink for heavy metals" [13] and are also one of the main sources of material contributing to street dusts. The soil survey of Coventry revealed that whilst roadsides and industrial soils carried relatively high concentrations of metals, private gardens figured amongst the highest concentrations for Pb and Cu, and was second highest for Cd and Ni. However, whilst some of the metals in soil exceeded background levels, none exceeded CLEA SGV or ICRCL trigger concentrations. This suggests that whilst contaminated soil may well be finding its way onto the streets, with Cd, and to a certain extent, Ni appearing to have a close association with the soil (Table 3), and Ni concentrations relatively high in the airborne particulates, there must be other sources for Zn, Cu, and Pb. Nickel has been identified as a "marker species" for fossil fuel burning [12], with Zn and Cu being characteristic of incineration and Pb associated with traffic, but also incineration and road dust. This study has differentiated Ni from the other metals which may be reflected in their different sources.

The European Air Quality Directive (Council Directive 96/62/EC) at present focuses on "pollutants of concern" of which Pb is the only metal. Other metals such as Cd and Ni have been included in Directive 2004/107/EC (2005), but there does not appear to be legislation to cover the elements with the highest concentrations found in the present study: Zn and Cu. Pb in particular, but also Cd in Coventry dusts and particulates now appear to have declined to lower values.

References

1. Charlesworth SM, Everett M, McCarthy R, Ordóñez A, De Miguel E (2003) A comparative study of heavy metal concentration and distribution in deposited street dusts in a large and a small urban area: Birmingham and Coventry, West Midlands, UK. Environ Int 29:563–573

2. Charlesworth SM, Foster IDL (2005) Gamma emitting radionuclide and metallic elements in urban dusts and sediments, Coventry, UK: implications of dosages for dispersal and disposal. Min Mag 69:759–767
3. Harrop DO (2002) Air quality assessment and management – a practical guide. Spon Press, London
4. Goodman GT, Roberts TM (1971) Plants and soils as indicators of metals in the Air. Nature 231:287–292
5. Adamo P, Giordano S, Vigiani S, Castaldo Coblanchi R, Violante P (2003) Trace element accumulation by moss and lichen exposed in bags in the city of Naples (Italy). Environ Pollut 122:91–103
6. Giordano S, Adamo P, Sorbo S, Vinglani S (2005) Atmospheric trace metal pollution in the Naples urban area based on results from moss and lichen bags. Environ Pollut 136:437–442
7. Steinnes E (1995) A critical evaluation of the use of naturally growing moss to monitor the deposition of atmospheric metals. Sci Tot Environ 160: 243–249
8. Goodman GT (1979) Airborne heavy metal pollution. In: Bromley RDF, Humphrys G (eds.) Dealing with dereliction: the redevelopment of the Swansea Valley. University College of Swansea, Swansea
9. Macklin M (1992) Metal pollution of soils and sediments: a geological perspective. In: Newson MD (ed.) Managing the human impact on the natural environment: patterns and processes. Belhaven Press, London
10. Defra (Department of Environment, Food, and Rural Affairs) and EA (Environment Agency) (2002) Assessment of risks to human health from land contamination: an overview of the development of guideline values and related research. Report CLR7
11. ICRCL (Inter-Departmental Committee on the redevelopment of Contaminated Land) (1983) Guidance on the assessment and redevelopment of contaminated land. ICRCL Paper 59/83, Department of the Environment, London
12. QUARG (Quality of Urban Air Group) (1996) Airborne particulate matter in the United Kingdom. Third report of the Quality of Urban Air Review Group. Department of the Environment, London
13. Baird C (1999) Environmental chemistry: heavy metals in soils, sewage and sediments. W.H. Freeman, New York

Size, morphological, and chemical characterization of aerosols polluting the Beijing atmosphere in January/February 2005

S Norra,[1] B Hundt,[1] D Stüben,[1] K Cen,[2] C Liu,[2] V Dietze,[3] E Schultz[3]

[1]Institute of Mineralogy and Geochemistry, Universität Karlsruhe (TH), Karlsruhe, Germany
[2]Institute of Geochemistry, China University of Geosciences, Beijing, China
[3]German Weather Service, Freiburg, Germany

Abstract

One of the most air-polluted cities in the world is Beijing, where PM2.5 was sampled on filters by a mini-volume sampler (200 L h^{-1}) and coarse particles on collection plates by means of the passive sampler Sigma-2. From 15 January to 5 February 2005, sampling was carried out in two modes, by collecting particles over periods of several days separated in day and night samples, and alternately during night and daytime in intervals of 12 h. The sampling site was located in north-west of Beijing. Automated microscopic image analysis of coarse particles showed significantly differing size distributions of particle fractions between 3 and 100 µm for day and night samples. The different size distributions could be attributed to varying conditions of atmospheric dispersion and particularities due to local traffic. Elemental carbon containing particles accounted for 10% to 30% of total coarse particles. Scanning electron microprobe analyses identified a wide range of different anthropogenic and geogenic particles. Average PM2.5 concentration during daytime was about 75µg m^{-3} in the first, 200 µg m^{-3}

in the second, and 64 µg m^{-3} in the third week. Peaks of daily PM2.5 concentrations reached more than 300 µg m^{-3} in the second week. Trace metals such as Pb, As, Cu, and Zn showed similar temporal courses. Several different sources of aerosols were identified by means of scanning electron microscopy (SEM). Backward trajectories indicated that highly polluted air masses reaching Beijing during the second week were advected from south. These air masses passed with low velocities industrial areas in the south and the whole city before reaching the sampling site, whereas air masses entering Beijing during the first and third week came from less industrialized regions in the north. Due to the specific spatial distribution of air pollution sources in the area of Beijing, trajectories could provide a simple approach to an air pollution forecast.

Introduction

During the last decade, epidemiological studies showed that atmospheric particles contribute to a decreasing life expectancy of man, especially in intensively polluted areas, such as cities. In the UK, 12,000 deaths are attributed to total PM10 and 7000 death to anthropogenic PM10 [1]. For total Europe, 288,000 deaths every year can be attributed to PM10 [2]. This was the reason that European Union (EU)-wide threshold values for PM10 were set in 2005. Recently, research is focused, in particular, on PM2.5 since there is more and more evidence that negative health effects can be attributed to this fraction [3].

Beijing is regarded as one of the most air-polluted cities in the world. About 15 million inhabitants are affected by seasonal occurring dust storms and a mixture of different anthropogenic air pollution sources, such as power plants, traffic, heavy and chemical industry, and domestic heating. Although the Beijing Municipality as well as the central government is increasingly paying attention to the atmospheric pollution of China's capital and has carried out many countermeasures, the efficiency of these countermeasures are still affected by a limited knowledge about the spatial distribution and temporal behaviour of involved sources.

The Beijing Environmental Protection Bureau currently monitors PM10, SO_2, NO_2, and CO. The indexes of SO_2, NO, O_3, and CO dramatically exceed the National Air Quality Standards of China. Coal and natural gas are the main sources for heating in winter in Beijing and traffic is another important source of aerosol. However, the situation is characterized by the still widespread use of solid fuels, such as coal and lignite for private heating and industrial energy production causing far above-average particle emissions in comparison with oil and gas. Furthermore, local construction sites, coal heaps, and bare soils considerably contribute to particle pollution, in particular to a high coarse particle load. Additionally, the dust and aerosol load is also related to geogenic sources such as dust transport from deserts. During spring, in average four to five sand and dust storm events occur in northern China causing high loads of particulate matter (PM) in Beijing.

Sampling and methods

Sampling location and period

The area chosen for sampling of atmospheric particles was located in north-west of Beijing (Fig. 1) in the campus of the China University of Geosciences (CUG) near to a highly frequented street artery. For particle sampling, active and passive techniques were used. The sampling period was from 15 January to 5 February 2005. Partly, particles were sampled for several days interval, differentiating between day and night sampling;

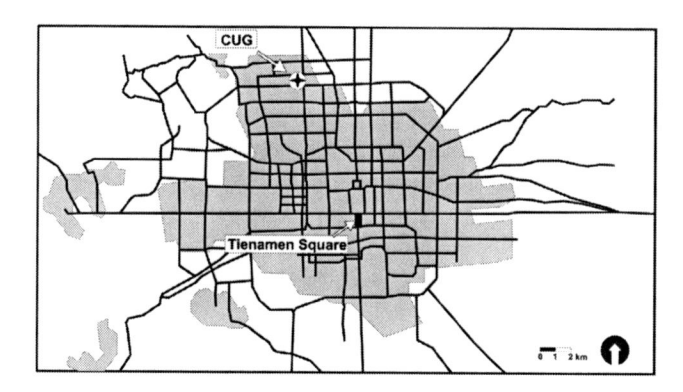

Fig. 1. Sampling location at China University of Geosciences (CUG)

partly, particles were sampled for 12 h intervals only, between 7 a.m. and 7 p.m. and 7 p.m. and 7 a.m. All samples were taken at 1.5 m height.

Passive sampling of coarse particulates (>3 μm)

Passive sampling of atmospheric particles larger than 3 μm in diameter by Sigma-2 technique is described in the technical guideline VDI 2129/4 [4]. The Sigma-2 sampler consists of a tube and a protective cap out of aluminium. The interior of the sampler contains a carrier with an inlaid sampling plate located at the bottom of the tube. Airborne particles are allowed to enter the tube through four rectangular windows in the cap and the tube, as well but twisted against one another. Due to this construction the inside of the sampler is widely protected against the impact of precipitation and wind. In the claimed interior of the tube, particles are settling down onto the highly transparent collection plate suitable for optical microscopy. The particles are fixed on the surface of the collection plate by a weather-resistant adhesive. Microscopic analysis provides a size fractioned number deposition rate of collected particles. By assuming spherical shape and unit density for all particles, a size-fractionated mass deposition rate of particles can be calculated. Finally, a size-fractionated particle mass concentration can be calculated from the mass deposition rate by approximating individual particle deposition velocity by its settling velocity assumed under the calmed conditions in the interior of Sigma-2. An automatic image analysis system consisting of an optical microscope, high-resolution CCD Camera and a PC-aided image analysis system was used in order to measure object-specific features of individual particles, such as projected area diameter and mean grey value. This approach provided a size-fractionated particle concentration of a transparent, typically mineral component and a black, typically elemental carbon containing component of the size interval from 3 to 100 μm optical equivalent diameter [5].

Active sampling of PM2.5

PM2.5 was collected with mini-volume samplers (Leckel, Berlin) at a pump rate of 200 L h^{-1}. Aerosols were sampled on cellulose nitrate filters (Sartorius) and quartz fibre filters (QF 20, Schleicher & Schüll, Dassel, Germany). Gravimetric analysis was performed after at least 48 h equilibration with a microbalance (Sartorius MP2) for three times. One quarter of the filter was used for chemical element analysis.

Chemical element analysis

For the analysis of trace metals, parts of the filters were digested in Teflon vessels using concentrated HNO_3 (Merck, sub-boiled), concentrated HF (Merck, p.a.), and concentrated $HClO_4$ (Merck, s.p.). Determination of elements was performed by inductively coupled plasma mass spectrometry (ICP-MS) PQ 3 (VG Elemental). Quality control was performed by additional analysis of reference material SRM 1648 (urban PM) acquired from National Institute of Standards (NIST), USA. Results of standard material are within ±10% of the certified values for element concentrations [6].

Scanning electron microscopy

Scanning electron microscopy (SEM) was carried out at the Laboratory for Scanning Electron Microscopy, Universität Karlsruhe. Adhesive collection plates and cellulose nitrate filters were sputtered with Ag to guarantee electrical conductivity. Particles on these plates were investigated with respect to morphology (image) and main element composition.

Results and discussion

Morphology of particles of Beijing atmosphere

Generally, particles collected during the sampling period could be classified as particles of anthropogenic sources, such as fuel combustion and traffic (e.g., fly ashes, soot, pure metallic particles, and tire wear), of biological sources (pollen and diatoms), and of geogenic sources (e.g., salts such as halite, sulphates and carbonates, quartz, silicates, and aluminosilicates). Some anthropogenic particles are presented in Figure 2. Although many geogenic particles were found on the filters and adhesive plates, anthropogenic particles contribute for a great portion to the overall particle load in the atmosphere of Beijing. Besides soot and fly ashes, many coal particles were identified. Geogenic particles comprise minerals such as quartz, amphiboles, calcite, and halite. Halite might be formed secondarily by crystallization from liquid phases in the atmosphere or might be an indicator for oceanic air masses transporting sea salt particles towards Beijing from the coast around 200 km east of Beijing. Only little portion of biogenic particles were found owing to the sampling period during winter with relatively low biological activity.

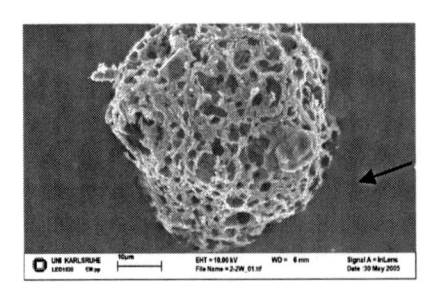

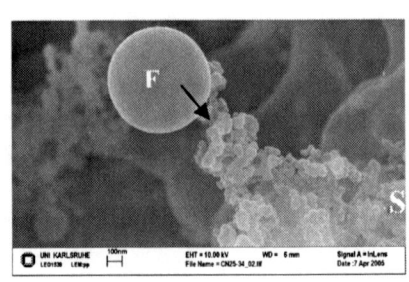

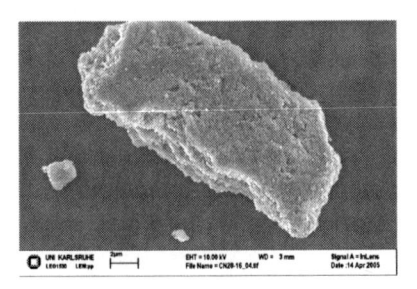

Fig. 2. Examples of Beijing atmospheric particulates of anthropogenic sources. A. Incompletely incinerated carbon particle carrying a pure sulphur particle (S) sampled in the period 31 January till 3 February. B. Fly ash particle (F) and soot (St) sampled in the period 17 January, 7 p.m., till 18 January, 7 a.m. C. Pure Cu particle collected on 23 January between 7 a.m. and 7 p.m.

Temporal course of coarse particulates (>3 μm)

Concentrations of a coarse total particle (CTP) component and a coarse black component (CBC) between 3 and 100 μm were determined for the period from 17 January, 7 p.m. to 22 January, 7 p.m. Adhesive collection plates were changed every 12 h. Concentrations of CTP varied between 170 and 400 μg m^{-3} from 17 January until 20 January and increased up to 530 μg m^{-3} on 22 January during daytime (Fig. 3).

CBC varied between 40 and 120 μg m^{-3} and counted for 10–30% of CTP. The small increase of CBC towards the end of the monitoring time is much less intensive as the increase of CTP. Small CBC particles of size 3–6 and 6–12 μm show higher concentrations during night than during daytime (Fig. 4).

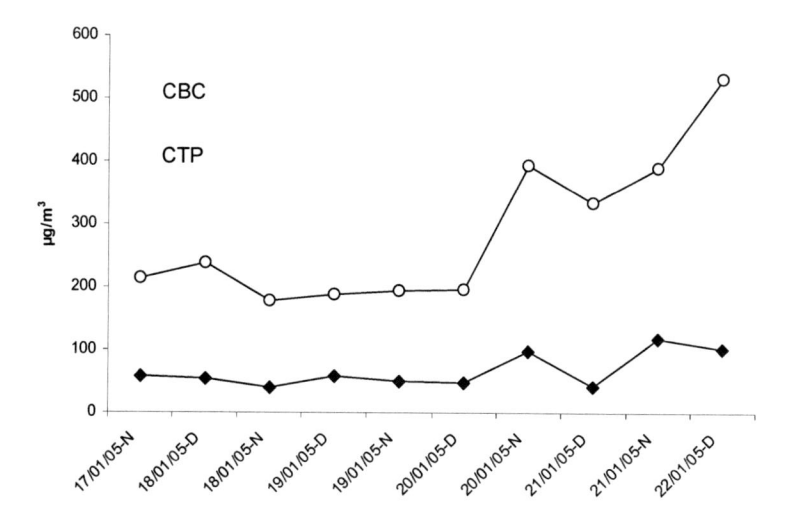

Size, morphological, and chemical characterization of aerosols polluting 173

Fig. 3. Temporal course of CTP and CBC in Beijing from 17 January, 7 p.m. to 22 January, 7 p.m. 2005

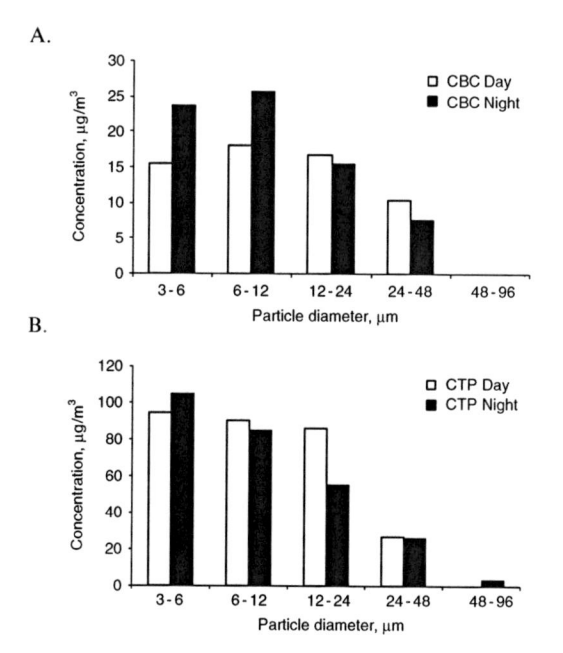

Fig. 4. Average mass concentrations of atmospheric particles (A. CBC and B. CTP) of different size classes collected during night and daytime from 17 January, 7 p.m., to 22 January, 7 p.m.

Two reasons are assumed to produce this pattern. First, during night-time in winter heating processes will be more intensive than during daytime owing to lower temperatures. Secondly, atmospheric mixing height typically is decreasing during night limiting particle dispersal to a shallow layer below an intensifying temperature inversion. Additionally, a particularity of heavy transport in Beijing may be responsible for this effect. Lorries are normally only allowed to enter Beijing at night to mitigate traffic jams during daytime. Lorries are not only emitting elevated amounts of soot, but are also an important source of coarse particles by resuspending road dust particles and producing an high amount of tire wear. CTP concentrations at day are more or less on the same level (about 90 µg m^{-3}) for the size intervals 3–6, 6–12, and 12–24 µm particle diameter, but rapidly decrease towards the largest fraction above 48 µm. At night, CTP concentrations of all size fractions uniformly decrease from about 105 up to less than 5 µg m^{-3}. According to often reported particle mass size distributions [e.g., 7], the coarse mode shows a maximum in particle concentration at about 10 µm particle diameter. The actual pattern can considerably differ from this idealized pattern due to the impact of local coarse particles [8]. At Beijing, particle concentration of 3–6 µm class is comparable with the concentration of the size class 6–12 µm in case of CBC and even higher in case of CTP reflecting a flat concentration maximum around 10 µm for CBC and a shallow shoulder reaching up to 20 µm particle diameter for CTP, respectively.

Temporal course of PM2.5

PM2.5 was sampled in two different modes. In the first mode, PM2.5 was sampled for several days during daytime and night-time. In the second mode, PM2.5 was sampled for 12 h periods. Daytime sampling always lasted from 7 a.m. to 7 p.m., night-time sampling from 7 p.m. to 7 a.m. During the first week (15–22 January), PM2.5 concentration at daytime accounted for about 75 µg m^{-3} and at night-time for around 90 µg m^{-3}. Average PM2.5 concentrations increased during the second week (22–27 January) to almost 200 µg m^{-3} showing no significant difference between day and night-time (Fig. 5). Between 31 January and 5 February, PM2.5 showed lowest average concentrations with 50 µg m^{-3} during night-time and 64 µg m^{-3} during daytime. Between the both periods last mentioned, PM2.5 showed intermediate concentrations, which were during night-time (162 µg m^{-3}) almost twice as high as during daytime (85 µg m^{-3}). Sampling of PM2.5 for short intervals of 12 h resulted in a more differentiated appearance. PM2.5 showed highest concentrations of 307 µg m^{-3} during

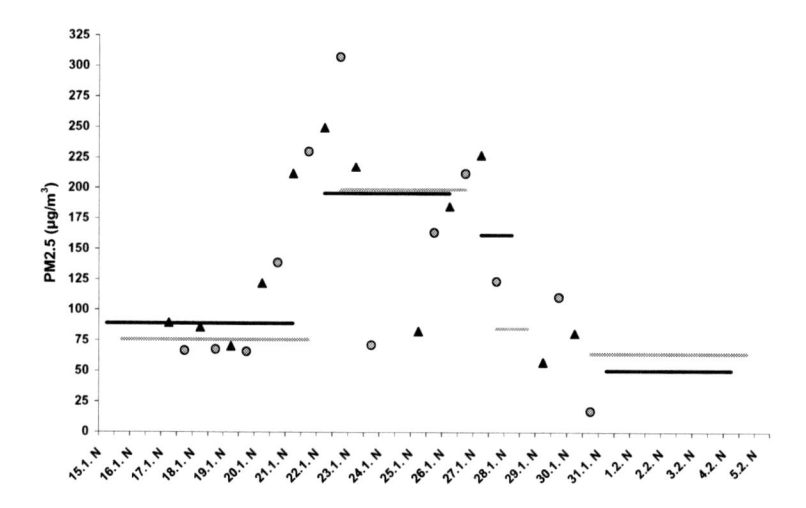

Fig. 5. Temporal courses of PM2.5 concentrations. Horizontal lines represent average concentrations of integrating sampling over specific periods of several nights and days; grey lines show PM2.5 concentrations at day; black lines indicate PM2.5 concentrations at night. Single symbols represent sampling periods of 12 h (day: 7 a.m. till 7 p.m.; night: 7 p.m. till 7 a.m.). Grey circles indicate PM2.5 concentrations during daytime; black triangles indicate PM2.5 concentrations during nighttime. Data of night-time sampling are plotted at the x-axis. Labelling of data of day time sampling is disabled.

daytime of 23 January. A second peak of high PM2.5 concentrations of 227 µg m^{-3} occurred in the night from 27 to 28 January.

The increase of PM2.5 concentrations towards 22 January corresponds with the temporal course of the fraction >3 µm. Modelling trajectories of air masses by means of National Oceanic and Atmospheric Administration (NOAA) hybrid Single-particle Lagrangian integrated trajectory (HYSPLIT) model showed that air masses reached Beijing from southern directions during periods of high concentrations of particles and from northern directions during periods of low concentrations of particulates. In periods when air masses reached the sampling station from south, those air masses are polluted by emissions from industry south of Beijing and by the diverse pollution sources of the city itself. When the air masses reached the sampling station from northern directions the air pollution was less since north of Beijing larger industrial complexes do not exist and the extension of the city itself is less towards north in comparison with the areas south of the sampling station. Furthermore, wind velocities of only about 2 m s^{-1} occurred during the first and second week indicating an elevated probability for inversion conditions. During the third week, wind velocities reached up to 6 m s^{-1} causing more intense turbulence and less residence time of air masses in the areas they pass. Thus, owing to processes of concentration of pollutants in the atmosphere, air pollution increases in cases of elevated residence times of air masses above Beijing and industrialized areas, whereas less air pollution occurs during periods of shorter residence times of air masses above areas of intensive emissions.

Higher PM2.5 concentrations at night during the period of 15 January to 22 January 2005, correspond to the results for coarser particles. Especially the finer fractions of coarse particles (CBC: 3–6 and 6–12 µm; CTP: 3–6µm) show higher concentrations at night than at day. However, this pattern can not be generalized as was observed for following sampling periods. In general, weather conditions (e.g., wind directions, inversion layers, temperature) and diverse urban activities (heating modes, construction activities, intensity of lorry traffic, etc.) can cause differences in air pollution and corresponding temporal course and size distributions, respectively. The specific mechanisms responsible for changes of Beijing aerosol pollution are still not fully understood and have to be investigated.

Temporal courses of trace element concentrations in PM2.5

Trace element concentrations (Cd, Cu, Zn, As, Pb) of particulates show a similar temporal course of concentration as total PM2.5. Again, two peaks of maximum concentrations were observed on 23 and 27 January 2005. Figure 6 exemplarily shows this course for Pb and As, both toxic elements for human beings. Lead reaches maximum concentration during the night from 27 to 28 January and as was found peaking during daytime of 27 January. In contrast to the temporal course of PM2.5, which showed its maximum concentration on 23 January, Pb and As show their maximum concentrations on 27 January.

Arsenic concentrations of PM2.5 varied from about 200 to 300 mg kg^{-1}. Lead concentrations varied from about 2000 to 3000 mg kg^{-1} till 30 January and from 1200 to 1700 mg kg^{-1} during the last sampling period. Concentrations of Pb, As, Cd, Cu, and Zn of PM2.5 were less during the first short period of high pollution (23 January) than during the second short period of high pollution (27 January). Aerosols sampled in the 12 h mode showed concentrations up to 2260 mg kg^{-1} Pb and 226 mg kg^{-1} as during the first and 3310 mg kg^{-1} Pb and 400 mg kg^{-1} as during the second pollution period. Thus, aerosols during both periods were slightly differently composed. With respect to harmful trace elements, the second period of pollution was the more toxic one.

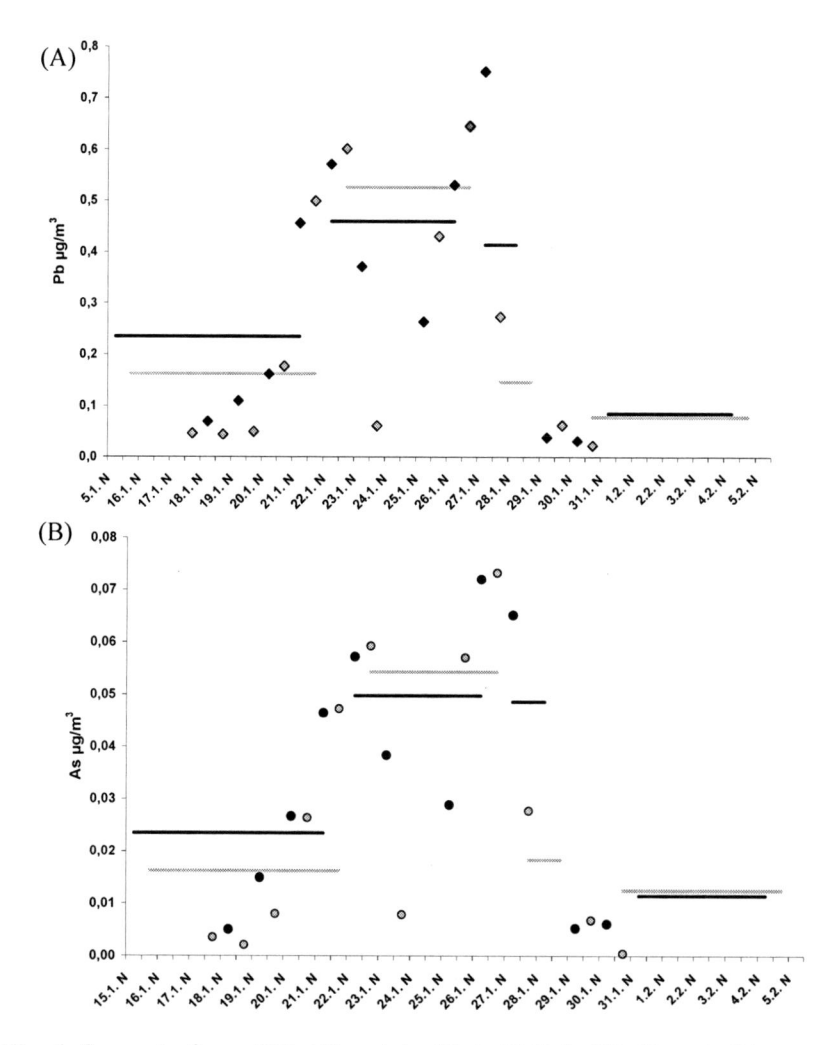

Size, morphological, and chemical characterization of aerosols polluting 177

Fig. 6. Concentrations of Pb (A) and As (B) in PM2.5 of Beijing air. Lines represent courses of integrating sampling over specific periods of several nights and days. Grey lines indicate element concentrations at day and black lines indicate element concentrations at night. Single symbols represent sampling periods of 12 h (day: 7 a.m. till 7 p.m.; night: 7 p.m. till 7 a.m.). Grey symbols indicate element concentrations during daytime; black symbols indicate element concentrations during night-time. Dates of night-time sampling are plotted at the *x*-axis. Labelling of dates of daytime sampling is disabled

Conclusions

China has not established air quality standards for PM2.5 yet. However, PM2.5 pollution at Beijing exceeded even the long-term standards for PM10 according to regulations of China of 250 µg m^{-3} and the EU (40 µg m^{-3}) for most days during the sampling period. During the first week of sampling, also standards for daily total suspended particles (TSPs) (China: 300 µg m^{-3}; EU: 150 µg m^{-3}) were already exceeded by the mass concentration only of the coarse particle fraction between 3 and 100 µm. Until now, no limit values exist for BC and the concentration of harmful chemicals in PM2.5 as well. Nevertheless, both variables show high concentrations in Beijing atmosphere and should be reduced as much and as quickly as possible because of expected adverse health effects these extreme concentrations mean to the population.

Pb and As concentrations with respect to PM2.5 fraction observed in Beijing are far higher than in other cities of the western hemisphere such as New York (Pb about 6 ng m^{-3}, as about 1 ng m^{-3}) [9]. He et al. [10] already observed in 1999 and 2000 at Beijing similar concentration ranges of PM2.5 and of Pb in the fraction of PM2.5 in comparison with this study. Some research already has been carried out to identify the sources causing Beijing air pollution [11]. World wide, those sources (fossil fuel combustion, secondary particles, particle resuspension, industry, biomass burning, etc.) are typically the same and only vary with respect to the portions they contribute to the pollution. Thus, additional to investigations on Beijing air pollution, further studies have to focus on the reasons for the not decreasing air pollution, which have to include urban planning and development aspects. Those studies may also contribute details to an identification of the specific reasons of different pollution levels. For example, lorry traffic and heating modes are processes, which can be regulated by urban management if their consequences to air pollution are comprehensively understood for the specific urban situation. A valuable contribution to a better understanding of the relation between effective sources, occurring pollution and ruling atmospheric conditions are provided by separate day and night-time sampling, even when only cost-effective, passive, or active techniques are used.

Furthermore, results of this study clearly demonstrate that high air pollution is typically related to air masses entering the Beijing area from southern directions with low wind velocities, whereas lower air pollution occurs in times when air masses reach Beijing from northern directions with elevated wind velocities. This relationship could be used to develop an air pollution forecast system for Beijing.

Acknowledgements

The authors express their gratitude to Mr. P Pfundstein and Mr. V Zibat (both Laboratory for Scanning Electron Microscopy, Universität Karlsruhe) for support of SEM. We also like to thank Mrs. C Mössner and Mrs. C Haug (both Institute of Mineralogy and Geochemistry, Universität Karlsruhe) for technical support with respect to dissolution of samples and measurements by ICP-MS. Special thanks go to Mr. M Fricker (German Weather Service), who gave many helpful advices for the installation of the samplers during the field campaign. Furthermore, the authors express their gratitude to Professor A Wiedensohler from the Leibniz-Institute for Tropospheric Research at Leipzig for some valuable discussions. Finally, we thank Deutsche Forschungsgemeinschaft (DFG) for financial support of this project.

References

1. Pearce D, Crowards T (1996) Particulate matter and human health in the United Kingdom. Energ Policy 24:609–619
2. BMU (2005) Feinstaub. Bundesministerium für Umwelt, Naturschutz und Reaktorsicherheit. Berlin
3. Lall R, Kendall M, Ito K, Thurston GD (2004) Estimation of historical annual PM2.5 exposures for health effects assessment. Atmos Environ 38:5217–5226
4. VDI 2119-4 (1997) Messung partikelförmiger Niederschläge, mikroskopische Unterscheidung und größenfraktionierte Bestimmung der Partikeldeposition auf haftfolienb, Probenahmegerät Sigma-2
5. Schultz E (1993) Size-fractionated measurement of coarse black carbon particles in deposition samples. Atmos Environ 27A:1241–1249
6. Hundt B (2005) Geochemische Untersuchungen von Stäuben aus Peking sowie Identifikation der Aerosolbelastungsquellen. Unpublished diploma thesis, Institute of Mineralogy and Geochemistry, Universität Karlsruhe
7. Maynard RL (2001) Particulate air pollution. In: Brimblecombe P, Maynard RL (eds.) The urban atmosphere and its effects. Imperial College Press, Danvers, pp 163–194
8. Schultz E, Alessandro M, Endlicher W (1999) A two years air pollution survey in Gran Mendoza, Argentina-analysis, spatial distribution and seasonal variations of traffic related pollutants. MERIDIANO-Revista de Geografia 7:102–118
9. Ito K, Xue N, Thurston G (2004) Spatial variation of PM2.5 chemical species and source-apportioned mass concentration in New York City. Atmos Environ 38:5269–5282

10. He K, Yang F, Ma Y, Zhang Q, Yao X, Chan CK, Cadle S, Chan T, Mulawa P (2001) The characteristics of PM2.5 in Beijing, China. Atmos Environ 35:4959–4970
11. Zheng M, Salmon LG, Schauer JJ, Zeng L, Kiang CS, Zhang Y, Cass GR (2005) Seasonal trends in PM2.5 source contributions in Beijing, China. Atmos Environ 39:3867–3976

Air pollution levels in two São Paulo subway stations

RK Fujii,[1] P Oyola,[2] JCR Pereira,[1] AS Nedel,[3] RC Cacavallo[1]

[1]School of Public Health, University of São Paulo, Av. Dr. Arnaldo 715, CEP 01246-904 São Paulo, SP, Brazil
[2]Mario Molina Centre, Santiago, Chile
[3]Institute of Astronomy, Geophysics and Atmospheric Sciences – Astronomy Department, University of São Paulo (IAG/USP) (support by CNPq)

Abstract

In megacities like São Paulo, which is responsible for approximately 16% of Brazil's gross national product (GNP), the subway serves as a major transportation mode, transporting 2.7 million people daily. The underground portion of the subway system is a confined space that may permit the concentration of contaminants either from the outside atmosphere or generated internally. The objective of this study is to evaluate the air quality of two of São Paulo's subway stations (Clínicas and Praça da Sé), by identifying the fungi species, the nitrogen dioxide (NO_2), BTEX (benzene, toluene, ethylbenzene, and xylene), and particulate material (PM_{10}) concentrations. Afterwards, a comparison of the internal pollutants concentration values with external atmosphere parameters is performed.

Introduction

The Metrô São Paulo is responsible for the operation and expansion of the subway transport, a system with high transportation and connection capacity inside the public transportation network in the metropolitan area. The

São Paulo subway system operates in 60.5 km of extension with 54 stations, 29 of which are underground, distributed in four lines, with 2.7 million passengers daily.

The contaminants concentration levels became an air quality marker for confined environments. This epidemiologic marker has been used to assess the ventilation system performance and to verify the external air renewal, with the support of Technical Reports and Referential Standards. Thus, assessment of CO, CO_2, NO_x, NO_2, suspended particles, volatile organic compounds (VOCs), and micro-organisms' concentration levels can be used to evaluate the ventilation ratio, indicating the air quality in these places.

This study aimed at evaluating the air quality of two of São Paulo subway stations (Clínicas and Praça da Sé), by identifying the fungi species, the nitrogen dioxide (NO_2), BTEX, and PM_{10} concentrations and monitoring data on temperature and relative humidity. Afterwards, a comparison of the internal pollutants concentration values with external atmosphere parameters was performed.

Methodology

Studied sites

Two stations were selected for the present study. The main characteristics taken into account in that selection were: architecture, depth, users/hour rate, and geographic localization.

1. CLÍNICAS (CLI) – Inaugurated on 12 September 1992, is located at Dr. Arnaldo Avenue, with 9510 m² of area, and has a 20,000 passengers/hour/peak capacity. It has access to the hospital complex of Clínicas and Emílio Ribas Hospital therefore having a comparatively high traffic of sick people.
2. PRAÇA DA SÉ (PSE) – Inaugurated on 17 February 1978, is located at the Praça da Sé, with 39,925 m² of area, and has a 100,000 passengers/hour/peak capacity. This station links two subway lines: 1 and 3 being, for this reason, the busiest station of the entire subway network.

The samples for fungi, NO_2, and benzene analyses were collected in the different levels of each studied station. In PSE station, the samples were collected in line 1 and line 3 platforms, access mezzanine, and the outdoor environment. In CLI station the sampling sites were platform, mezzanine, offices, and outdoor environment.

The samples collection was performed on a continuous sampling basis or specific sample collection during the busiest periods of both the morning and the afternoon.

Fungi samples

The air samples for microbiologic analyses were collected during the busiest morning and afternoon periods, in different levels of the station, and at the external surrounding area. The sample collection was performed by a 1-stage impactor with 28 L min^{-1} flow, 1.5 m from the ground level (respiratory zone) and the material was cultivated in a 4% Sabouraud dextrose agar culture medium.

After the sample collection, the plates were incubated at 28°C for 7 days in the laboratory. The fungi identification and the colony-forming units (CFU m^{-3}) quantification were made with an optical microscope, and the plates were stained with lactofenol blue [1].

The analyses considered the CFUs count by genus, the percentage of fungi genus by sample, and a second analysis quantified the CFU amount per cubic metre of collected air.

Gaseous pollutants

The gaseous pollutants of interest were nitrogen dioxide and BTEX, which had their concentrations measured by the following techniques.

DOAS continuous analyser

This measuring system allows for the monitoring of several pollutants present in the air in a continuous way. This can be selected at the instrument's operating programme according to their light-absorption characteristics at different wavelengths. These compound concentrations are determined by using the differential optical absorption spectroscopy (DOAS) technique, which is based on the Lambert-Beer's law. A DOAS system basically consists of a light transmitter that displays a long and fine beam, and a receiver that collects such light and sends it to an analysing device by means of fibre optics. Light transmission through fibre optics allows the analyser to be located some distance and to be kept indoors. Light absorption is produced when the beam crosses the air at some predetermined frequencies by means of the existing chemicals. The light being received at the analyser is processed by comparing it through references kept in the instrument and is finally assessed on the absorption produced, which is proportional to the averaged concentration along the measuring beam. The length of the optic trip can vary from some hundreds of metres to some kilometres and is conditioned by the compound being tested, among other parameters [2].

Hourly averages were calculated using the DOAS NO_2 concentration data to make comparisons with the passive sampler results at the same sampled period.

Passive samplers

Passive samplers have previously been used to BTEX and NO_2. They constitute an interesting complement to the continuous monitors, because they allow an inexpensive monitoring with a high space resolution. The measurement principle is that the ambient air is diffused passively through a filter covered with a certain absorbent material specific for the compound to be determined. In order to assure that the flow is laminar – though strong winds may blow – a stainless steel filter or Teflon mesh is placed at the open side of the sampler [3,4]. After the sampler has been exposed to ambient air concentrations, the BTEX and NO_2 contents are analysed at the laboratory by using gas chromatography for BTEX and flow injection analysis (FIA) for NO_2.

The NO_2 passive samplers were installed in pairs. Two sampling campaigns with a 30-day exposure period each – June 2005 and between August and September 2005 were carried out. There was only one campaign for the BTEX tubes, with one tube for each place of interest. These were installed during the second NO_2 tubes campaign.

Particulate material

The DataRAM4 continuously monitors the real-time concentration and median particle size of airborne dust, smoke, mist, and fumes. The instrument also displays air temperature and humidity. With appropriate particle discriminators, the unit provides high-reliability measurements correlated with PM_{10}, $PM_{2.5}$, $PM_{1.0}$, and breathable fractions. This instrument monitors the concentration of fine particulates in ambient air by a combination of aerodynamic size pre-selection, dual-wavelength nephelometry, and concurrent relative humidity sensing. With its self-calibrating internal filter and automatic zeroing, the DataRAM4 determines median particle size down to 0.05 μm regardless of concentration. The monitoring campaign was carried out, for each station, with two instruments during a 7-day period in June 2005.

Temperature and relative humidity

These parameters were registered by the DataRAM4 in the indoor environment of each station. For the outdoor temperature and relative humidity parameters, data from a case study released by the same research group were used [5].

Results and discussion

Fungi

A total of 40 air samples were collected for fungi analyses, 8 being from the stations' outdoor environment, and the other 32 from the different levels of each station. There were 681 CFU isolated by the mycologic exam, with major occurrences of *Cladosporium* sp., *Alternaria* sp., *Penicillium* sp., *Candida* sp., and *Aspergillus* sp., as shown in Table 1.

Low-toxicity regular fungi were found in the analysed samples, though it is cautious to point out that, in sensitized individuals, they can cause allergic or toxic reactions. In each station the fungi percentiles for indoor and outdoor environment were quite similar, as well as the values for the indoor environment for the two stations.

There are neither indoor nor outdoor biological air contamination standard levels, because even having a low air micro-organism concentration it could contain high pathogenicity species. Low air micro-organism concentration levels does not necessarily mean healthy and clean environment.

The fungi genus incidences found in this study are quite similar to those observed by Awad [6], although it is important to point out that the subterranean station in Cairo has an air conditioning system.

Table 1. Indoor and outdoor fungi percentiles in Clínicas and Praça da Sé subway stations, São Paulo, 2005

Fungi	Clínicas station Indoor (%)	Clínicas station Outdoor (%)	Praça da Sé Station Indoor (%)	Praça da Sé Station Outdoor (%)
Alternaria sp.	18.4	27.7	16.8	10.8
Aspergillus sp.	3.9	6.0	2.3	1.1
Candida sp.	3.9	2.4	2.7	1.1
Cladosporium sp.	52.7	37.3	52.7	61.3
Curvularia sp.	3.4	2.4	2.0	1.1
Penicillium sp.	15.0	8.4	14.1	7.5

Gaseous pollutants

The results obtained with passive samplers indicated that the outdoor environment pollution has an influence over the indoor stations' environment and tunnels. Those pollutants are transported by the ventilation system or captured by the station access area, which makes the indoor and outdoor concentration values similar.

Table 2 shows the results obtained for nitrogen dioxide (NO_2) in each station's indoor and outdoor environments.

The maximum and minimum values shown in Table 2 for the indoor environment of the PSE occurred in the line 1 platform, which suggests some external agent influence during the station's cleaning or maintenance process.

Hourly averages were calculated and comparisons with the passive sampler results at the same sampled period were used.

Figure 1 shows the comparisons between the DOAS NO_2 concentration data and the passive sampler values, which substantiate the feasibility of the passive sampler technique. The BTEX passive sampler's data indicate low concentration values, as seen for NO_2, and we concluded that the presence of those two gaseous pollutants in the indoor environment of the stations is due to the concentrations on the outdoor atmosphere. The use of this sampler category allows for a regional mapping indicating the BTEX behaviour at different points. Table 3 presents the BTEX concentration levels with one sample collected per referred area.

Table 2. Nitrogen dioxide (NO_2) average, maximum and minimum concentration levels, standard deviation, and number of measurements at Clínicas and Praça da Sé station's indoor and outdoor environments, São Paulo, 2005

	Clínicas station		Praça da Sé station	
	Outdoor ($\mu g\ m^{-3}$)	Indoor ($\mu g\ m^{-3}$)	Outdoor ($\mu g\ m^{-3}$)	Indoor ($\mu g\ m^{-3}$)
Average	62.81	68.01	77.28	76.56
Max.	87.08	76.95	89.56	142.00
Min.	21.91	52.09	43.38	6.53
Std.	23.35	7.80	17.30	32.66
No.	10	8	6	18

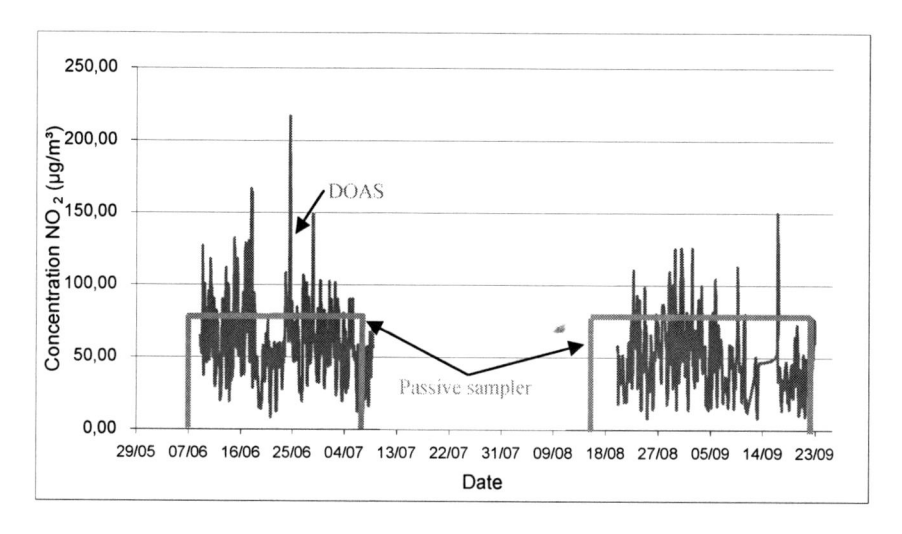

Fig. 1. Comparisons between the DOAS NO$_2$ concentration data and the passive sampler values, São Paulo, 2005

Indoor and outdoor concentrations of NO$_2$, benzene, toluene, and xylene were similar, indicating that the observed values are connected to the external environment atmosphere. In this case the subway station acts as an external contaminants receptor.

Table 3. Benzene, toluene, ethylbenzene, and xylene concentration distribution according to sampled areas, São Paulo, 2005

Sampler area	Benzene (µg m^{-3})	Toluene (µg m^{-3})	Ethylbenzene (µg m^{-3})	p- + m-Xylene (µg m^{-3})	o-Xylene (µg m^{-3})
FSP/USP[a]	6.2	21.2	4.4	8.6	2.7
CLI platform	5.7	22.7	4.7	9.1	2.8
CLI mezzanine	5.6	22.2	5.6	10.1	3.0
CLI outdoor	5.7	22.8	4.7	9.3	2.9
PSE line 1	4.1	16.8	2.7	5.3	1.9
PSE mezzanine	4.4	17.8	2.6	5.1	1.6
PSE outdoor	4.3	21.3	2.9	5.7	1.9
PSE line 3	3.8	18.0	2.7	5.2	1.7

[a]School of Public Health – DOAS site [5].

Table 4. PM_{10} average concentration levels, standard deviation, and limits observed in Clínicas and Praça da Sé stations, São Paulo, 2005

	CLI platform ($\mu g\ m^{-3}$)	CLI mezzanine ($\mu g\ m^{-3}$)	PSE line 1 ($\mu g\ m^{-3}$)	PSE line 3 ($\mu g\ m^{-3}$)
Average	312.4	243.9	150.9	124.2
Standard	145.01	112.03	54.94	47.00
Max.	742.1	585.8	297.2	234.2
Min.	38.1	28.6	50.7	34.4

Particulate material

As shown in Table 4, the particulate material (PM) concentration levels are high. The CONAMA's year average air quality standard for this pollutant is 50 $\mu g\ m^{-3}$ [7].

Many studies that investigate PM also observed high concentration levels, which characterize the subway transport system as a particle generator due to the spinning parts attrition, incoming external PM, and air flux inside the tunnels, associated with the impracticability of a ventilation system installation with the capability to remove this sort of pollutants [8–10].

Another feature that must be better known is the PM composition, which could help in the identification of those particles' toxicity as well as their source – external environment atmosphere or subway system itself.

Temperature and relative humidity

Station's temperature and relative humidity are related to the ventilation system activity, with 72% of the removed heat from the tunnels being due to mechanic ventilation and air motion, and 28% being due to ground heat conduction [11].

The temperature and relative humidity data (Table 5) collected by DataRAM4 simultaneously with the PM_{10} samples, showed a high temperature and low humidity in the Clínicas station. This condition makes the air drier, contributing to the high PM concentrations. On the other hand, this condition is unsuitable for micro-organism development.

The temperature for line 1 of Praça da Sé station was highly correlated with that for line 3 platform ($r = 0.83$), while its association with the outdoor environment was lower ($r = 0.66$).

In the Clínicas station we observed a low correlation between the mezzanine and platform temperatures ($r = 0.42$), which was very similar to the correlation between the platform and the outdoor environment ($r = 0.44$).

Table 5. Temperature and relative humidity averages per sampled area, São Paulo, 2005

Sampler area	Temperature (°C) Average	Standard	Relative humidity (%) Average	Standard
CLI mezzanine	33.8	0.55	35.6	2.13
CLI platform	33.4	1.08	37.8	3.21
CLI outdoor	19.1	3.42	75.8	17.06
PSE line 3	28.6	0.79	44.9	5.74
PSE line 1	27.7	1.48	50.1	3.76
PSE outdoor	18.2	3.76	78.3	18.22

There was no association between the temperatures of the mezzanine and the outdoor environment ($r = 0.10$).

In the Clínicas station, internal temperature is extremely high and has a particular behaviour. In addition, its dynamics is not affected by external environment parameter variations. It is important to point out that the mechanical ventilation system of the station is still undergoing installation. Another fact that must be highlighted is that the data collection instrument was not properly located in the Clínicas station (very close to the station's ceiling).

Conclusions

The present study completed its main purpose of evaluating the air quality of the Clínicas and Praça da Sé subway stations in the city of São Paulo.

Although high pathogenicity fungi were not found in the analysed samples, it is reasonable to recommend that periodic analyses be performed in different sites, as most of them have appropriate conditions for pathogenic fungi development. A preventive procedure that should be adopted is the use of disinfectant cleaning solutions, even for the tunnels which are periodically cleaned.

For gaseous pollutants, the NO_2 measurements presented similar values when compared to those obtained by DOAS equipment. This study was the first opportunity of measuring BTEX in São Paulo city, being therefore an important reference.

Concerning the comparisons between the two stations, Praça da Sé station showed a better condition concerning temperature and humidity when compared to the Clínicas station, most likely because there is a functioning ventilation system.

Considering the substances and micro-organisms focused in this study, PM_{10} must have the main attention. Although in this study it was possible to know that its concentration levels are extremely high, further studies are necessary to assess what substances compose this PM and their toxicity.

References

1. ANVISA (2003) RE No. 9 16/01/03 Ministry of Health. National Agency of Sanitary Surveillance (ANVISA)
2. Edner H, Ragnarson P, Spaennare S, Svanberg S (1993) Differential optical absorption spectroscopy (DOAS) system for urban atmospheric pollution monitoring. Appl Optics 32:327–333
3. Fermer M, Svanberg P (1998) Cost-efficient techniques for urban- and background measurements of SO_2 and NO_2. Atmos Environ 32:1377–1381
4. Hangartner M (1990) Diffusive sampling as an alternative approach for developing countries. Swiss Federal Institute of Technology, Zürich, Switzerland. VDI Berichte Nr. 838, pp 515–526
5. Nedel AS, Cacavallo RC, Oyola P, Artaxo P (2005) Uso integrado de ferramentas para otimização e manejo de informação ambiental: Estudo de caso na cidade de São Paulo, Brasil. Ciência e Natura special volume December 2005.
6. Awad AHA (2002) Environmental study in subway metro stations in Cairo, Egypt. J Occup Health 44:112–118
7. CONAMA (1990) Environment National Committee. National committee of Environment. Standard Air Quality. RE Conama/No. 003 de 28/06/1990.
8. Johansson C, Johansson P (2003) Particulate matter in the underground of Stockholm. Atmos Environ 37:3–9
9. Branis M (2006) The contribution of ambient sources to particulate pollution in spaces and trains of the Prague underground transport system. Atmos Environ 40:348–356
10. Aarnio P, Yli-Tuomi T, Kousa A, Mäkelä T, Hirsikko A, Hämeri K, Räisänen M, Hillamo R, Koskentalo T, Jantunen M (2005) The concentrations and composition of and exposure to fine particles ($PM_{2.5}$) in the Helsinki subway system. Atmos Environ 39:5059–5066
11. Rosa ES, França FA (1995) Estimates of heat's transfer between soil and tunnels of subway systems. COBEM, Belo Horizonte, Brazil. Special vol.:1–4

Air quality nearby different typologies of motorways: Intercomparison and correlation

CS Martins, F Ferreira

Department of Sciences and Environment Engineering, College of Sciences and Technology, New University of Lisbon, Portugal

Abstract

A study has been carried out to evaluate the air quality impact under different typologies of motorways in Portugal. The monitoring of atmospheric pollutants with the use of indicative and continuous simultaneous measurement methods was the strategy to evaluate the conformity with National/European air quality legislation. This paper gathers results obtained in two major air quality-monitoring campaigns. Meteorological and traffic data were also collected. Benzene showed a decreasing linear dispersion behaviour for all the considered motorways and seems to be the best descriptor to evaluate the impact on air pollution through indicative measurements.

Introduction

The main purpose of this paper is to evaluate the road traffic impact on air quality under different typologies of motorways in Portugal: A1 North, A2 South, and A5 Cascais that links Lisbon to Porto, Faro, and Cascais, respectively (Fig. 1). The first is marked by a pendular traffic with two significant peaks along the day, but with more heavy-duty diesel traffic than the others. The second has a peculiar characteristic, meant for expressive traffic peaks associated with a seasonal increase to/from Algarve, a

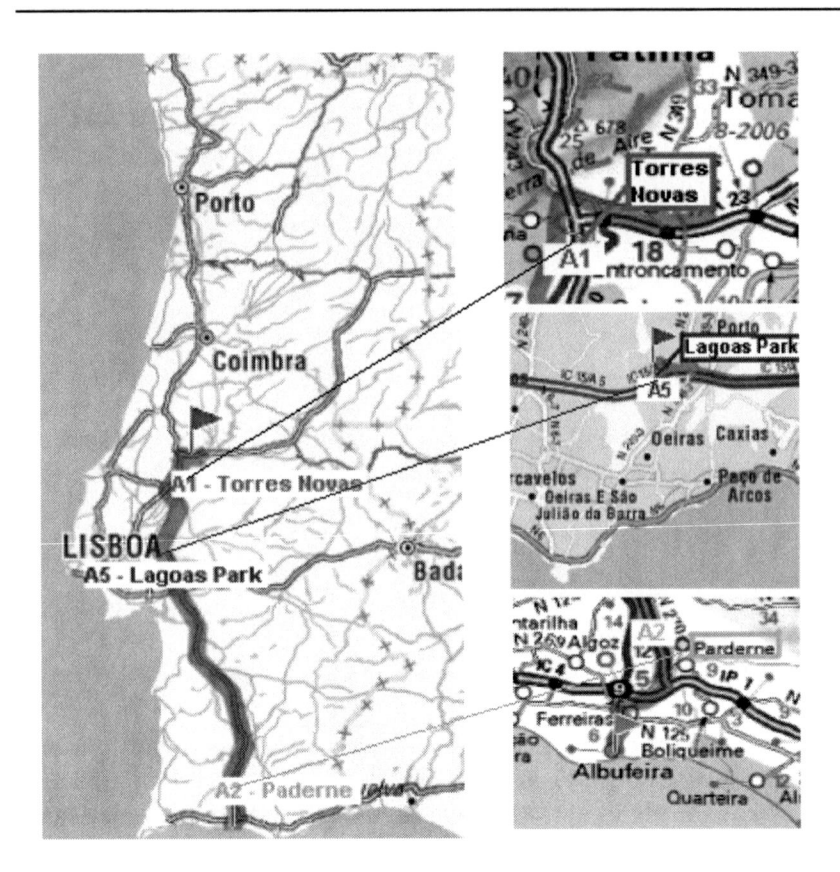

Fig. 1. Geographical points location map

much requested tourist area. The last one has similar characteristics to the first but with much higher average daily traffic [1].

The Directive 96/62/CE mainly aims to maintain and improve the air quality through the establishment of long-term quality objectives, common criteria, and methods, on the European Union (EU) level for the air quality assessment in order to get information on ambient air quality and ensure that it is made available to the public [2]. Therefore, the obtained results were also compared with the current legislation in order to evaluate the impact in the nearly population exposure due to the vehicles circulation in those motorways.

This paper gathers results obtained in two major air quality-monitoring campaigns, accomplished between 8 March and 10 May (first campaign), and between 28 June and 18 August (second campaign), 2004. Both campaigns were made during 3 weeks each one.

Three sampling points distributed along three main sections of each motorway were selected, namely: Santarém (S) – Torres Novas (TN) for A1, Oeiras (O) – Carcavelos I for A5, and Almodôvar (ALM) – São Bartolomeu de Messines (SBM) for A2.

Experimental – Method

The two campaigns implied the continuous measurement of gases (CO and NO_2) and particles (PM_{10}). Also, the weekly averaged concentrations of NO_2 and C_6H_6 with diffusion sampling tubes placed at some particular points, previously chosen [3]. Meteorological parameters were measured due to their influence in the dispersion of atmospheric pollutants, namely wind speed and wind direction, rainfall, temperature, and humidity [4]. A meteorological station was always placed in the same location where pollutants were monitored in the vicinity of the Snif-Airlab (mobile laboratory of the New University of Lisbon) and close to the places where the indicative measurement with diffusion tubes was taking place [5]. Figure 2 presents a scheme with the location of the measuring equipment.

The rank of diffusion tubes at different distances (10, 25, 50, 75, and 100 m) of the berm and at both sides of the motorway, aimed to provide a better understanding of the dispersion pattern of the pollutants along those areas. The chosen points had looked for to be representative of the population exposure in their vicinity to the pollutant concentrations as also conditioned by the needed two requests: the existence of plain land in both

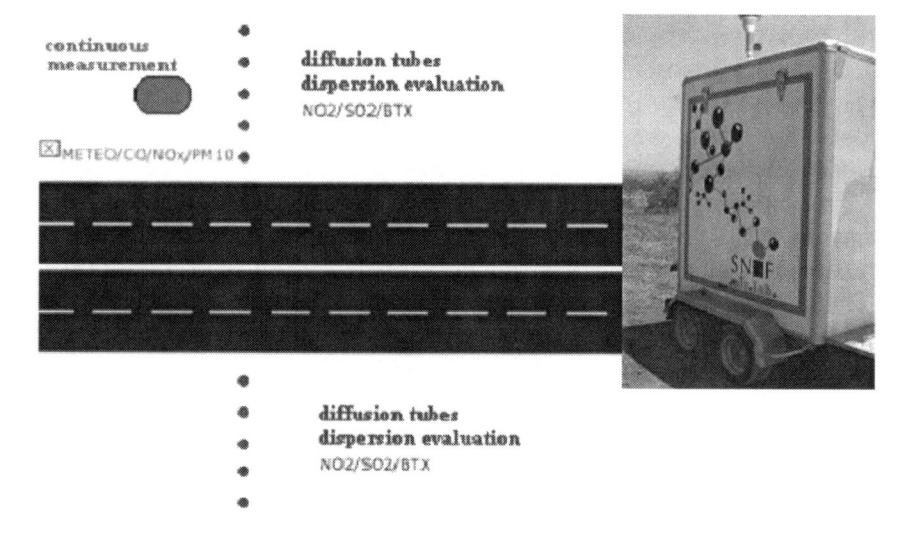

Fig. 2. Project of installation of the measuring equipment and Snif Lab

sides of the road, for the indicative measurement, and on the other hand, the existence of electric energy to the Snif analysers for the continuous measurement.

The monitoring of pollutant gases was made using passive Radiello diffusion tubes for sampling of NO_2 and BTX (benzene, toluene, and xylene). Toluene and xylene are pollutants not currently legislated in terms of air quality. This equipment allows collecting an integrated pollutant sample for one specific exposure period. The results correspond to the average concentration for the chosen period of exposure. The exposure time of the diffusers in the sampling points was 1 week.

The pollutants CO and NO_2 were measured using a chemiluminescence continuous measurement technique with integrated average concentrations of 15 min, which were compiled in hourly average values. For CO, taking into account the legislated 8-h base averaged value, this integrated period was used to allow a comparison with the air quality limit value foreseen in the Council Directives 1999/30/EC and 2000/69/CE.

The monitoring of particles was carried through continuous beta radiation equipment. The Directive 1999/30/EC indicates the method of reference for sampling and measuring of particulate matter (PM_{10}), which passes through a size-selective inlet with a 50% efficiency cut-off at 10 μm aerodynamic diameter) as the one described in the European norm EN 12341. This norm considers the gravimetric method as the reference method. However, this legislation allows the use of other methods if demonstrated that their results are equivalent to the ones of the reference method. Therefore, a correction factor of 1.18 to the PM_{10}-collected data (correction factor for traffic stations) was applied on the basis of intercomparison studies performed by the Portuguese Institute for the Environment. Averaged concentrations of 15 min were collected and integrated as hourly averaged values. Since the legislation establishes daily and annual limit values, the hourly data were then integrated on a daily basis to aim the respective comparison with the Directive 1999/30/EC.

The meteorological conditions influence the concentration and the distribution of the atmospheric pollutants [6]; therefore, the measurement of the meteorological parameters was performed in all sampled locations. A portable meteorological station measuring wind direction and speed, rainfall, temperature, and sun radiation, was used for this purpose. Those parameters were measured at height of 5 m above ground level and 25 m far motorway berm. The collected data are on hourly basis.

The simultaneous use of these two methods (indicative and continuous) is sufficient to verify the conformity with the legislation, as well as evaluating and comparing the correspondent obtain concentration levels in the different studied motorways. However, salient that for each method the obtained values are only representative of the sampling period.

Results and discussion

The results obtained were compared with the current EU legislation. The comparison was made with the legislative values for the year 2005 (limit value – LV and/or limit value + margin of tolerance – LV + MT), assuming different time basis (hourly, octo-hourly, or daily), depending on the pollutant. Also, it was made an analysis with the purpose of evaluating the road traffic impact on air quality under three different typologies of motorways associated with the distinct use of those.

Continuous measurement

Considering the analysis of the concentration behaviour of the several pollutants monitored, it was proceeded to the intercomparison of measured values, in hourly basis. Figure 3 shows the evolution of hourly concentrations of CO that occurred during the first campaign in the three different motorways and it shows that for the Lagoas Park (A5), those concentrations are always higher compared with the other evaluated motorways. Nevertheless

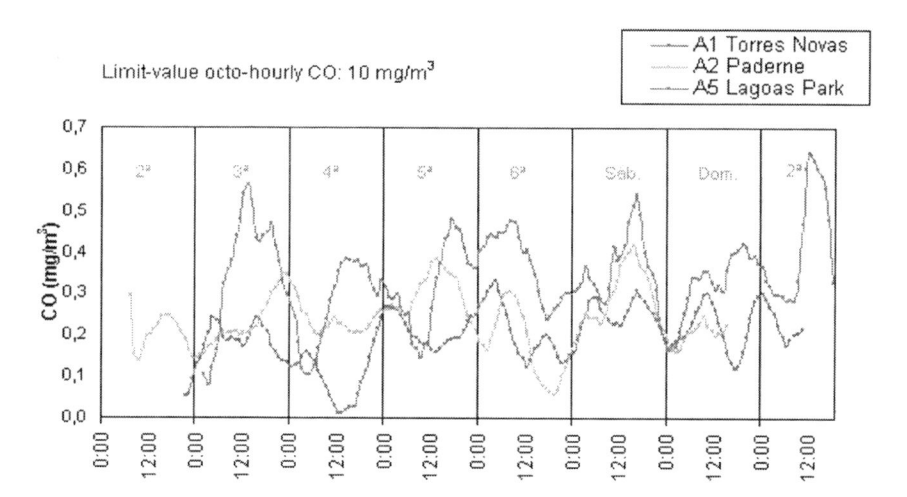

Fig. 3. CO hourly concentrations under different typologies of motorways (first campaign)

Torres Novas (A1) and Paderne (A2) seem to have more or less the same concentrations levels, however through distinct trend along the day. Relative to daily analysis some concentration peaks have a coincident trend, even so with different concentration levels; on the other hand occasionally such behaviour is completely divergent. Matching the first and second campaigns, the concentrations levels are similar once the trend lines were in analogous baseline.

This pollutant is well correlated with traffic flow and those concentration variations could be associated with the affluence of vehicles through this road section.

Those values can be considered as very low taking into account the correspondent threshold value of 10 mg m^{-3} (octo-hourly basis). Figure 4 shows the evolution of hourly concentrations of NO_2 occur during the first campaign in the three different motorways and like CO, this pollutant had similarly behaviour in what concerns to the highest measured values in Lagoas Park (A5) and also concerning with the general occurred peaks at the beginning or at the end of the day. This could be explained because of the correspondent increment on traffic marked notoriously by two flow peaks along the day. Differentiate levels of concentration along all weeks were observed for all the three motorways, but with more expression for

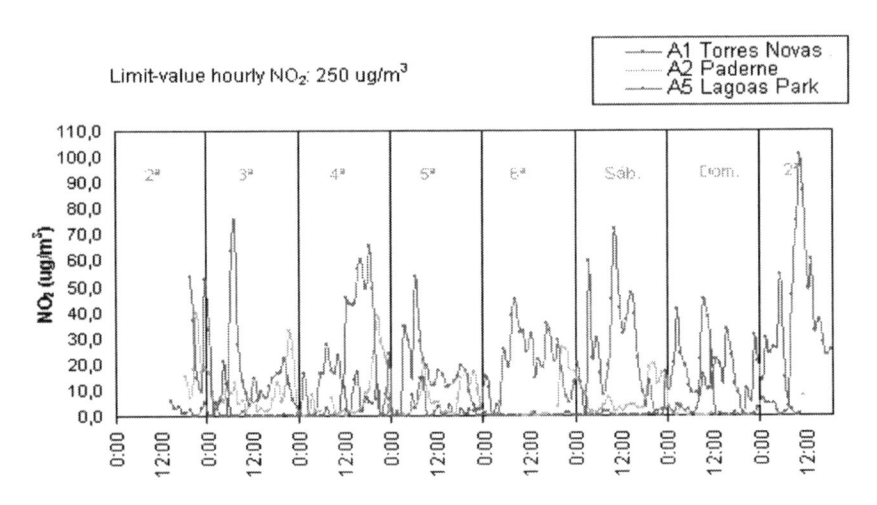

Fig. 4. NO_2 hourly concentrations under different typologies of motorways (first campaign)

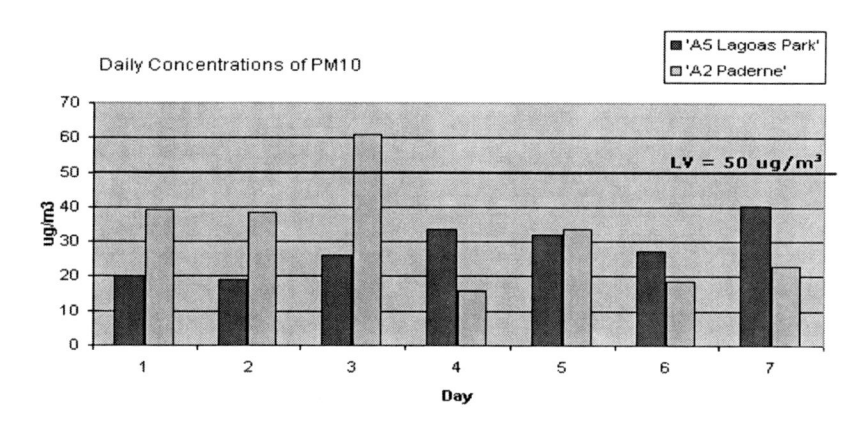

Fig. 5. PM$_{10}$ daily concentrations under different typologies of motorways (first campaign)

A5 than the others. This could be associated to the distinct traffic flow verified during both monitoring periods. Measured concentrations are under the hourly threshold value for NO$_2$ (250 μg m^{-3}).

Concerning PM$_{10}$, these had the opposite performance to the one verified for the gases, once the highest concentrations were observed in the Paderne (A2).

This comparison was made only for A2 and A5 motorways, due to technical problems with the analyser during the sampling campaign in Torres Novas (A1). No exceedances to the daily limit value were observed with an exception for the A2 (Fig. 5).

Indicative measurement

The next NO$_2$ weekly averaged concentrations are related with the indicative measurement made in Torres Novas (A1), Lagoas Park (A5), and Paderne (A2) (Fig. 6).

Those concentration levels were relatively different concerning the motorway, but always with low observed values. It is possible to verify, mainly for the first campaign, a decreasing linear pollutant dispersion behaviour with the distance to the road, especially until 50 m.

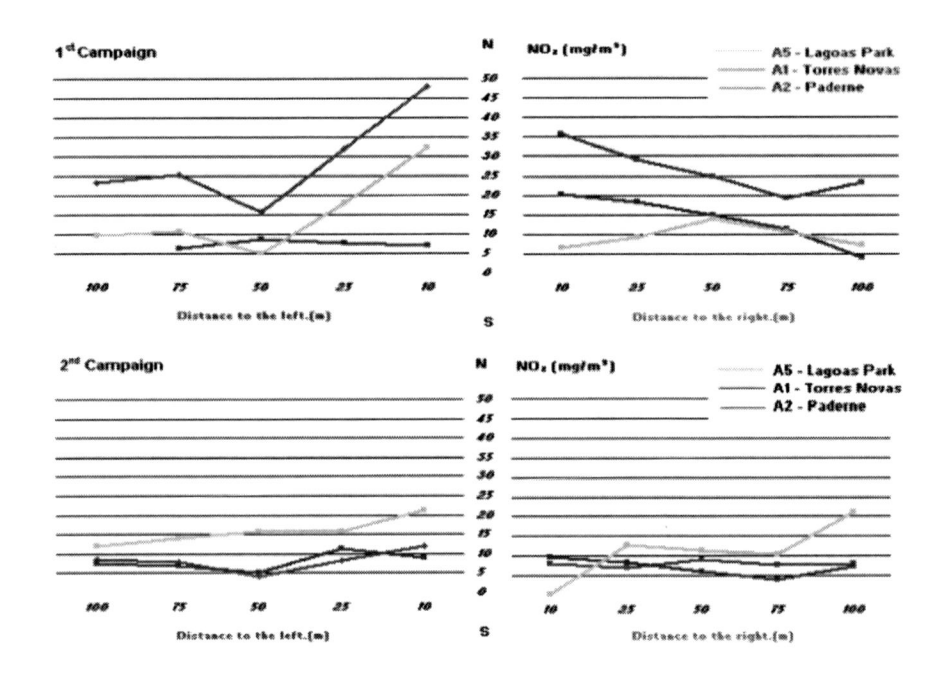

Fig. 6. NO$_2$ weekly concentrations under different typologies of motorways (first and second campaigns)

Concerning to the second campaign, the pollutant behaviour in what concerns to the A5 motorway was random and it was not possible to obtain linear pollutant dispersion. The wind variable conditions can explain not only the linear trend but also the error associated with this measurement methodology. NO$_2$ concentration values are higher at the first campaign than those from the second one. This is related with the different number of vehicles circulating in each campaign. Nevertheless, the measured values never exceeded the correspondent limit value. There is approximate linear dispersion behaviour of the registered concentration values with the distance from the road. The NO$_2$ concentration between campaigns, show a similar relation with the concentrations registered by the continuous measurement, mainly in the second campaign, which had met higher measured values close to twice more, relatively to those of first campaign. The weekly averaged concentrations of C$_6$H$_6$ could be observed in Figure 7.

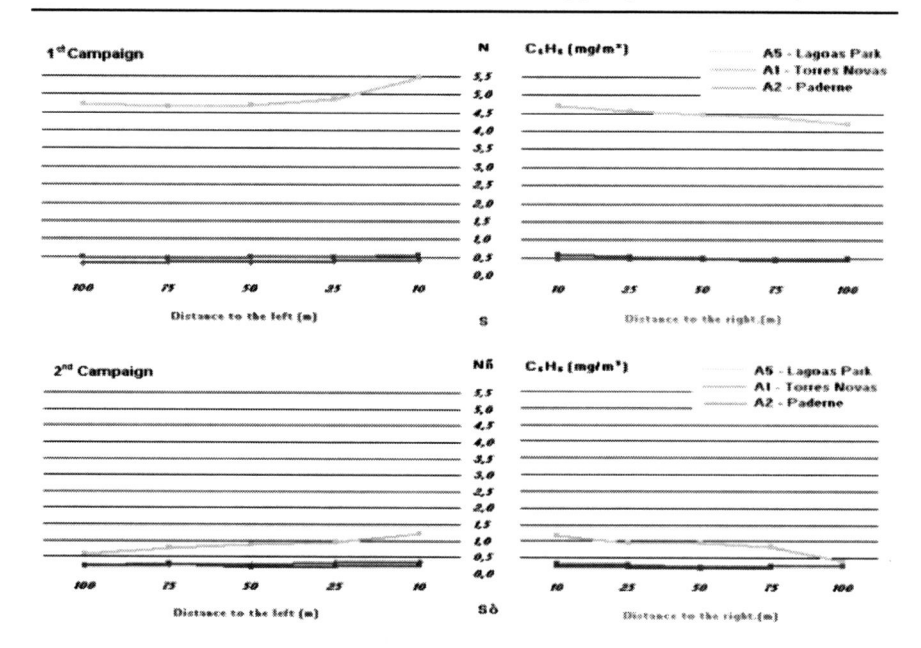

Fig. 7. C_6H_6 weekly concentrations under different typologies of motorways (first and second campaigns)

Benzene (C_6H_6) measured values are low and similar for the two campaigns, with an exception for Lagoas Park in the A5 motorway, once their values were much more higher than the observed in the other two locations at A1 and A2 motorways. This could be explained because of about 3600 vehicles of hourly average traffic in that motorway, against nearly 1500 vehicles at the A1 and A2 motorways. The comparison of the concentrations through both measurement campaigns also indicate that the value levels were slightly higher in the first campaign although, for the A5 location there was a very big difference between those concentrations levels, just because of the nearly half vehicles that circulated during the second campaign period.

It is possible to observe the presence of a clear dispersion pattern of benzene pollutant during some different distances from the motorways and for both sides of the road.

A direct comparison with the legislation is not possible because measured benzene concentrations correspond to weekly averages and not annual averages.

Conclusions

In general, the concentration values obtained along the two campaigns for all the pollutants were low, except for benzene.

Regarding the measurements of CO and NO_2, very low concentrations values were observed, but with some significant variations along the day, however, quite below the threshold value for the protection of human health.

The monitoring air quality data for A2 South only indicates potential problems for particulate atmospheric pollution. High concentration levels were measured close to the motorway and possibly associated to the PM_{10} direct emission through the vehicles exhaustion as to the resuspension from the vehicles circulation.

C_6H_6 values presented decreasing linear dispersion behaviour mainly until 50 m distance from the motorway. For NO_2 pollutant this occurrence is not so clear. The NO to NO_2 oxidation time could explain this, since it is flowed from the pipe. Benzene seems to be a better pollutant to characterize traffic pollution nearby motorways. Knowledge about the traffic counts for the campaign periods allowed a better explanation of the motorway impact in terms of air quality, through the relationship established with certain observed peaks. However, the notorious cyclic pattern described, under the three motorways, by two systematic peaks in a number of vehicles every day, although with distinct levels of hourly average traffic, it allows to see a non-direct correspondence to the measured concentrations trend. These non-linear behaviours can be explained once the distinct meteorological conditions occurred over those several monitored points as also by the punctual occurrences of calm situations during the respective sampling periods. Additionally, the winds did not assume a systematic dominant direction; therefore, those points were not always in such a position that allowed catching the pollutant mass source from the traffic roads.

The monitoring of atmospheric pollutant through two different measurement methods (indicative and continuous measurement) was sufficient to evaluate the conformity with the air quality legislation. The obtained results through several campaigns that had taken place in those motorways, allowed congregating some useful information that could be used by stakeholders in order to support the air quality assessment in motorways.

References

1. EIA (1998) Tecninvest, T 61204 – Study No. 1814, A2-South motorway, Section Almodôvar – Algarve Longitudinal Road
2. Castro HA, Gouveia N, Escamilla-Cejudo JA (2003) Methodological issues of the research on the health effects of air pollution. Rev Bras Epidemiol 6(2):São Paulo
3. Ferreira F, Tente H, Torres P, Mesquita S, Santos E, Boavida F, Jardim D (2001) Background levels of sulphur dioxide, nitrogen dioxide and ozone in Portugal. International conference measuring air pollutants by diffusive sampling
4. Borrego C, Tchepel O, Monteiro A, Miranda AI, Barros N (2002) Influence of traffic emissions estimation variability on urban air quality modeling. Water Air Soil Pollut: Focus 2:487–499
5. Miranda AI, Conceição M, Borrego C (1993) A contribution to the air quality impact assessment of a roadway in Lisbon. The air quality model APOLO. Sci Tot Environ 134:1–7
6. Rao KS, Gunter RL, White JR, Hosker RP (2002) Turbulence and dispersion modeling near highways. Atmos Environ 36/27:4337–4346

Assessment of air pollution in the vicinity of major alpine routes

P Suppan,[1] K Schäfer,[1] J Vergeiner,[2] S Emeis,[1] F Obleitner,[2] E Griesser[2]

[1]Institute for Meteorology and Climate Research (IMK-IFU), Forschungszentrum Karlsruhe, 82467 Garmisch-Partenkirchen, Germany
[2]Institute of Meteorology and Geophysics, University of Innsbruck, 6020 Innsbruck, Austria

Abstract

The Alpine environment as a sensitive region is heavily influenced by major traffic routes. Within the project ALPNAP (monitoring and minimization of traffic-induced noise and air pollution along major Alpine transport routes) the integration of advanced scientific methods of monitoring and simulating the air quality distribution and noise propagation is a key aspect for the impact analysis on human health. A methodology for measurement strategies and model simulations for air pollutants will be demonstrated for the Brenner traverse. First, results of a field measurement campaign give detailed insights of the complexity of the atmospheric conditions and the distribution of air pollutants in the Inn valley.

Introduction

Road traffic emissions combined with unfavourable meteorological conditions like calm winds and low inversion layers have severe impacts on the natural environment as well as on the socio-economic development of regions in the vicinity of major Alpine traffic routes. Increasing amounts of heavy-duty vehicles and passenger cars force the increase of traffic emissions, which continually violate the EU air quality thresholds of, e.g., PM_{10} and/or NO_2 not only in this region.

ALPNAP is an ongoing European Union (EU) research project focusing on corresponding effects along several major transit routes across the European Alps. Assisting regional authorities with appropriate output, a unique cooperation of scientists within the Alpine region was set-up to better assess and predict the spatial and temporal distribution of air pollution and noise propagation close to major Alpine traverses.

In a first step, a measurement campaign in the lower Inn valley was designed to determine cross-valley air pollution and meteorological information, as well as vertical profiles to determine flow regimes (valley, slope winds), mixing layer height, stability in the boundary layer, and emission sources at specific locations. Covering the major part of the winter season, this data will be used for enhanced analysis as well as for the set-up and validation of corresponding models merging the wealth of different remote sensing and in situ measurements. In a second step, air quality simulations will be carried out, in order to receive detailed three-dimensional (3D) information of the distribution of air quality parameters.

Methodology

Measurements

In the frame work of ALPNAP, a field campaign in the Inn valley was performed between November 2005 and February 2006. The campaign was based on existing background information and was focused on air pollution and noise strain in the lower Inn valley. The target area was centred on Schwaz/Vomp (47°20′N, 11°41′E, 540 m a.s.l.), where exceedances of NO_2 and PM_{10} are frequently recorded.

From the viewpoint of air pollution induced by inner Alpine traffic, the most important goals for the field campaign are outlined below:

- Study the spatial and temporal variation and distribution of air pollution concentrations within a cross section of the valley induced by inner Alpine traffic.
- Determine the dependence of the observed air pollution patterns on meteorological parameters like wind, stability or mixing height, etc.

- Evaluate the spatial representativeness of routine air pollution networks in the area as a function of distance to the main sources and altitude.
- Evaluate the best use of an existing slope temperature profile for mixing height and stability analyses.
- Use the field data for the set-up and validation of analysis of dispersion models used in the area of investigation.

Based on the existing routine air pollution measurements, a two-way approach was introduced for the field phase. Permanent air pollution and meteorological instrumentation including a SODAR (vertical profiles of acoustic backscatter intensity and Doppler signal, as well as derived wind and turbulence parameters and mixing layer heights), a ceilometer (vertical profiles of optical backscatter intensity and derived mixing layer heights), an open-path differential optical absorption spectroscopy (DOAS), a long-path air pollution information at different paths), three automated weather stations (AWS), and 10 passive samplers for NO_2 provide consistent data sets of the temporal development of the meteorology and air pollutants throughout the winter. The location of the instruments as well as some images of the measurement devices are shown in Figure 1. During high air

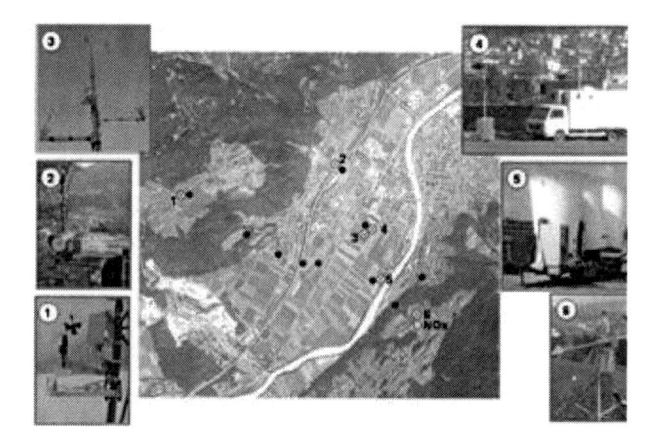

Fig. 1. Target area Schwaz/Vomp and instrumentation. Note the river Inn and the A12 highway (closer to the northern slopes). Images of the measurement devices clockwise from bottom left: (1) AWS Vomperberg on power pylon, (2) DOAS receiver or emitter unit, (3) 10 m AWS Schwaz, (4) measurement van with PM_{10} device, (5) SODAR, and (6) AWS Arzberg. At location 5 a ceilometer was set up additionally. Tiny circles denote NO_2 passive sampler locations

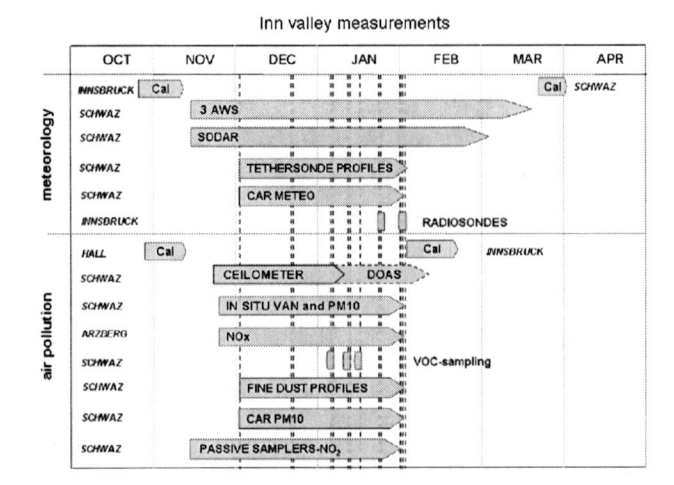

Fig. 2. Operation periods of the different measurement instruments. Cal denotes calibration phase. The dashed lines represent additional measurements, where tethersonde profiles, car traverses, and VOC sampling were performed

pollution episodes, additional measurements were performed. These include measurements with a tethered balloon to determine the vertical structure of dust particles, temperature, humidity, and wind. Mobile car traverses yielded PM_{10} and meteorological information and sampling of volatile organic compounds (VOCs) at selected spots. The operation periods of the different measurement instruments are given in Figure 2.

Modelling

Additional to air quality measurements also the simulation of the horizontal and vertical distribution of air pollutants on a high temporal and spatial resolution is a key aspect. In this study it should be demonstrated how to utilize advanced methods to simulate the concentrations of airborne pollutants, such as NO, NO_2, different VOCs, and PM, as a function of the emissions within such a complex topography. Furthermore, the evaluation of the consequences of specific emission reduction measures (e.g., traffic ban on certain vehicles as a function of time of day and year, future development of traffic) will be considered.

Therefore, a model hierarchy from the microscale to mesoscale has to be set up to calculate traffic emissions, to simulate meteorological and air pollution parameters for supporting further studies on the health impact

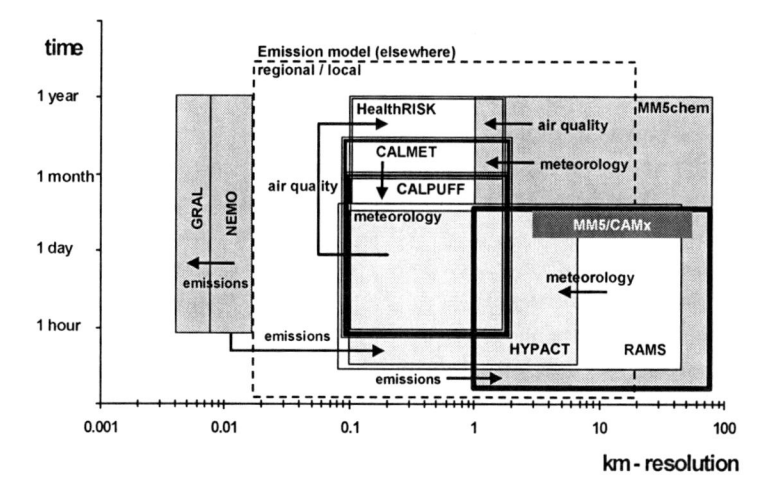

Fig. 3. Model chain and modelling tools used by the ALPNAP project partners and their possible spatial and temporal field of application

in an Alpine environment. In Figure 3, the hierarchy of the models starting from the traffic emission model system NEMO/GRAL, to forecast tools like like RAMS/HYPACT to the online-coupled mesoscale Chemistry-Transport Model MCCM with a spatial resolution of less than 100 m and up to 100 km, respectively, is shown. The simulation period varies between hours and more than 1 year for the calculation of annual mean thresholds of NO_2, O_3, and PM_{10} according to the European air quality directives for the whole Brenner traverse from Rosenheim/Germany to Verona/Italy.

Results

Generally, the NO concentrations are clearly dominated by the traffic volume. Thus, the highest values are found at the highway (DOAS, 570 m a.s.l.) followed by the valley floor in 800 m distance to the highway (Schwaz, 540 m a.s.l.) and at the southern slope (Arzberg, 720 m a.s.l.). The exceedance at the valley floor as compared to the slope station is basically related to a stable layering of the valley atmosphere more or less during all the time. Outstanding high pollution episodes were found during: 20–24 December 2005, 09–22 January and 25 January–02 February 2006, when the detailed structure of the valley atmosphere was captured by the additional intensive measurements periods.

Moreover, the data indicate that the ratio of concentrations near the highway to the background is systematically higher for NO_2 and NO by more than a factor of 2. On a local scale the data allow, e.g., to demonstrate the impact of nearby construction work imposing on the CO, NO, and NO$_x$ measurements. As a well-known feature on the other hand, a dominant "new year" peak in PM$_{10}$ concentration at midnight (31 December 2005/01 January 2006) was found.

These features are exemplified for the 10-day period from 6 to 15 January 2006, which was characterized by stable conditions within the valley boundary layer. A very regular and undisturbed weather pattern is revealed by Figure 4. The almost undisturbed cycle of short wave radiation indicates the prevailing clear sky conditions. Very cold (dense) air at the valley bottom leads to inversion conditions most of the time (temperature gradient is positive, warmer air at AWS Arzberg 180 m a.g.l. than at AWS Schwaz), except for almost isothermal conditions in the early afternoon. The according wind pattern shows a typical winter-time valley regime. Outflow from south-west, usually stronger in the second half of the night (more cold air has formed in the upper Inn valley), except for some early afternoon periods with very weak valley inflow.

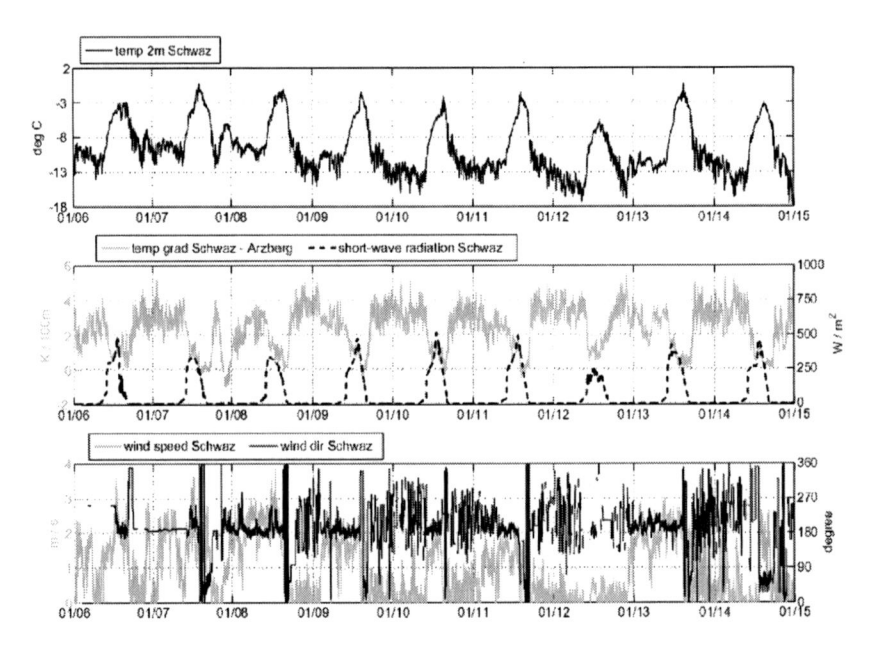

Fig. 4. Weather conditions during the period from 6 to 14 Jan 2006. 2 m temperature at AWS Schwaz in °C (*top*), Temperature gradient between AWS Schwaz and Arzberg in K 100 m^{-1} and short wave radiation at AWS Schwaz in W m^{-2} (*middle*), and wind speed in m s^{-1} and direction in deg at AWS Schwaz (*bottom*)

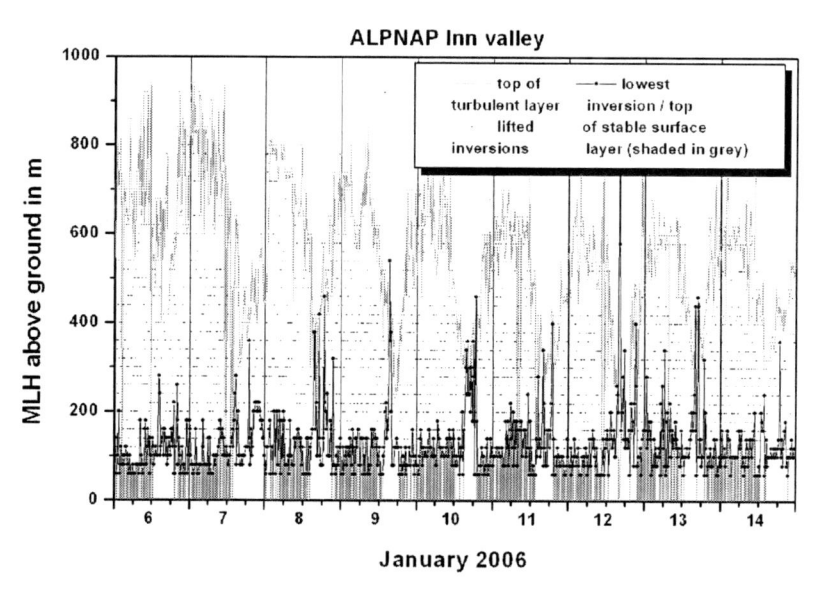

Fig. 5. Diurnal variation of the vertical thermal structure of the valley boundary layer from SODAR acoustic backscatter and vertical velocity variance observations. Upper grey line shows top of turbulent layer. Small bars indicate multiple lifted inversions, grey shading from below marks a stable surface layer. Small bullets signify the mixing layer height defined as the top of the stable surface layer or – if absent – the height of the lowest lifted inversion

The temporal variations of air pollution concentration (CO, PM_{10}, and NO_2) at the valley ground are clearly dominated by the mentioned weather conditions and emissions. As for the NO_2 and CO emissions, the main source is road traffic at the highway. The NO_2 and CO maxima in the morning and in the evening correspond to the local traffic maxima. In the early afternoon mixing height rises (see thin lower line in Fig. 5) and the concentrations decrease (Fig. 6), whereas the ozone concentrations increase. During midnight the CO concentrations show a maximum and PM_{10} and NO_2 concentrations show a minimum. As outlined in Table 1, the daily mean values of PM_{10} and NO_2 are increasing within this 10-day period. This effect is attributed to the accumulation of pollutants due to a stable and high pressure-dominated Grosswetterlage with an inefficient air mass exchange.

In Arzberg, the NO_2 concentrations are generally lower than at the valley bottom. Only in the beginning of the period during well-mixed conditions in the early afternoon the concentrations are on the same level as at the valley ground.

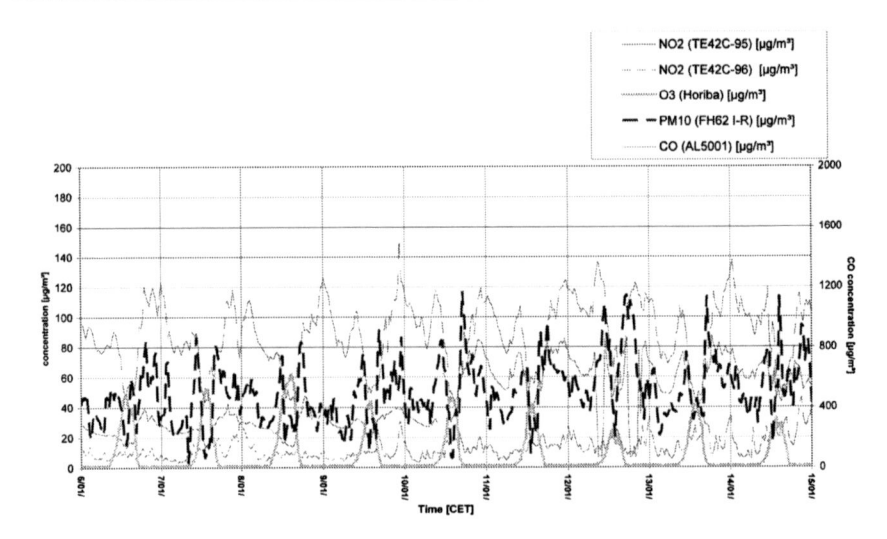

Fig. 6. Time series of NO_2, O_3, PM_{10}, and CO at the station Schwaz and NO_2 at Arzberg (instrument TE42C-96)

Table 1. Daily mean concentrations of NO_2 in the valley (Schwaz), NO_2 on the southern slope (Arzberg), and PM_{10} at Schwaz between 6 and 15 Jan 2006

Date	NO_2-Schwaz ($\mu g\ m^{-3}$)	NO_2-Arzberg ($\mu\ m^{-3}$)	PM_{10}-Schwaz ($\mu g\ m^{-3}$)
06 Jan	24.9	8.8	43.8
07 Jan	27.1	11.6	42.7
08 Jan	28.2	10.3	42.2
09 Jan	30.7	10.4	44.8
10 Jan	48.7	11.7	52.3
11 Jan	63.9	12.9	51.7
12 Jan	54.1	15.8	65.8
13 Jan	61.7	16.6	52.3
14 Jan	61.1	20.5	57.4
15 Jan	65.7	28.3	80.2

The typical evolution of the polluted boundary layer is shown for 13 January 2006. The acoustic remote sensing technique of the SODAR detects vertical stratification (Fig. 7a) and vertical wind profiles. The optical backscatter from a ceilometer is an indication of the aerosol density (Fig. 7b). An inversion in the lowest 100 m persists until midday, when solar

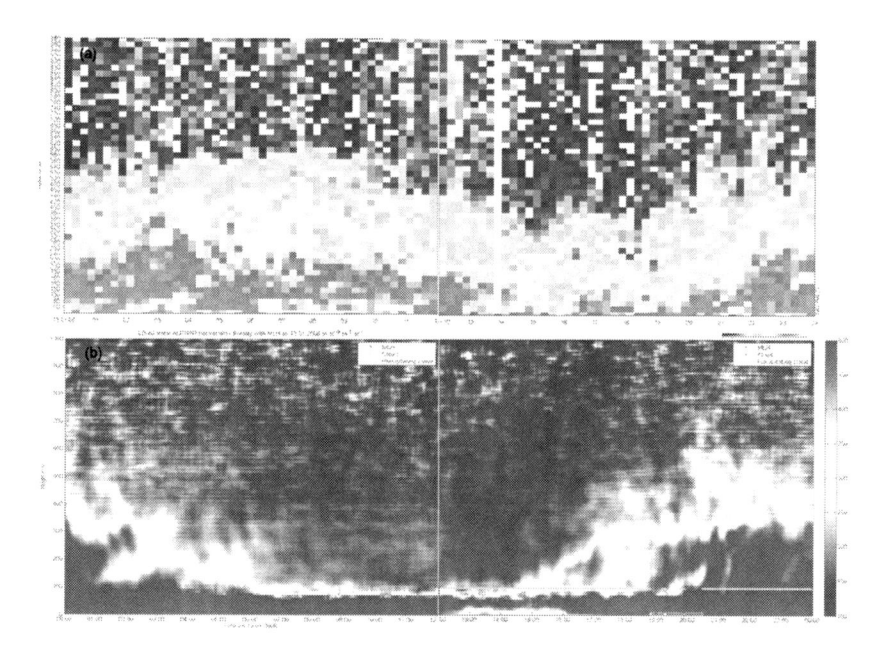

Assessment of air pollution in the vicinity of major alpine routes 211

Fig. 7. Diurnal variation of vertical structures of the valley boundary layer on 13 January 2006. (a) Acoustic backscatter intensity from SODAR observations. The darker grey shading near the surface signifies the most stable thermal stratification. (b) Optical backscatter intensity from ceilometer observations. The whitish and grey shades indicate high aerosol concentrations; the black areas near the surface signify moist mist or even fog

radiation leads to a short break up in the early afternoon before reforming after sunset. The ceilometer reveals that the polluted layer is restricted to about 80 m a.g.l. before noon, when the intensity decreases from the bottom while still present below the sharp inversion layer.

A more detailed picture at a given time can be seen from the tethersonde ascents depicted in Figure 8. The left graph shows the vertical structure around 1030 CET. The beginning of the marked inversion around 90 m a.g.l. marks the top of the polluted layer, with PM_{10} values gradually decreasing with height from 70 µg m^{-3} to background values close to zero. After 2 h, radiation has heated up the surface layer, leaving a more pronounced inversion layer at 140 m a.g.l. Consequently, the dust particles where vertically redistributed leading to more uniform values of around 50 µg m^{-3}. Hence, the surface exposure is reduced due to a deeper and more uniformly distributed aerosol layer.

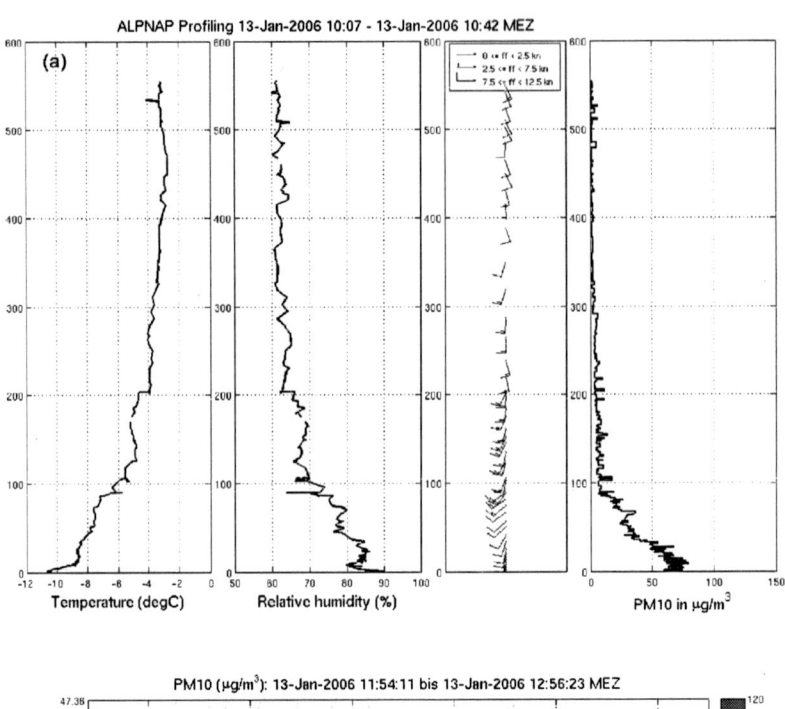

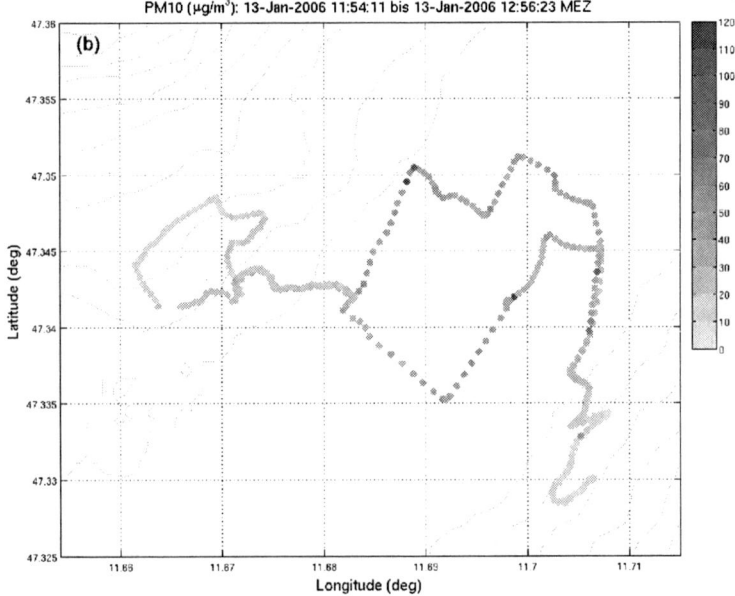

Fig. 8. Tethersonde ascents on 13 Jan 2006 at (a) 1007–1042 CET, (b) 1154–1234 CET The four panels show temperature in °C, relative humidity in %, wind vanes and PM$_{10}$ in μg m^{-3}

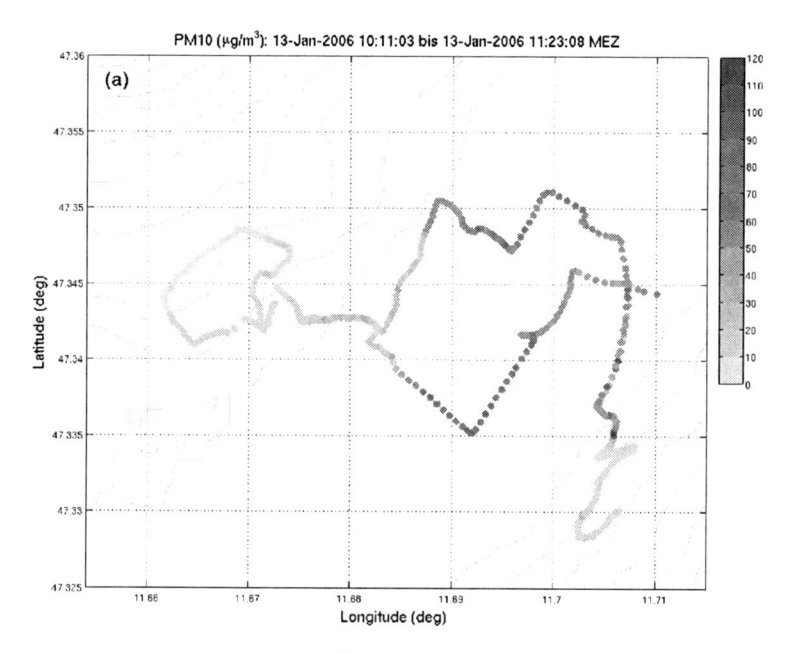

[Figure: PM10 (µg/m³): 13-Jan-2006 10:11:03 bis 13-Jan-2006 11:23:08 MEZ, showing latitude vs longitude plot with concentration scale 0-120]

Fig. 9. PM_{10} concentration (µg m^{-3}) on 13 Jan 2006 from car measurements between 1011 and 1123 CET (*left*) and between 1154 and 1256 CET (*right*) including GLOBE30 topography. Note the decrease in concentration on the valley bottom towards noon due to the better mixing conditions

The mobile measurements of the PM_{10} concentrations give detailed information of the distribution at the valley bottom and on both slopes up to 200 m a.g.l. (Fig. 9). In the morning PM_{10} levels at the valley bottom are close to 80 µg m^{-3} and at the slopes close to 10 µg m^{-3}. Towards noon, as already seen in the tethersonde graph, the concentration in the valley decreases due to lower emissions. Moreover, the lowest layers show a more evenly distribution, while in levels above the PM_{10} concentrations are nearly unchanged.

Conclusions

A methodology for the assessment of the air quality within Alpine valleys was introduced. Both, measurements and simulations will be used to extend our knowledge about the distribution of air pollutants in the inhomogeneous terrain of a valley and the impact on human health.

In a first review a brief description of the air quality situation within the cross section of the Inn valley near Innsbruck was summarized. First results of the measurements show the influence of weather conditions and traffic emissions to the overall air quality situation within the valley. Their qualitative interpretation confirms that stable meteorological conditions without relevant vertical and horizontal exchange are the main reasons for the observed NO_2 and PM_{10} threshold exceedances during winter 2006. However, the detailed evolution indicates a number of complex interactions linking, e.g., small-scale topography, diurnal variation of sources (traffic volume), and the relevant meteorological parameters (solar insolation, inversion strength, wind, and synoptic conditions), respectively.

However, even detailed measurements like these cannot reflect the full temporal and spatial variability of the complex flow regimes or the horizontal and vertical distribution of chemical parameters in alpine valleys. Future numerical simulations will contribute there and the corresponding validation process will greatly benefit from the available data. Thus, setting up different models in a most sophisticated way will enable for process oriented studies, model intercomparisons and impact studies.

Further information about the ongoing work can be obtained at the ALPNAP website http://www.alpnap.org/.

Acknowledgements

The project ALPNAP is implemented through financial assistance from funds of the European Community Initiative Programme "Interreg III B Alpine Space". We like to thank Andreas Krismer, Herbert Hoffmann, and Carsten Jahn for their contributions to this work. This project has received European Regional Development Funding through the Interreg III B community Initiative.

The relative impact of automobile catalysts and Russian smelters on PGE deposition in Greenland

S Rauch, J Knutsson

Water Environment Technology, Chalmers University of Technology, 412 96 Göteborg, Sweden

Abstract

Elevated platinum group elements (PGEs) concentrations have been reported at Summit in Central Greenland. Automobile catalysts and Russian smelters have been suggested as potential sources of PGEs at Summit, but the relative importance of these sources still needs to be determined. Here, we provide a source characterization based on relative PGE concentrations and air mass modelling. This study suggests that both catalysts and smelters are potential sources of PGE in Greenland, but that the contribution of smelters is relatively limited. The larger contribution from catalysts is probably the result of higher emission rates related to the number of vehicle equipped with a catalyst. The presence of automobile catalysts-derived PGE in Greenland suggests that automobile catalysts are resulting in widespread contamination.

Introduction

Platinum group elements (PGEs) are among the least abundant elements in the earth's crust. However, emission of Pt, Pd, and Rh from automobile catalysts [1] is resulting in elevated concentrations of these elements in the urban and roadside environment [2–4]. Whereas emitted PGE were

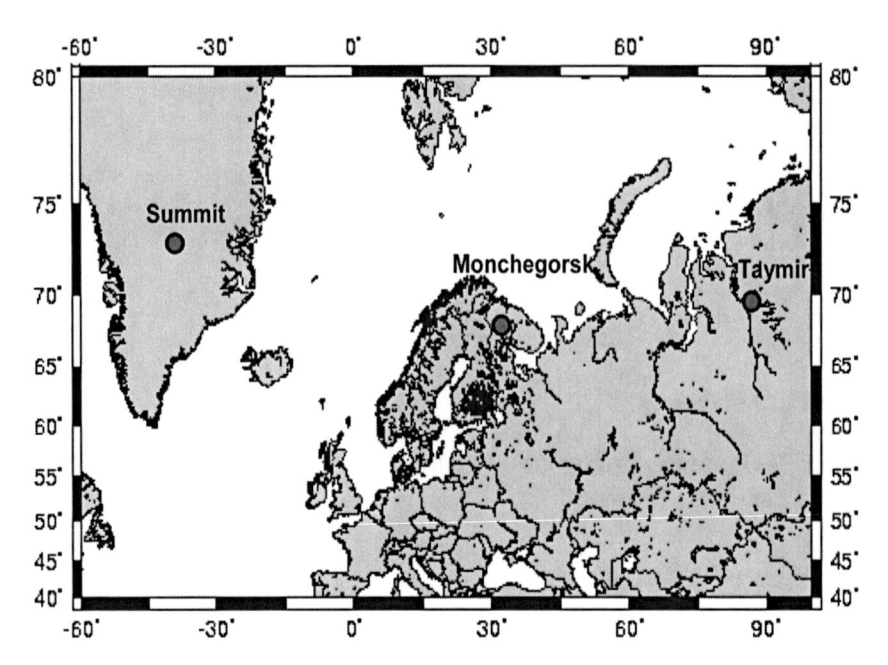

Fig. 1. Location of the Summit site in Central Greenland and Russian smelters (Moncehgorsk and Taymir) for which dispersion has been modelled

believed to remain in the roadside environment, the recent finding of increasing PGE concentrations at Summit in Central Greenland [5,6] has raised concern over the possible widespread dispersion of these metals. Because of the rapid increase, potential natural sources are unlikely to be responsible for the observed increase. Until now, two possible anthropogenic sources of PGE in Greenland have been discussed, i.e., automobile catalysts [5] and Russian smelters [7]. Because fine particles have relatively long atmospheric residence time and can be transported over long distances, the occurrence of PGE fine particles in automobile exhaust might explain elevated PGE concentrations in Greenland [8]. Russian smelters are important PGE emitters [9,10] and their location above the Arctic Circle (Fig. 1) makes them a potential source of PGE in Greenland.

At present, there is no agreement on the relative contribution from these two sources and the source of PGE at Summit still needs to be determined. To determine the relative importance of these sources to PGE deposition in Greenland, we have used relative PGE composition in Greenland snow and an atmospheric circulation model in both backward and forward modes.

Characterization of PGE deposition in Greenland

Elevated PGE concentrations were observed at Summit in Central Greenland in the mid-1990s [5]. Platinum, Pd, and Rh deposition rates in Greenland in the mid-1990s were estimated to be approximately 0.08 ± 0.04, 0.2 ± 0.1, and 0.011 ± 0.005 µg m^{-2} year^{-1} [8], respectively, based on PGE concentrations [5,6] and snow accumulation rates at Summit.

Comparison of relative Pt, Pd, and Rh concentrations at Summit and in materials characteristic of automobile emissions and Russian PGE production provide information relevant to PGE source identification at Summit. The plot of Pt/Pd versus Pt/Rh (Fig. 2) reveals differences in the relative composition of catalyst-derived materials and smelter-derived materials with an overlap. Deposition at Summit has a composition similar to that of urban samples, but has a low Pt/Pd resulting in the deposition at Summit also partially belonging to the overlap zone. Although the comparison indicates that both automobile catalysts and smelters are viable sources of PGE at Summit, automobile catalysts appear to be a more important source (Fig. 2).

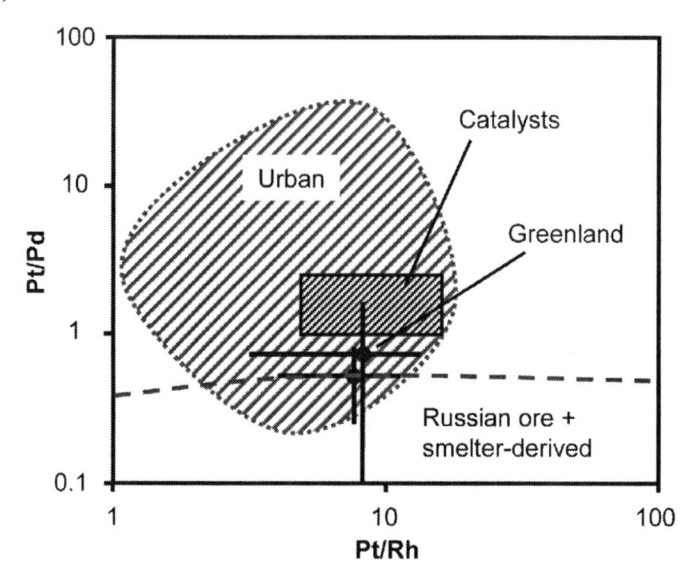

Fig. 2. Relative PGE composition in recent Greenland ice compared with the composition of urban samples, catalysts, Russian ores, and samples in the vicinity of Russian smelters (Adapted from [8].)

Modelling

The hybrid single-particle Lagrangian integrated trajectory (HYSPLIT) model was developed by the Air Resources Laboratory (ARL) (National Oceanographic and Atmospheric Administration, USA) and the Australian Bureau of Meteorology to assist in determining the impact of catastrophic pollution events and planning an effective response. The model is a combination of Lagrangian and Eulerian modelling approaches, i.e., advection and diffusion are calculated separately (Lagrangian framework) and concentrations are determined on a fixed grid (Eulerian approach) [11,12].

Two independent modelling approaches were considered to determine the source of PGE at Summit. Because automobile catalysts are a diffuse source of PGE, it is difficult to model the fate of emitted particles. The possibility for catalyst-emitted particles to reach Summit was determined using back trajectories with Summit as an endpoint. In contrast, smelters are localized emission sources and it is therefore possible to model the dispersion of emitted particles. The dispersion of PGE emitted from two Russian smelters was modelled in the forward mode.

Back trajectory modelling

The origin of the air mass over an area can be determined using atmospheric models and it is possible to determine the source of air pollutants from long-range transport. The back trajectory approach provides an interesting approach for the study of PGE emitted from catalysts. If trajectories ending at Summit pass over urban areas, there is a good chance that fine particles can be transported from the urban areas to summit. In contrast, if trajectories pass over smelting areas in Russia, there is a high probability that PGE are transported from the smelters to Greenland. Although this approach is not quantitative, it can provide an indication of the source of PGE at Summit. Ten-day back trajectory calculations were performed for a 1-year period. Calculations were based on vertical velocity and were performed using the National Centre for Environmental Prediction (NCEP, USA Reanalysis weather data. Trajectories were found to originate or pass over both smelters and urban areas (Fig. 3) with no clear tendency for a specific source area.

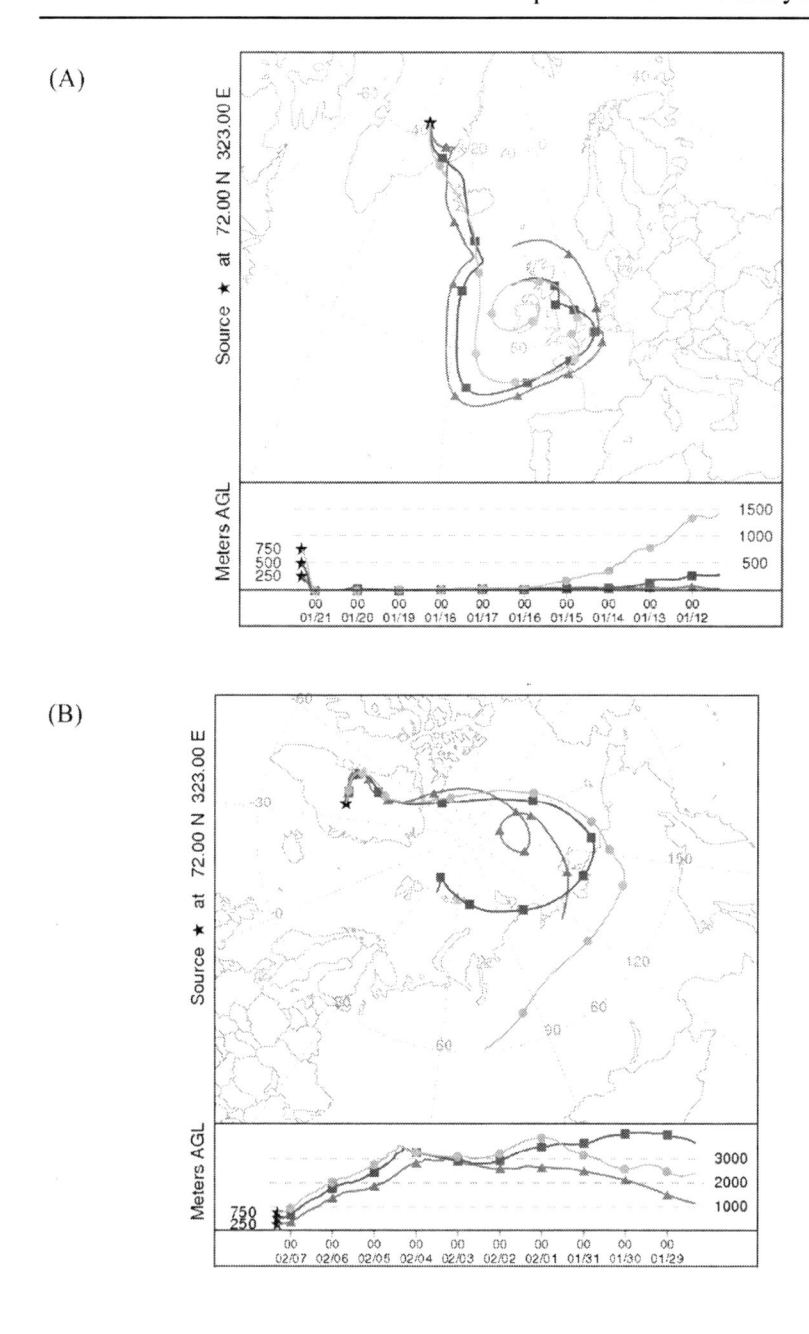

Fig. 3. Examples of 10-day atmospheric trajectories ending at Summit at three different altitudes (triangles: 250 m; squares: 500 m; circles: 750 m). (A) Trajectory passing over urban areas; (B) trajectories passing over northern Russia

Dispersion modelling

In addition to back trajectory calculations, HYSPLIT enables the modelling of atmospheric dispersion from a point source and can predict deposition at a specific location. Two smelters (Fig. 1) were selected as PGE emitters. Simulation was performed for a 100-day period using reanalysis weather data. Because PGE emission rates from Russian smelters and emitted particle sizes are not available, the following assumptions were made:

- Two emission scenarios, i.e., low and high emission, were selected for input into the model. The scenarios are based on production numbers, emission factors for Cu and Pb.
- Twenty-five percent of emitted particles (particle mass) were assumed to have a diameter of 0.5–5 μm, the remaining particles were assumed to have a diameter of 15 μm.

Validation of the model and input parameters was performed by comparing the deposition estimates with measured deposition rates at a location approximately 8 km from the Monchegorsk smelter [13]. Estimates from the model with the high emission scenario are within 20% of the measured deposition, indicating that the input parameters and the model accurately describe the dispersion of PGE from the smelter.

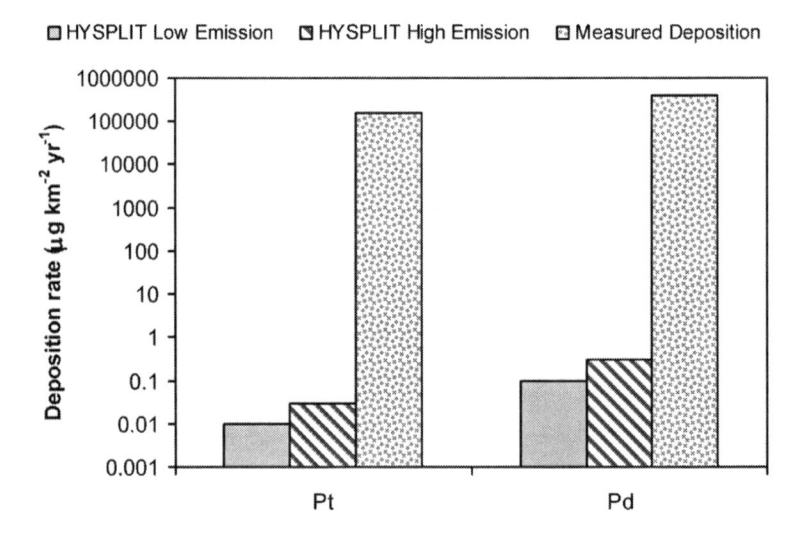

Fig. 4. Comparison of deposition rates at the Summit. HYSPLIT estimates include emission from both Monchegorsk and Taymir. Measured deposition estimated from [8] estimated using reported deposition and snow accumulation rate

The results show that the deposition of Pt and Pd from PGE emission by the two smelters is in the order of 0.01–0.1 μg Pt m^{-2} year^{-1} and 0.1–1 μg Pd m^{-2} year^{-1} (Fig 4). Measured deposition is at least 100,000 times larger than deposition estimated using HYSPLIT. Although the modelling results are expected to be affected by errors, the magnitude of the difference with measured PGE deposition at Summit demonstrates that the contribution of Russian smelters is limited in comparison with other sources.

Because back trajectories did not point to a specific source area and the contribution from smelters is limited, deposition at Summit is attributed to a higher total emission rate for catalysts than for smelters, possibly due to the large number of vehicles equipped with a catalyst.

Conclusions

The study presented here shows that smelters are a relatively minor contributor to PGE deposition in Greenland. Automobile catalysts is the only other known atmospheric source of PGE and catalysts are therefore expected to be responsible for the observed increase in PGE deposition in Greenland, possibly due to the relatively large number of vehicles equipped with a catalyst. This study supports a widespread contamination by catalysts. Contamination is expected to continue in the near future owing to the growing number of vehicles worldwide and the implementation of catalyst-requiring legislation in developing countries.

Acknowledgements

The authors gratefully acknowledge collaboration and discussion on PGE dispersion within the AGS-funded project on PGE dispersion. The authors also gratefully acknowledge the NOAA ARL for the provision of the HYSPLIT transport and dispersion model and READY website used in this publication (http://www.arl.noaa.gov/ready.html).

References

1. Moldovan M, Palacios MA, Gomez MM, Morrison G, Rauch S, McLeod C, Ma R, Caroli S, Alimonti A, Petrucci F, Bocca B, Schramel P, Zischka M, Pettersson C, Wass U, Luna M, Saenz JC, Santamaria J (2002) Environmental risk of particulate and soluble platinum group elements released from gasoline and diesel engine catalytic converters. Sci Tot Environ 296:199–208

2. Gomez B, Palacios MA, Gomez M, Morrison GM, Rauch S, McLeod C, Ma R, Caroli S, Alimonti A, Schramel P, Zischka M, Pettersson C, Wass U (2002) Platinum, palladium and rhodium contamination in airborne particulate matter and road dust of European cities. Risk assessment evaluation. Sci Tot Environ 299:1–19
3. Rauch S, Hemond HF, Peucker-Ehrenbrink B (2004) Recent changes in platinum group element concentrations and osmium isotopic composition in sediments from an urban lake. Environ Sci Technol 38:396–402
4. Tuit CB, Ravizza GE, Bothner MH (2000) Anthropogenic platinum and palladium in the sediments of Boston harbor. Environ Sci Technol 34:927–932
5. Barbante C, Veysseyre A, Ferrari C, Van de Velde K, Morel C, Capodaglio G, Cescon P, Scarponi G, Boutron C (2001) Greenland snow evidence of large scale atmospheric contamination for platinum, palladium, and rhodium. Environ Sci Technol 35:835–839
6. Barbante C, Boutron C, Morel C, Ferrari C, Jaffrezo JL, Cozzi G, Gaspari V, Cescon P (2003) Seasonal variations of heavy metals in central Greenland snow deposited from 1991 to 1995. J Environ Monitor 5:328–335
7. Renner R (2001) Catalyzing pollution. Environ Sci Technol 35:138A–139A
8. Rauch S, Hemond HF, Barbante C, Owari M, Morrison GM, Peucker-Ehrenbrink B, Wass U (2005) Importance of automobile catalysts for the dispersion of platinum group elements in the Northern Hemisphere. Environ Sci Technol 37:8156–8162
9. Niskavaara H, Kontas E, Reimann C (2004) Regional distribution and sources of Au, Pd and Pt in moss and O-, B- and C-horizon podzol samples in the European Arctic. Geochem Explor Environ Anal 4:143–159
10. Boyd R, Niskavaara H, Kontas E, Chekushin V, Pavlov V, Often M, Reimann C (1997) Anthropogenic noble-metal enrichment of topsoil in the Monchegorsk area, Kola peninsula, northwest Russia. J Geochem Explor 58:283–289
11. Draxler RR, Rolph GD (2003) HYSPLIT (hybrid single-particle Lagrangian integrated trajectory) model access via NOAA ARL READY website (http://www.arl.noaa.gov/ready/hysplit4.html). NOAA Air Resources Laboratory, Silver Spring, MD.
12. Rolph GD (2003) Real-time environmental applications and display system (READY) website (http://www.arl.noaa.gov/ready/hysplit4.html). NOAA Air Resources Laboratory, Silver Spring, MD
13. Gregurek D, Melcher F, Niskavaara H, Pavlov VA, Reimann C, Stumpfl EF (1999) Platinum-group elements (Rh, Pt, Pd) and Au distribution in snow samples from the Kola Peninsula, NW Russia, Atmos Environ 33:3281–3290

Cultural heritage stock at risk from air pollution

J Watt,[1] E Andrews,[1] N Machin,[1] R Hamilton,[1] S Beevers,[2] D Dajnak,[2] X Guinart,[3] P de la Viesca Cosgrove[3]

[1] Centre for Decision Analysis and Risk Management, School of Health & Social Sciences, Middlesex University, Enfield EN3 4SF, UK
[2] Environmental Research Group, King's College London, SE1 9NH, UK
[3] Departament de Medi Ambient de la Generalitat de Catalunya, Diagonal 523-525, 08029 Barcelona, Spain

Abstract

This paper reports a methodology for assessing the risk of damage caused to key cultural heritage buildings because of exposure to air pollution. Both corrosion and soiling damage are considered. Emissions inventories linked to a dispersion model allow the production of pollutant (SO_2, O_3, PM_{10}) contour maps across a city. Combining this information with dose–response functions allows the rate of building material damage to be mapped. The location of key cultural heritage properties can be added to these maps, allowing the risk to these building to be evaluated. UN Educational, Scientific and Cultural Organization (UNESCO) World Heritage Sites in London and Barcelona are used to demonstrate the methodology.

Introduction

Cultural heritage properties (historic buildings, works of art, medieval glass, archaeological treasures, etc.) are valued by society, but are at risk of damage due to exposure to air pollution. The two main forms of damage are *corrosion* (this was principally due to SO_2 exposure in the past but increasingly other pollutant effects are becoming more important as SO_2

levels fall) and *soiling* (principally due to exposure to particulate air pollution). Cultural heritage damage presents a special cause for concern because, unlike most ecosystems, building materials have no ability to recover naturally and the historic and artistic importance of the objects may be related to the preservation of the original materials, whereas materials in modern buildings may be replaced more readily. A number of research programmes, mostly supported by the European Commission, have examined the scientific basis for exposure and response and this has resulted in the availability of dose–response functions relating a number of materials to pollutant exposure [1]. These materials can be subdivided in the following categories depending on the availability of quantitative or qualitative data on effects of environmental parameters on degradation:

- Category 1: Dose–response functions are available
- Category 2: Damage functions might be obtained from, e.g., maintenance data
- Category 3: Limited knowledge on environmentally affected degradation

Dose–response and damage functions model the physical damage done by pollution but this is only the first necessary step in establishing the cost incurred. If policy or management practice is to be changed to increase protection of heritage, it is desirable that some measure of the extent of the risk is available. For regional, national or international scale air quality policy it is desirable to be able to estimate the cost implications involved and this involves estimation of the amount of material exposed. This is complex for heritage objects and will not be considered here. On the other hand, more local actions by heritage managers may not require such detail. It is likely that managers will have a good estimate of the amount of different material used in the construction of their object, but will be less familiar with the air pollution situation and the resultant implications for air quality. Thus it may be sufficient to have an indication of the likely damage at the location of the object – a risk map. We present this approach here and report on studies conducted in London and in Barcelona, which relied on bringing together two databases:

1. The identification and location of cultural heritage buildings
2. The spatial distribution of those air pollutants which can cause damage to these buildings

In this case, the material composition is not collated, but the distribution of heritage objects spatially is related to the calculated damage functions (and potentially cost functions) to show how heritage is put at risk. Owners and managers, who have access to data on materials for their buildings, will be able to locate them on the map and learn about potential damage and policy makers will be made aware of the potential impact.

Methods

Stock at risk for heritage objects

The identification and location of cultural heritage buildings occurs at international, national, and local levels. At the international level, UN Educational, Scientific and Cultural Organization (UNESCO) provides a list of cultural heritage with universal importance. The UNESCO list [2], identified by the World Heritage Committee as having outstanding universal value, includes 812 properties with 628 cultural, 160 natural, and 24 mixed properties. Almost half of the properties are located in Europe.

The list is not sufficient for a comprehensive stock-at-risk evaluation since a huge number of valuable objects and monuments are not placed on the list for a variety of reasons, for example, a reluctance to provide the commitment to maintenance that is implied by inclusion. In addition, a number of sites represent collections of objects such as a historic city centre. At national and local level, most countries in Europe have an established procedure for identifying historic buildings and other valuable objects and monuments. For example, in the UK, the Department for Culture, Media and Sport is required to compile lists of scheduled monuments (nationally important archaeological remains) and also buildings with special architectural or historic interest, in order to provide guidance for local planning authorities. The administration of the listing system is carried out by English Heritage and properties identified are called *listed buildings* [3]. Buildings which are chosen for the list are classified as grade 1, grade 11*, or grade 11. The stock of these buildings in England and Wales, and in London, is shown in Table 1.

Table 1. Listed buildings in England and Wales

Building type	Description	No. of building UK	London
Grade 1	Buildings of exceptional interest	13,000	584
Grade 11*	Buildings which are particularly important, of more than special interest	26,000	1356
Grade 11	Buildings with special interest which warrant every effort being made to preserve them	461,000	16,569

The City of Westminster contains 21% of the listed buildings and also three of the four UNESCO sites in London. The properties in London and Barcelona are:

London
Westminster Palace
Westminster Abbey
St. Margaret's Church
Tower of London

Barcelona
Palacio Guell
Casa Mila
Palaude la Música Catalana
Hospital de Sant Pau

Some of the properties are shown in Figure 1.

Hospital de Sant Pau, Barcelona   Tower of London

Fig. 1. UNESCO sites in Barcelona and London

Results and discussion

Case study: London

The modelling work undertaken used receptor-based or statistical modelling techniques, combined with the use of commercially available models. Receptor-based techniques have become more robust because of the increased availability of measurements in London for the development of empirical and statistical modelling. In particular, the rapid growth of the London Air Quality Network (LAQN) since 1993 has provided a wealth of high-quality air pollution data for many different pollutants. The increased availability of these data has led to increasingly sophisticated model developments as well as providing new insights into the underlying mechanisms of urban air pollution [4]. The model was run for the year 2002 using Heathrow hourly sequential meteorological data. All sources contained within the London Atmospheric Emissions Inventory (LAEI) were used. For road transport emissions, the new LAEI 2002 emissions were used. For the large combustion processes (known as part A processes) within and outside London, a detailed modelling approach was undertaken, using specific process information updated to 2002. All other emission sources were taken from the LAEI 2001 and assumed that there was no change between 2001 and 2002. The model runs for 2002 predict the annual average SO_2, NO_2, and O_3. The resulting total SO_2 concentration is shown in Figure 2. In terms of a general trend across London, high concentrations of 7–9 µg m^{-3} SO_2 arise to the east of London and these decrease to 4–5 µg m^{-3} in incremental bands to west London. The influence of the part A sources in the east Thames area can be clearly seen. The addition of transport sources of SO_2 has increased concentrations, particularly along the M25 motorway which lies along the outer edge of the map. A comparison with measurement sites in London showed that the model predictions were good, with, on average the model predictions being 6% higher than the measured concentrations. Errors (model – measured) are both positive and negative and 90% of the concentrations are within ±2.6 µg m^{-3}. It is worth noting that annual mean concentrations in central London are very slightly reduced from surrounding areas, this reduction reflects the absence of large industrial sources in this area.

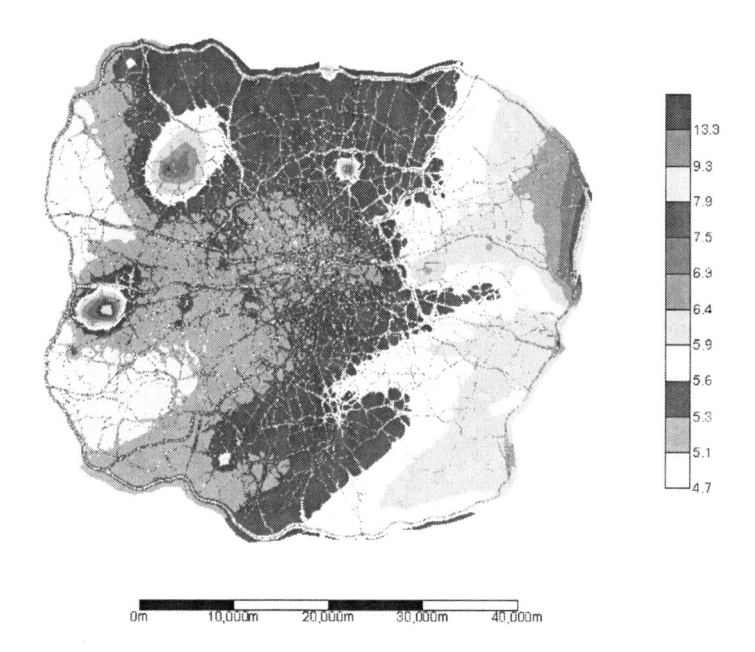

Fig. 2. Total annual SO_2 mean (µg m^{-3}) for London – 2002 emission scenario

To assess the corrosion implications, ozone maps were also produced using ozone equations derived from Tidblad et al. [5], and these maps combined with the dose–response functions were run for the year 2002 using information from Heathrow hourly sequential meteorological data. The prediction results are illustrated in Figure 3 as detailed maps of London at a spatial resolution of 20 × 20 m^2 for an area up to and including the M25 motorway. It can be seen that the material loss data shows high copper corrosion in outer London and lower copper corrosion in central London. This is representative of the complex combination of both O_3 and SO_2 concentrations. The low ozone concentration in central London, suburban "town" centres, Heathrow Airport and at roadside has been counterbalanced by the high SO_2 concentrations located in Heathrow Airport, the eastern edge of London and along main roads such as the M25 as seen.

Soiling is determined by PM_{10} concentrations and the London Emissions Inventory contains all the relevant emissions data to enable a PM_{10} contour map to be produced. This is shown in Figure 4.

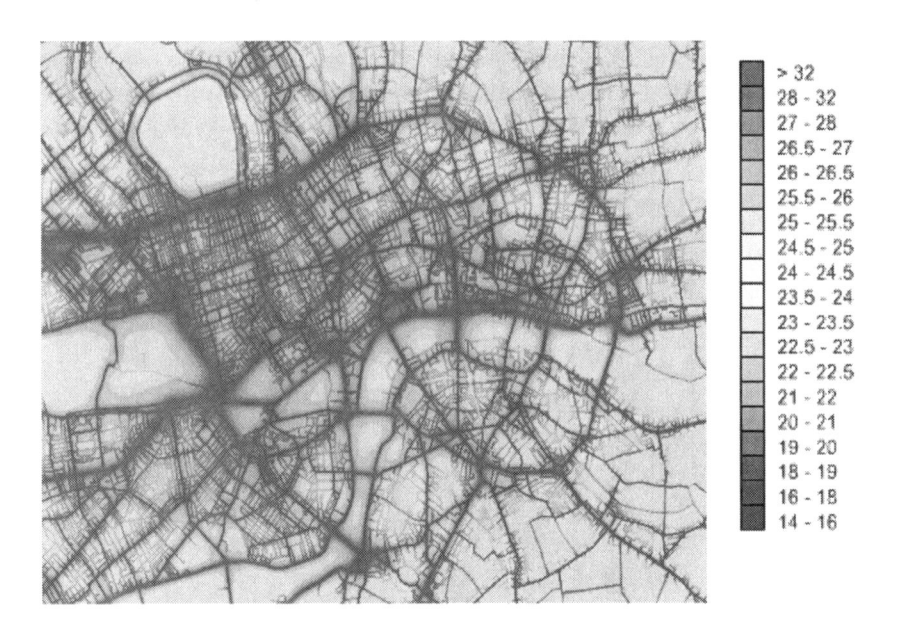

Fig. 3. London mapping showing the material loss for copper in 2002 (in g m^{-2} and in percentage of background corrosion)

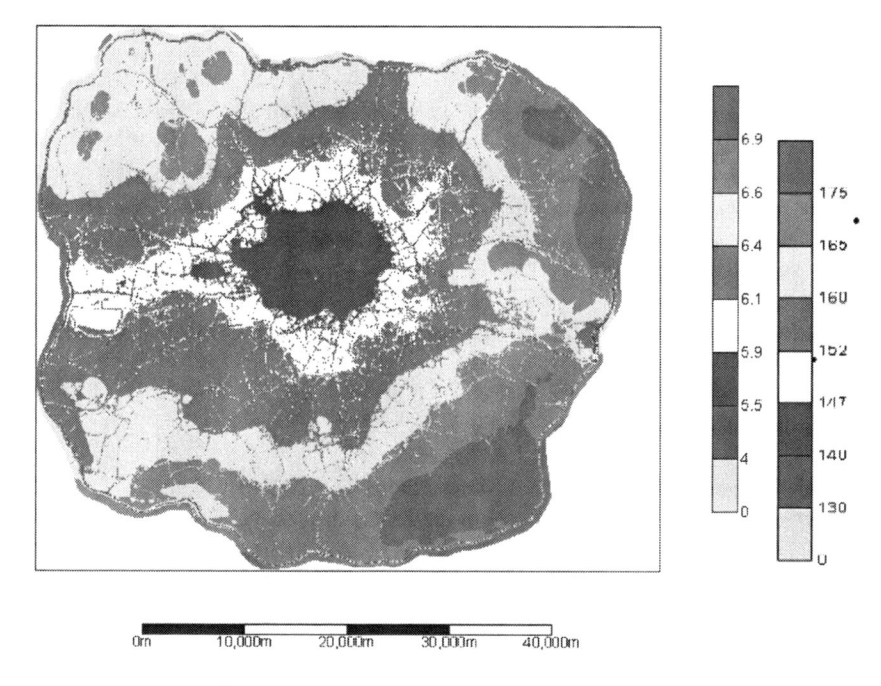

Fig. 4. PM$_{10}$ (μg g^{-1}) levels in London, 2006 [6]

Results from the Multi-Assess programme [7] provided dose–response functions for soiling. For painted steel, the result was:

$$\Delta R = R_\text{O}[1-\exp(-C_\text{PM0} \times T \times 5.9\times 10^{-6})]$$

where C_PM10 is in μg m^{-3} and T is in days.

Application of a dose-response function to any location on the map allows the soiling damage at that location to be calculated.

Estimated material damage at the London UNESCO sites

Pollution levels (annual mean)		5.5 μg SO$_2$ m^{-3}
		30 μg PM$_{10}$ m^{-3}
Material damage	Corrosion: copper	Material loss: 5.5 g m^{-2} per annum
		Surface recession rate: 0.61 μm per annum
	Soiling: painted steel	47% loss in reflectance over 10 years or 27% loss in reflectance over 5 years

This information, combined with economic information on costs and benefits, statutory requirements and information on public attitudes, allows those responsible for the management of the building to establish an appropriate maintenance schedule.

Case study: Barcelona

Compared to London, Barcelona experiences lower levels of SO$_2$, but higher levels of PM$_{10}$. This is a reflection of the relatively small number of major energy-consuming industrial plants within the city but the large amount of traffic present on non-motorway roads (carreteres secundàries) present in the city. The impact of emissions from these roads is shown in Figure 5.

Estimated material damage at the Barcelona UNESCO site, Hospital de Sant Pau

Pollution levels (annual mean)		5.0 µg SO$_2$ m^{-3}
		40 µg PM$_{10}$ m^{-3}
Material damage	Corrosion: copper	Material loss: 5.1 g m^{-2} per annum
		Surface recession rate: 0.57 µm per annum
	Soiling: painted steel	57% loss in reflectance over 10 years or 35% loss in reflectance over 5 years

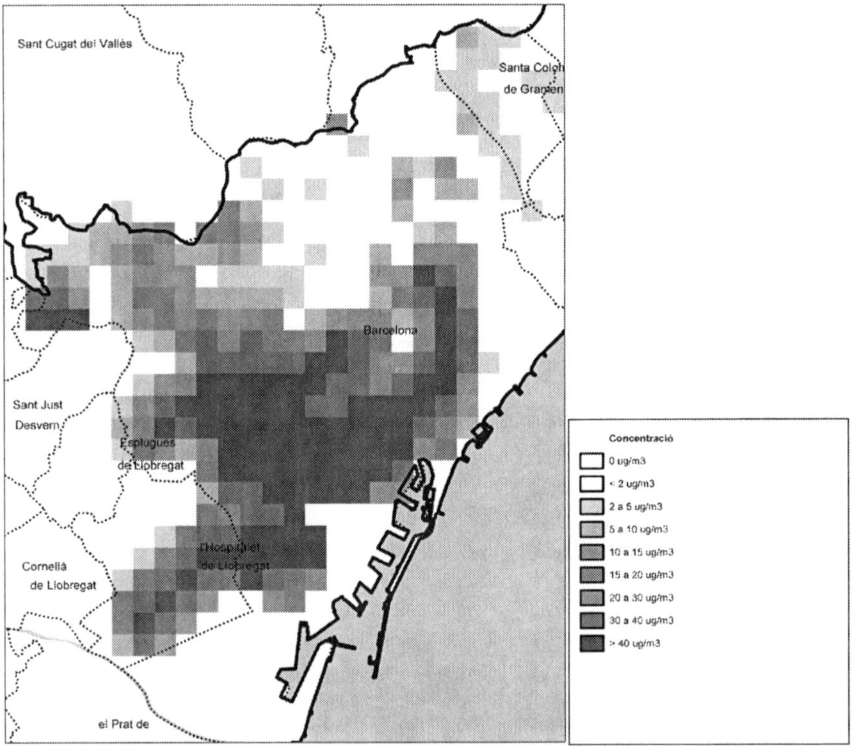

Fig. 5. PM$_{10}$ levels in Barcelona, 2006

Conclusions and future research

This paper has demonstrated that the risk to heritage of damage from air pollution may be illustrated for managers of heritage and policy makers by the production of risk maps that combine damage functions and heritage location. The spatial distribution of air pollutant concentrations is generally well known for major cities in Europe. The identification and location of the major heritage properties, as identified in the UNESCO list, is widely available. However, the identification and location of additional heritage properties at national and local levels is much more variable. Detailed risk maps are being produced for a number of cities and regions within the European Union (EU)-sponsored Cult-Strat programme [8].

The materials composition of the properties is generally not known and without this information it is not possible to quantify the total amount of damage caused by air pollution nor is it possible to quantify the costs of maintenance, including conservation, repair, and cleaning. This information is useful for development of another type of policy – air quality. The cost–benefit methods used to underpin this have leant heavily on health impacts, but materials damage is also included and it is therefore important to start to obtain an accurate inventory of heritage material at risk and the cost implications at national and European scales.

Material composition and stock of modern buildings can be modelled by estimating the amount of different materials in typical types of building and then estimating the numbers of each type in a given region. This relies on that fact that, within limited geographical areas and over restricted timescales, buildings of a given type are frequently very similar to each other. Heritage buildings are typically much more diverse since they may be survivors of earlier periods and valuable since they are relatively rare and/or may be unique or unusual at the time of construction. A series of small-scale studies are being undertaken to evaluate the potential of using identikits for some types of heritage or to estimate heritage building stock by inspection. These will be reported elsewhere. The Cult-Strat project also has, as one of its major objectives, the development of a methodology which can be employed to provide this information.

Three case studies are being conducted as key components of this research:

- London and Barcelona will continue to develop the approaches described in this paper, and also undertake an audit of materials in selected zones of the cities.
- Paris: a study is concentrating on the central part of Paris, adjacent to the banks of the River Seine between the Ile Saint Louis on the eastern

side and the Eiffel Tower on the western side. The study will produce an inventory of cultural heritage stock, map the stock, apply dose–response functions, and identify the zones at risk [9].
- Madrid: this study is considering all monuments, monumental sites, buildings, and sculptures included in the Madrid City Regional Office list and exposed to the atmosphere. A total of 260 immovable objects and 1300 movable objects are included in the study and a database of material composition has been established. The study will be extended to the greater Madrid region [10].

Acknowledgements

John Watt and Ron Hamilton are grateful to the European Commission for funding provided to the projects Multi-Assess and Cult-Strat. All partners in the above projects contributed to the themes developed in this paper. In particular, the authors wish to record their appreciation to Vladimir Kucera, Johan Tidblad, Stefan Doytchinov, Tim Yates, Roger Lefevre, Anda Ionescu, and Daniel de la Fuente.

References

1. Tidblad J, Kucera V, Mikhailov AA, Henriksen J, Kreislova K, Yates T, Stöckle B, Schreiner M (2001) UNECE ICP Materials: Dose-response functions on wet and dry deposition effects after 8 years of exposure. Water Air Soil Pollut 130:1457–1662
2. United Nations Educational, Scientific and Cultural Organisation (UNESCO), World Heritage (2006) http://whc.unesco.org/
3. English Heritage, Research and Conservation, Heritage Protection (2006) http://www.english-heritage.org.uk/
4. Greater London Authority, Air Quality Research (2006) http://www.london.gov.uk/mayor/environment/air_quality/research/index.jsp
5. Tidblad J, Mikhailov AA, Henriksen J, Kucera V (2004) Improved prediction of ozone levels in urban and rural atmospheres. Protection Metals 40:67–76
6. London Air Quality Network (2006) http://www.londonair.org.uk/london/asp/BOC.asp
7. Swedish Corrosion Institute (KIMAB) (2006) http://www.corr-institute.se/MULTI-ASSESS/
8. European Commission, Research (2006) http://ec.europa.eu/research/fp6/ssp/cultstrat_en.htm
9. Ionescu A, Lefevre R (2006) Personal communication
10. de la Fuente D (2006) Personal communication

III. Contaminated Environments and Remediation

Organic contaminants in urban sediments and vertical leaching in road ditches

A-M Strömvall,[1] M Norin,[2] TJR Pettersson[1]

[1]Water Environment Technology, Department of Civil and Environmental Engineering, Chalmers University of Technology, SE-412 96 Göteborg, Sweden
[2]NCC Construction Sverige AB, Göteborg, Sweden

Abstract

This is a study of the environmental impact of organic contaminants emitted from urban traffic and road infrastructure in Göteborg, Sweden. The vertical leaching of organic contaminants in road ditches has been investigated, as well as their occurrence in storm water sediment, urban soil, and shallow groundwater.

A total of 80 specific organic contaminants were analysed in a storm water sediment sample, and of these as many as 40 specific organics were identified. The concentration of total semi-volatiles, alkylbenzenes, aliphatics, 4-nonylphenols, total of mono- and di-nonylphenol ethoxylates, carcinogenic polycyclic aromatic hydrocarbons (PAH), diethyl hexylphthalate (DEHP), and several brominated flame retardants, were all analysed in high concentration. Depth profiles, in clay, clay/sand, and sand road ditches, at four places along highway E20, were analysed for a total of 40 specific organic compounds. In the soil profiles, total semi-volatiles (<2300 mg kg^{-1} dw) and carcinogenic PAH-16 (<1.0 mg kg^{-1}dw) were identified. In one of the clay/sand profiles, total semi-volatiles were analysed in decreasing concentrations but until a depth of 1 m, and the carcinogenic PAH-16 until 1.5 m. The relative composition of the specific PAH-16 indicates rubber tyres, vehicle exhausts, and asphalt materials to be the main sources of PAH contamination.

Even in urban shallow groundwater, total semi-volatiles and carcinogenic PAH-16 were identified in remarkably high concentrations.

The occurrence of total semi-volatiles and carcinogenic PAH deep in road ditches, and the high levels in urban groundwater, show the need for efficient construction of road ditches and treatment of road runoff to prevent contamination. The high levels of total semi-volatiles in all samples show that most of the contaminants occurring in urban environments are still unidentified compounds with unknown environmental effects.

Introduction

In modern industrial society, tens of thousands of different organic chemicals have been brought into use as products in large quantities, or as additives in materials and products. Thousands of organic substances, most of them still unidentified, are also generated unintentionally as contaminants in the various manufacturing or combustion processes. Many of these compounds had never occurred on earth before man began to produce them, and many of them are highly toxic, carcinogenic, suspected as hormone disruptors, and able to accumulate in food webs. Knowledge concerning the toxic organic contaminants in the urban soil and water environments is poor, and research in the area is now highly desired for sustainable management of the urban soils and water flows.

One of the major pathways of pollutants in urban areas is the storm water system, collecting a mix of substances from, i.e., traffic, building materials, and atmospheric fallout. In the storm water systems, mainly heavy metals and organic substances are transported to the receiving waters. The sources and transport of many heavy metals have been investigated for decades and are well known today. This is however not the case for organic pollutants. Some measurements of, i.e., polycyclic aromatic hydrocarbons (PAHs) have been conducted [1–4], but the knowledge on most priority organic pollutants in storm water is poor. More than 600 compounds are potentially occurring in urban storm water, and 11 of these have been selected and prioritized as high-risk and toxic pollutants, after a chemical hazard identification and assessment tool (CHIAT) procedure [5,6].

The most important sources of toxic organic pollutants in the urban environment are the traffic, besides leaching from construction and building materials. There are thousands of partially known substances in car exhausts, which are the major sources of hydrocarbons in the road environment [7]. Vehicle exhaust, unburned petrol and diesel fuel, rubber tyre particles, vehicle construction materials and paints, lubricating oils, the road environment's concrete, and the asphalt material itself are all probable emission sources of PAH and other semi-volatile contaminants leaching from roads [3,8–13]. Many of the identified semi-volatile contaminants are combustion by-products or additives in vehicle or road construction materials.

Knowledge of the content and possible release of hazardous substances from building material is scarce, but the content and leaching behaviour of such compounds in concrete constituents and paints on wooden facades have recently been presented [14]. Nonylphenol ethoxylates (NPE) and nonylphenols (NP) are widely used as air-entraining admixtures in concrete [15], and these compounds could leach from concrete to the environment. NPE and NP have also been widely used as additives in paints, asphalt, lubricants, stabilizers, petroleum demulsifiers, fungicides, bactericides, dyes, drugs, adhesives, rubber chemicals, phenolic resins, and plasticizers. Surfactants, such as NP, are genotoxic and persistent compounds with the ability to bioaccumulate, and are suspected hormone disruptors [16,17]. Phthalates are essentially used as plasticizers in the production of polymeric materials [18]. DBP and DEHP are used as a binding agent in rubbers, plastics, and surface coatings, and as a high boiling solvent. DBP is also readily used as an additive to bitumen [19], and could be leached from asphalt materials, which have been used for paving [9,20]. Phthalates are easily released into the environment and are ubiquitous environmental pollutants [21]. Phthalates are also commonly used substances in urban areas, and potential sources such as lined panels, roofs, surface treatment of vehicle under-bodies, cables, hoses, jointing materials, adhesives, and paints have been pointed out for these contaminants in Sweden.

This study has focused on finding new organic contaminants in the urban water cycle through chemical analysis of urban storm water sediment and soil. The concentrations of total hydrocarbons and PAH have also been analysed in soil from road ditches, to assess the potential vertical leaching of organic contaminants from road runoff. Also surface soil and groundwater have been collected at different places in Göteborg, and analysed as reference samples for total hydrocarbons and PAH.

Experimental method

Urban storm water sediment

The sediment in the Järnbrott storm water pond was selected for screening of urban organic pollutants, and to be used as a reference for an urban area with high degree of contamination. The goal was to identify as many organic contaminants as possible. The Järnbrott storm water pond has an urban catchment area with traffic, housing, and small-scale industrial activities [22, 23]. The pond has a dry-weather area of 6200 m^2 and a volume of 6000 m^3, and the depth then varies from inlet to outlet between 0.5 and 1.6 m. The sediment sample was taken with a core sampler in the middle of the pond, and sent to a certified environmental laboratory for chemical analysis. Based on findings from the literature, a total of 80 specific organic contaminants were selected for the chemical analysis. The groups of contaminants analysed were volatile aliphatic and aromatic hydrocarbons, volatile chlorinated aliphatics and aromatics, semi-volatile hydrocarbons, PAHs, phenols, alkyl phenols and alkylphenol ethoxylates, phthalates, and brominated flame retardants.

Soil and water – road ditches

After a survey of local soil and bedrock maps, and discussions for possibilities of soil sampling, two locations along highway E20, with different soil composition, were selected for the study of vertical leaching of organic contaminants. The locations were Partille, with clay and clay/sand ditches, and Jonsered with sand ditches. The traffic on E20 at Partille, 10 km from Göteborg centre, is characterized by slow-moving dense traffic and queues during rush hours. At Jonsered, 20 km from Göteborg centre, the traffic on E20 is more high-speed. The traffic load at Partille was estimated to be about 40,000 vehicles per day, and at Jonsered 30,000 vehicles. The selected road ditches treat the motorway runoff through infiltration in the ditches. At Partille, the ditches were cleaned from contaminants, through excavation of the top surface soil layer, 3 years before this study.

The soil samples were collected through an auger sampler at two places, with a distance of 60 m between each, in the ditches at both Partille (L1 and L2) and Jonsered (M1 and M2). Soil samples were taken and depth profiles created. At Partille, the depth profile L1 had the following depths: 0, 0.3, 1.1, 1.5, 1.9 m, and L2 had 0, 0.3, 0.5, 0.7 m; at Jonsered, M1 had 0 and 0.4 m, while M2 had 0, 0.4, 1.0, 1.5 m. Sampling tubes for collection of groundwater were installed in the holes after the soil sampling. The depth of the groundwater tube at Partille was 1.5 m, and at Jonsered 1.6 m. At Jonsered, groundwater samples were successfully collected, but at Partille it was not possible to collect any groundwater, due to the dense clay soil structure in the ditches. Water and sediment samples were also collected in a nearby creek at Jonsered, for use as a reference sample.

During the sampling, soil profile L1 in Partille consisted of a topsoil layer until a depth of around 0.3 m, sand till a depth of 1.0 m, surface clay till 1.5 m, and pure clay till 1.9 m. The L2 profile had the topsoil in the surface layer, but the surface clay starting already at 0.3 m, and thereafter pure clay till a depth of 0.7 m. At the Jonsered M1 profile the surface layer was topsoil, and from a depth of 0.15 m the sand started until a depth of 1.5 m. In the profile M2, the surface layer was topsoil with groundwater at the surface, and the sand started at 0.15 m till a depth of 1.5 m.

At Partille and Jonsered, a total number of nine and six samples, respectively were analysed for loss of ignition (organic content), semi-volatiles as chlorobenzenes, sum of hydrocarbons, sum of PCBs, and sum of PAHs, but also all the specific EPA PAH-16. Volatile aromatics and aliphatics, volatile chlorinated aliphatics and aromatics, and phthalates were also analysed in the surface soil samples in soil profile L1 (Partille) at depths 0 and 0.3 m, and M1 and M2 (Jonsered) at 0 and 0.4 m. At Jonsered, also the reference groundwater, creek water, and sediment were analysed for loss of ignition (organic content) and total/dissolved organic carbon (TOC/DOC), semi-volatiles such as chlorobenzenes, sum of hydrocarbons, sum of PCBs and sum of PAH, and all the specific EPA PAH-16. For the water samples, also phenols and volatile aromatics and aliphatics were included in the analysis.

Soil and groundwater – city areas

As city area reference samples, surface soil (at 0 and 0.25 m) as well as groundwater were taken at three places in the city of Göteborg: Järntorget, Backaplan, and Packhusplatsen. The land use at the sites is: major city street (Järntorget), shopping centre (Backaplan), and minor city street (Packhusplatsen). The soil samples were taken through tubes pounded into the soil until a depth of 30 cm. The shallow groundwater samples were collected in groundwater collection tubes already installed at the places, through pumping from the wells. The wells were pumped dry the day before sampling to get refreshed groundwater. Surface soil samples were also collected in the area of Partille (Sävedalen), situated 500 m from E20 and with a land use of urban forest. All these reference samples were analysed for loss of ignition or TOC/DOC, semi-volatiles such as chlorobenzenes, sum of hydrocarbons, sum of PCBs, and sum of PAHs, and the specific EPA PAH-16. Selected samples were also analysed for NP and its mono- and di-ethoxylates, but also for specific phthalates.

Chemical analysis

All sediment, soil, and water samples were collected in glass containers or bottles, and sent for chemical analysis to a certified environmental laboratory for chemical analysis following standardized methods. The water and sediment samples were sent immediately after collection, but some of the soil samples were deep-frozen, and were sent for chemical analysis first after reports of concentrations from the more surface layers.

Results and discussion

Järnbrott storm water sediment

A total of 80 specific organic contaminants were analysed in the Järnbrott storm water sediment sample, see Table 1, and of these as many as 40 specific organic contaminants were identified. The total semi-volatiles, alkylbenzenes, aliphatics, 4-nonylphenols, total of mono- and di-nonylphenol etoxylates, carcinogenic EPA PAH-16, DEHP, and several brominated flame retardants, were all found in high concentration.

The concentration of total semi-volatile hydrocarbons was as high as 12,000 mg kg^{-1} and of this only 0.5% was identified and quantified as specific compounds. The result shows that most of the organic contaminants in the storm water sediment are still unknown compounds with potential environmental effects.

All of the 16 specific US EPA–PAH were identified in high concentration (21 mg kg^{-1}) in the sediment, and the sum of carcinogenic PAH of 5.3 mg kg^{-1} exceeds the Swedish guideline value of 0.3 mg kg^{-1} for contaminated sites with sensitive land use, but not the guideline value of 7 mg kg^{-1} for less less sensitive sites [24].

The concentration of 4-nonylphenol was 3.1 mg kg^{-1} and exceeds the Canadian quality criteria guideline value for NP and its ethoxy- lates in sediments of 1.4 mg kg^{-1} [25]. The concentrations of 4-NF-diethoxylate and 4-NF-monoethoxylate were also high with values of 1.5 and 5.3 mg kg^{-1}, respectively. Also 4-octylphenol and 4-OF-ethoxylates were identified in the sediment. Of the phthalates analysed, di-2-ethylhexyl-phthalate (DEHP) was identified in the sediment sample in a concentration of 23 mg kg^{-1}. Eight specific compounds in the group of brominated flame retardants were also identified, where the pentabromo- and tetrabromo- diphenylether compounds were found in concentrations of around 10 μg kg^{-1}. Many of the

Table 1. Organic contaminants identified in the Järnbrott storm water sediment

Chemical parameter/analysis method	Chemical	Conc.	Guideline value
Loss of ignition, % dw		16	KM[a]/MKM[b]
Semi-volatiles, mg kg^{-1} dw/GC/MS	Hydrocarbons	12,000	
	PAHs	9.0	
	Phenols	0.31	4/40[c]
	Chlorobenzenes and PCBs	<1	0.02/7[c]
Volatiles, mg kg^{-1} dw/GC/MS	Toluene	0.52	10/35[c]
	Ethylbenzene	0.40	12/50[c]
	Xylenes	2.5	15/70[c]
	Some alkylbenzenes	40	–
	Naphthalene	3.2	
	Indane	0.35	1.4[d]
	Aliphatics > C8–C10	34	100/350[c]
US EPA PAH-16, mg kg^{-1} dw/HPLC	Acenaphtene	0.16	
	Fluorine	0.52	
	Phenanthrene	3.2	
	Anthracene	0.52	
	Fluoranthene	5.5	
	Pyrene	5.1	
	Benzo(a)anthracene	1.0	–
	Chrysene	0.65	
	Benso(b)fluoranthene	1.0	–
	Benso(k)fluoranthene	0.53	–
	Benso(a)pyrene	1.2	
	Ideno(123cd)pyrene	0.17	
	Benzo(ghi)perylene	0.96	
	Dibenz(ah)anthracene	0.72	–
	ΣCarcinogenic PAH	5.3	0.3/7[c]
	Σ16 EPA–PAH	21	20/40[c]
Alkylphenols and their ethoxylates, mg kg^{-1} dw/HPLC	4-Nonylphenol (NP)	3.1	
	4-NP-monoethoxylate	5.3	
	4-NP-diethoxylate	1.5	Sum of all: 5.7/14[d]
	4-Octylphenol (OP)	0.16	
	4-OP-monoethoxylate	0.051	
	4-OP –diethoxylate	0.011	
Phthalates, mg kg^{-1} dw/GC/MS	Di-(2-ethylhexyl) phthalate	23	59[e]
	Other specific phthalates	<0.80	–
Brominated flame retardants, µg kg^{-1} dw/GC/MS	Tetrabromobisphenol-A	5.2	–
	Tetrabromodiphenylether	8.8	–
	Tetrabromodiphenylether 47	8.5	–
	Pentabromodiphenylether	11	–
	Pentabromodiphenylether 100	1.2	–
	Pentabromodiphenylether 99	9.8	–
	Hexabromodiphenylether	3.1	–
	Heptabromodiphenylether	2.4	–

[a]KM: sensitive land use
[b]MKM: less sensitive land use
[c]Generic guideline values for contaminated soils in Sweden
[d]Canadian Soil Quality Guidelines for protection of Environmental and Human Health
[e]Dutch guideline value for sediments

organic contaminants analysed were in concentrations below the detection limits of the methods.[1]

Road ditches

Depth profiles, in clay, clay/sand, and sand road ditches, at four places along the motorway E20, were analysed for a total number of 40 specific organic compounds. Of these only total semi-volatile hydrocarbons and PAH were found in concentrations high enough for quantification to be presented (see depth profiles in Fig. 1). In the surface soil in the road ditches, total semi-volatiles up to 2300 mg kg^{-1} in profile L2 (Partille), and carcinogenic EPA PAH-16 up to 1.0 mg kg^{-1} in profile L1 (Partille), were identified. In the clay profile L2, total semi-volatiles were identified in decreasing concentrations down to a depth of 1 m, and the carcinogenic PAH-16 in the clay/sand profile L1 until a depth of 1.5 m. These findings of high levels deep in the soil profiles, indicating vertical transport of organics in the soil, were not expected since organic contaminants are hydrophobic compounds and strongly bonded to organic particles in soil. A possible transport of the organics could be through adsorption to organic particles or colloids, small enough to be transported vertically in the ditches with the infiltration of road runoff. In the M1 and M2 profiles (Jonsered), the levels of organic contaminants were low (Fig. 1).

PAH urban areas

In the reference surface soil samples, taken in the centre of Göteborg, total semi-volatiles and carcinogenic PAH-16 were also analysed in high concentrations (for selected samples see Table 2). Even in the shallow urban groundwater samples, taken in central Göteborg, total semi-volatile hydrocarbons and carcinogenic PAH-16 were identified in remarkably high concentrations, up to 1400 μg L^{-1} and up to 0.4 μg L^{-1}, respectively. The results show that the concentration of carcinogenic PAHs in shallow urban groundwater exceeds

---

[1] Benzene, aliphatics >C5–C8, MTBE, acenaphtylene, 4-NP-triethoxylate, -tetraethoxylate, 4-OP-triethoxylate, -tetraethoxylate, dimethyl-, diethyl-, di-*n*-propyl- di-*n*-butyl-, di-isobutyl-, dipentyl-, dicyclohexyl-, butylbenzyl-, di-*n*-octyl-phthalate, hexabromocyclodecane, octa-, nona-bromodiphenylether, decabromobiphenyl, -diphenylether, 1,1,1-trichloro-, 1,1,2,2-tetrachloro-, 1,1,2-trichloro-, 1,1-dichloro-, 1,2-dichloroethane, 1,2-dichloropropane, *cis*-1,2-dichloro-, *trans*-1,2-dichloro-, trichloroethene, dichloro-, trichloro-methane, vinylchloride, monochlorobenzene, and dichlorobenzene.

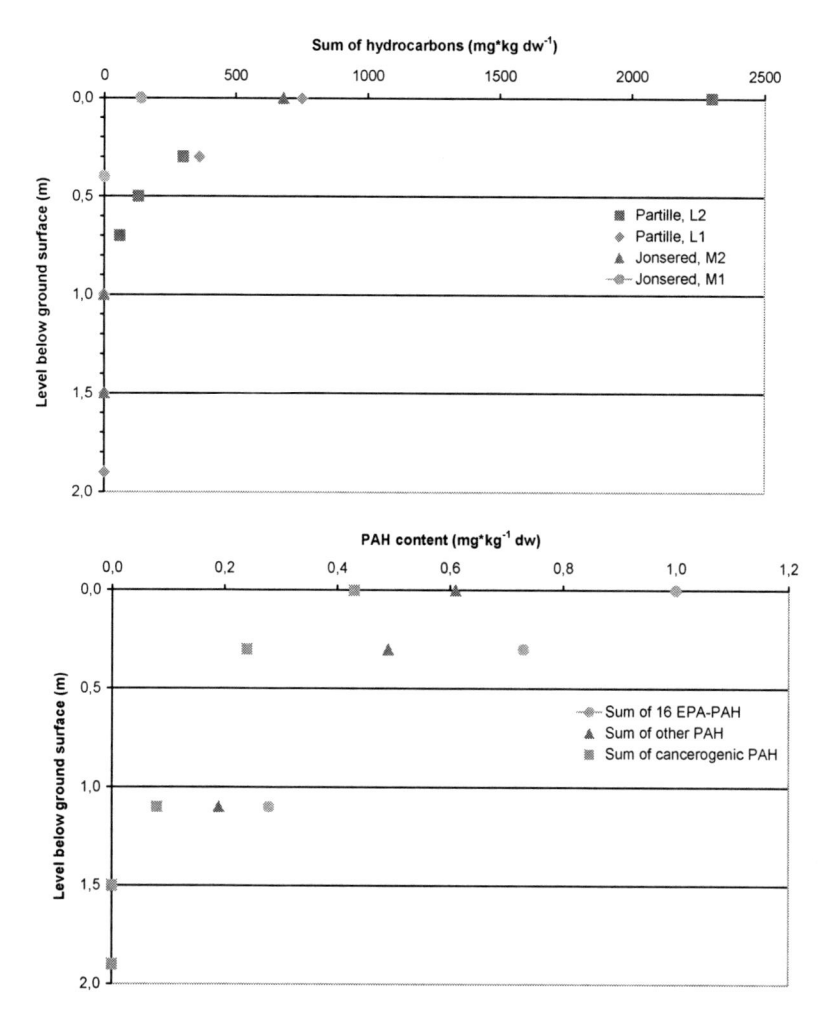

Fig. 1. Vertical soil profiles for concentrations of total semi-volatile hydrocarbons and polycyclic aromatic hydrocarbons (PAH) (profile L1) in road ditches along road E20

the Swedish guideline value, 0.2 µg L^{-1}, for contaminated groundwater at petrol stations [26].

In Table 2, the relative composition of PAH-16 in the Järnbrott storm water sediment, and soil from surface road ditches, are compared with the percentage in road runoff sediment at Gårda [8], urban surface soils, and urban shallow groundwater. Three of the most dominant specific PAH in several samples were phenanthrene (9–18%), fluoranthene (13–26%), and pyrene (12–26%), indicating that tyre rubber [12,13] and diesel exhaust

Table 2. The relative composition of PAHs in selected soil, storm water sediment, and shallow groundwater samples taken in Göteborg

Location	Soil (%)					Sediment (%)		GW (%)
	Partille	Järn-torget	Backa-Plan	Säve-dalen	Järn-brott	Gårda		Järn-torget
Naphthalene	n.d.	n.d	n.d.	n.d.	n.d.	3.0		3.0
Acenaphtylene	n.d.	9.0	n.d.	n.d.	n.d.	n.d.		4.2
Acenaphthene	n.d.	n.d.	n.d.	n.d.	0.7	0.6		n.d.
Fluorene	3.4	n.d.	1.3	n.d.	2.4	1.0		4.9
Phenanthrene	13	5.9	12	8.1	15	18		9.0
Anthracene	3.0	0.9	1.1	3.1	2.4	4.1		2.3
Fluoranthene	21	13	20	16	26	14		17
Pyrene	12	13	16	16	24	26		15
Benzo(a)anthracene*	9.2	6.9	10	6.1	4.7	3.9		7.4
Chrysene*	5.4	4.1	8.2	10	3.0	12		5.2
Benso(b)fluoranthene*	4.1	8.6	7.8	13[e]	2.5	4.2		6.7
Benso(k)fluoranthene*	3.4	3.8	3.8		5.6	0.8		3.3
Benso(a)pyrene*	9.6	7.9	7.8	7.1	5.6	2.1		8.1
Ideno(123cd)pyrene*	7.5	12	6.5	6.1	3.4	3.7		6.7
Benzo(ghi)perylene	6.1	15	7.4	5.1	4.5	3.0		6.7
Dibenz(ah)anthracene*	n.d	1.0	n.d.	n.d.	0.8	3.0		n.d.
ΣPAH*[a], mg kg^{-1} dw	0.4	1.3	1.0	0.44	5.3	2.7[e]		0.37[d]
ΣPAH-16[b], mg kg^{-1} dw	1.0	2.9	2.3	0.98	21	9.2		0.98[d]
rel. ΣPAH*[c],%	40	45	43	45	25	29		38

*Carcinogenic PAH
[a]ΣPAH* = sum of carcinogenic PAH
[b]ΣPAH-16 = sum of sixteen PAH analysed
[c]Rel. ΣPAH* = ΣPAH* divided by ΣPAH-16
[d]in µg L^{-1}
[e]sum of benso(a)fluoranthene and benso(k)fluoranthene
GW: groundwater

[28] could be important sources of pollution. In the soil sample taken at Järntorget, in central Göteborg, also benzo(ghi)perylene (15%) was a dominant PAH, and indicative for petrol exhaust [27,28]. The relatively high amount of chrysene at Sävedalen (10%) and in the Gårda sediment (12%), indicate diesel exhaust as well as leaching and surface abrasion of bitumen from road asphalt [9,12]. The relative amount of carcinogenic PAH was higher in the soil samples (40–45%) than in the storm water sediment samples (25–38%). These results could be explained by the higher molecular weight of the carcinogenic PAH, and thereby a higher

hydrophobicity of these PAH with consequent higher affinity to the soil particles in road ditches. The non-carcinogenic PAH are more hydrophilic and thereby more easily transported to storm water sediments.

Conclusions

The occurrence of total semi-volatile hydrocarbons and carcinogenic PAH-16 deep in road ditches, and the high levels in urban groundwater, show the need for efficient construction of road ditches and efficient treatment of road runoff water in urban and traffic-related areas to prevent contamination. Furthermore, the occurrence of PAH in soil at a great distance (500 m) from the highway shows that the contaminants are spread over a large area, far from the road. The high levels of total semi-volatiles in all samples show that most of the contaminants occurring in urban environments are still unidentified compounds with unknown environmental effects. The relative composition of the specific PAH-16 indicates rubber tyres, asphalt materials, and vehicle exhausts to be the main sources of PAH contamination. The occurrence of alkylphenols and alkylphenol ethoxylates, phthalates, and brominated flame retardants in the Järnbrott sediment should be further investigated with the goal of finding sources of pollution.

Acknowledgements

The authors are grateful to Erik Carlsson and Thomas Johansson who performed all practical work in field. The members of the SVR, Swedish Civil Engineering National Organization, the Road Division Group are all gratefully acknowledged for initiating the project. The SBUF, Development Fund of the Swedish Construction Industry, provided financial support for the study.

References

1. Krein A, Schorer M (2000) Road runoff pollution by polycyclic aromatic hydrocarbons and its contribution to river sediments. Wat Res 34:4110–4115
2. van Metre PC, Mahler BJ, Furlong ET (2000) Urban sprawl leaves its PAH signature. Environ Sci Technol 34:4064–4070
3. Takada H, Tomodo O, Ogura N (1990) Determination of polycyclic aromatic hydrocarbons in urban street dusts and their source material by capillary gas chromatography. Environ Sci Technol 24:1179–1186

4. Mikkelsen PS, Hafliger M, Ochs M, Jacobsen P, Tjell JC, Boller M (1997) Pollution of soil and groundwater from infiltration of highly contaminated stormwater – a case study. Water Sci Technol 36: 325–330
5. Baun A, Eriksson E, Aabling T, Ledin A, Mikkelsen PS (2004) Screening tool for problem and hazard identification of xenobiotic organic compounds in stormwater. Environment and Resources DTU, Technical University of Denmark (submitted)
6. Ledin A, Aabling T, Baun A, Eriksson E, Mikkelsen PS (2004) CHIAT – chemical hazard identification and assessment tool, Miljö och Resurser DTU, Danmarks Tekniske Universitet
7. Ball DJ, Hamilton RS, Harrison RM (1991) The influence of highway related pollutants on environmental quality. In: Hamilton, RS, Harrison, RM (eds.) Highway pollution. Elsevier, Amsterdam
8. Pettersson TJR, Strömvall A-M, Ahlman S (2005) Underground sedimentation systems for treatment of highway runoff in dense city areas. Tenth international conference on urban drainage, Copenhagen/Denmark, 21–26 August 2005
9. Norin M, Strömvall A-M (2004) Leaching of organic contaminants from storage of reclaimed asphalt pavement. Environ Technol 25:323–340
10. Andersson ÅC, Strömvall A-M. (2001) Leaching of concrete admixtures containing thiocyanate and resin acids. Environ Sci Technol 4:788–793
11. Sadler R, Delamont C, White P, Connell D (1999) Contaminants in soil as a result of leaching from asphalt. Toxicol Environ Chem 68:71–81
12. Lindgren Å (1998) Road construction materials as a source of pollutants. Ph.D. thesis, Department of Environmental Engineering, Luleå University of Technology
13. Rogge WF, Hildemann LM, Mazurek MA, Cass GR (1993) Sources of fine organic aerosol. 3. Road dust, tire debris and organometallic brake lining dust: roads as sources and sinks. Environ Sci Technol 27:1892–1904
14. Togerö Å (2004) Leaching of hazardous substances from concrete constituents and painted wood panels. Ph.D. thesis, Department of Building Technology and Building Materials, Chalmers University of Technology, Göteborg, Sweden
15. Rixom RM; Mailvaganam NP (1999) Chemical admixtures for concrete, 3rd edn. E&FN Spon, London, p 437
16. Teles M, Gravato M, Pacheco M, Santos MA (2004) Juvenile sea bass biotransformation, genotoxic and endocrine responses to β-naphthoflavone, 4-nonylphenol and 17β-estradiol individual and combined exposures. Chemosphere 57:147–158
17. Ruthann AR, Steven JM, Paul WG, Gang S, Brody JG (1998) Identification of alkylphenols and other estrogenic phenolic compounds in wastewater, septage, and groundwater on Cape Cod, Massachusetts. Environ Sci Technol 32:861–869
18. Harris CA, Henttu P, Parker MG, Sumper JP (1997) The estrogenic activity of phthalate esters in vitro. Environ Health Perspect 105:802–811
19. Zielinski J, Bukowski A, Osowiecka B (1995) An effect of polymers on thermal stability of Bitumen. J Therm Anal 43:271–277

20. Norin M (2004) Subsurface environmental impact in urban areas. Ph.D. thesis, Department of Geo-engineering, Chalmers University of Technology, Göteborg, Sweden
21. Soto AM, Sonnenschein C, Murray MK, Michaelson CL (1998) Estrogenic plasticizers and antioxidants. In: Eisenbrand, G. (ed.) Hormonally active agents in food symposium, October 1996. Universität Karlsruhe, Kaiserslautern, pp 142–66
22. German J (2003) Reducing stormwater pollution. Ph.D. thesis, Department of Water Environment Transport, Chalmers University of Technology, Göteborg, Sweden
23. Pettersson TJR (1999) Stormwater ponds for pollution reduction. Ph.D. Thesis, Department of Water Environment Transport, Chalmers University of Technology, Göteborg, Sweden
24. Swedish Environmental Protection Agency (2002) Contaminated sites. Report 5053
25. Canadian Council of Ministers of the Environment (2001) Canadian sediment quality guidelines for the protection of aquatic life; nonylphenol and its etoxylates. CCME, Canada
26. Swedish EPA (1997) Development of generic guideline values, Report 4639
27. Larsen RK, Baker JE (2003) Source apportionment of polycyclic hydrocarbons in the urban atmosphere: a comparison of three methods. Environ Sci Technol 37:1873–1881
28. Harrison RM, Smith DJT, Luhana L (1996) Source apportionment of atmospheric polycyclic aromatic hydrocarbons collected from an urban location in Birmingham. UK Environ Sci Technol 30:825–832

The use of an epiphyte (*Tillandsia usneoides* L.) as bioindicator of heavy metal pollution in São Paulo, Brazil

AMG Figueiredo,[1] CA Nogueira,[1] B Markert,[2] H Heidenreich,[2] S Fränzle,[2] G Liepelt,[2] M Saiki,[1] M Domingos,[3] FM Milian,[4] U Herpin[5]

[1]Instituto de Pesquisas Energéticas e Nucleares, IPEN-CNEN/SPAv. Prof. Lineu Prestes 2242, CEP 05508-000, São Paulo, SP, Brazil
[2]Internationales Hochschulinstitut Zittau, Markt 23, D-02763, Zittau, Germany
[3]Instituto de Botânica, Av. Miguel Stefano No. 3687, CEP 04301-902, São Paulo, SP, Brazil
[4]Universidade de São Paulo, Instituto de Física, R. Do Matão, Trav. R, 187, 05508-900, São Paulo, SP, Brazil
[5]Universidade de São Paulo, NUPEGEL/ESALQ, CP 9, 13418-900 Piracicaba, SP, Brazil

Abstract

In the present work, *Tillandsia usneoides* L., an epiphytic bromeliad, was used as bioindicator of atmospheric metal pollution in São Paulo, Brazil, the biggest city in South America. *Tillandsia* samples were collected from an unpolluted area and were exposed bimonthly at ten sites of the city with different pollution levels and at a control site. Seven trace metals (Ni, Co, Cu, Cd, Pb, V, and Sb) were analysed in the plants by inductively coupled plasma mass spectrometry (ICP-MS) thereafter. The results indicated that Co, Ni, Cd, and V can be attributed to industrial sources while Cu can be associated to both vehicular and industrial sources. Sb is suggested to be influenced mainly by vehicular sources. For Pb, no evident sources could be identified so far as it was spread evenly along the monitoring sites.

Introduction

Biomonitoring of air pollution with plants has been a common practice for many decades. In general, bioindicators can be defined as organisms that can be used for the identification and qualitative determination of human-generated environmental factors, while biomonitors are organisms mainly used for the quantitative determination of contaminants. Bioindicators can be classified as being sensitive or accumulative. Bioaccumulation is the result of the equilibrium process of biota compound intake/discharge from and into the surrounding environment [1].

Spanish moss (*Tillandsia usneoides* L.) and many other species of this genus have been used recently as bioaccumulators, indicators, and monitors of airborne elements [2–4]. Air quality studies have shown that geographic variations in mineral concentrations in plants of *T. usneoides* L. were often correlated with the proximity to aerosol sources, e.g., the ocean, roads, mines, power plants, soils, and urban centres [5].

The city of São Paulo is one of the biggest cities in the world. The metropolitan region of São Paulo (MRSP) has a population of about 18 million people, with about 8000 km^2, with severe environmental problems due to the atmospheric emissions of about 2000 highly pollutant industries and emissions from about 7.8 million motor vehicles [6]. Serious environmental and health problems have specially been observed in the region due to particulate matter (PM$_{10}$) with varied composition [7].

In previous studies [8,9], samples of *T. usneoides*, taken from an unpolluted area, were exposed for 8 weeks at different sites of the city of São Paulo, in order to evaluate the potentiality of this species as a bioindicator of atmospheric metal pollution in São Paulo. Instrumental neutron activation analysis (INAA) was used to analyse trace elements in the plants. This analytical technique allowed the determination of 21 elements, such as As, Ba, Cr, and Zn, and notable correlations of Zn and Ba to vehicular sources and of Cr to industrial emissions sources could be identified. However, in spite of the multi-elemental analysis characteristics of INAA, this analytical technique is not suitable for the determination of Pb, Ni, and Cd, which are also relevant elements in pollution studies. Thus, in the present study, seven trace metals (Co, Ni, Cu, Cd, Pb, V, and Sb) were analysed in the *T. usneoides* samples by inductively coupled plasma mass spectrometry (ICP-MS), providing a more complete description of the atmospheric metal pollution in São Paulo.

Sampling and exposure sites

T. usneoides samples were sampled at a small farm with natural vegetation, located in the city of Mogi das Cruzes, about 70 km to the north-east of São Paulo city. The area has low industrialization and traffic influence and consequently low impact of PM_{10} and metal pollution. All samples were collected in the same area in order to guarantee same origin conditions during the experiments.

In this study, ten monitoring sites next to automatic monitoring stations operated by the government agency of air quality control (CETESB) were chosen in MRSP to transplant the samples of *T. usneoides* (Fig. 1). Seven sites were situated in the city of São Paulo and three in the extended city (Santo André, São Caetano, and Mauá), having different levels of PM_{10} and metal pollution. Four exposure sites were located in the downtown area, representing a car shift region, where 20% of the total number of vehicles are not allowed to run one day of the week, by the number of the license plate (PI, DP, CC, and IB; Fig. 1). The transplantation of plants was also performed at Mogi das Cruzes, which served as the control site. Each exposed sample was composed of 5 g of plants, tied by Teflon strings to a gyrator apparatus (six samples per apparatus, 1.5 m above the soil), which turned

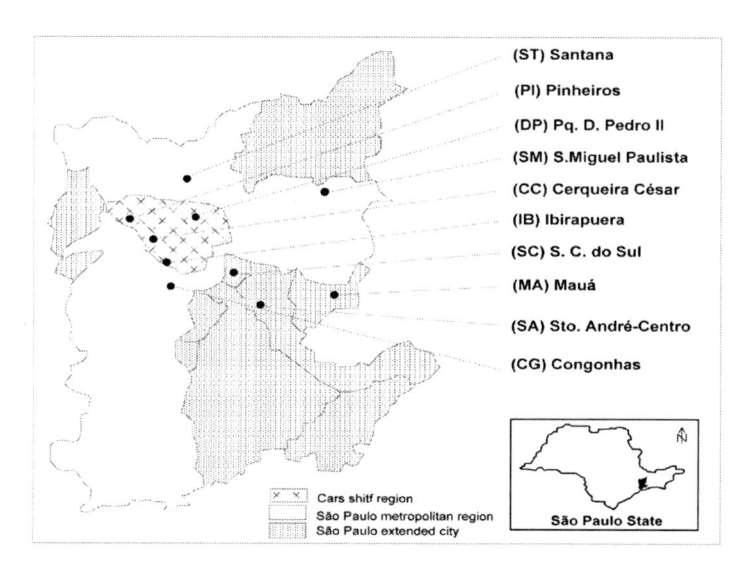

Fig. 1. Exposure sites of *Tillandsia usneoides* plants in the metropolitan region of São Paulo. Car shift region: area where 20% of the total number of vehicles are not allowed to run one day of the week by the number of the license plate

with the wind so that homogenous contact with air contaminants was guaranteed. The samples were submitted to exposure for 8 weeks and, after exposure time, were substituted by new samples allowing to performing eight transplantation experiments (A = April–May/2002; B = June–July/2002; C = November/2002–January/2003; D = February–April/2003; E = April–May/2003; F = June–July/2003; G = September–October/2003; H = November/2003–January/2004).

Experimental

T. usneoides samples were freeze-dried without washing and homogenized using an agate vibratory micro-mill. At the IHI Zittau laboratories, the *Tillandsia* samples were digested by microwave-assisted dissolution using nitric acid and hydrogen peroxide (MLS microwave oven; PTFE reaction vessels). The measurements were done by ICP-MS, using a Perkin–Elmer–Elan DRC. To evaluate the accuracy of the data, three biological reference materials were analysed: peach leaves (NIST SRM 1547), tea (GBW 08505), and white cabbage (BCR 679). The results obtained presented accuracy better than 15% in relation to the certified and recommended values.

Results and discussion

Figures 2A–G show the enrichment of the element concentrations in *T. usneoides* exposed at the monitoring sites in relation to the concentrations measured in plants from the control site during the different monitoring periods. This relation, in percentage, was calculated by:

$$RC_E = (CE_A - CE_C / CE_C) \times 100 \qquad (1)$$

where:

RC_E = Enrichment (%) of the concentration of the element E
CE_A = Concentration of the element E in the sample
CE_C = Concentration of the element E in the control sample

For Co and Ni, there was a significant enrichment in São Miguel Paulista (SM), compared to the other sites (Figs. 2A and B). In this area, there is a metal processing plant, which produces about 16,000 t year^{-1} of Ni and 600 t year^{-1} of Co, which may indicate that Co and Ni contents of *T. usneoides* are related to emissions from the mentioned metal plant.

(A)

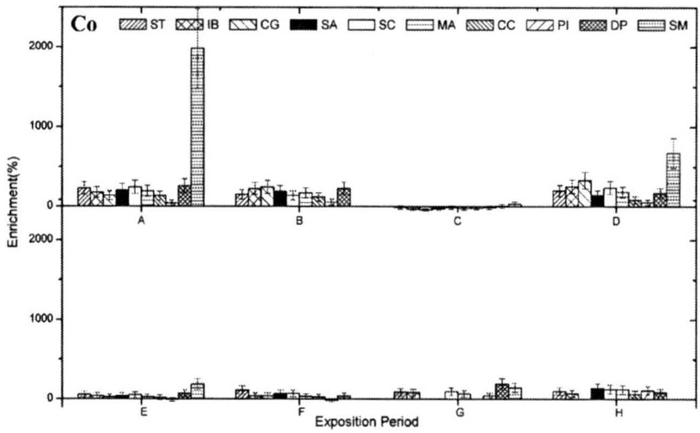
Co
ST IB CG SA SC MA CC PI DP SM
Enrichment(%)
Exposition Period
A B C D E F G H

(B)

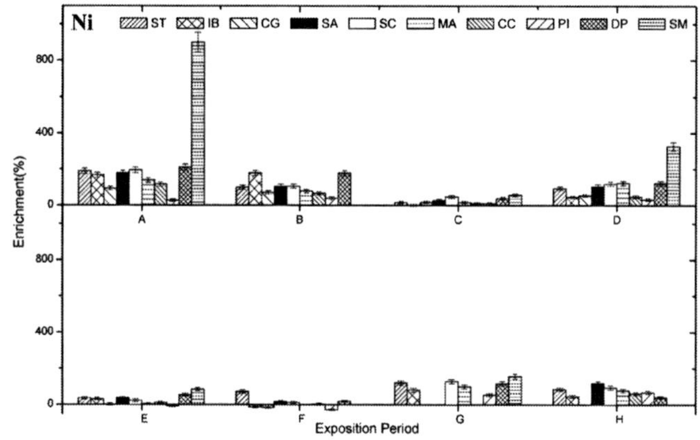
Ni
ST IB CG SA SC MA CC PI DP SM
Enrichment(%)
Exposition Period
A B C D E F G H

(C)

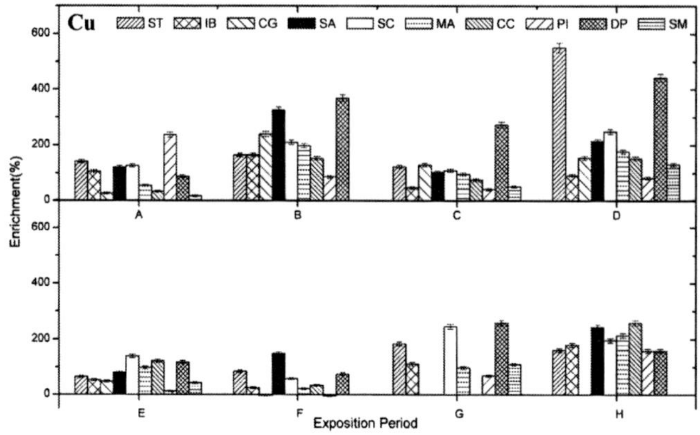
Cu
ST IB CG SA SC MA CC PI DP SM
Enrichment(%)
Exposition Period
A B C D E F G H

(D)

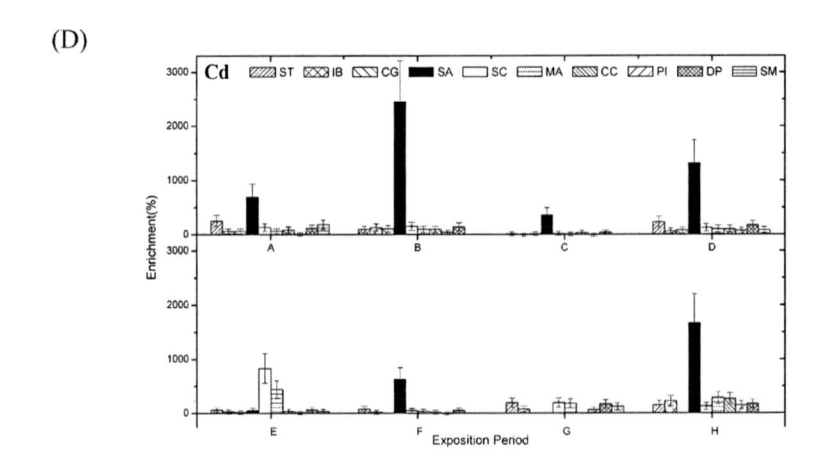

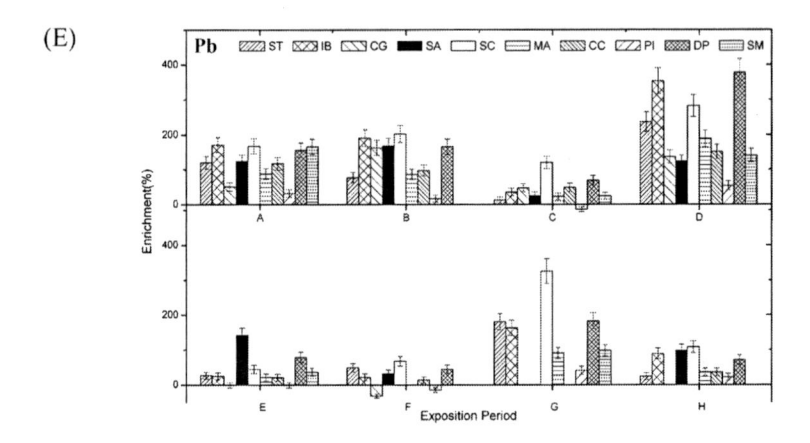

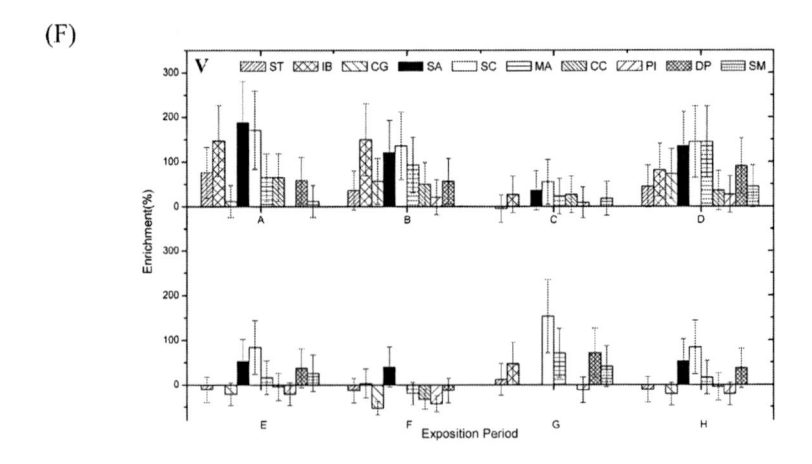

(E)

(F)

(G)

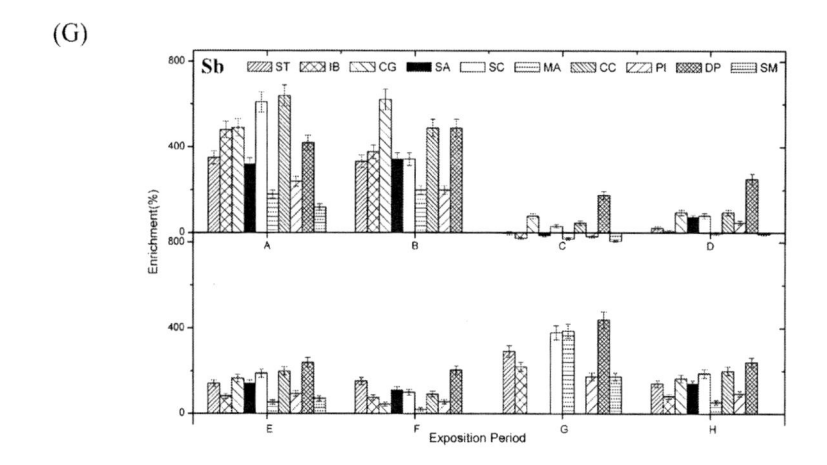

Fig. 2. Enrichment of the elements (%) in *T. usneoides* in relation to the exposure sites and exposure periods. (A) Co; (B) Ni; (C) Cu; (D) Cd; (E) Pb; (F) V; (G) Sb. Exposure period: A = Apr–May/2002; B = June–July/2002; C = Nov/2002–Jan/2003; D = Feb–Apr/2003; E = Apr–May/2003. F = June–July/2003; G = Sept–Oct/2003; H = Nov/2003–Jan/2004. Stations: ST = Santana; IB = Ibirapuera; CG = Congonhas; SA = Santo André; SC = São Caetano; MA = Mauá; CC = Cerqueira César; PI = Pinheiros; DP = Parque D. Pedro; SM = São Miguel

For a better understanding of the behaviour of the other elements analysed, a cluster analysis was applied by using Statistica software and the dendrogram obtained from the data analysis is presented in Figure 3.

The dendrogram shows that V and Cd are in one group. Vanadium is considered a toxic element and is associated with industrialized areas [10].

This element has increased during recent years in urban environments due to industrial dusts. In this study, in fact, high levels of vanadium were observed in Santo André (SA), São Caetano (SC), and Mauá (MA), which are highly industrialized (Fig. 2F). The highest enrichments of Cd were also observed in the same sites (Fig. 2D), and may be associated with the industrial emissions.

T. usneoides showed high concentration of copper in Parque D. Pedro (DP), downtown, near big avenues with high volume of traffic, and in SA (Fig. 2C), with high density of metallurgic industries, indicating industrial and vehicular sources probably caused by the abrasion of car brakes.

Antimony was highly concentrated in plants exposed in DP, and is suggested to be mainly associated with vehicular sources.

Pb (Fig. 2E) is still widespread in spite of lead-free gasoline in Brazil being available since 1983. However, fuel containing lead continues to be

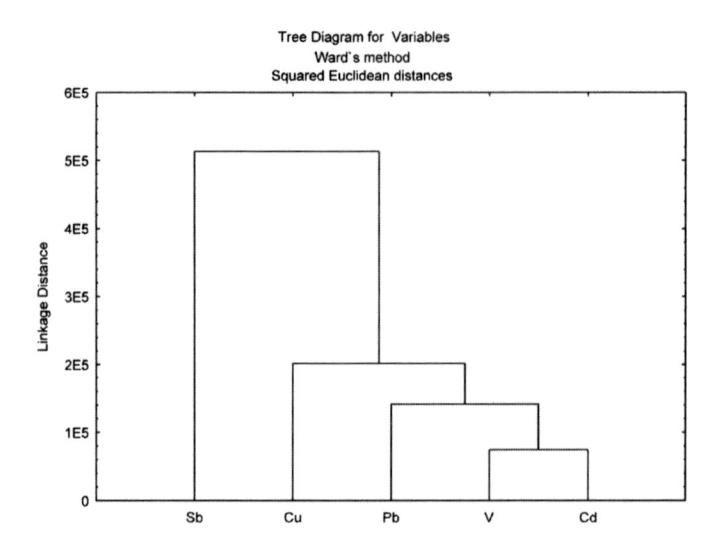

Fig. 3. Dendrogram obtained from the data of Pb, Cd, Cu, V, and Sb

used in aircraft and helicopters (São Paulo has intense helicopter traffic), which may explain this behaviour.

Conclusions

Traffic-related elements such as Cu and Sb presented high concentrations in exposure sites near heavy traffic avenues (cars, buses, and trucks) and may be associated to vehicular sources. For Cd, V, Co, and Ni, the highest contents were related to industrial zones and can be associated to the presence of anthropogenic emission sources.

Acknowledgements

The authors thank the Brazilian financial agencies FAPESP, CAPES, and CNPq for financial support and CETESB for allowing the use of the monitoring stations.

References

1. Conti ME, Cechetti G (2001) Biological monitoring: lichens as bioindicators of air pollution assessment – a review. Environ Pollut 114:471–492
2. Brighigna L, Ravanelli M, Minelli A, Ercoli L (1997) The use of an epiphyte (*Tillandsia* caput medusae morren) as bioindicator of air pollution in Costa Rica. Sci Total Environ 198:175–180
3. Calasans CF, Malm O (1997) Elemental mercury contamination survey in chlor-alkali plant by the use of transplanted Spanish moss, *Tillandsia usneoides* (L.). Sci Total Environ 208:165–177
4. Pignata ML, Gudiño GL, Wannaz ED, Plá RR, González CM, Carreras HA, Orellana L (2002) Atmospheric quality and distribution of heavy metals in Argentina employing *Tillandsia capillaris* as a biomonitor. Environ Pollut 120:59–68
5. Husk GJ, Weishampel JF, Schlesinger WH (2004) Mineral dynamics in Spanish moss, *Tillandsia usneoides* L. (Bromeliaceae), from Central Florida, USA. Sci Total Environ 321:165–172
6. CETESB (2005) Relatório de qualidade do ar no Estado de São Paulo – 2004. Série Relatórios
7. Molina MJ, Molina LT (2004) Megacities and atmospheric pollution. J Air Waste Manage Assoc 54:644–680
8. Figueiredo AMG, Saiki M, Ticianelli RB, Domingos M, Alves ES, (2001). Determination of trace elements in *Tillandsia usneoides* by neutron activation analysis for environmental biomonitoring. J Radioanal Nucl Chem 249:391–395
9. Figueiredo AMG, Alcalá AL, Ticianelli RB, Domingos M, Saiki M (2004). The use of *Tillandsia usneoides* L. as bioindicator of air pollution in São Paulo, Brazil. J Radioanal Nucl Chem 259:59–63
10. Wannaz ED, Carreras HA, Pérez CA, Pignata ML (2006). Assessment of heavy metal accumulation in two species of Tillandsia in relation to atmospheric emission sources in Argentina. Sci Total Environ 361:267–278

On-line matrix separation for the determination of PGEs in sediments by ICP-MS

A De Boni,[1] W Cairns,[2] G Capodaglio,[1,2] P Cescon,[2,1] G Cozzi,[1] S Rauch,[3,4] HF Hemond,[4] C Boutron,[5] C Barbante[1,2]

[1]Department of Environmental Science, University of Venice, Italy
[2]Institute for the Dynamics of Environmental Processes – CNR, Venice, Italy
[3]Chalmers University of Technology, Goteborg, Sweden
[4]Massachusetts Institute of Technology, Cambridge, USA
[5]Laboratoire de Glaciologie et Géophysique de l'Environnement – UMR CNRS/Université Joseph Fourier, Grenoble, France

Abstract

The use of commercially available cation exchange cartridges for the on/off-line removal of the major interferences of platinum group elements (PGEs) during analysis by inductively coupled plasma mass spectrometry (ICP-MS) was investigated and the method was validated using standard solutions and a certified reference material (BCR-723), road dust. The developed method was then applied to the analysis of dated sediment cores taken from the Venetian Lagoon. The results obtained showed enrichment for the PGEs when compared to the mean crustal values, and were in good agreement with the results found by other research groups for sediments from different parts of the globe.

Introduction

More and more restrictive regulations to control air pollution from emissions from cars became one of the most important aspects of the development and research activity on internal combustion engines in the last few decades.

After-treatment technologies have been developed to enable new vehicles to comply with the emission standards established by the latest legislation.

The main pollutants from vehicle exhaust are hydrocarbons (HC), formed by the incomplete combustion of the fuel, carbon monoxide (CO) and nitrogen oxides (NO$_x$); their emissions from cars can be reduced by more than the 90% through the employment of catalytic converters, whose main active components are Pt, Pd, and Rh, often referred to as platinum group elements (PGEs).

Unfortunately, the ever increasing employment of the catalytic converters has lead to an increasingly widespread emission of PGEs into the environment; in fact, it is well known that, as a consequence of mechanical and thermal abrasion, these elements are released into the environment in amounts that are influenced by the speed of the automobile, type of engine, type, and age of catalyst [1,2]. A great number of studies have dealt with the dispersion and accumulation of PGEs in several environmental matrices (see for instance [3] and references listed herein). Of course, to assess the dispersion of PGEs into the environment, it is important to obtain time series of their changing occurrence in environmental archives, such as snow/ice and sediments. Studies of the levels of PGEs in snow and ice by Barbante et al. [4,5] in samples from the northern hemisphere have demonstrated that the levels have dramatically increased since the widespread introduction of catalytic converters. This data trend is also seen in studies on sediments from a lake in North America [6].

It is also known that a fraction of the PGEs released by cars is bio-available [7], so it has become really important to have a reliable analytical technique for the accurate determination of PGEs in environmental matrices. The determination of PGEs in environmental matrices is made very difficult by their very low concentrations and requires the use of very sensitive analytical techniques, such as inductively coupled plasma mass spectrometry (ICP-MS). In spite of the high sensitivity of the ICP-MS technique, the direct determination of PGEs is sometimes hampered by the presence of strong chemical interferences. For this reason it is necessary to introduce either a mathematical correction method [8,9] or a matrix separation technique for the removal of these interferences. In the case of geological samples, fusion with sodium peroxide followed by tellurium co-precipitation [10] or NiS fire-assay [11] has been successfully used, but the extensive sample handling, the high blank levels and the incomplete recovery of PGEs do not allow an easy application to environmental samples. Other suitable methods for the determination of PGEs employ the preconcentration of the analytes [12–15] or matrix separation [16–20] using ion chromatography or micro-columns for the removal of the interferences. The latter method exploits the property of PGEs to form anionic chlorocomplexes in HCl media [21], which are unable to interact with the cationic surface of the resin and pass through the column, while interferences, present mainly as positive species, are retained on the resin itself.

In this work the same property of PGEs has been exploited by using much smaller cation exchange columns to carry out the on/off-line separation of the matrix. The study of the retention of the major interferences and the elution of the PGEs has been carried out and the method was validated by analysing a certified reference material (BCR-723), road dust. The developed method was then applied to the analysis of dated sediment cores taken from the Venetian Lagoon.

Experimental – Method

Instrumentation

The method development and the optimization of the matrix separation were carried out on an Agilent 7500i inductively coupled plasma quadrupole mass spectrometry (ICP-QMS) (Agilent Technologies, Yokogawa Analytical Systems, Tokyo, Japan) fitted with the Integrated Sample Introduction System (ISIS), consisting of two extra peristaltic pumps and two 6-port valves under full computer control, an ASX510 autosampler (CETAC Technologies, Omaha, USA) and a PFA double pass spray chamber (CPI International, the Netherlands), thermostated to 2°C, fitted with the standard Agilent V-groove low flow nebulizer (Agilent Technologies). The method validation and the sample analysis were carried out both on the Agilent 7500i ICP-MS, fitted with a PolyPro concentric nebulizer (ESI, Omaha, USA) and on an Element2 inductively coupled plasma sector field mass spectrometry (ICP-SFMS) (Thermo Finnigan MAT, Bremen, Germany) working in low resolution mode, fitted with a PFA microflow nebulizer (ESI, Omaha, USA).

The strong cation exchange (SCE) cartridges used in this work, were Maxi Clean cartridges (Alltech Associates Inc., Deerfield, IL, USA), which consist of a styrene divinylbenzene polymer with a sulphonic acid functional group (particle size range, 45–150 μm, molecular exclusion limit, 1000 Da, exchange capacity, 2.0 meq mL^{-1}).

Reagents

High purity deionized water (18 MΩ cm^{-1} resistivity) (Purelab Ultra, Elga, High Wycombe, UK) and high purity acids (Suprapur grade, Merck,

Darmstadt, Germany) were used throughout and all solutions were prepared in a class 1000 clean room under a class 100 laminar flow bench.

Standards of the PGEs (Pt, Pd, and Rh) and the main interferences were prepared from single element standards and a multielement standard, respectively (CPI International, the Netherlands).

BCR-723 urban road dust certified reference material (IRMM, Geel, Belgium) was used for quality control.

Sample collection

Six duplicate sediment cores were taken from near by the bridge that connects Venice to the mainland at distances from the bridge of 1 m (core pair 1), every 10 m (core pairs 2–5), and at ~200 m (core pair 6). For the sample collection a tube with a 7 cm internal diameter was used. The plastic tubes were closed with plastic film and then capped both at the bottom and at the top, and placed vertically in order to avoid any disturbance of the core.

Sample preparation

The sediment cores were kept frozen in a cold room at –20°C until sample preparation, which was carried out while they were still frozen. Starting from the top they were cut into 1 cm slices for the first 8 cm and then into 2 cm slices for the rest of the core. From each slice the inner part of 4 cm diameter was taken, weighed, and dried to constant weight in an oven set at 105°C. The sediment was homogenized with a pestle and mortar and sieved through a 2 mm sieve.

A 50 mg sample of sediment was digested in a microwave digester (Milestone FKV, Ethos 1600, Sorisole, Italy) using the operating parameters described in Table 1 and the following mixture of reagents: 4 mL H_2O, 3 mL HCl, and 1 mL HNO_3. The digest was then cooled, made up to 25 mL and was stored at 4°C prior to analysis.

Table 1. Operating conditions for microwave-assisted acid digestion of 50 mg of sediment or BCR-723 certified reference material

Step	Power (W)	Time (min)
1	250	10
2	400	10
3	650	8
4	400	10
5	250	10

About 50 mg of the certified reference material BCR-723 urban road dust was digested in the same manner to control the accuracy of the methodology.

Preparation of the strong cation exchange cartridges

Before use, the SCE cartridges were conditioned by passing ultra-pure water for an hour, and then washed with 1% (v/v) hydrochloric acid for 10 min at a flow rate of ~0.7 mL min^{-1}. The retention efficiency of the column for the matrix was tested and the acid strength of the digest optimized before use [22].

Between two samples the micro-column was washed with 30% (v/v) hydrochloric acid plus 1% (v/v) hydrofluoric acid for 20 min, and then reconditioned with 1% (v/v) hydrochloric acid for 10 min at a flow rate of ~0.7 mL min^{-1}.

Analytical procedure

To monitor the signal during analysis a time-resolved analysis acquisition was chosen. During each experiment the following phases were followed as the signal was acquired: initially a 1% (v/v) hydrochloric acid solution (blank solution) with no column in-line was analysed until a steady state signal was reach; after the reaching of a stable signal for the blank solution, the sample capillary was switched to the sample solution to obtain a response plateau of the analytes without the column in-line; the micro-column was inserted in-line during the analysis of the sample solution; signal monitoring was then stopped and the column regenerated as described above.

Results and discussion

Analysis of the BCR-723 urban road dust reference material

To study the retention and elution dynamics of the main interferents and of the PGEs, 50 mg of road dust reference material (BCR-723 urban road dust) was digested as described above, then cooled, made up to 25 mL and diluted 1:1 with water before matrix separation.

For this the micro-columns were coupled on-line to the ICP-QMS. This is made possible by using the ISIS.

Figures 1 and 2 show the results obtained respectively for the main interferences and for PGEs by analysing the road dust reference material.

In Figure 1 the blank solution signal can be seen between 0 and ~160 s; at ~160 s starts the signal of the sample before the insertion of the column, which takes place at ~240 s. After the insertion of the micro-column the signal drops (at ~270 s) due to the elution from the micro-column of the acid solution used to condition the micro-column itself. After the washing out of the clean acid solution, a steady state signal is reached as the major interferences are retained by the micro-column.

Figure 2 shows the blank solution signal between 0 and ~160 s, the sample peak between ~160 and 240 s and the drop in the signal at ~270 s due to the elution from the micro-column of the conditioning solution. From ~320 s the signal starts to increase because the PGEs are not retained by the micro-column. Pd and Rh show a drop in signal, indicating that without interferences removal, they are subjected to heavy interferences. From Figure 2 it could seem that Pd is retained by the micro-column because the drop in the signal is very big, but actually the mean signals before (~13 cps) and after (~30 cps) the insertion of the micro-column are significantly different. On the contrary, the Pt signal is unchanged before and after matrix removal, probably due to the fact that when HF is not used in the digestion mixture, Hf, the main interference for Pt, remains undigested (as most of the Hf in the samples comes from the ceramic substrate of the catalytic converters)

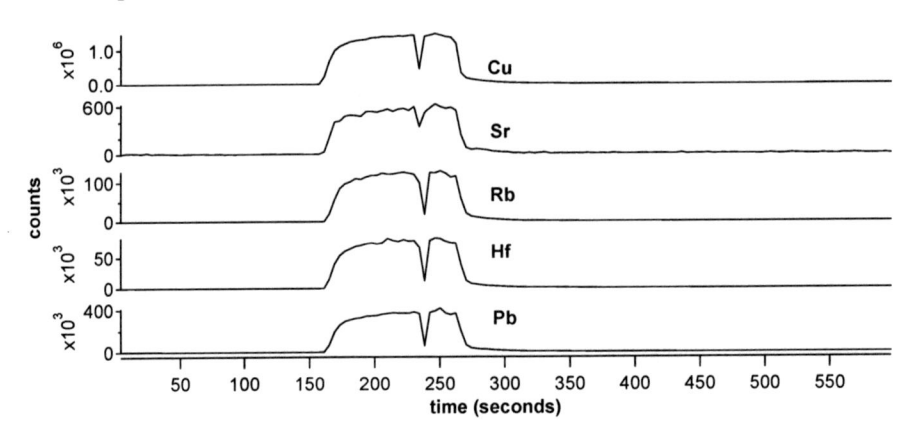

Fig. 1. Signal obtained for the major interferents in a 50 mg sample of BCR-723 road dust certified reference material digested in 4 mL H_2O, 3 mL HCl, and 1 mL HNO_3, made up to 25 mL and diluted 1:1, before and after matrix separation at a flow rate of 0.7 mL min^{-1}

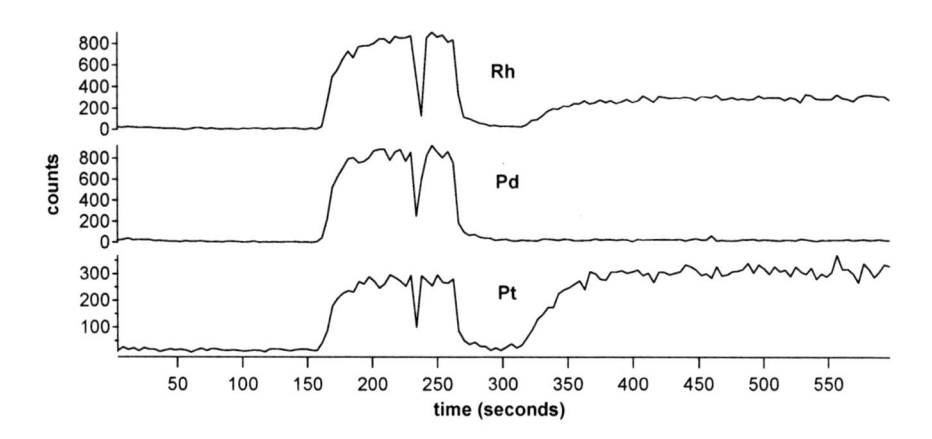

Fig. 2. Signal obtained for the PGEs in a 50 mg sample of BCR-723 road dust certified reference material digested in 4 mL H$_2$O, 3 mL HCl, and 1 mL HNO$_3$, made up to 25 mL and diluted 1:1, before and after matrix separation at a flow rate of 0.7 mL min^{-1}

and the small fraction that is mobilized is unable to form anionic complexes that normally elute with the PGEs and cause an increase in the Pt signal.

Quantitative analysis of the BCR-723 urban road dust reference material

The method validation was carried out by the analysis of a sample of the road dust reference material (BCR-723 urban road dust). About 50 mg of road dust reference material (BCR-723 urban road dust) was digested as explained before, then cooled, and made up to 25 mL. A standard additions calibration was done: standard additions of 0, 5, 13, 26, and 52 ng L^{-1} of Rh, 0, 5, 11, and 17 ng L^{-1} of Pd, and 0, 27, 53 and 107 ng L^{-1} of Pt were carried out before the matrix separation. The standard solutions were carried out before the matrix separation. The standard solutions were prepared in 20% (v/v) aqua regia. The interferences removal was done off-line and the calibration solutions were analysed with both ICP-QMS and ICP-SFMS.

The results obtained are summarized in Table 2 together with the method detection limits obtained with the ICP-QMS by 20 repetitions of a blank digest solution after passing through a micro-column.

Good calibration curves and good agreement between the results and the certified values were obtained, except for Pd with the ICP-SFMS. This is probably due to the presence of residual interferences that can be ameliorated by further optimization of the ICP-SFMS method.

Table 2. Results obtained from the analysis of the urban road dust reference material (BCR-723) and detection limits (3 × SD of the blank) obtained by 20 repetitions of a blank digest solution after passing through a micro-column

Element		Concentration found (ng g^{-1})	Certified value (ng g^{-1})	Detection limit in the solid (pg g^{-1})
ICP-SFMS	Rh	14.9 ± 0.1	12.8 ± 1.2	
	Pd		6.0 ± 1.8	
	Pt	79.4 ± 1.6	81.3 ± 3.3	
ICP-QMS	Rh	14.3 ± 0.1	12.8 ± 1.2	0.15
	Pd	4.7 ± 0.4	6.0 ± 1.8	0.62
	Pt	80.7 ± 0.4	81.3 ± 3.3	0.24

Analysis of sediments cores from the Venetian Lagoon

The developed method was applied to the analysis of dated sediment cores taken from the lagoon of Venice. It is well known that inside the city of Venice there are no cars, they are all stopped in a big car park at the entrance of the city. The bridge that connects Venice with the mainland has a very high volume of traffic (up to 30,000 vehicles per day cross the bridge when traffic volume is at its highest). This gave us the idea of collecting some sediment cores close to the bridge to see if an increase in the PGEs' concentration was present compared with the mean value for these elements in the Continental Crust.

Three sediment cores have been analysed up to now: the first, the third, and the sixth collected southward from halfway along the bridge. Both instruments have been employed to analyse the sediment cores, and a standard additions calibration method was used. The results obtained for Rh and Pt with the ICP-QMS were close to or lower than the detection limits, whereas, as already seen when analysing the road dust reference material (BCR-723), the best results for Pd were obtained with the ICP-QMS. For these reasons, the results obtained for Pt and Rh with the ICP-SFMS and those obtained for Pd with the ICP-QMS are presented in Table 3. In Table 3 a comparison between the obtained PGEs concentrations in the sediments and those of the continental crust are shown; the mean values obtained for each core are reported in this table with the ranges of values found in each core in brackets underneath.

Table 3 shows that the results obtained for Pd and Rh are much higher than those of the continental crust, whereas for Pt the increase is less significant, even if some samples are much higher than the crustal value. However,

Table 3. Comparison between the mean concentrations of PGEs (ng g^{-1}) in the continental crust and those obtained in sediments of the Venetian Lagoon

		Rh	Pd	Pt
Earth crust		0.06	0.4	0.4
Sediments	Core no. 1	0.96	4.81	0.59
		(0.062–3.21)	(1.61–9.15)	(0.024–3.96)
	Core no. 3	1.96	6.47	1.19
		(0.56–7.10)	(1.43–30.1)	(0.33–10.04)
	Core no. 6	5.32	7.48	0.97
		(1.19–43.64)	(1.62–34.07)	(0.44–6.30)

we have a high confidence in our results as they are in good agreement with those obtained by other research groups for river and surface sediments [3].

Conclusions

This study has shown that commercially available reusable cation exchange micro-columns are able to remove the interferences that prevent the direct determination of PGEs by ICP-MS. The main advantages of using these small columns are that they can be reconditioned and used many times; moreover, they can work with small sample, elution, and regeneration volumes. In this way the matrix separation becomes simpler and less time consuming.

The method has been successfully applied to an urban road dust reference material (BCR-723); good agreement between the results and the certified values were obtained, even if the ICP-SFMS method has still to be optimized for the determination of Pd.

Moreover, the method has been applied to the analysis of Venetian Lagoon sediment cores. The results obtained for Rh and Pt with the ICP-QMS were close to or below the detection limits, and the calibration curve for Pd obtained with the ICP-SFMS was ideal, so the employment of both both instruments was necessary.

Acknowledgements

This research was funded by the Alliance for Global Sustainability, Project 2227. The financial support of the Agence de l'Environnement et de la Maîtrise de l'Energie (contract no. 0566C0096) and of Istituto Nazionale della Montagna (IMONT) are also kindly acknowledged.

References

1. Moldovan M, Palacios MA, Gomez MM, Morrison G, Rauch S, McLeod C, Ma R, Caroli S, Alimonti A, Petrucci F, Bocca B, Schramel P, Zischka M, Pettersson C, Wass U, Luna M, Saenz JC, Santamaria J (2002) Environmental risk of particulate and soluble platinum group elements released from gasoline and diesel engine catalytic converters. Sci Total Environ 296:199–208
2. Artelt S, Kock H, Konig HP, Levsen K, Rosner G (1999) Engine dynamometer experiments: platinum emissions from differently aged three-way catalytic converters. Atmos Environ 33:3559–3567
3. Ravindra K, Bencs L, Van Grieken R (2004) Platinum group elements in the environment and their health risk. Sci Total Environ 318:1–43
4. Barbante C, Veysseyre A, Ferrari C, Van de Velde K, Morel C, Capodaglio G, Cescon P, Scarponi G, Boutron C (2001) Greenland snow evidence of large scale atmospheric contamination for platinum, palladium, and rhodium. Environ Sci Technol 35:835–839
5. Barbante C, Schwikowski M, Döring T, Gäggeler HW, Schotterer U, Tobler L, Van de Velde K, Ferrari C, Cozzi G, Turetta A, Rosman K, Bolshov M, Capodaglio G, Cescon P, Boutron C (2004) Historical record of european emissions of heavy metals to the atmosphere since the 1650s from alpine snow/ice cores near Monte Rosa. Environ Sci Technol 38:4085–4090
6. Rauch S; Hemond HF (2003) Sediment-based evidence of platinum concentration changes in an urban lake near Boston, Massachusetts. Environ Sci Technol 37:3283–3288
7. Sures B, Zimmermann S, Messerschmidt J, Von Bohlen A (2002) Relevance and analysis of traffic related platinum group metals (Pt, Pd, Rh) in the aquatic biosphere, with emphasis on palladium. Ecotoxicology 11:385–392
8. Gomez MB, Gomez MM, Palacios MA (2000) Control of interferences in the determination of Pt, Pd and Rh in airborne particulate matter by inductively coupled plasma mass spectrometry. Anal Chim Acta 404:285–294
9. Moldovan M, Gomez, MM, Palacios MA (1999) Determination of platinum, rhodium and palladium in car exhaust fumes. J Anal At Spectrom 14:1163–1169
10. Jin X, Zhu H (2000) Determination of platinum group elements and gold in geological samples with ICP-MS using a sodium peroxide fusion and tellurium co-precipitation. J Anal At Spectrom 15:747–751
11. Oguri K, Shimoda G, Tatsumi Y (1999) Quantitative determination of gold and the platinum-group elements in geological samples using improved NiS fire-assay and tellurium coprecipitation with inductively coupled plasma-mass spectrometry (ICP-MS). Chem Geol 157:189–197
12. Benkhedda K, Dimitrova B, Goenaga Infante H, Ivanova E, Adams FC (2003) Simultaneous on-line preconcentration and determination of Pt, Rh and Pd in urine, serum and road dust by flow injection combined with inductively coupled plasma time-of-flight mass spectrometry. J Anal At Spectrom 18:1019–25

13. Cantarero A, Gomez MM, Càmara C, Palacios MA (1994) On-line preconcentration and determination of trace platinum by flow-injection atomic absorption spectrometry. Anal Chim Acta 296:205–211
14. Moldovan M, Gomez MM, Palacios MA (2003) On-line preconcentration of palladium on alumina microcolumns and determination in urban waters by inductively coupled plasma mass spectrometry. Anal Chim Acta 478:209–217
15. Hidalgo MM, Gomez MM, Palacios MA (1996) Trace enrichment and measurement of platinum by flow injection inductively coupled plasma mass spectrometry. Fresenius J Anal Chem 354:420–423
16. Ely JC, Neal CR, O'Neill J, Jain JC (1999) Quantifying the platinum group elements (PGEs) and gold in geological samples using cation exchange pretreatment and ultrasonic nebulization inductively coupled plasma-mass spectrometry (USN-ICP-MS). Chem Geol 157:219–234
17. Bruzzoniti MC, Cavalli S, Mangia A, Mucchino C, Sarzanini C, Tarasco E (2003) Ion chromatography with inductively coupled plasma mass spectrometry, a powerful analytical tool for complex matrices. Estimation of Pt and Pd in environmental samples. J Chromatogr A 997:51–63
18. Jarvis I, Totland MM, Jarvis KE (1997) Determination of the platinum-group elements in geological materials by ICP-MS using microwave digestion, alkali fusion and cation-exchange chromatography. Chem Geol 143:27–42
19. Meisel T, Fellner N, Moser J (2003) A simple procedure for the determination of platinum group elements and rhenium (Ru, Rh, Pd, Re, Os, Ir and Pt) using ID-ICP-MS with an inexpensive on-line matrix separation in geological and environmental materials. J Anal At Spectrom 18:720–726
20. Mukai H, Ambe Y, Morita M (1990) Flow injection inductively coupled plasma mass spectrometry for the determination of platinum in airborne particulate matter. J Anal At Spectrom 5:75–80
21. Ely JC, Neal CR, O'Neill JA, Jain JC (1999) Quantifying the platinum group elements (PGEs) and gold in geological samples using cation exchange pretreatment and ultrasonic nebulization inductively coupled plasma-mass spectrometry (USN-ICP-MS). Chem Geol 157:219–234
22. Cairns W, Varga A, DeBoni A, Limbeck A, Barbante C, Capodaglio G, Boutron C, Cescon C (2006) Micro-column matrix separation for simultaneous platinum group element determination in a road dust certified reference material by ICP-MS (submitted)

Determination of PGE and REE in urban matrices and fingerprinting of traffic emission contamination

M Angelone,[1] F Spaziani,[1] C Cremisini,[1] A Salluzzo[2]

[1]ENEA C.R. Casaccia, V. Anguillarese 302, 00060 Roma, Italy
[2]ENEA C.R. Portici, Località Granatello, 88055 Portici-Napoli, Italy

Abstract

The results of a comparative investigation on urban and natural matrices (soil, road, and tunnel dust) from Latium (central Italy) are used to evaluate the influence of catalyzed traffic emission on the distribution of Pt and rare earth elements (REE). Normalized REE data show that REE distribution pattern in urban soils is strictly related to natural background. Road dust, although collected directly on asphalt and so richer in catalyst particles, do not show at the moment any significant difference: the results evidence that REE in urban samples substantially maintains their original geological mark. In addition, laser ablation inductively coupled plasma mass spectrometry (LA-ICP-MS) technique is utilized to identify and analyse catalyst fragments in samples: in fact, the association of platinum group element–convoy-group element (PGE–Ce) or PGE– lanthanum (La), as high intensity peaks found in urban matrices, reveals the presence of catalyst particles.

Introduction

The study of platinum group elements (PGE) and rare earth elements (REE) distribution in urban environments has become extremely interesting in the last decades, owing to the increasing use of these elements in car catalysts. In the catalyst PGE directly promote the conversion of harmful gases contained in the exhaust fumes (unburned hydrocarbons, nitrogen oxides, and carbon monoxide) to N_2, CO_2, and H_2O. REE, instead, mainly represented in this context by Ce, La, and Nd, stabilize the catalyst support (a γ-alumina based honeycomb) and enhance the oxidation of pollutants. The thermal and mechanical wear of catalyst, however, causes the release of fine particles, emitted in the order of few ng km^{-1}, enriched in PGE and REE: about 66% of them have a diameter >10 μm, while 21% ranges from 3.1 to 10 μm and the remaining 13% is <3.1 μm [1]. Catalyst particles are mainly accumulated in the roadside sites and contribute to the composition of urban dust (a heterogeneous material in which components are differentiated for morphological, chemical, and physical properties).

From a geochemical point of view, and despite the complexity of the urban environment, it is useful to introduce single and multi-element approaches as "tracers" to distinguish the different sources of pollution. For instance, PGE could give direct evidence for traffic emission contribution, Zn and Cd for tyre degradation and building corrosion, while the association of Co, Mn, and V reveals domestic emission. Instead, particles derived from natural soil weathering could be recognized by typical REE pattern and ratios, similar to those of earth crust or geological backgrounds.

The aim of this paper is to study the composition of some urban matrices, collected along the main roads and tunnels of Latium (the Italian region comprising Rome), with particular regard to PGE and REE, to assess the influence of vehicular circulation.

Experimental – Method

Sampling and pre-treatment

Urban samples (soils, road dust, and tunnel dust) were collected in municipal areas of Rome (inhabitants: 2,533,000; city area: 1285 km^2) and Viterbo (inhabitants: 53,300; city area: 406 km^2). We purposely selected these two cities with the aim to compare the result of a metropolis (Rome) and of a small town (Viterbo). Natural soils were collected from rural areas of Latium, including soils derived from volcanic materials, sedimentary rocks, and quaternary deposits. Soil and dust samples were dried at 50°C and sieved at 2 mm. Dust samples were further sieved to obtain the fraction <63 μm, since this is the sample's portion with the highest environmental relevance [2].

Apart from these environmental matrices, in order to better characterize the major pollution source operating in urban sites, we also analysed two used autocatalysts taken from two of the most sold cars in Europe. It is also important to specify that we have chosen the two major catalyst types, a ceramic one and a metallic one, with the aim to point out any differences concerning their PGE and REE contents. The external wrapping of each catalyst was removed and the internal honeycomb structure was grounded and stored.

Sample dissolution for total quantification

Samples were mineralized through a microwave-assisted acid dissolution based on a mixture of ultra-pure reagents (HNO_3–H_2O_2–HF–$HclO_4$). Samples were then heated near to dryness and the residues were dissolved in 1% HNO_3. The analytical procedure was verified on four certified reference materials (CRMs): WGB-1, BCR-723, NIST-2556, NIST-2557 for PGE, and GSP-2 for REE determination. The results obtained on CRMs were in very good agreement to the certified values.

Samples were quantitatively analysed by ICP-MS (Perkin–Elmer–Elan-6100). Spectral interferences were mathematically corrected [3]. However, in many cases, interferences on Pd and Rh were too intense to obtain a reliable determination for these two elements.

PGE–REE association for urban traffic fingerprint

The recognition of catalysed traffic contribution in urban matrices can be obtained through the study of elemental association in solid samples. This objective, however, is impossible to reach by means of conventional ICP-MS analysis. This technique in fact, requires the introduction of a solution, but the necessary dissolution procedure obtains only the total amounts of analytes in the whole sample. Laser ablation inductively coupled plasma mass spectrometry (LA-ICP-MS), instead, working directly on solid samples, can identify catalyst particles emitted with exhaust fumes [4]. In fact, the internal honeycomb structure of each catalyst has always a typical association of a high Ce (and often also La) concentration combined with high concentrations of PGE. Such composition and contemporary association is not common in environmental matrices and represents a characteristic fingerprint to distinguish autocatalyst particles. LA-ICP-MS analyses (using a Cetac-LSX-200 joined to a Perkin–Elmer–Elan-6000) were performed on solid (no grinded) samples. To better identify the influence of catalysed traffic on urban matrices, a subsample of tunnel dust was further sieved to split the fraction <20 µm and 20–63 µm ranges.

Results and discussion

Pt in urban matrices

Results of Pt determination in urban soils and dusts are summarized in Table 1, where there are also listed the Pt background values obtained for Latium natural soils [5]. All urban samples show a Pt enrichment with respect to the soil background. According with other studies [6], the highest Pt concentrations were found in tunnel dust samples (due to limited air circulation and restricted atmospheric influence). The amount of noble metal is highly dependent on the traffic intensity and the tunnel age. Indeed, the latest realized or recent cleaned tunnels, "Castel S. Angelo" and "Collina Fleming", show lower Pt amounts with respect to the other tunnels.

PGE in autocatalysts

The two autocatalysts analysed show a drastic differentiation about PGE contents, clearly related to their dissimilar structure (Table 2). The Pt/Rh ratio is compatible to the one generally associated to autocatalysts used in Europe (~5). From the introduction of the early models, in the mid-1970s, to the most efficient Three-way types (TWC), in the 1980s, catalysts with different PGE contents have been produced to improve engine performances and to lower emissions. The amount of PGE in a catalyst, therefore,

Table 1. Pt contents in Latium urban and natural matrices (ng g^{-1})

Urban matrices	Pt
Rome urban soils	11.2 ± 4.1
Viterbo urban soils	10.3 ± 3.4
Rome tunnel dusts	344 ± 250
Viterbo road dusts	110 ± 26
Latium natural soils	3.8 ± 1.9

Table 2. Pt contents in two European autocatalyst (µg g^{-1}). Means of five replicates

Catalyst type	Pt	Pd	Rh
Ceramic-based	990 ± 21	307±3	218±2
Metallic-based	2424 ± 89	16±1	567±2

can give information on its history, considering as example that Pd came to the fore from 1989 onwards, Rh has been introduced from 1986, while the newest generation of autocatalysts contains mainly Pd. Consequently, the analysed metallic catalyst can be classified as an old TWC (pre-1989), heavily based on Pt, while the ceramic one as a more recent Pt–Pd TWC.

REE in urban matrices

Results of REE determination in urban matrices of Latium are summarized in Table 3. The normalization of REE concentrations with respect to a geological "reference" value is surely a useful tool to obtain a comparison among information from different sites and, sometimes, allows relating samples to "contamination" sources. This approach was applied also on our samples (Fig. 1) using at first the upper continental crust (UCC) average data [7] as reference values (Table 3). Nevertheless, because of the variability and of the wide range of REE concentration in the different parent materials, this normalization does not clearly evidence any possible weak enrichment. For this reason we also normalized our data with the geological background of the studied areas in (average of the available data on selected natural soil of Latium) Figure 2. However, the light enrichments

Table 3. REE mean contents in Latium urban and natural matrices ($\mu g\ g^{-1}$)

REE	Rome urban soils	Rome tunnel dusts	Viterbo urban soils	Viterbo road dusts	Latium sediment soils	Latium volcanic soils	Latium geological background	UCC [7]
La	102 ± 4	51 ± 16	103 ± 15	88 ± 13	29 ± 13	99 ± 3	75 ± 31	30
Ce	295 ± 17	106 ± 39	198 ± 27	175 ± 24	49 ± 17	301 ± 17	192 ± 105	64
Pr	25 ± 1	11 ± 4	22 ± 3	19 ± 3	6.8 ± 2	25 ± 1	17 ± 8	7.10
Nd	64 ± 7	42 ± 15	87 ± 16	71 ± 11	28 ± 12	62 ± 4	57 ± 17	26
Sm	16 ± 0.8	7 ± 36	12 ± 2	10 ± 1	5.3 ± 2	16 ± 1	10 ± 5	4.50
Eu	3.9 ± 0.1	1.7 ± 0.8	2.8 ± 0.5	2.6 ± 0.3	1.4 ± 0.5	3.4 ± 0.6	2.3 ± 1.1	0.88
Gd	15 ± 0.8	6.3 ± 3	11 ± 2	10 ± 1	6.3 ± 3	15 ± 1	9.8 ± 4.7	3.80
Tb	1.3 ± 0.3	1.2 ± 0.1	1.4 ± 0.2	1.2 ± 0.2	1 ± 0.1	1.2 ± 0.1	1.1 ± 0.2	0.64
Dy	7.8 ± 0.4	3.4 ± 2	6.5 ± 1	5.7 ± 0.6	4.6 ± 2	8.2 ± 0.4	6 ± 2.4	3.50
Ho	1.1 ± 0.1	1.2 ± 0.1	1.1 ± 0.1	0.9 ± 0.1	0.6 ± 0.1	1.1 ± 0.1	0.9 ± 0.1	0.80
Er	3.7 ± 0.2	1.5 ± 0.7	3.2 ± 0.4	2.9 ± 0.3	2.7 ± 1.6	4.1 ± 0.1	3.2 ± 1.3	2.30
Tm	0.4 ± .01	0.3 ± 0.01	0.4 ± 0.05	0.3 ± 0.04	0.2 ± 0.02	0.4 ± 0.02	0.3 ± 0.04	0.33
Yb	2.7 ± 0.2	1.1 ± 0.5	2.6 ± 0.3	2.4 ± 0.3	2.2 ± 1	3.1 ± 0.4	2.6 ± 1	2.20
Lu	0.4 ± .02	0.2 ± 0.08	0.4 ±0.05	0.3 ± 0.05	0.3 ± 0.20	0.5 ± 0.07	0.4 ± 0.2	0.32
ΣLREE	505.9	218.7	424.8	365.6	119.5	506.4	353.3	132.5
ΣHREE	32.4	15.2	26.6	23.7	17.9	33.6	24.3	13.89
ΣREE	538.4	233.9	451.4	392.2	137.4	540	377.6	146.4

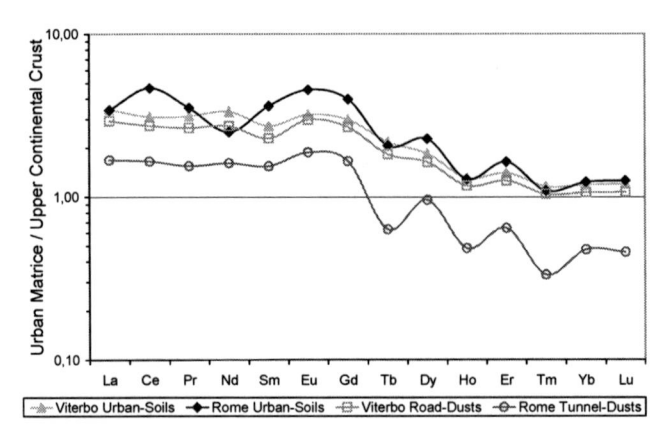

Fig. 1. Latium REE pattern with respect to UCC

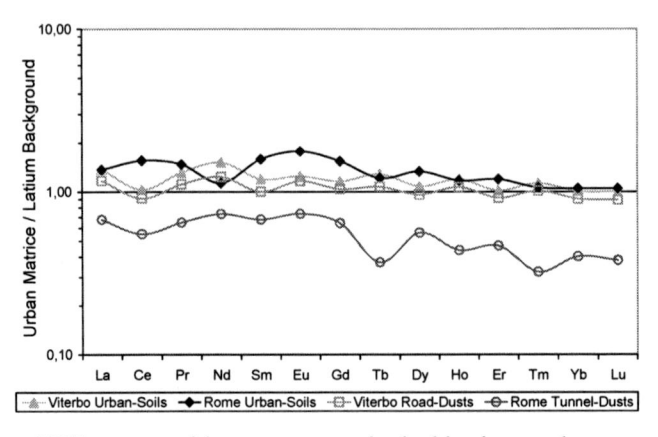

Fig. 2. Latium REE pattern with respect to geological background

found in this way in Latium urban soils and road dust cannot be directly and entirely attributed to the vehicular traffic, since in urban environment both natural and anthropic sources contribute to the REE contents. Besides, road dusts of Viterbo show a REE pattern very similar to urban soils of the same area, evidencing that road dusts mainly derive from soil particles transported by local winds. Tunnel dusts of Rome, instead, also showing a pattern enough similar to the others, are depleted in REE with respect to both their geological background and urban soils, as a result of the lower degree of interaction and particles exchange between tunnel and the outdoor environment (this type of dust, in fact, is mainly composed by particles derived from the decomposition of the structural materials of tunnels).

The relative concentration sequences of REE in the studied environmental matrices are very similar to those of related parent materials as reported in Table 3. In Rome urban soils the concentration sequence is Ce > La >

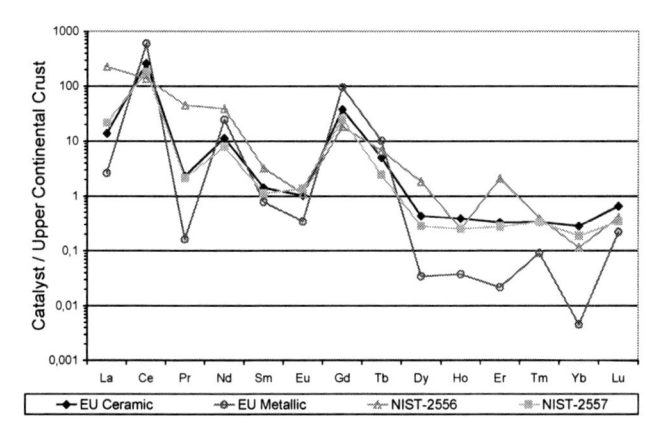

Fig. 3. Catalysts REE pattern with respect to UCC

ND > Pr > Sm > Gd > Dy > Eu > Er > Yb > Tb > Ho > Tm > Lu, substantially coinciding with that of tunnel dusts (with the slight difference Tb = Ho > Yb). A similar sequence characterize also Viterbo urban soils and road dust, where the only differentiation is that Er > Eu. These sequences similarity means that, despite anthropic contribute, urban soils still preserve their original REE distribution.

REE in autocatalysts

Data on REE concentrations in autocatalysts are reported in Table 4. The normalization with respect to the UCC (Fig. 3) evidences characteristic patterns related to differences in manufacturing technique. These patterns can be used, through a comparison with the ones of known catalysts, to describe and rapidly identify the typology of a catalyst. In fact, our ceramic European autocatalyst has a pattern strictly similar to NIST-2557, therefore we can declare that these two catalyst are based on an equivalent structure. The metallic European autocatalyst, instead, is clearly different to the others, since it is not based on a ceramic structure (like the ones mentioned above) and because the NIST-2556 is based on an aged configuration (called Pellet) never adopted in Europe.

Multi-element approach

Among urban matrices, dust can be considered the most relevant for the human health, since the particles <10 μm can be easily inhaled. This fraction that represents about 35% of the catalyst fragments emitted with exhaust fumes can enter into contact with peoples and especially children (breathing, skin exposition, etc.).

Table 4. REE mean contents in autocatalysts ($\mu g\ g^{-1}$)

REE	Ceramic	Metallic
La	412 ± 07	80 ± 2
Ce	16650 ± 374	38440 ± 930
Pr	16 ± 0.6	1.1 ± 0.2
Nd	295 ± 21	634 ± 15
Sm	6.4 ± 0.1	3.5 ± 0.7
Eu	0.9 ± 0.01	0.3 ± 0.001
Gd	142 ± 4	359 ± 13
Tb	3.2 ± 0.1	6.5 ± 0.7
Dy	1.51 ± 0.01	0.12 ± 0.01
Ho	0.31 ± 0.01	0.03 ± 0.01
Er	0.75 ± 0.06	0.05 ± 0.01
Tm	0.11 ± 0.01	0.30 ± 0.01
Yb	0.62 ± 0.01	0.01 ± 0.001
Lu	0.21 ± 0.01	0.07 ± 0.01
ΣLREE	17380	39160
ΣHREE	149	366

The urban environment promotes selective enrichment of some elements with respect to natural background. It is possible to evidence such phenomena by introducing an enrichment factor expressed as (EF) = $[M]_{us}[R]_{ns}/[M]_{ns}[R]_{us}$, where M is the considered element in urban sample (us) or Latium natural soils (ns), while R is the element considered as reference.

In this work, the selected reference element is La, because it is included in the catalyst composition but, at the same time, is less abundant than Ce. Results are reported in Figure 4, where it is evident the Pt enrichment testifying the autocatalyst-derived pollution. The log (EF) of REE results overall negative in Viterbo road dusts, while several positive values has been observed in tunnel dusts of Rome. Again, it is the aforesaid "enclosed" state of a tunnel that is responsible for this differentiation. Only Nd, among the key REE in the autocatalyst structure, show a positive enrichment, even though light, in both cases.

Traffic pollution fingerprint

A first application of LA-ICP-MS was carried out on a road dust sample. Some intense peaks of Ce and PGE are so elevated to shift the magnitude

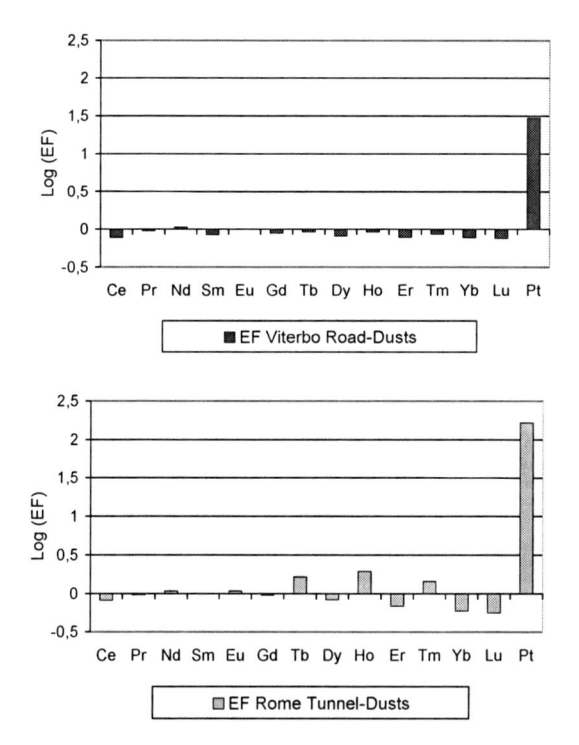

Fig. 4. Enrichment factors for Pt and REE, intended with respect to La, in Latium dusts

of the y-axis to such a level that the baseline signals are practically invisible. This is a first clue to recognize catalyst particles, because it suggests that the sample has particles with peculiar and very high concentrations of Ce and PGE (signal intensity is proportionally to analyte concentration). Moreover, the peaks appeared exactly at the same time, and this means that those high concentrations had been found in the same part of the sample; this is our second evidence, because the simultaneous presence of Ce and PGE in the same particles is not common (unless the particles derive from a catalyst). The definitive proof was obtained from LA analyses performed on our catalyst samples: the signal intensities of Ce and PGE, in fact, had an order of magnitude comparable to those found in road dust (taking in account that catalyst sample produce a more intense signal because it derive from the entire honeycomb structure, while catalyst particles in urban samples derive essentially from wash-coat fragments). This validate that the peaks found in road dust can undoubtedly be associated to catalyst particles.

A similar analysis was performed on tunnel dust, and each of the two fractions previously separated (<20 μm and 20–63 μm) were analysed. The

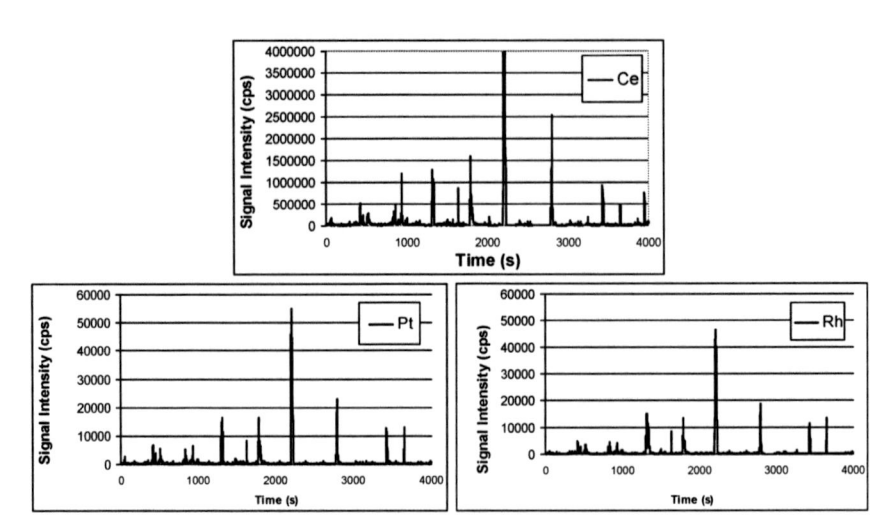

Fig. 5. Scanning laser ablation (LA) on a Rome tunnel dust (fraction 20–63 μm)

20–63 μm fraction was characterized by more Ce–PGE peaks than the <20 μm fraction (Figs. 5 and 6), suggesting that catalyst particles accumulated in urban samples have probably a dimension >20 μm. However, this is only preliminary information and further investigations must be made.

Finally, from some tests executed on dust samples purposely enriched by crushed catalysts, we verified that, apart from Ce and PGE association, also other elemental association, such as La–PGE, can be used (in some

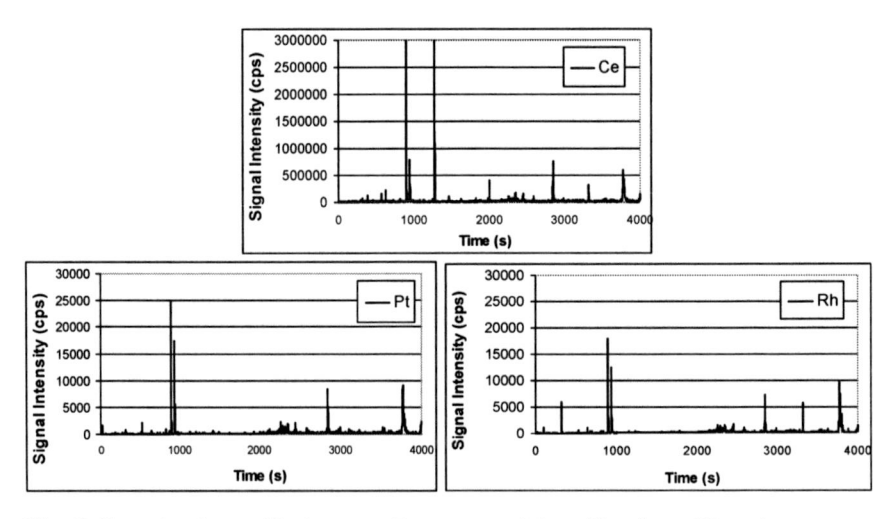

Fig. 6. Scanning laser ablation on a Rome tunnel dust (fraction <20 μm)

cases) to track catalyst particles in urban samples. However, considering that La is generally less abundant than Ce in the autocatalyst particles, its signal in LA-ICP-MS scanning is not always discernible from the signal produced by the background (geological particles, etc.).

Conclusions

A simple approach to study soils, road, and tunnel dust in Rome evidences a slight Pt pollution linked to autocatalyst emissions. Traffic-related REE enrichments are not yet clearly identifiable, because of high background values and heterogeneity of the other anthropic inputs. However, it has been demonstrated that LA-ICP-MS can show the presence of autocatalyst particles in urban samples, analysing both their content of PGE and of REE, and allows to overcome the limits inherent in the acid dissolution of solid samples.

References

1. Artelt S, Koch H, König HP, Levsen K, Rosner G (1999) Engine dynamometer experiments: platinum emissions from differently aged three-way catalytic converters. Atmos Environ 33:3559–3567
2. Gómez B et al. (2002) Levels and risk assessment for humans and ecosystems of platinum-group elements in the airborne particles and road dust of some European cities. Sci Total Environ 299:1–19
3. Angelone M, Nardi E, Pinto V, Cremisini C (2006) Analytical methods to determine palladium in environmental matrices. In: Zereini F, Alt F (eds.) Palladium emission in the environment. Springer, Berlin, pp 245–291
4. Rauch S, Morrison GM, Moldovan M (2002) Scanning laser ablation-ICP-MS tracking of platinum group elements in urban particles. Sci Total Environ 286:243–251
5. Cinti D, Angelone M, Masi U, Cremisini C (2000) Platinum levels in natural and urban soils from Rome and Latium (Italy): significance for pollution by automobile catalytic converters. Sci Total Environ 293:47–57
6. Ravindra K, Bencs L, Van Grieken R (2004) Platinum group elements in the environment and their health risk. Sci Total Environ 318:1–43
7. Taylor SR, McLennan SM (1995) Model of growth of continental crust through time. In: John Victor Walther (2005). Essentials of Geochemistry. Jones & Bartlett, Sudbury, MA.

Sorption behaviour of Pt, Pd, and Rh on different soil components: results of an experimental study

J Dikikh, J-D Eckhardt, Z Berner, D Stüben

Institute of Mineralogy and Geochemistry, University of Karlsruhe, Germany

Abstract

Experiments were carried out to investigate the role of the different soil minerals for adsorption, fixation, and remobilization of automobile catalyst emitted platinum group elements – PGE (Pt, Pd, Rh). It was shown that the adsorption capacity of the investigated soil minerals (kaolinite, Mn/Fe-oxides, quartz, feldspar, calcite) shows large differences mainly depending on specific surface and surface loading. The pH value plays an important role for the buffer capacity of calcite. Kaolinite and Mn/Fe-oxides are characterized by high-specific surface and variable surface loading, expressing the highest adsorption capacity and also stronger bonding for the PGE.

Individual PGE show differences in respect of their adsorption behaviour. Platinum was adsorbed slower and to a much lower extend than Pd and Rh. Pd is the most easily removable element, even though it was adsorbed to a relatively high amount by all investigated minerals.

The experiments demonstrated that mineralogical composition and geochemical conditions of soil exert a decisive influence on fixation and remobilization of catalyst emitted PGE.

Introduction

The use of platinum group elements (PGE) Pt, Pd, and Rh in automobile catalytic converters lead to their increasing emission into the environment. Meanwhile it was demonstrated that the catalyst emitted PGE are widely distributed not only in roadside soil, but also in plants, animals, and even in outlying areas such as Greenland ice [1–6]. The environmental relevance of the mobility of PGE is well documented [7,8]. However, many processes concerning mobility, transformation or uptake by plants and animals are not understood.

Soil is the first natural material getting in contact with catalyst emitted PGE and it is the interface to the living environment. Therefore the aim of our investigations was to obtain fundamental data concerning the behaviour of PGE in soil, focussing on the role of different soil minerals with regard to adsorption, fixation, and mobilization of PGE. Differences and mutual effects of the minerals as well as the strength of PGE bonding were investigated. Further on specific characteristics of Pt, Pd, and Rh were studied.

Experimental set-up and analytical methods

Experiments were carried out with five minerals, each having specific characteristics with respect to surface loading, specific surface, or chemical behaviour. The minerals were treated with low-concentrated PGE solutions in batch and column experiments, which will be discussed in another paper. In leaching experiments using different natural and near-natural waters, the strength of bonding of PGE on the materials from batch experiments was evaluated.

Minerals

Quartz was provided by Gebrüder Dorfner GmbH & Co (Hirschau/Germany). The grain-size distribution is very narrow in the range of 0.063–0.2 mm, the specific surface area (BET) is about 600 cm² g^{-1}.

Feldspar provided by Amberger Kaolin Werke Eduard Kick (Hirschau/Germany) is a potassium feldspar with a grain size distribution ranging from 0.063 to 2 mm (85% between 0.2 and 0.63 mm), and a specific surface area (BET) of about 3000 cm² g^{-1}.

A very pure kaolinite with a specific surface area (BET) of about 157,000 cm² g^{-1} was provided by Gebrüder Dorfner GmbH & Co (Hirschau/ Germany).

Calcite from Deutsche Terrazo (commercial brand Juraperle TWA) has a grain size of 0.2–1.8 mm and a specific surface area of 200 cm² g^{-1}.

Mn/Fe-oxides were provided by the waterworks of Karlsruhe. The material forms during purification of drinking water as coating of Mn- and Fe-oxides/-hydroxides on small quartz grains. The grain size ranges between 0.2 and 2 mm, with most grains in the size of 0.63–2 mm. The material is very porous and has a specific surface area (BET) of ~466,000 cm² g^{-1}.

Batch experiments

About 5 g of each mineral or mineral mixture were treated with 50 mL of 1 µmol L^{-1} PGE solution, prepared from 1000 µg mL^{-1} Pt and Pd AAS standards (Alfa, USA), and RhCl$_3$·nH$_2$O (Merck, Germany), by diluting with double distilled water; pH was adjusted to 4 with nitric acid. The solutions were sampled at specific time intervals in order to assess the amount of adsorbed PGE. The concentration of PGE was analysed with high resolution inductively coupled mass spectrometry (HR-ICP-MS).

Leaching experiments

Leaching experiments were carried out with materials used in the batch experiments with individual minerals. Samples were dried at 40°C. About 1 g of each mineral was treated separately with 10 mL of humic water of mountain moor lake (DOC 36.7 mg L^{-1}; pH 4.5), molten snow (pH 4.0), and ammonium nitrate solution (1 M, pH 4.7). Ammonium nitrate extraction is a German standard method to estimate the plant-available amount of heavy metals in soils.

The samples were shaken for 24 h with humic water and molten snow and 2 h with ammonium nitrate solution, respectively. Before analysing by HR-ICP-MS, samples were filtrated and stabilized with 100 µL HNO$_3$. The desorption rate was calculated relative to the amount of PGE adsorbed by each mineral during the batch experiments.

Procedural blank solutions were prepared in order to assess blank values and minimize matrix effects.

Analytical methods

PGE solutions were measured with HR-ICP-MS (Axiom, VG-Elemental) using external calibration in the range between 0.5 and 5 µg L^{-1} for Pt, Pd, and Rh, prepared from a 10 mg L^{-1} multi-element ICP-MS standard solution (Spex-Certi Prep. Inc., USA). Platinum was quantified by recording and averaging the signals for the isotopes ^{194}Pt, ^{195}Pt, and ^{196}Pt, while for Pd the isotopes ^{104}Pd, ^{105}Pd, ^{106}Pd, ^{108}Pd were used. Solutions from the batch experiments were measured at a mass resolution of 500 (m/Δm) for all isotopes, whereas solutions from the extraction experiments were analysed at a resolution of 4500 for Pt and 9000 for Pd isotopes and Rh, respectively.

Each reported value is the average of three separate runs. As internal standard Ir was used.

Results

Batch experiments with distinct mineral phases

The batch experiments with separate minerals yielded specific results in respect of both, minerals and PGE (Fig. 1). Platinum is removed quantitatively only by adsorption onto kaolinite, the other minerals show less adsorption, ranging between 40% (feldspar) and 70% (calcite).

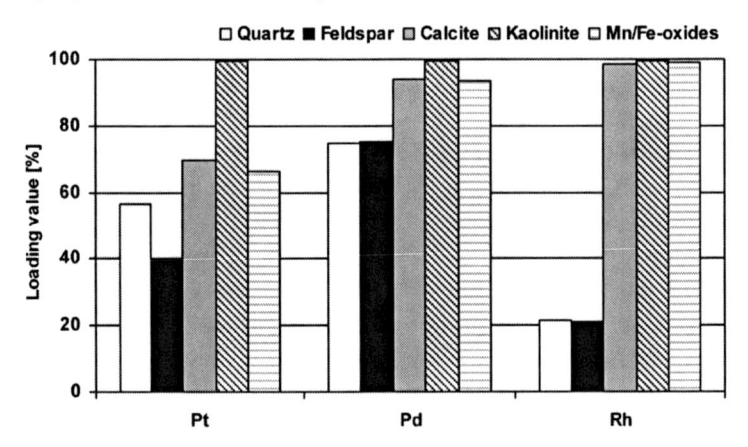

Fig. 1. Diagram showing the results of the batch experiments with individual mineral phases. Percent loading values are calculated from concentrations in solution at the end of the experiment relative to the starting concentration

Palladium and Rh are almost quantitatively adsorbed by calcite, kaolinite, and Mn/Fe-oxide, and up to 75% (Pd) and 20% (Rh), respectively by quartz and feldspar.

Summarizing the adsorption properties of the minerals relative to PGE, quartz, and feldspar show lower adsorption capacity as compared to calcite, Mn/Fe-oxide, and kaolinite. Similarly, Skerstupp et al. [9,10] reported about higher adsorption capacity for Fe oxides and the clay mineral montmorillonite.

Batch experiments with mixtures of mineral phases

Batch experiments with mineral mixtures were aimed to figure out mutual influences of the minerals investigated.

Pair-wise mixtures of all minerals (except for quartz, which is more or less inert), composed similar to the ratio of the minerals in a typical local soil, were prepared in six possible combinations. Additionally a "simulated soil" corresponding to the local soil composition was prepared including all minerals (Table 1).

The results show that Pt is adsorbed only to a minor portion (30–90%) compared to Pd and Rh by most of the mineral mixtures. Palladium and Rh were adsorbed to more than 90% by all mineral mixtures inclusively by the "simulated soil".

Figure 2 shows a very different behaviour for the investigated mixtures. The adsorption capacity of each mixture is in the range of the average of the adsorption capacities obtained from the experiments with individual minerals (Fig. 1). Mixtures with kaolinite or Mn/Fe-oxides display higher adsorption capacities, while feldspar and particularly calcite lead to a lower adsorption of PGE by the respective mixture.

The "simulated soil" displays an adsorption capacity, which in average reflects the characteristics of the individual minerals.

Table 1. Composition of the "simulated soil" (SS)

Minerals	Abbr.	Content in %
Mn/Fe-oxides	Mn	2
Kaolinite	K	4
Feldspar	F	15
Calcite	C	7
Quartz	–	72

288   J Dikikh et al.

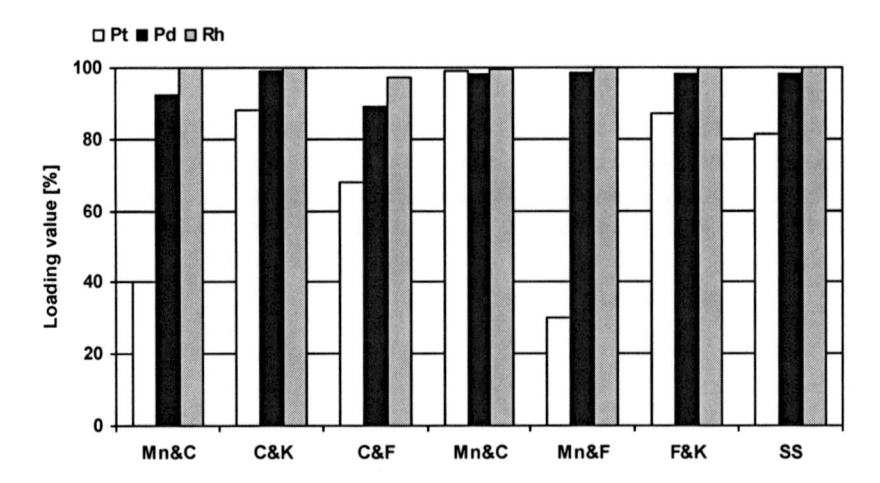

Fig. 2. Diagram showing the results of the batch experiments with mineral mixtures. Percent loading values are calculated from concentrations in solution at the end of the experiment relative to the starting concentration. Abbreviations of mineral mixtures see Table 1

These results clearly demonstrate that under the conditions of our experiments minerals do not mutually influence each other in respect of their adsorption properties for PGE.

Temporal trends of PGE adsorption

During batch experiments the solutions were daily sampled. The relative amount of PGE adsorbed after different time intervals indicates, that the temporal evolution of adsorption depends much more on the characteristics of the individual element than on the composition of the mineral mixture, because in general, all mineral mixtures show a similar time pattern in the adsorption behaviour of Pt, Pd, and Rh, respectively (Fig. 3).

Palladium was adsorbed by more than 80% by all mineral mixtures during the first day. At the end of the experiments after 50 days, between 90% and almost 100% of Pd was removed from the solutions by different mineral mixtures.

Rhodium showed the highest adsorption rate, with nearly 100% removal from the solution by all mineral mixtures. Most part of the Rh was adsorbed already during the first week of the experiments.

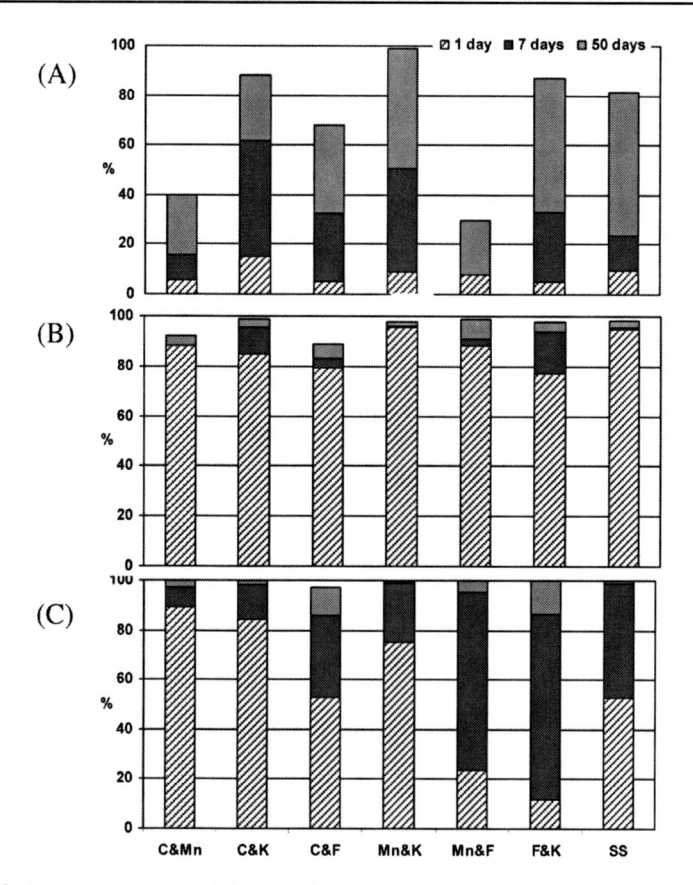

Fig. 3. Adsorption of Pt, Pd, and Rh by different mineral mixtures at different time intervals. Pt (A); Pd (B); Rh (C). For abbreviations of mineral mixtures see Table 1

Platinum showed the lowest and slowest adsorption properties. Depending on the mineral mixture, Pt was removed between 30% (Mn/Fe-oxides/feldspar mixture) and nearly 100% (Mn/Fe-oxides/kaolinite mixture) from the solution. After 1 day only a small amount of Pt was adsorbed by all mineral mixtures, most of Pt was adsorbed during the time interval between 7 and 50 days. Mixtures containing kaolinite always showed the highest adsorption capacity for Pt.

Leaching experiments

In general, PGE-leaching experiments with molten snow and ammonium nitrate show similar patterns for the five minerals investigated, with Pd being

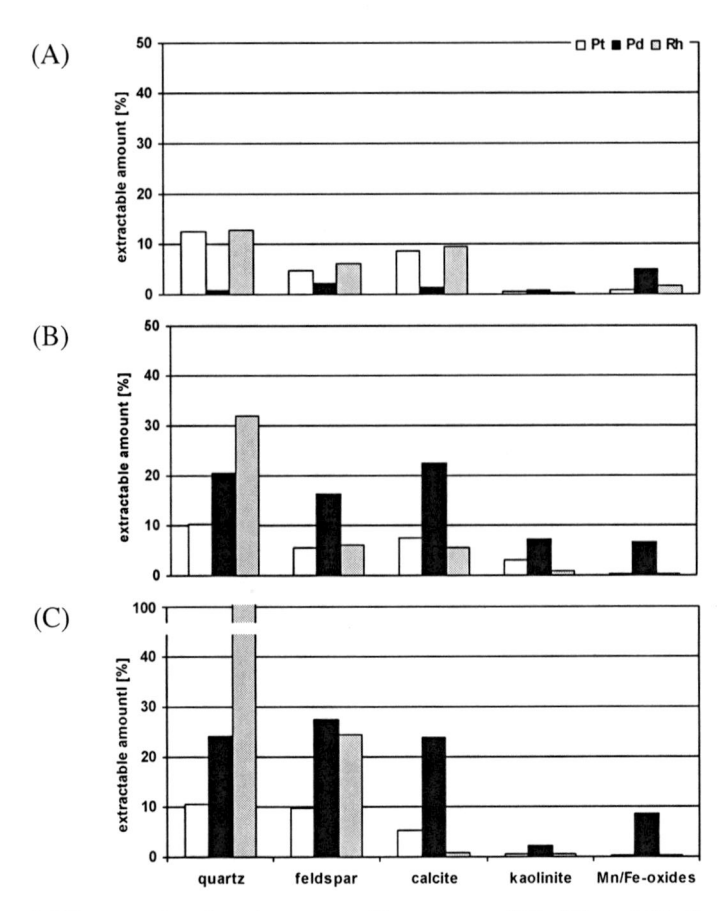

Fig. 4. Diagram showing the extractable amount of Pt, Pd, and Rh from minerals treated with PGE solutions during batch experiments: (A) extraction with humic water; (B) extraction with ammonium nitrate; (C) extraction with molten snow

easily desorbed than Pt and Rh (Figs. 4B and C). The only exception is in the relative amount of Rh released from quartz. This feature might be due to very less Rh retained by quartz during the batch experiments, which lead to a relatively high desorption ratio.

Leaching experiments carried out with humic water display contrasting PGE patterns, with Pd being less removed from the mineral surfaces into the solution than Pt and Rh (Fig. 4A). Probably, PGE (especially Pd) are not directly adsorbed by the respective minerals, but are bond by humic acids, which coat the mineral surfaces.

Obviously, minerals with the lowest adsorption capacity during the batch experiments (quartz and feldspar) also have the weakest bonding to

PGE. Kaolinite and Mn/Fe-oxides display not only the highest amount of adsorbed PGE, but also the strongest bonding.

Discussion

The results show evident differences in the adsorption behaviour of the three investigated PGE, which is due to the pH dependent speciation of these element in solution, with complexes of different size, structure, and loading [11–15].

The nature of PGE complexes in solution and the zero point of charge of mineral surfaces is largely dependent on pH [11,16,17]. The adsorption of PGE by quartz and feldspar in order Pd > Pt > Rh reflects the degree of hydration of PGE chlorocomplexes [17]. The experiments were carried out at a starting pH of 4, at which Pd-chlorides are strongly hydrolyzed as compared to Pt- and Rh-complexes. Further on, Pt in solution was present in form of Pt(IV)- and Pt(II)-complexes, with the bivalent Pt complex being easily hydrolyzed than Pt(IV)- and Rh(IV)-complexes. Because in contrast to PGE chlorocomplexes, hydrolyzed PGE complexes have a positive loading, their adsorption by the negative surfaces of quartz and feldspar result in the observed order.

In experiments including kaolinte, pH increased to neutral values. In this pH range PGE chlorocomplexes are hydrolyzed and completely adsorbed. At pH above 4 the surface loading of kaolinite becomes more and more negative, so that the adsorption capacity of mineral substrates with kaolinite for positive hydrolyzed PGE complexes increases.

At slightly alkaline pH values like in experiments with Mn/Fe-oxides, hydroxo-complexes occur. Pd complexes are positive in loading, while those of Pt and Rh are negative. The influence of pH on surface loading is especially important in the case of Mn/Fe-oxides. In neutral pH range (~5 to ~8.5) Mn oxides have a negative and Fe oxides a positive surface loading. Due to this both positive and negative PGE complexes can be adsorbed from the solution.

In batch experiments including calcite, pH was also slightly alkaline, but in this case, it is much likely that PGE precipitated as hydroxides or carbonates.

The high adsorption capacity of kaolinite and Mn/Fe-oxides is due to their high surface area and permanent surface loading; additionally Mn/Fe-oxides have a high porosity. Batch experiments with mineral mixtures show that the amount of adsorbed PGE is primarily controlled by the mineral phase with the higher adsorption capacity. Consequently, in the mixture of kaolinite and Mn/Fe-oxides Pd, Rh, and Pt were adsorbed to nearly 100%.

Calcite led to an increase of pH in all mineral mixtures, resulting in two effects: on the one hand to precipitation of carbonates, while the increasing negative surface loading of kaolinite and Mn/Fe-oxides resulted in a higher affinity for positively loaded PGE complexes.

Because specific surface area, surface loading, and pH are the most important parameters for the adsorption of PGE, it can be concluded that the immobilization of PGE in the "simulated soil" is controlled by kaolinite, Mn/Fe-oxides and calcite.

Conclusions

Batch and extraction experiments demonstrate that different soil minerals have specific adsorption characteristics for PGE with distinct behaviour for Pt, Pd, and Rh.

Specific surface, surface loading, and for calcite, also alkaline buffer capacity, are the main factors controlling the adsorption capacity of the minerals investigated. Additionally the minerals show differences in the strength of adsorption, corresponding to their adsorption capacity. Based on these characteristics minerals may be divided into two groups. Quartz and feldspar are characterized by low adsorption capacity and only weak bonding while kaolinite, Mn/Fe-oxides, and in part also calcite have relatively high adsorption capacities and stronger bonding for the PGE. Kaolinite and Mn/Fe-oxides are characterized by high specific surface and variable surface loading. In contrast adsorption on calcite is mainly controlled by pH, due to its higher buffer capacity.

It could be shown that the individual PGE have striking differences in respect of their adsorption behaviour. Platinum is adsorbed to a much lower extend than Pd and Rh. Additionally, kinetic effects lead to a considerably slower adsorption of Pt compared to Pd and Rh.

Among the PGE investigated, Pd is most readily adsorbed by all of the considered minerals. At the same time, it is also the element, which is most easily removed from the mineral surfaces by water. Our data support earlier presumptions that Pd is the most mobile of the catalyst emitted PGE [2,8].

Rhodium is adsorbed to the highest amount among the three investigated PGE. Its bonding strength is comparable to that of Pt as shown by the leaching experiments.

The experiments demonstrated that the mineralogical and geochemical composition of soil exerts a decisive influence on the fixation and mobilization of PGE. Clay-minerals, Mn/Fe-oxides and carbonates are components, which in conjunction with high pH values increase the bonding capacity of the soil for PGE.

References

1. Ek KH, Morrison GM, Rauch S (2004) Environmental routes for platinum group elements to biological materials – a review. Sci Total Environ 334–335: 21–38
2. Jarvis JK (2001) Temporal and spatial studies of autocatalyst-derived platinum, rhodium, and palladium and selected vehicle-derived trace elements in the environment. Environ Sci Tech 35:1031–1036
3. Moldovan M, Rauch S, Gomez M, Palacios MA, Morrison GM (2001) Bioaccumulation of palladium, platinum and rhodium from urban particulates and sediments by the freshwater isopod *Asellus aquaticus*. Wat Res 35:4175–4183
4. Platinum (1991) Environmental health criteria: 125. WHO. Also at http://www.inchem.org/documents/ehc/ehc/ehc125.htm
5. Palladium (2002) Environmental health criteria: 226. WHO. Also at http://www.inchem.org/documents/ehc/ehc/ehc226.htm
6. Ravindra K, Bencs L, Grieken R (2004) Platinum group elements in the environment and their health risk. Sci Total Environ 318:1–43
7. Fliegel D, Berner Z, Eckhardt JD, Stüben D (2004) New date on the mobility of Pt emitted from catalytic converters. Anal Bioanal Chem 397:131–136
8. Moldovan M, Palacios MA, Gomez MM, Morrison G, Rauch S, McLeod C, Ma R, Caroli S, Almonti A, Petrucci F, Bocca B, Schramel P, Zischka M, Pettersson C, Wass U, Luna M, Saenz JC, Santamaria J (2002) Environmental risk of particulate and soluble platinum group elements released from gasoline and diesel engine catalytic converters. Sci Total Environ 296:199–208
9. Skerstupp B, Frank G, Urban H (1994) Experimentelle Untersuchungen zur Adsorption von Chlorokomplexen und Kolloiden der Platingruppenelemente (PGE) auf Montmorilloniten, mit besonderer Berücksichtigung der Bestimmung der adsorbierten Spezies. Beiträge zur Jahrestagung, DTTG, Regensburg
10. Skerstupp B, Zereini F, Urban H (1995) Adsorption of platinum and palladium on hydrous ferric oxide (HFO) – an investigation by TXRF and XPS. Ber Dtsch Mineral Ges 7:234
11. Azaroual M, Romand B, Freyssinet P, Disnar JR (2001) Solubility of platinum in aqueous solutions at 25°C and pHs 4 to 10 under oxidizing conditions. Geochim Cosmochim Acta 65:4453–4466
12. Cotton FA (1982) Anorganische Chemie. Weinheim, Germany
13. Middlersworth JM, Wood, SA (1999) The stability of palladium (II) hydroxide and hydroxy-chloride complexes: An experimental solubility study at 25–85°C and 1 bar. Geochim Cosmochim Acta 63:1751–1765

14. Tait CD, Janecky DR, Rogers PSZ (1991) Speciation of aqueous palladium (II) chloride solutions using optical spectroscopy. Geochim Cosmochim Acta 55:1253–1264
15. Wood SA (1991) Experimental determination of the hydrolysis constants of Pt(II) and Pd(II) an 25°C from the solubility of Pt and Pd in aqueous hydroxide solutions. Geochim Cosmochim Acta 55:1759–1767
16. Scheffer F, Schachtschabel P (1998) Lehrbuch der Bodenkunde, 14 auflage. Enke, Stuttgart
17. Wagner M, Salichow A, Stüben D (2001) Geochemische Reinigung kleiner Fließgewässer mit Mangankiesen, einem Abfallsprodukt aus Wasserwerken (GreiFMan). Abschlussbericht zum BMBF-Vorhaben. Universität Karlsruhe
18. Nachtigall D (1997) Verfahren zur Bestimmung von Platinspezies in anorganischen und biologischen Systemen. Bio- und Umweltanalytik; Fraunhofer-Institut für Toxikologie und Aerosolforschung

Reactive soil barriers for removal of chromium(VI) from contaminated soil

A-M Strömvall,[1] M Norin,[2] H Inanta[3]

[1] Water Environment Technology, Department of Civil and Environmental Engineering, Chalmers University of Technology, SE-412 96 Göteborg, Sweden
[2] NCC Construction Sverige AB, Göteborg
[3] Environmental Resources Management – ERM, Jakarta, Indonesia

Abstract

The aim of this project was to find effective reactive materials as use in soil-bed barriers, for remediation of soil contaminated with chromium at Stallbacka industrial area in Sweden. Materials with different reduction/adsorption capacities of Cr(VI)/Cr(III) were tested in laboratory and in a field pilot-scale experiment. Concentrations of total Cr and Cr(VI) in the soil, highly contaminated with ferrochrome slag, were exceeding the guideline values for contaminated sites in Sweden.

Zero-valent iron (Fe^0) filling, FeSO$_4$·7H$_2$O, Na$_2$SO$_3$, field pine bark, modified pine bark, pine sawdust, and sphagnum peat were tested in batch or columns in mixture with the contaminated soil. All the materials, except peat, showed a good ability to reduce Cr(VI) in the batch experiments, and were chosen for further dynamic studies in columns. Iron sulphate and sodium sulphite were both shown to have a good ability to quickly reduce Cr(VI) in the columns, but the use might result in leaching of Fe and SO$_4^{2-}$ to surface and groundwater. For field bark it took a longer time to reduce/adsorb the same amounts of chromium, but it was functional for a longer time.

Reactive soil-bed barriers were constructed in field: soil with embedded layers of FeSO$_4$, pine bark underlying the soil, and soil without any reactive material layer. The iron sulphate was determined not to be suitable for the soil treatment, due to the high percentage of coarse materials in the soil texture, and thereby a quick washout of FeSO$_4$ during the water infiltration. The field reactive soil barrier with pine bark was proven to be effective in reducing Cr(VI), and also had the capacity to adsorb both total and dissolved chromium leaching from the contaminated soil.

Introduction

The Swedish EPA has identified almost 80,000 sites as potentially contaminated in Sweden, and the cost for remediation of the 1500 most contaminated sites is estimated to be 45 billion SEK. Similar, or even more severe, situations are present throughout Europe and North America. Due to the large number of sites and the high costs, there is a strong incentive for finding cost-efficient solutions that are still environmentally sustainable and provide reduced human and ecological risks. The remediation techniques used in Sweden are to a large extent based on *ex situ* methods, where the contaminated soil is excavated, transported, and treated, and combusted or disposed. In a sustainable perspective, cost-effective, and more environmentally sustainable, *in situ* methods to treat contaminated soil and groundwater at site are preferable.

In the USA and Canada [1,2], and in Germany [3], there have now been several implementations of the permeable reactive barrier (PRB) technology, for *in situ* treatment of contaminated groundwater. The PRB techniques have proven that passive reactive barriers can be cost-effective and efficient approaches to remediate groundwater contaminated with chlorinated hydrocarbons and chromate [4].

The toxicity of chromium is dependent on the oxidation state [5,6], and although Cr is an essential element for humans, the hexavalent Cr(VI) is toxic, mutagenic, and carcinogenic. Chromium mobility is dependent on the solubility in water, soil moisture, or other carrying fluid, and Cr moves at a rate essentially the same as the groundwater [7]. Chromate, CrO_4^{2-}, has a high solubility in soils and groundwater, and tends to be mobile in the environment [8]. In contrast, the reduced form, Cr(III), has a limited hydroxide solubility and forms strong complexes with soil minerals and humics. The pH and reduction–oxidation (redox) potential are the two measurable parameters that determine the speciation of Cr and thereby the mobility [9]. Chromium is predominantly ionic, Cr(III), at a pH of <3. At pH above 3.5, hydrolysis in water systems result in the presence of Cr(III) hydroxyl species as $CrOH^{2+}$, $Cr(OH)_2^+$, $Cr(OH)_3^0$, and $Cr(OH)_4^-$. At high redox and oxic conditions, Cr is present as Cr(VI) in the form of chromate, CrO_4^{2-} at higher pH, and at lower pH as hydrogen chromate $HcrO_4^-$ or dichromate $Cr_2O_7^{2-}$.

Stallbacka industrial area in Trollhättan has been identified as a contaminated site with high priority in Sweden. Apart from other contaminants, high concentrations of chromium have been analysed, which originate from the iron ore process industry used to produce stainless steel. During the steel production process a residual product of ferrochrome slag was formed. The slag was dumped and spread over a large area at the industrial site, causing widespread Cr contamination within soils and natural waters. According to a pollution investigation, the concentration of total Cr in the soil reaches up to 60,000 mg kg^{-1}. Hexavalent Cr was identified with concentrations in the soil of up to 1000 mg kg^{-1}, and in groundwater ~700 µg L^{-1}. Based on the guideline value for contaminated soil with less sensitive land use in Sweden, the levels of Cr in soil and groundwater exceed the guidelines [10].

In this project, suitable reducing agents or natural adsorbents for use in reactive barriers for future full-scale field remediation at Stallbacka have been investigated. Several different chemical reducing agents and natural adsorbents were tested as reactive barrier materials in laboratory experiments, and also in reactive soil-bed barriers in a pilot-scale field experiment at Stallbacka. The reactive materials were selected on the basis of findings from a literature survey.

Experimental methods

The slag-contaminated soil

The slag-contaminated soil used in the laboratory tests was randomly collected from the backfilling soil used in the field experiments at Stallbacka. The soil was air-dried at room temperature and sieved; particles <2 mm were selected and, after careful mixing, analysed and used for the batch and column tests. For characterization of the soil, four samples were randomly collected from the mixture. The pH values, conductivity, moisture content, loss of ignition, concentrations of total Cr, Cr(VI), and other elements were analysed. In order to determine the Cr content in the backfilling soil for the field experiment at Stallbacka, a total of 27 samples, nine for each soil-bed, were collected for chemical analysis. Also a total of 21, seven for each bed, were collected from the original soil during digging of the cavities, and analysed as reference samples for the contamination in undisturbed soil at Stallbacka. The soil samples were analysed for grain size distribution, loss of ignition, bulk density, moisture, pH, total Cr, and other elements, dissolved Cr and Cr(VI).

Reduction/adsorption materials

Based on a literature survey the following materials were used for the laboratory experiments:

- Ferrous sulphate heptahydrate, $FeSO_4 \cdot 7H_2O$ [8,11,12]:
- $Cr(VI) + 3Fe(II) \rightarrow Cr(III) + 3Fe(III)$ as
  $HcrO_4^- + 3Fe^{2+} + 3H_2O + 5OH^- \rightarrow 4\,Cr_{0.25}Fe_{0.75}(OH)_{3\,(s)}$ or
  $CrO_4^- + 3Fe^{2+} + 8H_2O \rightarrow 3Fe(OH)_{3\,(s)} + Cr(OH)_{3\,(s)} + 4H^+$
- Iron filling, Fe^0 [11]:
  $Cr(VI) + Fe^0 \rightarrow Cr(III) + Fe(II)$ or $Fe(III)$
  $Cr(III) \rightarrow Cr(OH)_{3\,(s)}$ – pH dependent.
- Sodium sulphite, Na_2SO_3 [6,11,13]:
- $6H^+ + 2HcrO_4^- + 3HSO_3^-$ (excess) $\rightarrow 2Cr^{3+} + 2SO_4^{2-} + S_2O_6^{2-} + 6H_2O$ or
  $5H^+ + 2HcrO_4^-$ (excess) $+ 3HSO_3^- \rightarrow 2Cr^{3+} + 3SO_4^{2-} + 5H_2O$
- Field pine bark and pine bark modified by heating [7,8,14]:
  $Cr(VI) + org_{bark} \rightarrow Cr(III) + ox\text{-}org_{bark}$, then Cr(III) adsorbed through complex bonding to polyhydroxy and polyphenol groups in tannin.
- Sphagnum peat [7,14,15]:
- $Cr(VI) + org_{peat} \rightarrow Cr(III) + ox\text{-}org_{peat}$, then Cr(III) adsorbed through chelating by carboxylate and hydroxyl groups in humic and fulvic acids.
- Pine sawdust [14]:

$Cr(VI) + org_{sawdust} \rightarrow Cr(III) + ox\text{-}org_{sawdust}$, then Cr(III) adsorbed in the sawdust through complex binding to polyhydroxy and polyphenol groups in lignin.

The characteristics of heavy metal content, organic content, moisture, and grain size distribution were determined for the sphagnum peat, pine barks, and sawdust.

Batch tests

The batch tests were carried out as a pre-test for selection of materials for the column experiments, but also for determination of Cr leaching from the contaminated soil. About 10 g of each of the active materials were introduced together with 20 g of soil into Pyrex glass beakers. A liquid-to-solid (L/S) ratio of 20 was used; thus 600 mL of nanopure water was introduced to each beaker, except for the beaker with only soil, which had an L/S of 30. All mixtures were agitated by plastic paddles, in closed beakers, at a speed of 200 rpm for 24 h. After completion of the tests, the solids were settled down, pH and conductivity were measured, and water samples collected for chemical analysis of total Cr, Cr(VI) and other elements.

Column experiments

A set of column experiments was carried out in order to study leaching of Cr and other elements from the slag-contaminated soil in beds, together with the selected and reactive materials, under flowing conditions. It should be noted that the column test is a simplified method to study leaching, compared to the more complex environment that exists under real conditions.

The column tests were performed by using polyethylene columns with a length of 40 cm and an inner diameter of 7 cm. At the bottom of each column, a 3 cm thick layer of acid-washed coarse sand, with particles ~1 mm, was placed as a filter to prevent carry-over of the solid materials from the columns. A total of 400 g dry weight of contaminated soil, for each column, was either mixed with the reactive materials or packed in mixed layers with the soil. For Na_2SO_3 and $FeSO_4$ the columns were packed with a total of 50 g of the active materials in three layers with the soil between. The treated pine bark and pine sawdust columns needed a packing to avoid clogging: first a bed with all of the soil, then 50 g dry weight of each material, pretreated in nanopure water for 24 h, and in mixture with 50 g of coarse sand beneath the soil layer. The field bark was also pretreated in nanopure water, and then 100 g dry weight packed as a layer under the soil bed. The columns were loaded with deionized water in a down-flow mode, and with a L/S ratio around two. The flow was controlled by a well-adjusted pump, and kept the water tables in the columns constant through an overflow tube. Through carefully measuring the inflowing, excess and leachate volumes, the water volumes flushed through the column bed materials could be calculated. All columns were wrapped in aluminium foil to minimize biological and photo-induced reactions. At the beginning of the experiments, the interval of sampling was 5 or 10 min, depending on the columns' different effluent speed. As the experiments progressed, samples were collected less frequently. The effluents' pH, conductivity, and redox potentials were measured at the same time as sample collection. Total Cr, Cr(VI) and other elements were analysed in the water samples. Ferrous iron, Fe(II), in the leachates from the $FeSO_4$ column, and sulphate concentrations in leachates from both the $FeSO_4$ and Na_2SO_3 columns, were also analysed.

Reactive soil barrier at Stallbacka

Three soil-bed experiments were constructed at the contaminated site: one with only slag-contaminated soil, another with 200 kg of FeSO$_4$ placed in three layers within the soil, and the third with field pine bark in a 20 cm thick layer beneath the contaminated soil. The soil beds were constructed by digging 1.2 m deep, and 3.0 m × 3.0 m cavities. To avoid any influence from the surrounding soil, the cavities were isolated with PVC plastic. On the bottom 20 cm layers of pre-washed sand were introduced as filters. The soil to be used in the beds was the backfilling soil with a mixture of the crushed excavated soil. The thickness of the soil to be treated was 0.5 m in all three beds, and they were designed in three layers, each compacted by using a soil compactor during construction. To reduce a rapid infiltration flow along the sides of the soil beds, a granular layer of bentonite clay was used as an upper layer, around the sides. A water solution was infiltrated through each soil bed to enhance the leaching of the contaminants. The beds were infiltrated twice with 3 m^3 of water, corresponding to approximately 2 years of annual rainfall. The water flow for the infiltration was 0.55 L s^{-1}. Leaching water from each pond was collected in a plastic tank and followed by chemical analysis. The leachate samples were collected twice, and analysed for pH, redox potential, conductivity, total Cr, and other elements, dissolved Cr, Cr(VI), SO$_4^{2-}$, and total organic carbon (TOC).

Chemical analysis

The contaminated soils' moisture content and organic carbon were determined by following standard methods [16], the grain size distribution in accordance with a Swedish standardized methodology (SS 027123), and the bulk density following an ASTM method (D 1895 B9). The soil and water samples' pH, electrical conductivity, and redox potential were measured by a standard method [16].

The soil samples were microwave-digested in a mixture of H$_2$O$_2$/HNO$_3$, and all the water samples were microwave-digested with HNO$_3$ before analysis of Cr(VI) and total Cr. The concentrations of Cr(VI) were analysed in the digested liquid, after filtration through 0.45 µm membrane filters, by the reddish-purple, 1.5-diphenylcarbohydrazide chromate method [17]. For ICP-MS analysis of total Cr and other elements, rhodium was added as internal standard to all samples before digestion. Selected water samples were also analysed for the concentrations of TOC following a standardized procedure (SS-EN 1484), and for concentrations of sulphate and ferrous iron, Fe(II), following HACH methods. For quality assurance, two leachate samples from the laboratory, three samples of the backfilling soil, and two leachate samples from Stallbacka were sent to a certified environmental laboratory for inter-calibration of the total Cr and other elements, and Cr(VI) analysis methods. The results were in good agreement with the concentrations measured at the Chalmers laboratory.

Results and discussion

Slag-contaminated soil

Soil at Stallbacka: For the contaminated backfilling soil used in the reactive soil-bed barriers in the field, the grain size distribution showed that the soil was dominated by coarse gravel and sand particles. The soil texture was classified as loamy sand with a capacity to infiltrate water relatively fast; the soil moisture was 13%, pH around 10, the bulk density 1.4 g cm^{-3}, hydraulic conductivity 2.5×10^2 cm s^{-1}, and loss of ignition 14%. The average concentrations of total Cr and Cr(VI) in mg kg^{-1} were, respectively, in the untreated soil-bed 9000 and 8, in the bed with iron sulphate 9200 and 6, and in the barrier with bark 10,400 and 5. The concentration of total Cr was about 80 times higher than the Swedish guideline value for polluted soils, assuming that all Cr is in the form of Cr(III) [10]. The total Cr level of contamination is classified as very serious, and the concentration of Cr(VI) classified as moderately serious. In the undisturbed soil samples taken during digging of cavities, the average of total Cr of 4000 mg kg^{-1} was lower than in the crushed and mixed backfilling soil. The average content of Cr(VI) was 11 mg kg^{-1}, and higher than in the backfilling soil, and shows great heterogeneity of the content in the soil material used.

Soil for laboratory experiments: The average concentration of total Cr in the sieved soil samples, with particles <2 mm, and taken at Stallbacka, i.e., the backfilling soil used in the field test, was 7500 mg kg^{-1}. Even if sieved and carefully mixed, the chromium concentrations in the soil material turned out to be inhomogeneous with concentrations varying from 5700–9000 mg kg^{-1}. The average concentrations of the other metals analysed were low, except for nickel with 70 mg kg^{-1}. The pH value of the soil was 12 and due to the high concentration of calcium, 120,000 mg kg^{-1}, suggested to be present as CaO, Ca(OH)$_2$ or CaSiO$_4$. At pH 12 and under oxic conditions, the dominant Cr(VI) species in water solutions of the soil will be dissolved chromate (CrO$_4^{2-}$).

Batch tests

The concentrations of Cr and other metals in the solutions after the batch tests of soil in mixture with different reducing/adsorbing agents are presented in Table 1. All the reactive materials used were shown to have the capacity to reduce Cr(VI) to Cr(III). The capacity of the reducing agents was in the order Na_2SO_3 = modified pine bark = sawdust = $FeSO_4 \cdot 7H_2O$ > iron filling > field pine bark > sphagnum peat > soil. The differences in concentrations of total Cr show heterogeneity of the soil material used, rather than the materials' chromium adsorption capacities.

Sphagnum peat had a rather weak reduction and adsorption capacity at the high pH of 12, due to Cr(VI) in the form of negatively charged chromate CrO_4^{2-}, and Cr(III) as $Cr(OH)_4^-$. Peat is otherwise known as a material with an excellent capacity to adsorb metal cations [18,19], but has no capacity to bind negatively charged ions, especially not in solutions at high pH. For this reason the peat was excluded in the coming column experiments. The low concentration of total Cr in the beaker with $FeSO_4$ may be explained by co-precipitation of Fe(III)/Cr (III) hydroxides in the solution at pH 9 [20], and could be an explanation for the brownish-coloured precipitation observed. Both the treated pine bark and sawdust adsorbed most of the water in the beakers, which indicated possible trouble with clogging if used in real soil beds. In the modified pine bark beaker, the solution also had a reddish colour, which may indicate a release of organic contaminants, such as soluble phenols from the tannin-containing bark. At the high pH level of the solutions, the reactive materials seem to have a weak capacity to adsorb other metals. For the natural materials such as pine barks, sawdust, and peat, even a release of Pb and Zn is indicated by the results.

Table 1. Metal concentrations measured in leachates after batch tests

	Cr(IV) mg L^{-1}	Cr total mg L^{-1}	Ni mg L^{-1}	Pb mg L^{-1}	Zn mg L^{-1}	pH	Conductivity mS cm^{-1}
Soil[a]	2.8	3.2	0.020	0.0040	0.02	11	0.40
Sphagnum peat	1.2	3.0	0.040	0.020	0.09	12	0.30
Field pine bark	0.60	3.4	0.050	0.060	0.04	11	0.40
Treated[b] pine bark	n.d.	3.5	0.060	0.030	0.21	10	0.50
Sawdust	n.d.	3.9	0.070	0.050	0.26	10	0.80
Iron filling	0.10	2.1	0.020	0.0050	0.02	11	0.40
$FeSO_4$ $7H_2O$	n.d.	1.2	0.070	0.0020	0.01	9	2.6
Na_2SO_3	n.d.	5.1	0.040	0.0030	0.03	12	18

[a] soil = slag-contaminated soil
[b] treated = modified

Column tests

The accumulative amounts of total Cr and Cr(VI) leached from the columns, with the reactive materials in mixture with slag-contaminated soil, are presented in Figure 1. The leached amounts are correlated with the calculated pore volume on the *x*-axis. A pore volume is the volume of water required to replace water in a certain volume of saturated porous media. For total Cr the removal capacity for the materials was in the order $FeSO_4 \cdot 7H_2O$ > Na_2SO_3 > field pine bark > iron filling > soil > pine sawdust. The order for the materials' capacity to reduce Cr(VI) was the same (Fig. 1), but with the exception that Na_2SO_3 was more efficient than $FeSO_4$. The total amounts of total Cr and Cr (VI) leached from the soil column were 86 and 58 mg, respectively, and pH was kept stable around 11. As shown in Figure 1, the Cr(VI) available was leached out already after 140 pore volumes, but corresponding to only 2% of the total Cr(VI) present in the contaminated soil. If scaled up to field conditions at Stallbacka, it may take several hundred years of natural precipitation to leach out the available Cr(VI) if no remediation is applied at the site.

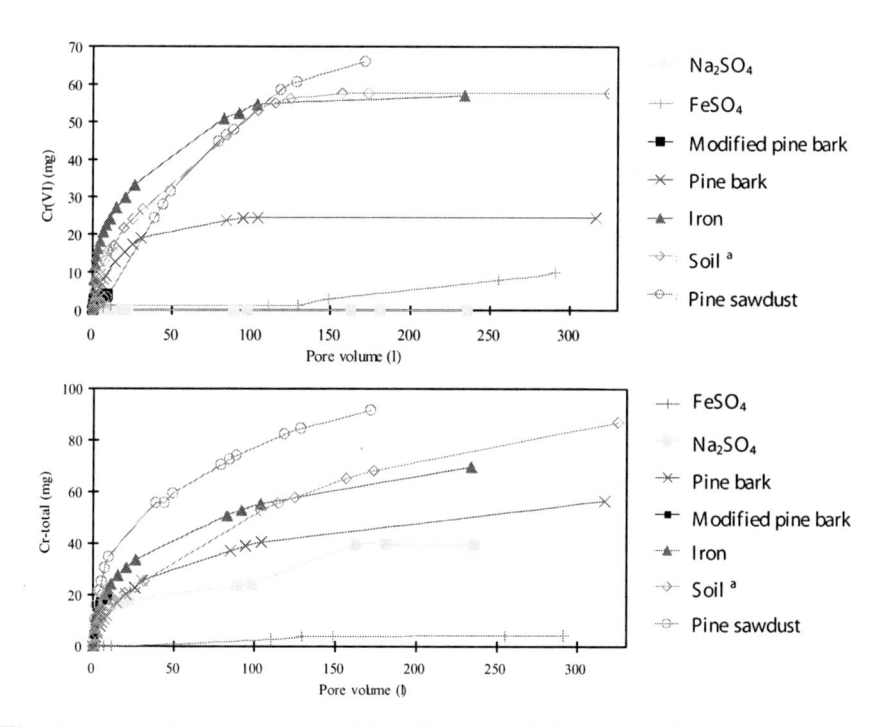

Fig. 1. Accumulative amounts of Cr(VI) and total Cr leached from the columns with different reactive materials ([a] soil = slag-contaminated soil)

From the column with FeSO$_4$, the total Cr leached was only about 5 mg but Cr(VI) was higher and 10 mg. These results reflect the error and different interferences in the analytical method used, especially for solutions with low chromium concentrations. The pH in this column increased from 6 to 11, and the redox potential was kept low but increased from −28 to +14 mV. The decrease of Cr(VI) in the leachates during the first 10 pore volumes was followed by a decrease in redox potentials, suggesting an efficient and fast reduction reaction between Fe(II) and Cr(VI). The following increase in redox potential and pH was consequently followed by an increase in release of Cr(VI), which may be explained by the competing and fast oxidation of Fe(II) by oxygen in water [21] or precipitation of Fe(OH)$_2$ at the higher pH level. This was also indicated by the precipitations formed in the columns, first a dark green felling of Fe$_2$O$_3$ and later orange/brown-coloured of Fe(III)/Cr(III) hydroxides. Serious leaching of sulphate and iron was measured in the leachates from the FeSO$_4$ column, and shows environmental side effects if used as a barrier material in the field. Sulphates could promote soil acidification, but could also enhance leaching of CrO$_4^{2-}$ through anion exchange [21].

Sodium sulphite was shown to be the best reducing material among the reagents tested in columns, and the Cr(VI) concentrations were under the detection limits in all the leachate samples analysed. The total Cr leached from the Na$_2$SO$_3$ column in a total of 40 mg was thus in the form of Cr(III) and, due to the high pH level from 12–11 and the low redox potentials increasing from −727 to +23 mV, in the form of dissolved Cr(OH)$_4^-$. The lack of organic materials and Fe(II) in this column made an environment unfavourable for oxidation and precipitation of Cr(III) in the leachates. High concentrations of SO$_4^{2-}$ were also analysed in leachates, and hence the same side effects as for FeSO$_4$ may be suspected if used in field barriers. From the pine field bark column a total Cr of 58 mg was leached, and 24 mg of this was in the form of Cr(VI). The reduction/adsorption of chromium was proven to be much slower and less effective than in the leachates from the FeSO$_4$ and Na$_2$SO$_3$ columns. The pH level was kept stable around 11, and the redox potentials decreased from +81 to +34 mV. The uptake of Cr(VI) by pine bark has complex mechanisms, where adsorption and redox reactions may proceed in parallel as in peat materials [15]. Different from the batch tests, the iron filling was shown to be less efficient in its ability to reduce Cr in the column, and resulted in the same capacity as soil without any reactive material. The low ability may be due to the fast flow speed in the column, and it is clear that iron filling needs a higher contact time to increase the Cr(VI) reduction capacity. The redox potential was also unfavourable, increasing from +6.8 to +41 mV.

From the pine sawdust column, a total of 90 mg Cr total was leached, and about 65 mg was in the form of Cr(VI). These levels of Cr leached are even higher than from the soil column without any reactive material. The pH was increasing from 5 to 11, but the redox decreased from +316 to +34 mV during the experiment, and thus the Cr was predominantly in the form of CrO_4^{2-} or $Cr(OH)_4^-$. As for pine barks and peat, the pine sawdust is expected to be most effective for adsorption of cations. At the high pH level, the results instead indicate a leaching of Cr from sawdust. For the treated pine bark it was only possible for 16 pore volumes, corresponding to 11 L, to flow through the column. Thereafter the column clogged, depending on the bark's high water-adsorption capacity, making this material unsuitable for use as filter material in a reactive bed. The concentrations of other metals in leachates from the columns were low, except for Ni, which leached in serious concentrations, especially from the $FeSO_4$ column.

Reactive soil barrier at Stallbacka

In Figure 2, the accumulative amounts of total Cr and Cr(VI) are presented for the constructed soil barriers at Stallbacka industrial area. The amount of total Cr leached from the reactive soil barrier with a layer of $FeSO_4$ was surprisingly highest, nearly 600 mg; from the bed with no treatment, it was lower at 150 mg. For the barrier with pine bark, the leaching of total Cr was much lower, around 25 mg. The pattern for dissolved Cr and Cr(VI), respectively, followed the same trends, with highest values in the $FeSO_4$ treated barrier 120 and 5 mg, in the untreated bed 80 and 2 mg, and the lowest values in the barrier with bark 10 and 0.5 mg. The high leaching of chromium from the reactive soil bed with iron sulphate may be explained by a fast flush-out of the $FeSO_4$ material from the bed, indicated by the high electrical conductivity, iron, and sulphate concentrations in the early leachates from this bed. The pH in the leachates from the untreated and $FeSO_4$ soil barrier was kept stable at 11–12, but for the leachates from the bark barrier more varied, and in the range 7–11. The redox potentials in the leachates from the untreated soil bed were rather stable, decreasing from 0 to –40 mV, and for the barrier with $FeSO_4$ increasing slightly from –40 to +40 mV. For the barrier with bark, the redox was steeply decreasing from +240 to +40 mV, showing an environment favourable for reduction of Cr(VI) to Cr(III). The concentration of TOC was 120 mg L^{-1} in the leachate from the bed with pine bark, and indicated a potential side effect of leaching of organic contaminants from the bark. The concentrations of iron and sulphate were, as expected, high in the leachates from the bed with $FeSO_4$.

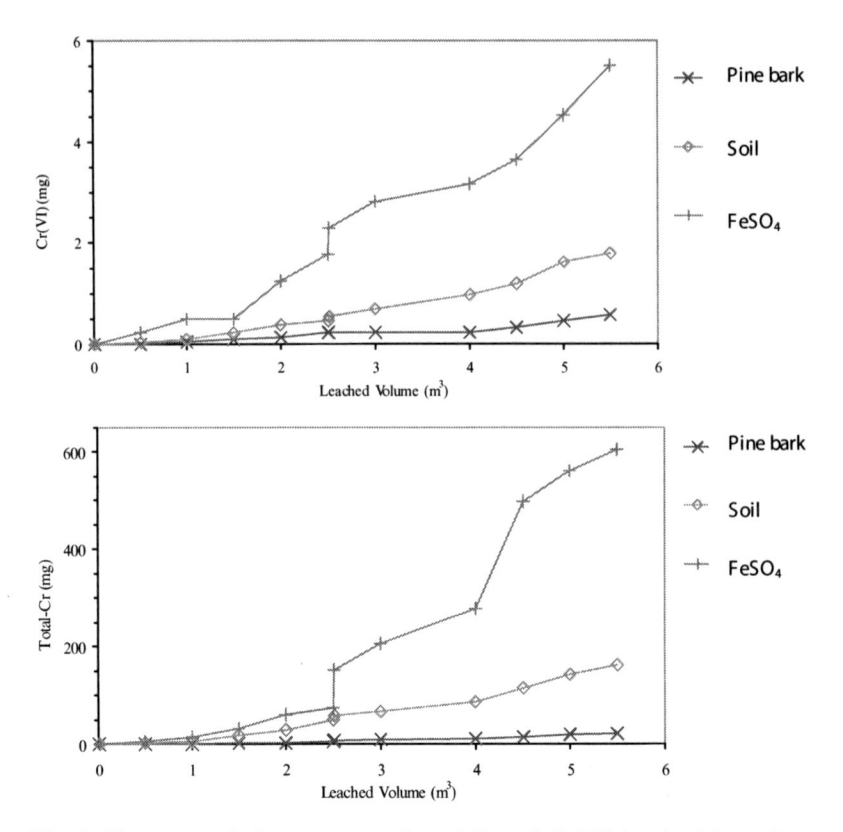

Fig. 2. The accumulative amounts of total Cr and Cr(VI) leached from the reactive soil-bed barriers at Stallbacka ([a] soil = slag-contaminated soil)

Conclusions

The contamination by total Cr and Cr(VI) in the polluted soil at the Stallbacka industrial area is serious, and remediation actions are highly desired. Introduction of an organic material such as pine bark has potential for use as a reducing agent for Cr(VI) and as an adsorbent for Cr(III), based on the results shown in this project. Barrier techniques where the soil could be treated without excavation and through direct injection of the active materials in the soil, or construction of a PRB for treatment of the groundwater, would be preferable.

The application of iron sulphate as an active material in reactive soil-bed barriers is not recommended as a remediation technique, because of the coarse soil texture. Furthermore, it was found that there is a difference in the results from laboratory columns and the field experiment, especially when iron sulphate was used.

Further studies need to be carried out in the field, particularly with mixed layers of Na_2SO_3 and pine bark or sphagnum peat. Sodium sulphite is suggested for the reduction of Cr(VI) to Cr(III) in the first flush, and pine bark/sphagnum peat for the secondary adsorption of Cr(III) working for a longer time period. The degradation rate of pine bark and sphagnum peat in the soil needs to be studied for determination of the barrier's vitality and lifetime.

Acknowledgements

The authors are grateful to Liang Yu, who performed all practical work in the laboratory experiments as well as parts of the fieldwork, and to Jesper Knutsson, who made the ICP-MS analysis. Anders Bank, environmental consultant, and Per Olsson of the Department of Contaminated Sites at the Regional Authority in Västra Götaland are both gratefully acknowledged for initiating the project. The Regional Authority of Västra Götaland provided financial support for the study.

References

1. Interstate Technology and Regulatory Council (2005) Permeable reactive barriers: lessons learned/new directions. PRB-4. Washington, DC. Also at www.itrcweb.org
2. Vidic RD (2001) Permeable reactive barriers: case study review, Technology evaluation report, TE-01-01. Ground-Water Remediation Technologies Analysis Center, Pittsburgh
3. Birke V, Burmeier H, Rosenau D (2003) Design, construction, and operation of tailored permeable reactive barriers. Pract Period Hazard Toxic Radioact Waste Manage 7:264–280
4. Powell RM, Puls RW, Powell PD (2002) Cost analysis of permeable reactive barriers for remediation of groundwater. In: Gavaskar AR, Chen ASC, Monterey CA (eds.) Proceedings of the international conference on remediation of chlorinated and recalcitrant compounds, 3rd edn., United States, May 20–23, 2002, 104–11. Battelle Press, Columbus, OH
5. Hansel CM, Wielinga BW, Fendorf S (2003). Fate and stability of Cr following reduction by microbial generated Fe (II). Department of Geological and Environmental Science, Stanford University, MFG, Inc., Fort Collins, CO
6. Palmer CD, Wittbrodt PR (1991) Processes affecting the remediation of chromium-contaminated sites. Environ Health Perspec 92:25–40

7. Palmer CD, Puls R (1994) Natural attenuation of hexavalent chromium in groundwater and soil. EPA/540S-94. US Environmental Protection Agency. Groundwater issue
8. Sass BM, Rai D (1987) Volubility of amorphous chromium(III)-iron(III) hydroxide solid solutions. Inorg Chem 26:2228–2232
9. Walter GR, Yiannakakais A, Chammas G (2001) Applying geochemical methods to remediate chromium contamination. Hydro Geo Chem, Inc, Tucson, AZ
10. Swedish Environmental Protection Agency (2002). Contaminated sites. Report 5053
11. U.S. EPA (2002) In situ treatment of soil and groundwater contaminated with chromium – technical resource guide. United States Environmental Protection Agency, EPA/625/R-005, October, 2000
12. Fetter CW (1999) Contaminant hydrogeology. 3rd edn. Prentice Hall, Upper Saddle River, NJ
13. Beukes JP, Pienaar JJ, Lachmann G, Giesekke EW (1999) The reduction of hexavalent chromium by sulphite in wastewater. Water SA 25:363–370
14. Bailey SE, Olin TJ, Bricka RM, Adrian DD (1999) A review of potentially low-cost sorbents for heavy metals. Wat Res 33:2469–2479
15. Sharma DC, Forster CF (1995) Column studies into the adsorption of chromium(VI) using sphagnum moss peat. Bioresour Technol 52:261–267
16. DSNR's Centre for Natural Resources (CNR) Wellington Laboratory (1999) Soil survey standard test method. Also at http://www.dlwc.nsw.gov.au/care/soil/soil_pubs/soil_tests/pdfs/mc.pdf
17. Standard methods for the examination of water and wastewater (1998) 20. edn., Lenore S. Clesceri et al. (eds.) American Public Health Association Board. California Environmental Protection Agency, New York
18. Kalmykova Y, Strömvall A-M, Steenari B-M (2006a) Alternative materials for adsorption of heavy metals and petroleum hydrocarbons from contaminated leachates. Environ Technol (submitted)
19. Kalmykova Y, Strömvall A-M, Steenarie B-M (2006b) Adsorption of Cd, Cu, Ni, Pb and Zn on sphagnum peat from solutions with low metal concentrations. J of Hazard Mat (submitted)
20. Seaman JC, Bertsch PM, Schwallie L (1999) In situ Cr(VI) reduction within coarse-textured, oxide-coated soil and aquifer systems using Fe(II) solutions. Environ Sci Technol 33:938–944
21. Geelhoed JS, Meeussen JCL, Roe MJ, Hillier S, Thomas RP, Farmer JG, Paterson E (2003) Chromium remediation or release? Effect of iron(II)sulfate addition on chromium(VI) leaching from columns of chromite ore processing residue, Environ Sci Technol 37:3206–3212

Cleaning of highway runoff using a reactive filter treatment plant – a pilot-scale column study

G Renman, M Hallberg, J Kocyba

Royal Institute of Technology, Stockholm, Sweden

Abstract

Removal of dissolved heavy metals in road runoff can be achieved by filtration through reactive materials. A four-column installation was set up at an existing treatment plant and used to examine different types of filter materials *in situ*. Two fractions of granulated activated carbon (GAC), Clinoptilolite and Polonite were investigated. The hydraulic loading was 1 m h^{-1} and the metal attenuation capacities were studied under unsaturated and saturated conditions. The relative effectiveness of the materials decreased in the order: GAC of fine fraction > GAC of course fraction > zeolite > Polonite. Aluminium, Fe, Mn, and Zn showed the highest concentrations in influent storm water and also showed elevated removal efficiencies.

Introduction

Runoff from urban catchment areas, especially from roads, parking lots, and build up areas, is highly polluted by such constituents as heavy metals (Cd, Cr, Mn, Co, Ni, Cu, Zn, V, Pb, etc.), organic compounds (such as polyaromatic hydrocarbons, polychlorinated biphenyls, and phenols)[1]. The contamination originates mainly from air deposition, brake linings, tyres, asphalt, petrol, oil, and also from building materials such as those used on roofs, and different galvanized materials, e.g., crash safety barriers [2].

A treatment system comprising of a retention pond followed by a reactive filter unit was developed in Sweden in 1998 [3]. A previous study on that system showed high efficiency of heavy metals removal for the same mineral and organic materials that were investigated in this paper [4]. These materials were natural and burned opoka (calcium silicate sedimentary rock manufactured as Polonite), clinoptilolite (type of zeolite), and pine bark. All of the tested materials showed promising metal reduction capacity after treatment of polluted storm water. Depending on the variant of tested substrate the average removal for, e.g., Cu was around 80% and for Zn between 63–79%.

The aim of this study was to investigate the hydraulic and metal adsorptive capacities of a number of filter materials in a particular environment. Granulated activated carbon (GAC) was used as reference material because of its recognized purifying capacity. A pilot-scale trial with columns was constructed during summer 2005 within an existing treatment system located at the island Lilla Essingen in the City of Stockholm. The system captures polluted storm water from parts of the traffic route E 4 (traffic load 120,000 vehicles per day), some local roads, parking lots, and roof water from a built-up area.

Fig. 1. A schematic cross section of the storm water treatment system at Lilla Essingen. 1: inlet-well, 2: retention pond, 3: two mechanical PIAB filters, 4: pump station, 5: two reactive filters, 6: Lake Mälaren. (After Stockholm Vatten AB)

Experimental

One of the reactive filter manholes in the treatment plant (Fig. 1) was used for the installation of the pilot-scale column experiment. Four transparent PVC tubes with internal diameter of 106 mm were used. Transparency of the columns facilitated the observation of hydraulic efficiency of water filtration through the material. Each column was 900 mm high and filled up to 500 mm with one of selected filter materials. Thus, the volume of each material was equal to 4.41 dm^3. Storm water, which was pretreated in a retention pond and by PIAB high-density polyethylene (HDPE) filters, was continuously pumped and equally distributed into every column. The hydraulic load was 1 m h^{-1} and the test was performed for 3 weeks under both unsaturated and thereafter under saturated conditions. Storm water was supplied by gravity through the columns. The column set-up is shown in Figure 2.

The filter materials comprised of two fractions of granulated activated carbon (NORIT GAC 830W, finer; NORIT PK3-5, coarser), Polonite (sieved fraction, 2–5 mm), and clinoptilolite, 1–2 mm). A description of the materials is given in Table 1. The treatment performance was based on water quality analysis of influent, untreated storm water, and effluent water discharged to a lake. Parameters and their respective instruments used for that analysis were turbidity (HACH 2100P), conductivity (FE 287 portable conductivity meter, EDT Instruments Company), pH and temperature (portable pH/mV/temperature meter, HACH), TOC and DOC (Multi N/C 3000 Analytic Jena AG according to SS-EN 1484-1), TOT-N (Multi N/C 3000 device, Analytic Jena AG), TOT-P (UV – 1601 SHIMADZU, according to SS 02 81 07-2), and concentration of the following heavy metals (Elan DRC-e ICP-MS): aluminium (Al), cadmium (Cd), chromium (Cr), copper (Cu), iron (Fe), lead (Pb), manganese (Mn), nickel (Ni), vanadium (V), and zinc (Zn).

Table 1. Physical parameters and pH of filter materials

Parameter	GAC 830 W	GAC PK 3–5	Clinoptilolite	Polonite
Particle size (mm)	0.60–2.3	62.8–5.0	1–2	2–5.6
Specific surface area (m^2 g^{-1})	1150	875	200–500	0.7
Density (kg m^{-3})	485	260	2250	700–980
pH	Alkaline	Alkaline	8–8.5	11.5–12.6

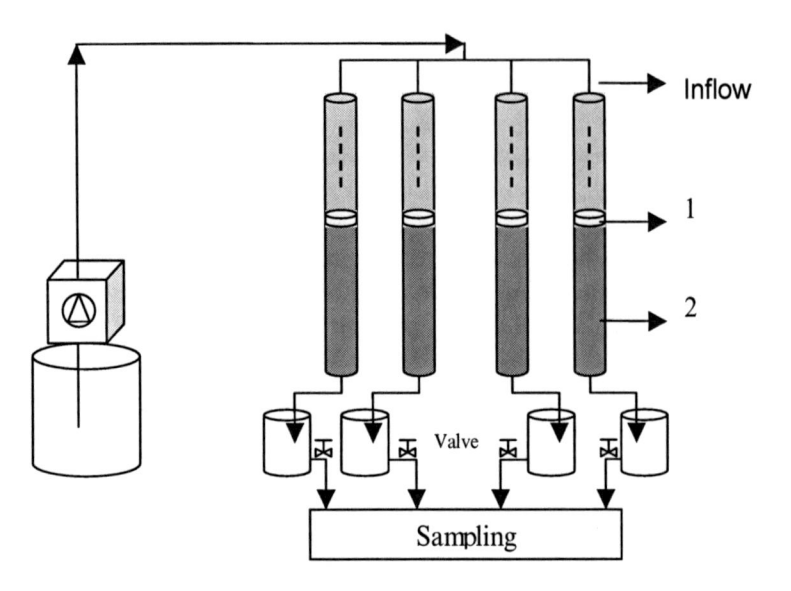

Fig. 2. Column set-up, field trial with downflow and unsaturated conditions

Results and discussion

Turbidity

During conditions of low influent turbidity in storm water, the highest efficiency was found for finer activated carbon (GAC 830W) that decreased turbidity from 4.4 FNU to a value oscillating at 1.4 FNU. Similar results were observed for clinoptilolite. The effluent from the column filled with Polonite and that filled with course GAC reached values of 2 FNU – 3 FNU, thus reduction rates remained rather low. Nevertheless, the reduction rates rose significantly after rain events with increasing turbidity of the storm water. When the influent water turbidity was around 17 FNU, the effluent values from columns with fine GAC and clinoptilolite were 2–2.6 FNU and 3–3.6 FNU, respectively. The reduction in the columns with course GAC and that with Polonite was about 64% and 50%, respectively.

Relatively low influent turbidity values measured at the initial period of the trial were caused by the sedimentation pretreatment unit during longer periods without storm events. Increased turbidity during runoff events clearly showed the mechanical filtration capacities due to small void size of the fine GAC fraction and the clinoptilolite.

Conductivity, pH

The conductivity of the influent storm water varied between 36 mS m^{-1} and 94 mS m^{-1}, and pH ranged from 7 to 7.5. High conductivity values were recorded for Polonite at the initial phase of the experiment (114 mS m^{-1}–118.5 mS m^{-1}). A possible reason was that calcium oxide (25% of the material content in dry mass) was washed out from the filter. Elevated pH values were also initially recorded in the effluent from the Polonite column. These ranged from 11.4 in the beginning of the experiment and dropped to 7.9 at the end of the test period. Effluent conductivity values did not vitally diverge from the influent value for the other materials.

Metal removal

Influent concentrations showed very low concentrations of Cd (<0.01 µg L^{-1}–0.03 µg L^{-1}), hence close to the detection limits of the instrument used for analysis. Low concentrations of Pb (0.8–5 µg L^{-1}), Cr (0.8–3 µg L^{-1}), and Ni (1–9 µg L^{-1}) were also found. One can therefore assume that the competitive role of those metals was significantly less in the adsorption process, and their reduction efficiency was consequently very low (see Table 2, unsaturated conditions). The results of other screened metals showed large differences in their removal by all tested substrates, which was mainly dependent on the concentrations of incoming water. The removal efficiency also differed for each filter material, though it still remains rather low for all media compared to some of the other studies [4–6].

All examined filter materials achieved good results for the reduction of such metals as Fe, Mn, Al, and Zn (total amount without partitioning into the colloidal and particulate fraction). These metals were present at highest concentrations in the analysed storm water during the whole period of sampling. Nonetheless, the highest removal efficiency was observed for activated carbon of smaller grain size (GAC 830W) and clinoptilolite, both under unsaturated and saturated conditions. The removal efficiency of Zn, Mn, Al, and Fe rose drastically with increasing concentrations of influent. Concerning Polonite this was most visible for the other analysed metals (Table 3). In other words, effluent concentrations stayed at the same lowered level or were diminished even more despite elevated concentrations of the influent. The effect of initial concentration on removal efficiency was also confirmed by previous studies [7,8].

Clinoptilolite exhibited the highest average removal capacity of Mn by 77.4%, while the finer GAC 830W achieved 72.3%. Copper and Cr uptake by clinoptilolite was, however, on average much lower (12–20%). It also

Table 2. Removal efficiencies by the filter media given in %

	Clinoptilolite		Polonite		GAC 830W		GAC PK3-5	
Element	Mean	SD	Mean	SD	Mean	SD	Mean	SD
Mn	80.4	11.7	52.8	26.3	83.7	12.2	68.8	13.3
Al	48.3	30.2	5.0	37.5	47.9	24.0	16.9	39.0
Fe	39.8	17.6	20.7	26.2	46.7	17.5	33.8	12.9
Zn	35.7	9.2	26.6	18.4	49.0	13.1	33.8	30.8
Pb	20.8	22.8	10.9	20.7	31.0	28.1	14.2	19.2
Cu	12.1	35.5	6.2	58.1	51.1	24.9	25.1	70.1
Cr	11.2	46.5	−246.6	132.2	27.2	58.3	21.4	48.7
V	4.2	42.7	−108.3	137.0	16.8	34.7	9.6	61.5
Ni	3.6	55.0	11.4	43.7	−49.4	96.7	10.9	76.2
Cd	−3.0	97.0	10.6	52.3	40.9	27.1	11.4	55.0

Negative values indicate release
n = 20; SD = standard deviation

showed a higher removal efficiency of Ni, Cd, and V under saturated flow. The average removal rate reached 21% for Ni, 20% for Cd, and 12% for V. This may indicate that the complete saturation of the filter bed resulted in a larger surface area that contributed to the treatment of the incoming storm water. Daily monitored concentrations in the clinoptilolite column effluent expressed in percentage of reduction revealed a maximum efficiency of Mn removal of 95.5% and a minimum of 46.7%, whilst for Fe the maximum amounted to 78%, and the minimum to 25%. The removal of Zn reached a maximum of 48% and a minimum of 4%.

Polonite removal rate of Fe, Mn, and Zn remained rather unchanged and did not depend on variations in influent concentrations. Polonite showed a high removal efficiency of metals despite its coarse fraction. The maximum removal rate of Mn was equal to 85%, while the minimum was 31% based

Table 3. Variations in metal removal efficiency by Polonite dependent on initial influent concentrations (C_{in})

Metal	C_{in} low (g L^{-1})	Removal (%)	C_{in} high (g L^{-1})	Removal (%)
Al	40–50	0	300–450	79
V	3–5	–	10–13	40–60
Cu	10–15	<25	22–42	68–74
Ni	1–3	0	6–9	50–77
Pb	1–3	0	>3	30–60
Cd	0.01	0	0.02–0.03	33–50

on the data obtained from everyday effluent concentration measurement of unsaturated flow. The corresponding data for saturated flow was 80.8% and 20%, respectively. In the case of Zn the removal efficiency was in the range of 17–50% for unsaturated flow and 6.3–51.6% for saturated flow. Concerning Fe the removal efficiency range was 13.6–48.6% and 6.3–54.1%, respectively. For such metals as Ni, Pb, Cd, Al, and Cu, Polonite performed poorly at low influent concentrations. However, a high removal efficiency of, e.g., nickel (50–77%) and copper (up to 60–70%) was reached whenever the influent concentrations increased (Table 3).

Release of Al was observed from the columns filled with Polonite and course GAC PK3-5 material during the initial test period of approximately 1 week.

The difference in removal performance was recognized as a result of the particle sizes of the materials. The sorption of the coarser GAC PK3-5 sorption was poorer than the finer GAC 830W for all investigated metals. The importance of particle size is significant due to the much larger surface area of reactive filters characterized by small particle diameter and thus a higher adsorption can be expected. However, both forms of GAC however performed also with broad range in removal rate.

The poor uptake can also be explained by short contact time of solution with filter material due to the high filtration velocity set up in this experiment. Metals bound to the particles of the storm water are mechanically removed with suspended solids during filtration. Assuming precipitation of metals hydroxide as a suggested removal mechanism for Polonite [6], one can conclude that the decrease of pH is a factor that could inhibit the process of uptake of dissolved elements.

Removal of organic carbon and nutrients

The concentration of TOC in incoming storm water was low or moderately high but mostly below 10 mg L^{-1}. The Polonite and clinoptilolite materials performed poorly, while both forms of activated carbon showed an expected reduction of organic matter. TOC removal was also observed to be decreasing over time due to the clogging phenomenon. Nonetheless, the finer GAC 830W decreased the TOC concentration to a value of around 4 mg L^{-1}, which is defined as very low or low concentrations according to the Swedish Environmental Protection Agency (SEPA) criterion. Coarser GAC PK3-5 achieved concentration levels ~7–9 mg L^{-1} in unsaturated conditions and up to 5 mg L^{-1} in constant saturated conditions. Thus, the difference cannot be found in either of the trials since the effluent concentrations are related to the levels of TOC in aqueous solution distributed into the pilot plant. However, the main part of organic carbon concentrations in the influent was the dissolved part hence the removal of these substances by activated carbon was the most efficient. Also it was confirmed that nutrients such as nitrogen and phosphorus appeared in the storm water in very low concentrations and were therefore very poorly removed by all tested materials.

Influence of hydraulic conditions

The removal efficiencies of the materials were similar in all experiments conducted under both unsaturated and saturated conditions. This can be explained by the hydraulic performance of the columns. The high hydraulic load probably caused saturation in parts of the columns aimed to be aerated under unsaturated flow. Therefore the aim to create two experiments with different hydraulic conditions can be assumed to have been unsuccessful.

Conclusions

Reaction kinetics, chemical equilibrium, association of metals with natural colloidal matter, and filtration velocity explain the differences in removal efficiencies of different contaminants. The design of interlinked reactive filter beds could be a possible solution to improve the removal efficiency, i.e., GAC for sorption demobilization of organic compounds, calcium silicate to destabilize colloidal components and precipitate selected metals, and natural zeolite to demobilize cationic metal species.

Acknowledgements

This research was a part of a joint project with the Stockholm Water Company (Stockholm Vatten AB) and the Swedish Road Administration. We would like to thank the Swedish Road Administration for the organization and installation of the pilot plant. Stockholm Water Company is acknowledged for the technical support and laboratory analysis.

References

1. Hares RJ, Ward NI (1999) Comparison of the heavy metal content of motorway stormwater following discharge into wet biofiltration and dry detention ponds along the London Orbital (M25) motorway. Sci Tot Environ 235:169–178

2. Sörme L, Lagerkvist R (2002) Sources of heavy metals in urban wastewater in Stockholm. Sci Tot Environ 298:131–145
3. Färm C, Renman G (1999) Proceedings of the 8th international conference on urban storm drainage, Sydney, Australia, 30 Aug–3 Sept, vol. 4, pp 1841–1846
4. Färm C (2003) Doctoral dissertation, Department of Public Technology, Mälardalen University
5. Ouki SK, Kavannagh M (1996) Department of Civil Engineering, University of Surrey, London
6. Kietlińska A, Renman G (2005) An evaluation of reactive filter media for treating landfill leachate. Chemosphere 61:933–940
7. Inglezakis VJ, Grigoropoulou H (2004) Effects of operating conditions on the removal of heavy metals by zeolite in fixed bed reactor. J Haz Mat B112:37–43
8. Athanasiadis K, Helmreich B (2005) Influence of chemical conditioning on the ion exchange capacity and on kinetic of zinc uptake by clinoptilolite. Water Res 39:1527–1532

Heavy metal removal efficiency in a kaolinite–sand media filtration pilot-scale installation

PJ Ramísio, JMP Vieira

Department of Civil Engineering, University of Minho, 4704-533 Braga, Portugal

Abstract

An experimental study was developed in order to evaluate the efficiency in retention of heavy metal from highway runoff using for filter bed current construction materials with known proprieties, sand and kaolinite. The control parameters for the experiments were: pH, conductivity, temperature, Zn, Cu, Pb, flow, and hydraulic head.

Preliminary results show that Zn is the most mobile metal with retention efficiency values decreasing to less than 50% in a 15-, 70-, and 110-day period, for the three different filter media. For Cu and Pb, and after a period of 260 days, the retention efficiencies obtained were above 70% and 40%, respectively, and above 90% in the sand and kaolinite media.

Introduction

A great number of different heavy metals and organic micro-pollutants appear as a result of automobile traffic. Cu, Pb, Zn, and Cd – sometimes extended to Ni and Cr – are considered the most important heavy metals associated with both mobile and stationary sources [1]. These metals are either dissolved in the storm water or bound to particulates [2,3].

The characterization and fractionation of metal elements in urban pavement runoff is important to develop an effective control strategy to immobilize metal elements before they leave the immediate vicinity of the roadway [4].

Another reason for fractionation study is that heavy metals can have an impact on environment and human health, especially in the dissolved form.

Unlike organics and xenobiotic compounds, metal elements are not degraded in the environment. As a consequence metal elements exert both a short-term toxicity impact, characterized by concentration or activity, and a long-term toxicity impact, characterized by mass accumulation [4].

Strategies such as infiltration methods, detention methods, vegetated filter strips, and wetlands have been proposed to control constituents in runoff from highways.

The two main active processes of immobilization of metals and nutrients in filtration systems are surface complexation and precipitation [3].

When infiltration ponds are used, the particles are retained by pond bottom layers, but dissolved heavy metals might contaminate aquifers. Even with detention ponds, where particles could settle, small particles and dissolved heavy metals contribute for groundwater contamination.

A decision with regard to the effectiveness of filtration or sedimentation is constrained by particle size gradation and density characteristics, as well as loading characteristics and specific site constraints. In contrast, if metal element loadings in urban drainage were mainly dissolved, effective *in situ* treatment would suggest the use of sorption and precipitation mechanisms [5].

Adsorption of metals to filter materials has been studied using different substrates, e.g., iron oxide-coated sand, natural zeolites, and granular activated carbon [3].

The immobilizing of dissolved metal complexes through sorption and particulate-bound metal elements through filtration mechanisms in an oxide-coated granular media trench has been reported. The field samples showed that Zn, Cd, Ni, and Cu mass were predominately in dissolved form, Pb and Cr were equally distributed between dissolved and particulate-bound, while Fe and Al were predominately particulate-bound [4].

The heavy metals desorption of is also relevant since by this process immobilized heavy metals can return to the dissolved form and contaminate groundwater. Desorption is a more difficult process than sorption. Sheidegger and Sparks [6] admitted that humic substances may be more responsible than clay minerals for sorption–desorption hysteresis. It is known that the diffusion of a chemical into micropores of organic matter and inorganic soil components is a very slow process and the non-labile form sorbed is difficult to desorb, being the labile form available for microbial attack. Therefore, sorption may be a reliable way to trap toxic pollutants in soil [2].

In developing an effective *in situ* treatment practice, feasibility must be evaluated based on characteristics of the waste loading and the physic-chemical mechanisms that can be utilized given the site, soil, and loading characteristics [5]. The development of an *in situ* retention technique, applying current construction materials with low variability proprieties, could help to develop a prototype solution for the reduction of heavy metal pollution from highway runoff.

The design of an adsorptive–infiltration technique is dependent of both the infiltration and sorption data. A research programme is in course to study those data and related processes and phenomena. This work reports the efficiency of dissolved heavy metal removal and physicochemical phenomena involved in the dual-media adsorptive filter.

Experimental details

The experimental installation consists of two reservoirs and three equal diameter cylindrical columns of transparent polyethylene. The layout of the installation is depicted in Figure 1.

The columns inner diameter is 172 mm. The bottom structure was constructed in order to allow a sliding movement at the end of the experiment. Downstream head control was established by means of a weir in order to easy measure the head loss through the filter media (Fig. 2).

In each column, different adsorptive filtration beds with 70 cm height where installed: a single sand medium in column C1; a dual media of 10% of kaolinite and 90% of sand in column C2; a dual media of 20% kaolinite;

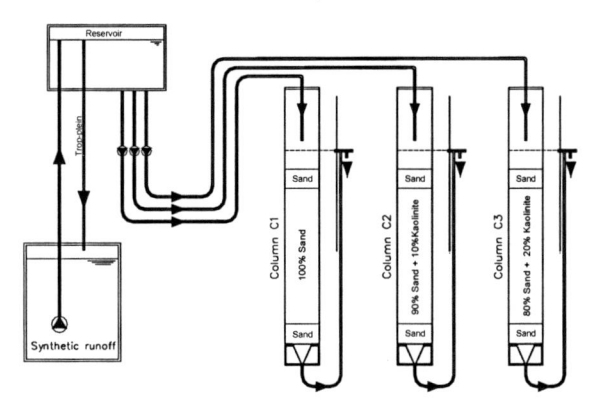

Fig. 1. Experiment layout

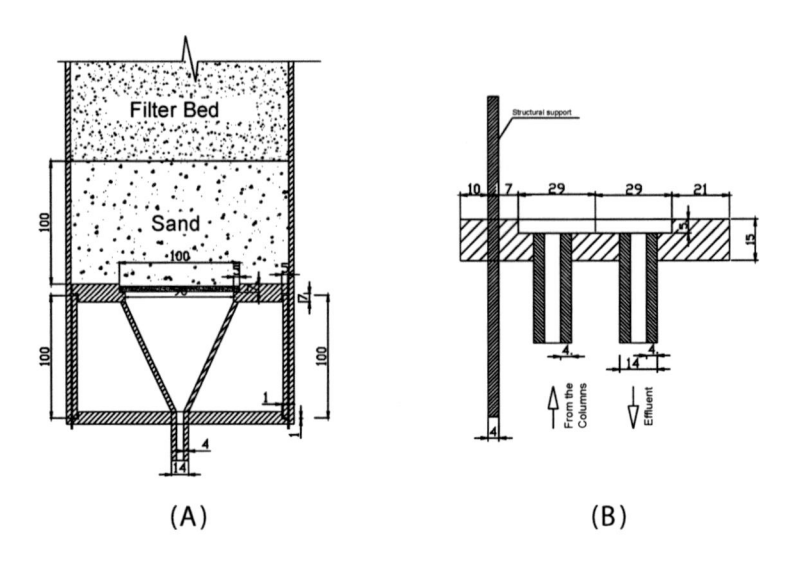

Fig. 2. Column details (units in mm): (A) Bottom retention structure; (B) Overflow weir

and 80% sand in column C3. For mechanical protection, a 10 cm sand layer was applied to the top and to the bottom of the filter bed. The filter media composition was defined in previous pilot-scale studies based on their hydraulic conductivity behaviour.

Each column was fed with a synthetic effluent by regulated flow pumps. Table 1 presents details of the filter material in the columns.

Kaolinite

Tables 2 and 3 present the chemical composition of the selected kaolinite, and the distribution of particle size.

Table 1. Filter bed weight (g)

	Column C1	Column C2		Column C3	
Layer	Sand	Sand	Kaolinite	Sand	Kaolinite
Top layer	3522.4	3522.4		3522.4	–
Filter media	24,705.8	22,235.2	24,705.8	19,764	49,411.7
Bottom layer	3522.4	3522.4		3522.4	–
Total	31,750.6	31,750.6		31,750.6	

Table 2. Kaolinite chemical composition analysis (x-ray fluorescence)

Composition	(%)
SiO_2	48.1
Al_2O_3	35.5
K_2O	2.0
Fe_2O_3	1.7
MgO	0.3
Na_2O	0.2
TiO_2	0.2
CaO	0.13
Ignition loss	12.0

Table 3. Kaolinite granulometry analysis (serigraph)

Particle size (μm)	(%)
<30	98 ± 1
<10	73 ± 3
<5	50 ± 4
<2	30 ± 4

Table 4. Typical clay CEC [7]

Mineral	Meq 100g^{-1}
Kaolinite	3–15
Illite	10–50
Chlorite	10–50
Vermiculite	00–200
Montmorillonite	80–200

Clay has high specific surface. Kaolinite has been reported to have specific surface from 15 to 50 m^2 g^{-1} that are small values when compared to montmorillonite (150–800 m^2 g^{-1}). The cation exchange capacity (CEC) defined as the maximum of cations that a clay mineral can exchange, are presented in Table 4.

Sand

The mineralogical analyses of the sand used for the top layer, filter bed, and bottom layer indicate the following mineral composition: 83% quartz; 13% feldspar; 4% mica; 0% kaolinite.

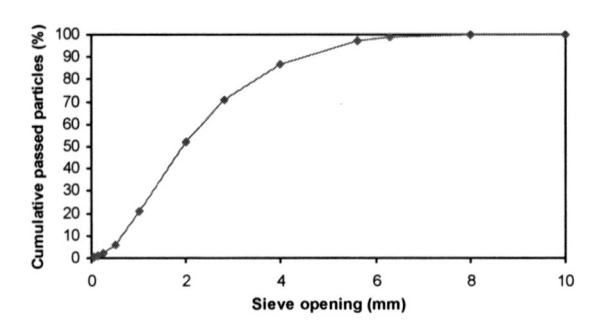

Fig. 3. Sand granulometry

The sand was sieved through different meshes and the granulometry distribution is presented in Figure 3.

Synthetic runoff

A synthetic highway runoff was obtained from the dilution of zinc acetate dehydrate ($(CH_3COO)_2$ Zn × $2H_2O$), copper (II) acetate monohydrate (($OOCCH_3)_2$ Cu × H_2O), and lead (II) acetate trihydrate (($CH_3COO)_2$ Pb × $3H_2O$). The maximum solubility value for these reagents at 20°C are: 430 g L^{-1}, 72 g L^{-1}, and 410 g L^{-1}, respectively.

Operational procedures

A constant effluent flow rate of 10 L d^{-1} was applied with different metal concentrations: 8.0 mg Zn L^{-1}, 1.0 mg Cu L^{-1}, and 0.4 mg Pb L^{-1}. After zinc breakthrough (at day 156), Cu and Pb concentrations were doubled.

The hydraulic load was 0.43 m d^{-1} (157 m $year^{-1}$).

The total organic carbon (TOC) concentration related to the acetate, in the original concentration, was 6.72 mg C L^{-1}, with the following contribution: Zn – 5.870 mg C L^{-1} (87.38%); Cu – 0.7553 mg C L^{-1} (11.24%), and Pb – 0.09266 mg C L^{-1} (1.38%).

Due to the long timescale of the experimental work, possible contamination of the synthetic runoff by atmospheric transport of small particles in the laboratory, biological growth, and/or other unexpected phenomena could be responsible for changes in dissolved heavy metal concentrations in the influent (e.g., biosorption or precipitation). Samples of recently prepared synthetic runoff and column influent were analysed in order to evaluate and quantify possible variations. Figure 4 depicts the results obtained during the experimental period.

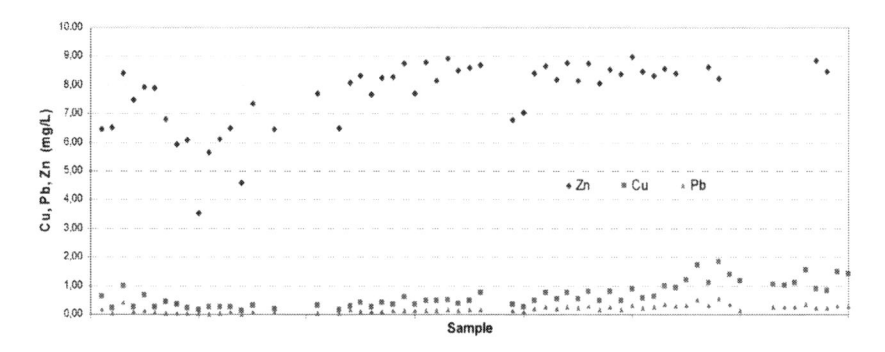

Fig. 4. Synthetic runoff and column influent concentrations

In each column, the dissolved heavy metal removal efficiency was calculated from the values obtained analysing simultaneously the dissolved heavy metal concentration in the inflow and the outflow.

The analyses were carried out according to standard methods [8]. Reagent-grade chemical products and ultra-pure water were used. All glassware coming in contact with the samples was washed with nitric acid and rinsed with deionized water.

After the sample was collected, the physicochemical parameters (pH, conductivity and temperature) as well as the hydraulic head were measured immediately. The sample was then filtered on a 0.45 μm porosity membrane, and acidified to pH less than 2 with HNO_3. This step was performed in order to measure the dissolved fraction. All the samples where then preserved at 4°C. Pb was analysed by graphite furnace atomic absorption spectrometry (GFAAS), while Zn was analysed by inductive coupled-plasma atomic emission spectrometry (ICP-AES). Cu was analysed by GFAAS or ICP. Detection limits were of 0.001 mg L^{-1} (GFAAS) and 0.01 mg L^{-1} (ICP-AES) for Cu, 0.002 mg L^{-1} (GFAAS) for Pb, and 0.01 mg L^{-1} (ICP-AES) for Zn.

Results and discussion

Samples of the synthetic runoff and column inflow and outflow were collected and analysed. The minimum, maximum, and average values for temperature, conductivity, and pH are presented in Table 5.

An electronic microscope observation was carried out in order to evaluate the particulate heavy metal retention in the filtration membrane, and the micro-organisms present in the synthetic runoff.

Table 5. Physicochemical results

Column/reservoir	Temperature (°C) Min.	Max.	Av.	Conductivity (μS) Min.	Max.	Av.	pH Min.	Max.	Av.
Inflow	16.4	25.8	21.3	90	132	102	6.93	7.85	7.5
C1 outflow	16.0	25.8	22.5	80	123	97	6.99	7.81	7.5
C2 outflow	16.2	25.6	22.5	77	115	93	7.11	7.90	7.6
C3 outflow	16.2	25.7	22.5	84	109	93	7.08	8.01	7.5

Figure 5 presents secondary electron images (SE) and backscattering electrons images (BSE) of particulate heave metal retained at the 0.45 μm porosity membrane. The particulate heavy metal was higher in the column inflow and with less extent in the columns outflow. The presence of lead occurred in the form of small spherical particles, while copper and zinc appeared normally together with a more homogenous spatial distribution.

An observation of the biofilm formed at the synthetic runoff retained in a filtration membrane is presented in Figure 6.

Results of dissolved heavy metal removal efficiencies in the three experimental columns are presented in Figure 7.

Preliminary results show that Zn is the most mobile metal with retention efficiency values decreasing to less than 50% in a 15, 70, and 110 day period, for columns C1, C2, and C3, respectively.

For Cu and Pb, and after 260 days, the retention efficiencies obtained were above 70% and 40%, respectively, in Column C1, and above 90% in the sand and kaolinite media columns.

The different retention efficiencies of Zn observed in these three columns suggest a direct relation to the filter bed characteristics, since the breakthrough period in column C2, with a 10% kaolinite filter bed, was almost half the breakthrough period in column C3, with a 20% kaolinite

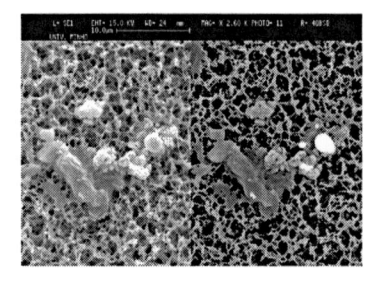

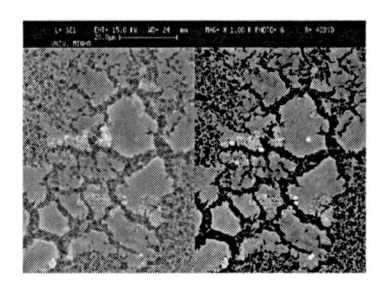

Fig. 5. Particle-bond heavy metal membrane retention

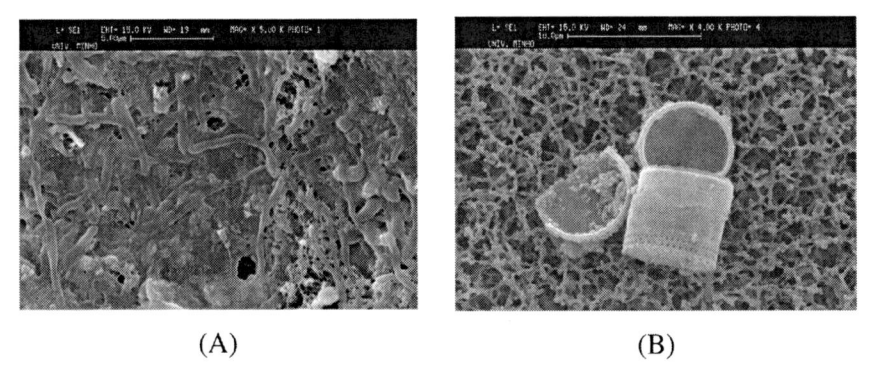

Fig. 6. Biofilm observation. (A) Mesh of micro-organisms; (B) diatom picture detail

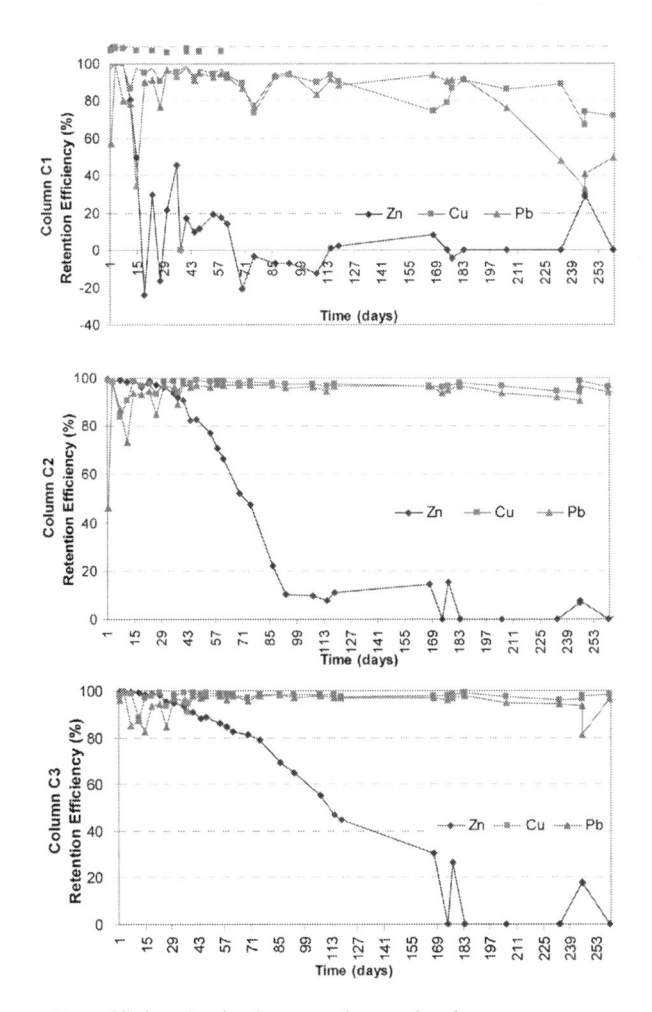

Fig. 7. Retention efficiencies in the experimental columns

filter bed. Although other processes may occur, these results indicate that the CEC is relevant and therefore sorption is the major phenomena.

The removal efficiencies of Cu and Pb are high, even for the sand filter bed. The efficiencies are not so dependent of the filter bed material, which may indicate that surface complexation and precipitation may have a more important role than in the retention of Zn.

At the end of this experiment, the filter bed will be removed and sliced to six layers. A homogeneous sample of each layer will be weighted and submitted to desorption. The desorption results and the retention efficiencies reported, will permit to more accurately estimate the main processes and phenomena involved as well as estimate the fraction of heavy metals that can be desorbed, and by this way contaminate groundwater.

Conclusions

The promising results obtained in this research work seem to be applicable as a technical solution for groundwater protection by retaining heavy metals from highway runoff.

Although many laboratory studies can be carried out, full scale results must be obtained to validate design parameters, since specific site characteristics such as rainfall characteristics, biological growth, temperature, and sediment granulometry have a major role in adsorptive filtration processes. This study is still running and the analytical data are now being evaluated in order to establish design parameters for a full-scale experimental station.

References

1. Hvitved-Jacobsen T, Vollertsen J (2005) Urban storm drainage pollution – concepts and engineering. Ph.D. course: process engineering of urban and highway runoff. Aalborg University, Aalborg, Denmark
2. Barbosa AE (1999) Highway runoff pollution and design of infiltration ponds for pollutant retention in semi-arid climates. Ph.D. thesis, Aalborg University, Aalborg, Denmark, 52 pp
3. Farm C (2002) Metal sorption to natural filter substrates for stormwater treatment – column studies. Sci Total Environ 298:17–24
4. Sansalone JJ et al. (1996) Fractionation of heavy metals in pavement runoff. Sci Tot Environ 189/190:371–378
5. Sansalone JJ (1999) Adsorptive infiltration of metals in urban drainage-media characteristics. Sci Tot Environ 253:179–188
6. Scheidegger AM, Sparks DL (1996) A critical assessment of sorption-desorption mechanisms at the soil water interface. Soil Sci 161:813–831

7. Gomes CF (1986) Argilas – O que são e para que servem. Fundação Caloust Gulbenkian (in Portuguese)
8. Standard Methods for the Examination of Water and Wastewater (1995) 19th edn. American Public Health Association, American Water Works Association, Water Environment Association

IV. Storm Water

Site assessment of road-edge grassed channels for highway drainage

M Escarameia,[1] AJ Todd[2]

[1]HR Wallingford, Howbery Park, Wallingford, Oxfordshire OX10 8BA, UK
[2]Atkins (formerly of TRL), Cornerstone House, Stafford Park 13, Telford TF3 3AZ, UK

Introduction

Worldwide, grass-lined channels (or swales) have long been used for the drainage of surface runoff from roads and motorways. Recently, the emphasis on sustainability and on minimization of impact on the environment have prompted the spread of swales as a drainage option which also provides flow attenuation and improvement of the quality of the discharged water. However, swales and other traditional grass-lined channels are typically too deep and steep-sided, from a vehicle safety view point, for use adjacent to carriageways.

A new environment-friendly drainage system was therefore developed by HR Wallingford (HRW) and Transport Research Laboratory (TRL) in collaboration with the Sports Turf Research Institute for the UK Highways Agency (HA). During an 8-year-long study, road-edge grassed channels were evaluated in terms of grass type specification, geometric characteristics, safety, hydraulic resistance, and performance on site through field trials. These trials are the subject of this paper and provided important information for the specification, construction, and maintenance requirements of grassed channels, as well as for their hydraulic capacity and environmental qualities. As a result of these very positive trials, the design document Advice Note HA119 [1] was produced giving guidance to highway drainage engineers on the design, construction, and maintenance of grassed channels.

Overview of development of grassed channels

In the UK, the drainage option currently preferred by the HA to collect surface runoff consists of triangular concrete channels built adjacent to the road edge with fin drains to collect subsurface flows. Grassed channels evolved as an environmentally improved version of concrete surface water channels. They are typically triangular in cross section, 2–3 m wide with mild side slopes of 1:5 or flatter and overall depth not exceeding 200 mm (Fig. 1).

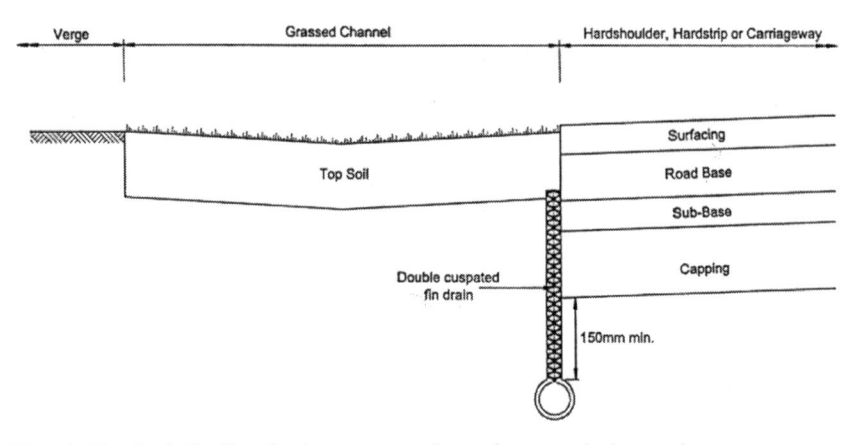

Fig. 1. Typical (indicative) cross section of grassed channel

Several key aspects were addressed before arriving at a fully viable solution (for full description see Escarameia and Todd [2]):

- *Separation of surface and subsurface flows*

Prevention of surface water infiltrating through grassed channels into the road construction layers can be achieved by means of a double cuspated fin drain between the road and the channel, draining into a carrier pipe.

- *Integration with the pavement and verge details*

Because grassed channels need to be wider than typical concrete channels, full account was taken of existing HA requirements concerning safety barriers, verge widths, installations of services and signs, etc.

- *Hydraulics of grassed channels*

Since existing methods for determining the hydraulic resistance of grass were not suitable for such shallow channels, a comprehensive full-scale testing programme was undertaken. This is described in detail in Escarameia et al. [3,4].

- *Buildability and identification of suitable grass types*

Methods of construction needed to be straightforward and capable of being integrated with the stages of pavement and road construction. The two most viable methods of lining the channels were identified as turfing and hydroseeding. Suitable grass types needed to be tolerant of damp conditions, be resistant to salt and damage by errand vehicles and be slow growing to minimize maintenance. Mixes dominated by perennial ryegrass and fescues are recommended for UK climatic conditions.

- *Durability and safety*

Another key factor in the design was durability, which depended on vulnerability to vehicle overrun, to weed invasion and silting up. Guidance was produced on maintenance to control weeds and grass height in order to ensure adequate conveyance. Tests carried out on a full-scale channel to evaluate construction methods and safety impact of the channels on vehicles, as well as vehicular damage to the channels, confirmed the feasibility of the channels and that safety performance was no worse than for concrete channels.

Field trials

Site selection and channel design

Several road schemes were identified as possible candidates but the construction programme of the A2–M2 Widening scheme in Kent fitted well with the research programme and provided good conditions in terms of space available, low risk of vandalism, and willingness from the contractor to experiment with a new drainage system. This was a design and build scheme where, initially, a kerb and sumpless gully drainage system had been considered. As well as building some 800 m of grassed channels based on designs provided by the research team, the carrier pipes and chambers for the conventional system were also laid as originally planned so that the system could be reinstated if the trials proved unsuccessful.

Description of monitoring sites and channels

The sites monitored were (Table 1):
1. Three crutches before bridge (London-bound): MP 44.4–44.3
2. Three crutches after bridge (London-bound): MP 42.2–42.1
3. Nashenden Valley (Coast-bound): MP 50.3–50.5

Note: MP stands for "marker posts"

Table 1. Characteristics of monitored sites

Site	Channel length (m)	Channel width (m)	Channel depth (m) – nominal	Channel slope (nominal)
1	43.9	2	0.150	1:75
2	94.0	3	0.150	1:66
3	147	2	0.150	1:35

The channels were lined with turf supplied by Turf Centre, Hampshire, consisting approximately of 60% fescue and 40% bent (Fig. 2). Samples of this grass mixture were checked by Smithsonian Tropical Research Institute (STRI) and considered adequate for purpose. A note with advice on procedures for the rapid establishment of the grass and maintenance recommendations was produced for the channel contractors.

The width of the grassed channels was determined by the available space on the verge of the road, from 2 to 3 m, whereas the channel depth was predefined at 150 mm for safety reasons. For the determination of the hydraulic resistance of the channels an average grass height of 50 mm was assumed, which corresponded to a Manning's resistance coefficient of approximately 0.075. In the specification provided by HRW/TRL it was

Fig. 2. View of site 3 from downstream

stated that each outfall grating ought to be protected by a 150 mm wide by 150 mm thick concrete surround. In order to provide a transition between the turf and the grated outfalls a surround of Grassgrid (Charcon) was installed around the gratings. The edge of the Grassgrid was depressed below the invert of the grassed channel by 50–75 mm to ensure that any vehicle wheel in the channel would not hit a rigid concrete edge.

Given the close proximity of groundwater abstraction points, the Environment Agency required that an impermeable membrane be placed underneath the grassed channels. The membrane selected was SGS Geosystems Ltd. (SGS)'s 1 mm gauge high-density polyethylene (HDPE).

Instrumentation

The monitoring of the grassed channels involved continuous measuring of rainfall and of water depths in the channels as a result of the runoff from the road and cutting. Two 0.2 mm tipping bucket rain gauges (Casella) were used to measure rainfall: one for sites 1 and 2 (due to their proximity to each other, Fig. 3), and another for site 3. Water levels in the channels were measured primarily by means of ultrasonic probes (manufactured by Pepperl and Fuchs). The probes were installed at two different locations in each monitored site: one close to the downstream end of the channel to

Fig. 3. Rain gauge for sites 1 and 2

capture maximum water depths and one further upstream. The ultrasonic probes were mounted on purpose-built chambers. Each chamber was connected by means of a flexible 20 mm diameter tube to a small rectangular collection channel covered with a perforated grating, which was placed at the invert of the grassed channel (Fig. 4). This set-up allowed the level of the runoff in the grassed channel to be transferred to the chambers, which acted as stilling wells, where the levels could then be read by the probes. During the monitoring programme problems were encountered related to

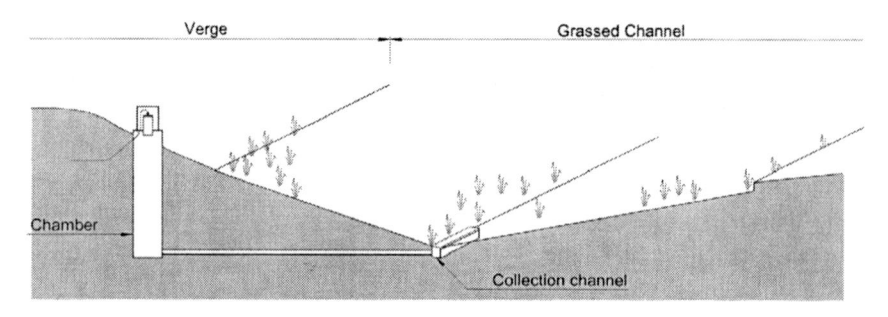

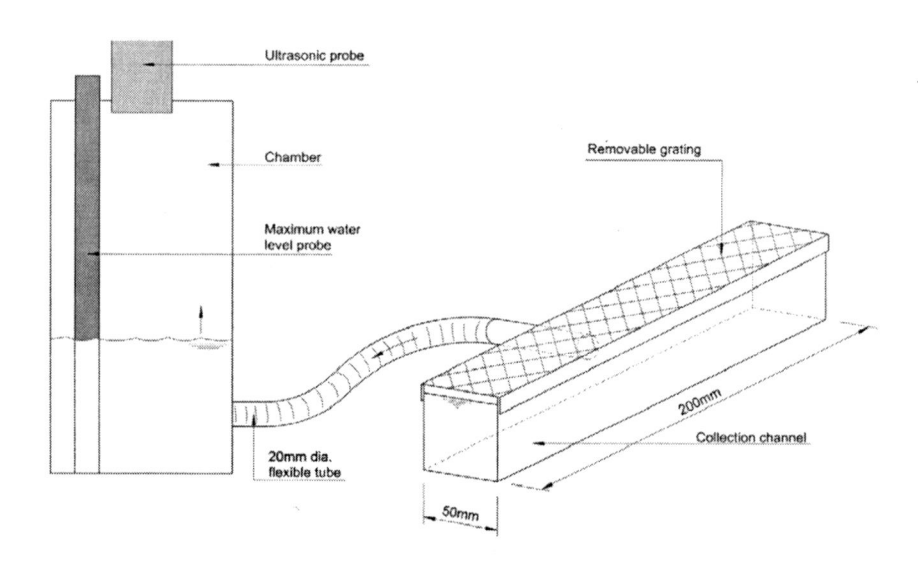

Fig. 4. Schematic of water depth probes and collecting chamber

malfunctioning of the ultrasonic probes on site and maximum water level probes were used as an alternative. With these probes, maximum levels are registered on a water-sensitive tape that discolours in contact with water. Data loggers (Datataker DT50, from Data Electronics Ltd.) were used to collect the data from the rain gauges and the water depth probes. The data loggers were triggered by a tip of the rain gauge (i.e., the start of a rainfall event) and were programmed to record rainfall and water depth data at 15 s intervals.

Other parameters such as the permeability of the underlying soil, grass height, channel geometry, and water quality were also measured at certain times during the monitoring trials.

Data collected

The trials began in early 2003, with field data collection starting in the beginning of April and completing at the end of July 2004. The data collection was hampered by difficulties associated with damage to the instrumentation (e.g., cables were regularly found to have been cut or to have damage to the linings) as well as equipment malfunction; however, useful data was obtained as described below.

Channel geometry

It was found that both the longitudinal slope of the channel invert and the cross-sectional shape of the channels remained fairly constant during the 16-month trial (with localized deepening in some places, which is thought to be a result of localized settlement due to lack of soil compaction rather than a result of erosion caused by the flow in the channels). There was some evidence of possible siltation at localized sections, but general conclusions were difficult to draw because of the effect of grass overgrowth on survey measurements. Occasional ruts caused by errant vehicles were easily repaired and did not compromise the integrity of the channels.

Hydrological, hydraulic, and permeability data

Rainfall information provided by the rain gauges was assessed to determine which storm events produced either significant amounts of rain or intense rainfall. Maximum peak rainfall intensities of 96 mm h^{-1} (or 0.027 L s^{-1} m^{-2}) were recorded, which corresponds to a return period higher than 1 year (and less than 5 years) for the location of the monitoring sites. These are the return periods usually considered for road drainage in the UK.

The volume of flow in the grass channel is affected by the rate at which water can soak into the channel bed and the time taken for the subsurface material to reach saturation. Permeability tests were therefore undertaken during different seasons at a number of locations within the trial channels using a falling head permeameter, thermometer, and a stop watch. The permeability measured during the trials was very variable depending on precedent weather conditions, ranging from 0.001 to 0.33 m s^{-1}.

It is well established that grassed channels have an increased ability to control pollutants when compared with non-vegetative systems. The limited information on water quality that was collected during the trials appears to confirm this, with low levels of heavy metals (below detectable limit for nickel) as well as of phosphate and nitrogen. The levels of hydrocarbon contents measured were also very low, much below the limits recommended in BS EN 858 [5] for discharge into watercourses.

Grass condition

With regard to the quality of the grass cover, it was found that regular inspections and maintenance are important to ensure a good quality for the grass. Invasion by large weeds was observed and, although the presence of weeds and dead patches did not impact on the hydraulic performance, they affected the visual appeal of the channels.

The quality of the grass was variable during the trials but generally speaking it was found that the turf provided adequate cover to the channels to prevent flow erosion. The grass survived quite well during the extremely hot summer of 2003, when temperatures as high as 30°C were registered in southern England. However, the turf at the sections where the water depth probes and chambers were installed appeared not to have ever fully recovered from installation of the instrumentation. The last metres of channel in all sites (between the downstream probe and the outfall) were the areas where the condition of the turf was the poorest, becoming fairly bare and/or overtaken by large weeds.

Areas of damage were also identified in sites 1–3 and repairs to the grassed channel had to be made in some cases. The damage was caused primarily by works machinery during installation of services in the verge beyond the channels. Repairs were, however, simple to carry out.

Analysis

The main objective of the analysis of the storm data collected was to check the applicability to the field of the design method that had been developed for the grassed channels based on laboratory tests. The analysis involved the following phases:

1. Development of a numerical model of runoff for prediction of water depth on grassed channels with time-varying rainfall
2. Processing of raw data
3. Determination of water depth profiles for the storm events monitored using the numerical model and comparison with the data collected.

A new numerical model, based on kinematic wave theory (this theory assumes the bed slope is equal to the friction slope S) was developed to compute the water depth at any required location along grassed channels with triangular cross sections. This model, Qgrass, was written in FORTRAN for a Windows environment with a Visual Basic interface.

The programme included the new resistance equation developed from laboratory data which was based on the Manning's equation and included a new roughness coefficient taking into account the dependence of the flow resistance on the term VR, the grass height H, and the type of grass.

The program Qgrass was run to determine the variation, at the location of the water level probes, of the water depth in the grassed channels during the storms selected for analysis. As a starting point, it was assumed that the catchment would produce 100% runoff. The results of the programme indicated that this assumption was valid as the agreement between the measured and calculated depths was very satisfactory. The numerical simulations captured very well the occurrence of the peaks of the storms and, in the great majority of cases, their intensity (see Figs. 5 and 6).

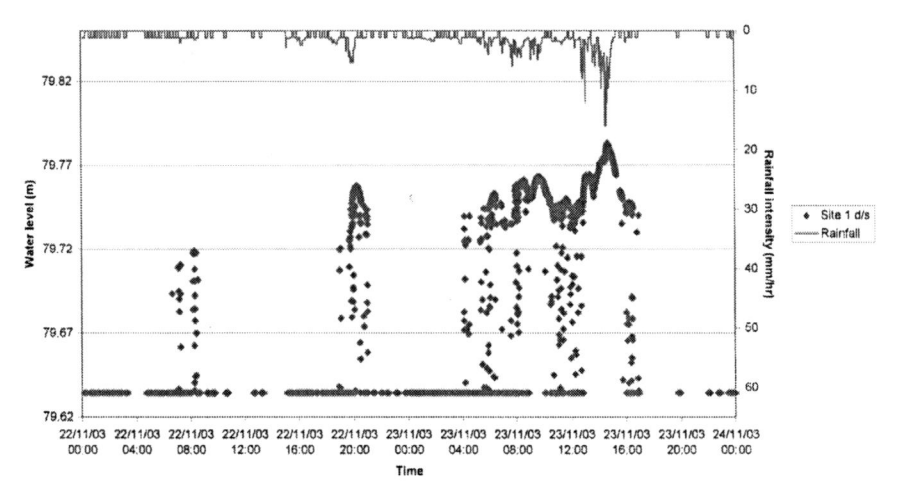

Fig. 5. Example of storm data collected (rainfall and water level in channel (22/11/03)

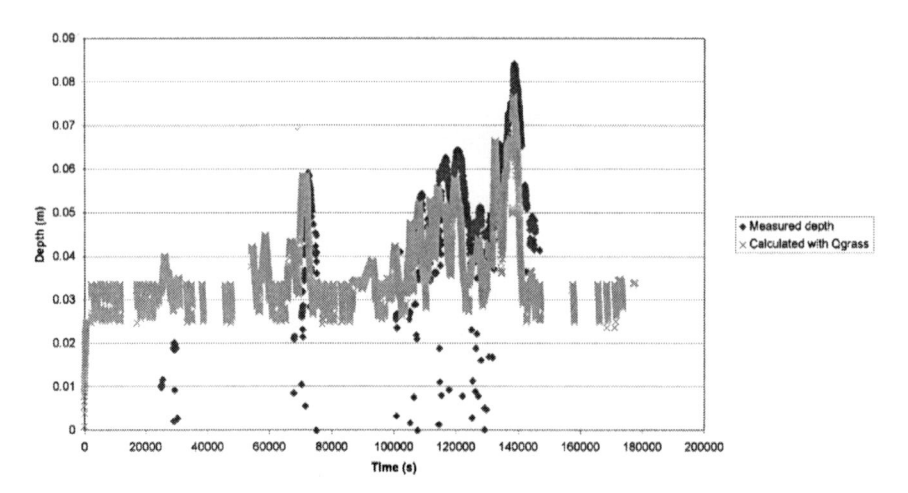

[Figure: plot of Depth (m) vs Time (s) showing measured depth and calculated with Qgrass]

Fig. 6. Example of comparison between measured and calculated water depths in grassed channel (22/11/03)

In none of the cases analysed did the calculated water depth exceed the channel depth (nominally 150 mm). Also, HRW was not aware of any reports of water on the hard shoulder or carriageway as a result of surcharge of the channels during storms. Overall, it can thus be concluded that the design procedure implemented in Qgrass was adequate.

Conclusions

From an environmental point of view the field trials carried out indicated that grassed channels are a good option for road drainage in that they provide "greening" of the verge without compromising their function as road runoff conveyance channels.

The quality of the grass was variable during the trials but generally speaking it was found that the turf provided adequate cover to the channels to prevent flow erosion. The durability of grassed channels cannot be dissociated from the level and frequency of maintenance they receive. Regular inspections and cutting (three times in the growing season in the UK) are essential to ensure the quality of the grass and flow capacity of the channels. Detailed surveys of the longitudinal profile and cross sections of the channels showed no significant signs of siltation or erosion during the 16-month long trials.

The permeability measured during the trials was very variable depending on precedent weather conditions, ranging from 0.001 to 0.33 m s^{-1}.

The design of grassed channels incorporates a fin drain at the edge of the carriageway to cater for infiltrated flows.

Limited water quality monitoring indicated that the channels have a beneficial effect when compared with concrete surface water channels. Work is soon to be commissioned to investigate the pollution removal ability of grassed channels.

Acknowledgements

The work described in this paper was funded by the UK Highways Agency. Special thanks are due to the HA Project Sponsor, Mr. S Santhalingam for his enthusiastic involvement in the work. Acknowledgement is also due to Mr. RWP May and Mr. Y Gasowski, formerly of HR Wallingford, for their contributions to the work, namely the numerical program Qgrass.

References

1. DMRB 4.2 (2006) Grassed surface water channels for highway runoff. HA119, vol. 4, Section 2, Part 9. The Stationery Office, UK
2. Escarameia M, Todd AJ (2006) Grassed surface water channels for road drainage. Project Report. HR Wallingford Report SR662, February 2006
3. Escarameia M, Todd AJ, May RWPM, Gasowski Y (2000) Grassed surface water channel. Interim Report. HR Wallingford Report SR572
4. Escarameia M, Gasowski Y, May R (2002) Grassed drainage channels – hydraulic resistance characteristics. Proc. ICE Water Marit Eng 154(4): 333–341
5. BSI (2002) Separator systems for light liquids (e.g., oil and petrol). BS EN 858, British Standards Institute, London

Evaluation of the runoff water quality from a tunnel wash

AE Barbosa,[1] J Saraiva,[2] T Leitão[1]

[1]Hydraulics and Environment Department, National Laboratory for Civil Engineering, Av. Do Brasil, 101, 1700-066 Lisbon, Portugal
[2]Buildings Department, Av. Do Brasil, 101, 1700-066 Lisbon, Portugal

Abstract

Tunnels are specific sections of roads, where the rainfall runoff process does not take place. Nevertheless, they are washed out periodically mainly for traffic safety reasons. The Portuguese National Laboratory for Civil Engineering (LNEC) monitored twice (in May and December 2003) the washout process of the 1570 m long IP2 Gardunha tunnel. The objective was to evaluate pollutant accumulation in the road pavement. Due to specific requirements from LNEC the wash operation was performed without the common detergents used to enhance the cleaning process. The annual average daily traffic (AADT), in 2003, was 9923. Measurements of flow and the collection of manual samples took place at the inlet of an oil separator tank that receives all the runoff from the tunnel, and discharges it into the local creek.

The first wash (1st wash) took place over two nights, with a total duration of 345 min, and an average flow of 1.6 L s^{-1}. The second wash (2nd wash) operation was shorter, with the duration of 103 min and an average flow of 4.2 L s^{-1}. The annual pollutant loads for metals, total suspended solids (TSS), oil, and grease and polycyclic aromatic hydrocarbon (PAH) were estimated based on wash volumes and pollutant concentrations.

Compared to other monitoring studies of highway runoff in Portugal, Gardunha tunnel wash water presented low pollutants content. Only an evaluation of the results taking into consideration the flow processes inside the tunnel atmosphere allowed an understanding of the data, being concluded that tunnel pavements accumulate much less traffic pollutants than ordinary road pavements, under similar operation conditions. Furthermore, pollutants emitted inside the tunnel will tend to build up outside, therefore increasing runoff concentrations in the road sections flanking tunnels. These results are important for the control of highway runoff impacts.

Introduction

In the framework of a research that the Portuguese National Laboratory for Civil Engineering (LNEC) has been carrying out for the Roads Institute, some roads were selected for runoff monitoring studies. The main objective was to gather information on road runoff (quality) characteristics in Portugal; therefore the case studies should represent different road sections and different climatic regions. Since tunnels are specific sections of roads never monitored before in Portugal one of the chosen sites was a tunnel segment located in central Portugal. Tunnels are confined areas not subject to the rainfall runoff process and therefore it should be likely to expect a higher pollutants accumulation inside them. These pollutants are usually released in a very short period, when tunnels are washed out, and the high pollutant concentrations may cause acute environmental impacts. Not only common road runoff pollutants are of concern; tensides, used to increase the removal of contaminants from tunnels, were reported by Rosedth et al. [1] as being acute toxic for aquatic organisms.

The references found in the literature, of studies concerning water quality from road tunnel washes, are from France [2] and from Norway [1,3,4]. Although it is not possible to compare results among the studies, due to the very different tunnel and tunnel use and maintenance characteristics, wash procedures, and monitoring methods, Table 1 presents some of the average pollutant concentrations obtained in those studies.

The Portuguese case study was the IP2 Gardunha tunnel (today part of the Highway A23). The tunnel is washed twice a year. Given the 6-month

Table 1. Concentration of the main pollutant found in the wash water from several tunnels

Sites	pH	Cond. ($\mu S\ cm^{-1}$)	Cd ($\mu g\ L^{-1}$)	Pb ($\mu g\ L^{-1}$)	Cu ($\mu g\ L^{-1}$)	Zn ($\mu g\ L^{-1}$)	TSS ($mg\ L^{-1}$)
Nordby[a]	7.41	930	1.94	93.6	260	2,600	2260
Frejus[b]	–	–	–	2750	–	8700	2960
Mont Blanc[b]	–	–	–	5200–15,000	–	26,500	5820–23,200
Chamoise[b]	–	–	–	3100	–	4800	2255
Les Monts[b]	–	–	–	12,100	–	9900	6678
Fourviere[b]	–	–	–	26,000	–	–	2354
Ringnes[c]	7.5–7.9	–	–	<0.516	11–28	119–7510	–
Nordby, Smihagen, Vassum[d]	–	–	–	170	680	13,800	3030

[a][4]
[b][2]
[c][3]
[d][1]

period of pollutants accumulation, it was expected that the monitoring results would surpass the concentrations observed in other roads. The monitoring of the washes of this 1570 m long tunnel took place in May and December 2003, and the results came as a surprise, being rather the opposite of what was expected. Further examination of the system, in collaboration with experts in tunnel atmospheric processes, leads to a more comprehensive understanding of the phenomena taking place, and permitted to elaborate a hypothesis that explains the results obtained at the Gardunha tunnel.

This paper presents the case study description, monitoring methods, results, and discussion. The tunnel atmospheric processes that support the hypothesis are explained as well. A more detailed description of the work done can be found in Barbosa and Antunes [5] and Barbosa et al. [6].[1]

---

[1] Both reports are written in Portuguese

Case study

Description of the site

The site selected is a 1570 m long tunnel located between 11 + 060 and 12 + 630 km of the road connecting Covilhã and Castelo Branco, known as IP2 Gardunha tunnel. The tunnel has an excavated area of 80 m^2 (55 m^2 of useful cross section) with a maximum burden of 235 m [7]. It is a single-gallery bidirectional tunnel, with a hydraulic diameter of 7.5 m and a mean slope of 1.7%. The axis of the tunnel is a straight line except for a gentle curve at about two-thirds of its length. The annual average daily traffic (AADT) of the road, in 2003, was 9923 vehicles of which about 20% are heavy duty. The tunnel paved cross section is of approximately 9 m.

The tunnel washout was monitored twice by LNEC. The first wash (1st wash) was in May 2003, over two nights, and the 2nd wash in December 2003, during one night. The water flowing at the tunnel pavement is drained to a longitudinal drain that directs the runoff to an oil separator tank (cf. Fig. 1) discharging into a local creek. Tunnel washes monitoring procedures.

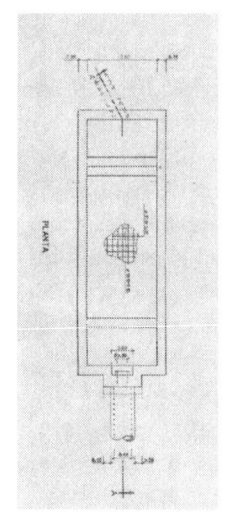

Fig. 1. View of the tunnel and of the collecting oil separator tank

The main objective of the washing and monitoring procedures described was to evaluate the pollutant loads accumulated in the tunnel road pavement during a 6-month period. For that purpose the wash operation was monitored at two different occasions (cf. Fig. 2). In accordance to LNEC request the common detergents applied to enhance the cleaning process, carried out regularly each 6 months, were not used.

The methodology adopted for the work has considered the analysis of two aspects of the runoff resulting from the tunnel wash: (1) the changes in flow and water quality along the wash process; and (2) the possibility to estimate the annual pollutant loads accumulated at the tunnel pavement.

The monitoring procedure consisted in registration of the water level, as well as collection of grab samples at the inlet of the oil separator tank, during the wash process. Figure 3 presents a section cut of the oil separator tank, and the ruler placed inside the tank for the water level measurements.

Fig. 2. View of the tunnel 2nd wash operation

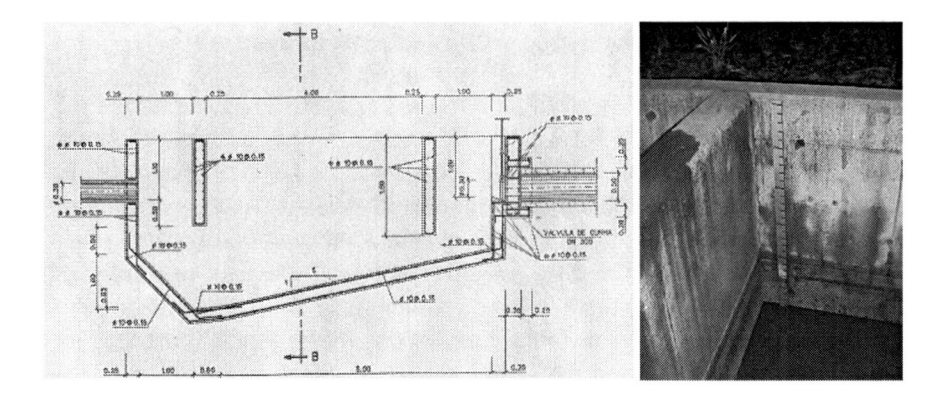

Fig. 3. Section cut of the oil separator tank and picture of the ruler for water level measurements

The water at the inlet of the oil separator tank was continuously monitored for the measurement of general physico-chemical parameters (pH, Eh, electric conductivity, and temperature) and samples were discretely collected (six samples for the 1st wash and eight for the 2nd wash) for the analysis of the following parameters: cadmium (Cd), lead (Pb), copper (Cu), zinc (Zn), iron (Fe), total suspended solids (TSS), oil and grease, and polycyclic aromatic hydrocarbons (PAH). The analysis have been done at a selected laboratory, according to accredited methods namely, the standard methods (SM) 3500 C for Cd, Pb, and Zn, and the SM 3500 D for Fe and Cu. Oil and grease analysis were accordingly to SM 55200 and TSS to SM 2540 D. The total PAH was based on gas chromatography, mass selective detection in the single ion monitoring mode (MSD/SIM); 16 PAH were measured, namely: acenaphthene, acenaphthylene, anthracene, benzo(a)anthracene, benzo(k)fluoranthene, benzo(b)fluoranthene, benzo(a)pyrene, benzo(g,h,i)perylene, chrysene, dibenzo(a,h), fluoranthene, fluorene, indeno(1,2,3,c,d)pyrene, naphthalene, phenanthrene, and pyrene.

The quality of the water used for the washes was also collected for analysis. The water from the 1st wash was groundwater, used by the road fire extinguishers system. The 2nd wash, as more water was needed in a shorter period of time, had an additional source of water. Water taken out from a local creek was transported to the site in a (metallic) fire engine tank with a capacity for 30,000 L.

Results and discussion

Results of samples analysis and *in situ* measurements

The 1st wash had a total duration of 345 min, and an average flow of 1.6 L s^{-1}. The second wash (2nd wash) operation was shorter and more intense, with 103 min of duration and an average flow of 4.2 L s^{-1}. These calculations were based on periodic measurements of the water level inside the oil separator chamber, being approximated. The first volume of water, equivalent to the water that filled the irregular bottom of the tank was not used for calculating flow but total volume. A summary of the quantity figures for the two campaigns is presented in Table 2 and Figure 4 illustrates the high flow variation. During the 1st wash (part I), there was a stop between 1 h 28 min and 2 h 20 min.

Table 2. Average flow and total volume for the two washing procedures

	Average flow (L s^{-1})	Total duration* (min)	Total volume (L)
1st wash	1.6	345	34,105
2nd wash	4.2	103	25,956

*Total duration of the washes, excluding the pause period that took place in the 1st wash

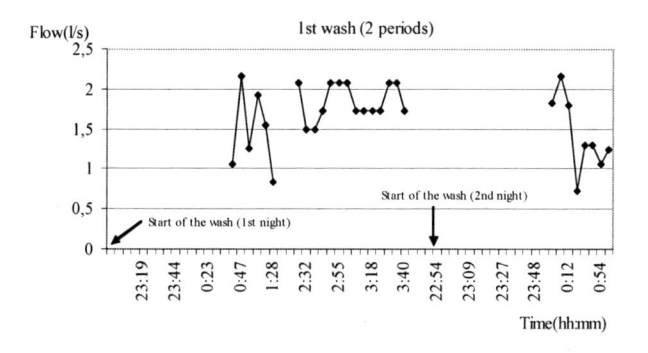

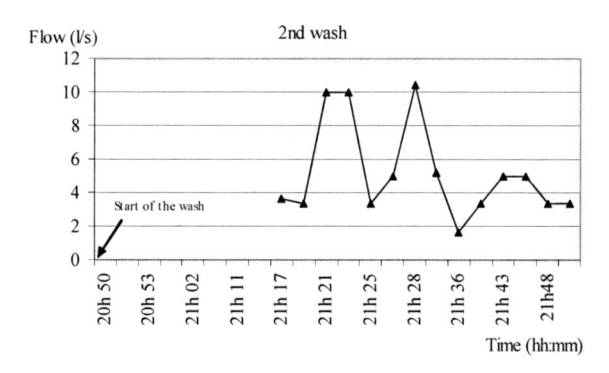

Fig. 4. Flow variation during the washes

The quality of the water used for the wash is presented in Table 3. The results – as averages, medians, maximums, and minimums – for the characterization of the water quality at the inlet of the oil separator chamber, are presented in Table 4. The quality parameters and pollutants were measured on the six and eight discrete samples taken, respectively, along the 1st wash and the 2nd wash. PAH in the 2nd wash were always below the detection limit (0.05 µg L^{-1}).

Table 3. Quality characterization of the water used for the washes of the tunnel

	1st Wash Hose	2nd Wash Hose	Fire engine tank
Parameter			
T (°C)	16.0	12.5	9.3
pH	9.1	8.1	7.3
Cond. (μS cm^{-1})	260.0	290.0	68.1
Cd (μg L^{-1})	<1	2.0	1.0
Pb (μg L^{-1})	2	<5	<5
Cu (μg L^{-1})	<2	<5	<5
Fe (μg L^{-1})	<50	120	1420
Zn (μg L^{-1})	73	370	40
TSS (μg L^{-1})	1.6	1.4	4.6

Table 4. Average, median, maximum, and minimum of the results obtained for the two washing campaigns

	1st Wash (six samples)				2nd Wash (eight samples)			
Parameter	Av.	Median	Max.	Min.	Av.	Median	Max.	Min.
T (°C)	15.8	15.8	16.4	15.5	10.4	10.6	11.1	9.7
pH	7.7	7.5	8.4	7.2	8.3	8.2	8.7	8.0
Cond. (μS cm^{-1})	514.0	450.0	683.0	380.0	367.9	346.0	650.0	232.0
Cd (μg L^{-1})	5	6	6	3	2	2	3	1
Pb (μg L^{-1})	48	46	92	20	35	32	65	13
Cu (μg L^{-1})	195	175	340	82	108	99	300	9
Fe (μg L^{-1})	5693	5515	11,980	1,710	4703	4600	7040	1,410
Zn (μg L^{-1})	962	700	2210	370	560	525	1150	230
TSS (mg L^{-1})	680.4	522.6	2182	88.9	1078.5	654.5	3003	227.0
Oil and grease (mg L^{-1})	40.4	29.0	125.2	7.0	50.4	47.6	91.8	27.8
PAH (mg L^{-1})	0.34	0.25	0.58	0.22	–	–	–	–

Calculation of annual pollutant loads at Gardunha tunnel

One of the objectives of the monitoring study was to calculate the annual pollutant load (based on 1 year of monitoring) for the most common road pollutants. Estimations of annual pollutant loads for metals, TSSs, oil

Table 5. Estimation of 2003 annual pollutant loads at the IP2 Gardunha tunnel

Pollutant	Pol. load	Pol. load	Annual pol. load	mg km^{-1} vehicle^{-1}	g ha^{-1} year^{-1}
Cd (g)	0.170	0.048	0.218	3.3E–05	0.15
Pb (g)	1.6	0.9	2.5	0.0004	1.78
Cu (g)	6.7	3.2	9.9	0.0015	6.99
Fe (g)	194.2	139.9	334.0	0.0505	236.41
Zn (g)	32.8	16.7	49.5	0.0075	35.00
TSS (kg)	23.2	32.1	55.3	83,647	39,126.1
Oil and grease (kg)	1.4	1.5	2.9	0.4353	2,036.0
PAH (g)	11.6	–	29.0	0.0044	20.5

and grease, and PAH are presented in Table 5. Values of measured accumulated traffic for the two periods were 2,075,978 (from the 18 December 2002 to 13 May 2003) and 2,133,797 vehicles (from the 14 May 2003 to the 18 December 2003). These figures were registered by the road managing company, SCUTVIAS.

Discussion of the results

It is noticeable (Table 2) that the 2nd wash must have provided, when compared to the first one, a more effective physical removal of the pollutants due to the higher flow. The total water volume for the two campaigns was used for the calculation of the total pollutant load accumulated at the tunnel pavement.

It is important to acknowledge that the water matrix is not typical for rain water; therefore, it should be taken into account to explain the results, in particular for pH, conductivity, and Fe. The reasons have to do both with the origin of the water and the delivery processes, namely through the hoses from the fire extinguishers and the fire engine (metallic) tank. Generally rain water presents slightly acidic pH (around 6), very low conductivity, and no trace of metals or suspended solids.

This is not the case of groundwater that usually has high conductivity and basic pH, as can be observed from Table 3. Water from the local creek has lower conductivity, and the presence of some suspended solids may be due to resuspension during the pumping. The Zn and Fe content are explained by the materials that were in contact with the water before and during the wash. This is clear for the 1420 µg L^{-1} of Fe found in the water from the fire engine tank. Such facts explain the observed higher conductivity in the 1st wash: 514 µS cm^{-1}, compared to an average of 368 µS cm^{-1} in the 2nd wash.

It was observed during the wash processes that the first volume of water to reach the oil separator chamber presented higher concentrations, especially of conductivity, TSS, Zn, Cu, and Pb. During the remaining of the process the samples exhibited results in the same order of magnitude. The cause for it should be the removal of particles accumulated on the longitudinal drain, flushed with the first runoff which flows through the entire drain before reaching the oil separator system.

The concentrations of the metals Zn, Cu, and Pb agree with the relative pattern found in many other national studies, that is: Zn >> Cu> Pb. It is noteworthy that the presence of oil and grease, and PAH are the highest ones ever registered in any road runoff measurement of such pollutants, including the case of a main national highway, with a traffic volume three times higher than IP2 [5]. Possible explanations for it could be the absence of sun light inside the tunnel, that prevents photodegradation, possibly lower dilution ratios when compared to runoff produced from rain events, and a more expedite conservation of samples after collection. The two latter may also be responsible for the observed Cd concentrations above the detection limit.

It was assumed that the non-detected PHA in the 2nd wash was due to summer high temperatures (the site can reach 36°C from July to August). This assumption was based on the common knowledge that phenomena such as biodegradation and volatilization are likely to increase their rate with temperature. There are also references in the literature to the effect of temperature in the emission of PAH. For instance, Tuominen et al. [8] measured polycyclic organic compounds generated from vehicle emission, both in laboratory tests and at a busy street in Helsinki. The tests were performed at –20°C and 20°C; the street monitoring took place during autumn and hard winter, and was concluded that the concentrations of polycyclic aromatic compounds was much higher in the cold samples than in the warm samples. For these reasons, the PAH load from 146 days before the 1st wash was converted in an equivalent load for 219 days representing the accumulation period for the 2nd wash. This value was used to calculate the annual load.

A general relevant conclusion is that, if the annual loads from other monitoring studies of highway runoff water in Portugal are compared to the ones from Gardunha tunnel, the latter has, comparatively, exceptionally lower pollutants loads. In Table 6 three examples are presented, for the highways A1, A2, and A6, with average daily traffics of, respectively: 30,299 (in 2002); 16,344 (in 2002), and 2918 (in 2004).

Table 6. Annual pollutant loads for three Portuguese roads and the IP2 tunnel (Adapted from [9].)

Pollutant	Annual pollutant load (g ha^{-1} year^{-1})			
	A1	A2	A6	IP2 (Gardunha)
Pb	101.2	24.4	13.8	
Cu	285.3	193.3	61.9	
Fe	6038.0	1922.3	2689.4	236.4
Zn	1329.5	1202.2	2632.4	35.0
TSS	704,525.5	42,556.5	149,423.0	39.1
Oil and grease	–	–	74,247.6	2036.0

Gabet [2] compared the pollution from wash water from tunnels with correspondent loads from open air roads. The author concludes that the average values for pollutant loads inside and outside tunnels are in the same order of magnitude. Nevertheless, Gabet [2] refers that Pb and Zn loads are lower inside tunnels, compared to open air and that hydrocarbons showed an opposite trend, being higher inside tunnels. These statements agree with the observations at Gardunha tunnel.

Atmospheric and tunnel flow processes: application to the case study

The mechanics of the flows and its relevance for the study

A key factor to further understand Table 6 results is to identify the specific mechanisms responsible for the transportation, deposition, and removal of particles inside tunnels and then compare them to those taking place in open air roads.

It is known that pollutants may be particulated or attached to particles and their dynamics are different in free and confined flows which may explain the differences found.

When the wind blows over a layer of loose particles spread on a horizontal, solid, flat surface, a relatively well-defined velocity is required to initiate the movement of the particles. Two basic mechanisms can be identified: drag, over the surface, and saltation.

The first is present when the drag forces on the particles due to the flow over the surface became higher then the friction forces between the surface and the particle. The second one occurs when the lift forces on the particles are greater than their own weight so that the trajectories of the particles iniate, on average, vertically and a strong dependence of the phenomenon on the exact properties of the supporting surface – e.g., smoothness, angle of friction, and inclination – is not to be expected. The opposite stands for the drag phenomenon for which the particles must be larger then the typical dimension of the roughness elements, and the inclination of the surface and its friction coefficient are of utmost importance.

The discussion of deposition of traffic particulated pollutants on road pavements of tunnels and its comparison with the deposition in open air lanes has to consider these two situations. The deposition or removal rate depends on the traffic (the same outside and inside the tunnel), on the atmospheric flow local velocities (unidirectional or bidirectional in the tunnel, depending on the traffic) and on the shear stresses induced by the flow over the pavement, defined as a rule by the friction coefficient, C_f. Atmospheric boundary layer (ABL) local characteristics and the mechanical ventilation of the tunnel determine the value of these parameters

The case of Gardunha tunnel

Based on existing data and measurements of IP2 site variables referred in the literature [10–12] the phenomena at Gardunha tunnel were modelled using the basic equations of fluid mechanics.

Further details, such as the discussion on the effect of the slope, the fact that the portals are not in a flat area; and on the effect of these factors on the local wind conditions, are presented elsewhere [13].

The measure of comparison between particle transports by flows inside and outside the tunnel is given by the relation between shear stresses, τ_w inside ($\tau_{w\,in}$), and outside ($\tau_{w\,out}$) the tunnel. This value was obtained through the following equation where index 0 is associated with tunnel conditions (axis of the tunnel as the reference height) and index 10 with ABL conditions (reference values 10 m above ground level):

$$\tau_{w\,in}/\tau_{w\,out} = C_f/C_{10}\left(U_0/U_{10}\right)^2 \sim 1.8$$

The result indicates that the "every day" shear stresses inside the Gardunha tunnel almost double the ones taking place outside, and accordingly the rate of removal of particles from the road pavement is higher inside the tunnel. This information is of great value to understand the results and trends obtained in the Gardunha tunnel wash monitoring study.

Conclusions

If a purely numerical relationship would be applied it is observed that doubling the calculated load concentrations at the Gardunha tunnel signify still comparatively lower values, though already in the same order of magnitude, for some of the pollutants (e.g., TSS load would be of 78,252.2 g ha^{-1} year^{-1} that compares to equivalent figures from Portuguese open air roads).

Such results are not surprising since it is known that pollutant accumulation or removal processes are not merely physical. Indeed, a complex set of physical, chemical, and biological processes take place changing the pollutants form, fate, and behaviour in the environment. The main conclusion of this paper is that tunnel pavements should accumulate much less traffic pollutants than ordinary road pavements, under similar operation conditions.

Additional investigation is necessary to further understand the phenomena, and the authors believe that this issue is important for the assessment of road runoff impacts and environmental protection, concerning sections of roads with tunnels. The fact that pollutants emitted inside the tunnel will tend to build up outside, therefore increasing runoff concentrations in the road sections flanking tunnels is a relevant information that was not seen in the literature.

Acknowledgements

LNEC is deeply grateful to SCUTVIAS Operation Director, J Garcia and his team for all the support and thoughtfulness demonstrated during the monitoring. The authors would like to thank Maria José Henriques and Maria João Moinante, from LNEC, and Pedro B Antunes, from ESTV, for support in the monitoring work. The authors are grateful to Torleif Bækken and Dr. Michel Legret for providing information concerning the Norwegian and the French studies.

References

1. Rosedth R, Amundsen CE, Snilsberg P, Langseter AM, Hartnik T (2003) Wash water from road tunnels – content of pollutants and treatment options. In: Reinosdotter K (ed.) International conference on urban drainage and highway runoff in cold climate

2. Gabet M (1991) Pollution des eaux de lavage des tunnels routiers. Revue Générale des Routes et Aérodromes, vol. 687, pp 57–64
3. Andersen S, Vethe Ø (1994) Mobilisation of heavy metals during tunnel maintenance. Sci Tot Environ 146:479–483
4. Bækken T, Åstebøl SO (1997) Tunnel wash water from the Nordby tunnel to the Årung river. Investigation of biological effects (in Norwegian). Report NIVA/Interconsult 1997
5. Barbosa AE, Antunes PB (2004) Águas de escorrência de estradas. Sistemas para minimização de impactes. Report 128/04-NRE/DHA, Laboratório Nacional de Eng. Civil, pp 66
6. Barbosa AE, Leitão T, Carvalho CR (2003) Águas de escorrência de estradas. Sistemas para minimização de impactes. Report 233/03-NRE/DHA, Laboratório Nacional de Eng. Civil, pp 130
7. Pinto da Cunha A (2002) Observação do túnel da Gardunha I (IP2). Final Report. Report 149/02 – NOS, Laboratório Nacional de Eng. Civil
8. Tuominen J, Pyysalo H, Laurikko J, Nurmela T (1987) Application of GCL-selected ion monitoring (SIM)-technique in analysing polycyclic organic compounds in vehicle emissions. Sci Tot Environ 59:207–210
9. Leitão TE, Barbosa AE, Henriques MJ, Ikävalko V, Menezes JT (2005) Avaliação e gestão ambiental das águas de escorrência de estradas. Final Report. Report 109/05 – NAS, Laboratório Nacional de Eng. Civil, pp 243
10. Saraiva JAG (1983) Wind effects on high rise buildings. Turbulence – Thesis, Laboratório Nacional de Eng. Civil, Lisbon, pp 262
11. Novenco (1997) Velocity and sound measurements in Tunnel Gardunha Portugal. Report Novenco, BV Nederlands, pp 18
12. Estanqueiro AIL et al. (eds.) (2005) Wind atlas of Portugal, Instituto Nacional de Eng. E Tecnologia Industrial, Lisbon, Portugal, 3 pp + CD.
13. Saraiva JAG, Barbosa AE. Shear stress induced by air flow over roads. Open roads and tunnels: a comparison (in preparation)

An investigation of urban water and sediment ecotoxicity in relation to metal concentrations

L Scholes, R Mensah, DM Revitt, RH Jones

Urban Pollution Research Centre, Middlesex University, Queensway, Enfield, EN3 4SA, UK

Abstract

Implementation of the European Union (EU) Water Framework Directive (WFD) has refocused attention on the impact of non-point source discharges on receiving water bodies. However, although the physico-chemical characteristics of urban runoff have been widely reported, its ecotoxicological impacts are less certain. There is currently considerable debate surrounding how the ecological status of water bodies should be determined and the all-encompassing environmental quality standards, required by the WFD, developed and complied with. To contribute to the debate surrounding these issues, the ecotoxicity of urban water and sediments were investigated in relation to concentrations of selected heavy metals in these environments. The Microtox test was used to evaluate the ecotoxicity of urban stream water, water samples in which sediments had been resuspended, porewater and sediments. Total concentrations of Zn, Cu, and Cd were determined in the same fractions. Data reported indicate that the toxicity of surface water is greater during wet weather events compared to dry weather conditions. Two different statistical approaches for assessing the associations between metals and ecotoxicity are described and the different interpretations provided are fully discussed.

Introduction

The European Union (EU) Water Framework Directive (WFD) requires member states to take a holistic approach to the management of aquatic ecosystems [1]. This marks a clear shift from current approaches to water quality management, which primarily focus on comparisons with physico-chemical water quality standards, to one which additionally integrates and addresses biological and hydromorphological aspects of aquatic ecosystems. As these components are temporally and spatially interlinked, the development of a detailed understanding of the behaviour and fate of pollutants within and between them is fundamental in enabling the development of a sustainable and integrated management approach.

An intercompartment relationship receiving particular attention is the sediment–water interface as these sediments have the greatest susceptibility to resuspension processes, as well as providing substrate and/or food sources for benthos [2]. This issue is of particular concern within the urban aquatic environment, where urban sediments act as a sink for organic and inorganic pollutants [3,4]. The ecotoxic effects of urban and highway runoff have been widely reported (see review by Baun et al. [5]). However, although well studied in relation to their physical and chemical characteristics, comparatively little is known about the ecotoxicology of urban sediments, either *in situ* or within the water column following a resuspension event [6,7]. The development of "Translating Relaying Internet Architecture Integrating Active Directories (TRIAD)" or combined approaches which integrate information on physico-chemistry, ecotoxicity, and ecology at a particular site is seen as a way forward [8]. However, current levels of knowledge are insufficient for the development of strategic water quality management strategies which can integrate biological, chemical, and physical aspects across different compartments of aquatic ecosystems. This study aims to contribute to this research need through an investigation of the ecotoxicity of urban stream water, sediment, porewater, and samples in which sediments have been resuspended, in relation to concentrations of Cd, Cu, and Zn in the same fractions.

Methodology

Site description and sampling programme

Samples were collected at six separate locations along an increasingly urbanized stretch of the Pymme's Brook, an urban watercourse in North

London (UK) which receives substantial surface water flows from the surrounding urbanized catchment area as well as discharges from combined sewer outlets (CSOs). At each sampling site, duplicate water and surface sediment samples were collected and *in situ* measurement of dissolved oxygen, temperature, and pH were carried out. Samples were collected on four separate occasions; twice during dry weather conditions and twice during wet weather conditions. Following collection, samples were transported to the laboratory and stored at 4°C prior to analysis by Microtox within 72 h and for metals as soon as possible thereafter.

Preparation of samples prior to analysis

Loosely associated surface water was allowed to drain from each sediment sample by placing sediments in tilted evaporation dishes. Sediments were sieved and the <4 mm size fraction collected and used in all subsequent experiments. Resuspension samples were prepared by shaking 50 g of sieved wet sediment sample in a flask with 250 mL of double-deionized water and gently shaking for 14 h to simulate the impact of a storm event (i.e., prolonged low intensity turbulence to compensate for a short intense storm event). The aqueous fraction was collected and centrifuged to remove particulate material. Porewater was extracted from sediment by a combination of applied pressure and vacuum filtration of a known mass of sediment through a Whatman 42 filter paper.

Microtox analysis

The ecotoxic effects of surface water, resuspension water, and porewater samples were evaluated in duplicate using the whole effluent test as described in the AZUR user manual [9], with the use of phenol as a standard reference material. The solid-phase test (SPT) was used as described in the AZUR user manual to determine in duplicate the ecotoxicity of sediment samples [9].

Metals analysis

Replicate water, resuspension water, porewater, and sediment samples were acidified with nitric acid, taken to dryness and then redissolved in 1% nitric acid. Concentrations of Zn and Cu were determined in each sample fraction using inductively coupled plasma-atomic emission spectrometry (ICP-AES), with Cd being reliably determined in sediment samples only using the same technique.

Results and discussion

Results of ecotoxicity analysis

An overview of the ecotoxicity determined in surface water, porewater, and resuspension water samples is provided in Figure 1 (data are presented as EC_{20} values as this proved to be the most realistic representation of ecotoxicity in the collected samples). In surface water samples, an EC_{20} value could only be determined in 1 of 12 samples collected during dry weather conditions as opposed to being measurable in 9 of the 12 samples collected during storm event conditions, indicating that wet weather flows have the potential to exert a greater toxic impact, as determined using the Microtox technique. Serial sample collection through the profile of a storm event by Marsalek et al. [10] reported that ecotoxicity as determined using various tests (including Microtox) displayed the "first flush phenomenon" typically reported during the physico-chemical monitoring of storm events. Although the timing of sample collection in the current study in relation to storm event profiles is unknown (sampling consisted of the collection of grab samples), results support the conclusion that storm events are associated with a greater ecotoxic, as well as physico-chemical, loading. Ecotoxic impacts were regularly detected in porewater and resuspension water samples during both dry weather and wet weather conditions (Fig. 1). This is thought to be associated with the fact that urban sediments are known to act as a long-term reservoir for pollutants, as opposed to surface waters which characteristically show greater temporal variation.

It is difficult to make direct meaningful comparisons with other ecotoxicity studies of urban runoff as data tend to be reported in a variety of formats (e.g., EC_{10}–EC_{50}) and end points (e.g., growth, reproduction, mortality) as well both the physico-chemical characteristics of the storm event and timing of sample collection in relation to storm event profile being unclear. However, the current study is consistent with storm water runoff studies which report ecotoxic responses ranging from no effect to substantial inhibition [5,11,12]. In the current study, EC_{20} values range from a maximum toxicity of 13% (recorded in a resuspension water sample) to having no discernable toxic impact (referring to samples in which the data did not permit the calculation of an EC_{20} value).

Ecotoxicity of resuspension water was significantly lower during wet weather conditions in comparison with dry weather conditions ($p = 0.011$; 2 sample t). This could indicate that the higher storm flows observed during sampling are flushing pollutants from the sediment or, alternatively, are transporting and depositing cleaner sediments. However, no significant

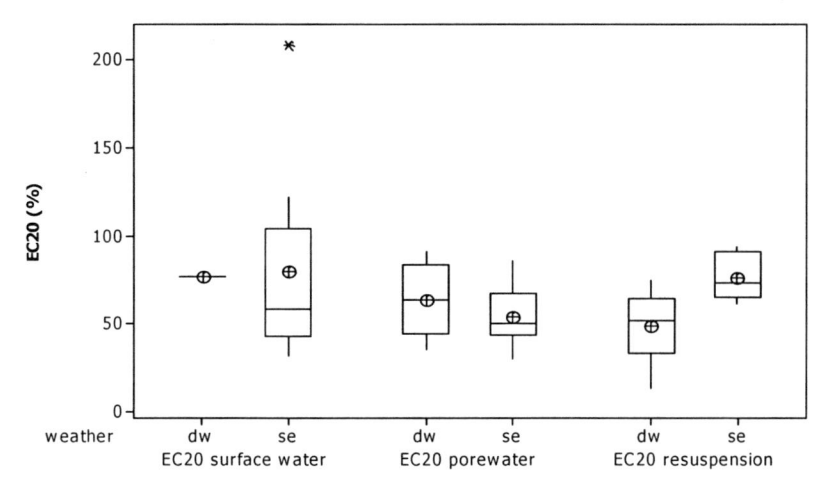

Fig. 1. EC_{20} values determined in surface water, porewater, and resuspension water samples: dry weather (dw) versus storm events samples (se) (%)

differences between the levels of toxicity determined in any of the water samples and sampling site (data not presented) or between surface water, porewater, or resuspension water sample fractions (Fig. 1) were found ($p > 0.05$; analysis of variance [ANOVA]).

Ecotoxicity of sediment samples

Only sediment samples collected during wet weather conditions were evaluated for their ecotoxicological impact using the Microtox SPT. This test involves bacteria coming into direct contact with sediment particles and associated pollutants, and is therefore of particular interest in the urban setting where the majority of pollutants are sediment-associated [13]. In terms of frequency of detection, SPT gave the highest response rate (100% of samples tested). In contrast, the detection rate was 75%, 67%, and 50% for the surface water, porewater, and resuspension water samples, respectively. Various authors have reported SPT as being more responsive than tests on, for example, sediment elutriates [14] and sediment porewaters [15]. However, because it was not possible to source a suitable reference sediment material within the time frame of the sampling programme, data collected within the current study cannot be directly used to support this finding. Although not possible to quantitatively evaluate the impact of urban pollution on sediment toxicity, it is feasible to evaluate sediment ecotoxicity semi-quantitatively by comparing SPT values with those reported in the literature (see Table 1).

Table 1. EC$_{50}$ values (g L^{-1}) for sediments determined using SPT in comparison with data from the literature

	Median	Range	
Current study	172.3	65–591	12
Storm water management pond[a]	5.3	0.8–79.9	19
Harbour sediment[b]	3.3	1.1–4.3	3
Sediment from an unpolluted site[b]	11.5	11–12	2

[a] [7]
[b] [16] (values expressed as dry weight)

The median sediment EC$_{50}$ value determined in this study is substantially higher (i.e., less toxic) than those reported in the literature and, for example, is over 10 times less toxic than sediment from a reported unpolluted site [16]. As the Pymmes Brook is known to be in receipt of both urban runoff and CSOs, the comparative lack of sediment toxicity determined in this study is surprising. However, sediment composition is known to heavily influence the toxicity of sediments as determined using Microtox [15] and a fuller evaluation of the sediment composition is required to enable data to be fully interpreted. In addition, these results also suggest a notable site specificity in both the environmental behaviour of pollutants and their ecotoxicological impact. This provides support for the use of a combined physico-chemical and ecotoxicological approach and also raises the possibility that results may need to be interpreted within a site-specific as opposed to a more generic context to enable their significance to be fully appreciated.

Results of metals analysis

Cu and Zn were detected in surface water, porewater, resuspension water, and sediment samples but due to the working limit of detection (LOD) for Cd (>0.6 µg L^{-1}) it was possible to reliably determine Cd concentrations in sediment samples only. Aqueous Cu and Zn concentrations tend to decrease in the order porewater > resuspension water > surface water (see Table 2). Concentrations of both metals were significantly greater in porewater samples than those determined in resuspension water or surface water samples ($p < 0.001$ in both cases; ANOVA) but differences between resuspension water and surface water samples were not significant. However, the higher median concentrations of Cu and, in particular, Zn in resuspension water samples support the proposal that resuspending sediments may result in the release of previously bound pollutants [17].

Table 2. Median, maximum, and minimum concentrations of metals in surface water, resuspension water, and porewater samples (µg L^{-1}) and sediment (µg g^{-1})

	Median	Min.	
Cu surface water	5.75	<0.75*	18.90
Cu resuspension water	9.55	<0.75*	70.90
Cu porewater	29.55	<0.75*	117.90
Cu sediment	27.80	5.60	152.20
Zn surface water	26.00	5.60	97.00
Zn resuspension water	43.90	13.80	189.50
Zn porewater	125.00	24.30	1529.60
Zn sediment	127.6	45.70	713.2
Cd sediment	0.70	<0.07**	1.50

*Working limit of detection on aqueous fractions (3 × SD on sample blank)
**Working limit of detection for sediment samples (3 × SD on sample blank/ assumed 5 g sediment mass)

Aqueous concentrations of Cu or Zn in any of the water fractions did not significantly vary according to sampling site location (data not presented). Concentrations of Zn in the surface water samples were found to be significantly higher in the storm event samples than during dry weather (2 sample t; $p = 0.010$). However, Cu concentrations did not show this trend. This difference in the behaviour of Cu and Zn could indicate that the Cu is more strongly bound to particulate materials, for example, particulate-associated organics [18] and underlines the complex nature and differential behaviour of pollutants within urban surface waters. In relation to sediment metal concentrations, there was no significant difference between concentrations determined during dry weather and wet weather conditions. Zn and Cu sediment concentrations were found to vary significantly according to site ($p = 0.005$ and 0.001, respectively; ANOVA) with concentrations of both metals determined at the site located furthest upstream being significantly lower than those recorded at various sites located further downstream, reflecting the fact that watercourse becomes increasingly urbanized along its length. However, Cd sediment concentrations did not vary in relation to the sampling site.

Investigation of associations between ecotoxicity and metals

To investigate if there were any associations between metal concentrations and ecotoxicity, data were examined using scatter plots and correlation.

An initial analysis appeared to demonstrate a very strong negative correlation between the Cd concentration in sediment and sediment ecotoxicity ($r = -0.940$, $p = 0.005$) and a strong negative correlation between Cd in sediment and the ecotoxicity of surface water ($r = -0.868$, $p = 0.025$) (Fig. 2). The data also indicate a moderate degree of negative correlation between the concentration of Cu in sediment and sediment ecotoxicity ($r = -0.663$, $p = 0.019$). In contrast, the data indicate a moderate positive correlation between Cu sediment concentrations and the ecotoxicity determined in porewater (Fig. 2). As decreasing EC values indicate increasing toxic effect, the negative correlation reported between Cu and Cd in sediments and the ecotoxicity of sediments and surface water (Cd only) suggests that as Cd and Cu concentrations increase, ecotoxicity also increases. The moderate positive correlation between Cu sediment concentrations and the ecotoxicity of porewater samples indicates that as Cu sediment concentrations increase, the ecotoxicity of porewaters decrease which could suggest that as the total sediment load of Cu increases, the equilibrium between particle-associated Cu and Cu in the porewater alters with, for example, initially bound Cu acting as a nucleation site around which further Cu is deposited. No evidence of a significant correlation between the concentration of Zn in sediment and ecotoxicity in sediment or surface water, or between Zn and Cu concentrations in aqueous fractions and the ecotoxic impacts determined in the same compartment was detected (data not presented).

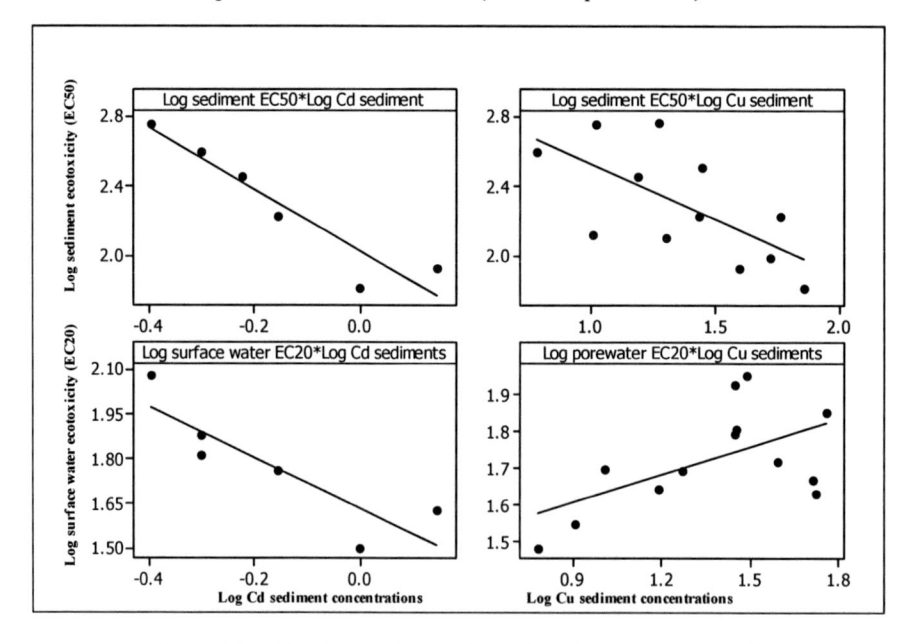

Fig. 2. Scatter plots of Cd and Cu sediment data versus ecotoxicity data of components where a significant correlation was identified

However, it is important to note that the above statistical analyses only include paired sample sets (i.e., situations when both a metal concentration and a toxicity value could be determined in the same sample) as it is not possible to plot or correlate data when only one of the parameters is detected. In relation to the surface water, porewater and resuspension samples, there were several occasions when Cu and Zn were detected, but ecotoxicity could not be determined within the same compartment and such results clearly need to be considered. The data was therefore subjected to an alternative statistical procedure involving the allocation of metal concentration values into one of two groups; metal concentrations when toxicity was recorded (toxic subset) and metal concentrations when no toxicity was detected (non-toxic subset). This procedure was carried out for each metal in each compartment on a site-by-site basis. If it is assumed that metals have no association with toxicity then the mean differences between the metal levels in the toxic and non-toxic subsets should be zero. A null hypothesis can therefore be constructed in which samples of mean differences are compared to a test mean of zero and tested for significant difference (1 sample t). Results of this analysis indicate that the toxic subsets tend to have higher metal concentrations than their non-toxic counterparts. For example, Zn and Cu concentrations in toxic subsets were higher in porewater and resuspension water samples in 20 of the 24 site-specific comparisons undertaken for these compartments. However, this difference was not found to be statistically significantly different (1 sample $t, p > 0.05$).

In relation to sediment metal and ecotoxicity data, there were six occasions when sediment ecotoxicity was determined, but Cd sediment concentrations were below the LOD. If the LOD is used to calculate a "surrogate" Cd sediment concentration and this value is used within the data set, analysis strongly suggests no correlation between these two parameters ($r = -0.097$). This is because a range of reported sediment ecotoxicity values have been allocated the same Cd sediment metal concentration, effectively cancelling out the strong negative correlation between these parameters ($r = -0.940, p = 0.005$) reported earlier. These results clearly indicate the importance of fully considering and reporting the approach used to analyse data as it may have a significant impact on the conclusions drawn.

In this paper, two approaches to analysing data are presented; the first involves the correlation of samples in which both toxicity and metals were detected, whereas the second approach also enables "non-toxic" samples and samples in which metals were below the LOD to be included within the statistical analysis. The application of both approaches is useful as both methods contribute towards developing a fuller understanding of factors influencing the ecotoxicological behaviour of urban sediments. It is noted that the sediment toxicity values determined in the current study were lower by over an order of magnitude than those reported in the literature. This information, together with determination of metals in some samples but no ecotoxicity (and vice versa), could suggest the presence of "threshold values" in relation to both metals and ecotoxicity around which relationships between parameters either alter or breakdown. Alternatively, it could reflect the complexity of the urban aquatic environment in which a diversity of pollutants interact in a number of antagonistic and/or synergistic ways which are difficult to predict. Although the second approach appears more robust in that it uses more of the data, the first approach is also useful as it may provide an insight into relationships between metals and ecotoxic responses at higher concentrations (i.e., in data sets when metals and ecotoxicity co-occur), revealing relationships which may not have been detected through the use of the second approach alone.

Conclusions

This study describes an investigation of the ecotoxicity of different components of urban watercourses (surface water, porewater, water samples in which sediments have been resuspended and sediment) in relation to concentrations of Cu, Zn, and Cd. Analysis of aqueous samples using EC_{20} values indicates that surface water has a greater toxic impact during wet weather as opposed to dry weather, although the appropriateness of categorising storm water ecotoxicity values using EC_{20} values, as well as the magnitude of the value at which action should be taken, has yet to be fully established.

The apparent correlations observed between metals and ecotoxicity which emerge on analysis of samples in which these parameters co-occur indicates an association between selected metals and ecotoxicity. However, use of an alternative statistical approach, which enables a more complete use of the data, does not support results generated using the first approach. This raises issues of data handling and interpretation in ecotoxicity studies in general, as well as illustrating the difficulties inherent in establishing relationships between toxicity and specific pollutant concentrations in field samples. Whilst it is acknowledged that ecotoxicity associated with urban aquatic environments is likely to be a function of many pollutants behaving synergistically and/or antagonistically and the results of the two approaches to analysing and presenting data indicate alternative scenarios, the strength of the correlations reported between metals and ecotoxic effects within samples where parameters co-occur could indicate the presence of "threshold values" where, for example, it may be possible to relate metal concentrations above a certain value to a predictable ecotoxic response. This issue is highlighted as a promising area for further research which may provide further insight into the environmental behaviour and ecotoxicology of pollutants within a complex and poorly understood environment.

References

1. EU WFD (2000) Directive 2000/60/EC of the European Parliament and of the Council of 23 October 2000 establishing a framework for Community action in the field of water policy. 2000/60/EC, 23 October 2000
2. Lepper P (2004) Towards the derivation of quality standards for priority substances in the context of the water framework directive. Fraunhofer-Institute Molecular Biology and Applied Ecology
3. Ellis JB, Revitt DM, Llewellyn N (1997) Transport and the environment: effects on water quality. J. Inst Water Environ Manag 11:170–177
4. Wilson C, Clarke R, D'Arcy BJ, Heal KV, Wright PW (2005) Persistent pollutants urban river sediments survey: implications for pollution control. Water Sci Technol 51:217–224
5. Baun A, Qualmann S, Eriksson E, Scholes L, Revitt DM, Seidel M, Mouchel JM (2005). Toxicity of stormwater samples from 4 different European catchments. Daywater Deliverable 4.5, www.daywater.org, verified 17 May 2006
6. Farm C (2003) Monitoring, operation and maintenance of detention ponds for road runoff. Proceedings of the international conference on stormwater management in cold climates, Portland, Maine, USA
7. Marsalek J, Anderson BC, Watt WE (2002) Suspended particulate in urban watershed ponds: physical, chemical and toxicological characteristics. Proceedings of 9ICUD, Oregon, USA. doi:10.1061/40644(2002)201
8. Smoulders R, De Coen W, Blust R (2004) An ecologically relevant exposure assessment for a polluted river using an integrated multivariate PLS approach. Environ Pollut 132:245–263
9. AZUR Environmental (1998) Microtox User Manual www.azurenv.com, Verified 17 May 2006
10. Marsalek J, Rochfort Q, Brownlee B, Mayer T, Servos, M (1999) An exploratory study of urban runoff toxicity. Water Sci Technol 39:33–39
11. Brent RN, Herricks EE (1999) A method for the toxicity assessment of wet weather events. Water Res 33:2255–2264
12. Schiff K, Bay S, Stransky C (2002) Characterization of stormwater toxicants from an urban watershed to freshwater and marine organisms. Urban Water 4:215–227

13. Lee P-K, Touray J-C, Baillif P, Ildefonse J-P (1997) In: Marselek J, Rochfort Q, Brownlee B, Mayer T, Servos M. An exploratory study of urban runoff toxicity. Water Sci Technol 3912:33–39
14. Loureiro S, Ferreira ALG, Soares AMM, Nogueira AJA (2005). Evaluation of the toxicity of two soils from Jales Mine (Portugal) using aquatic bioassays. Chemosphere 61:168–177
15. Doherty FG (2001) A review of the Microtox® toxicity test system for assessing the toxicity of sediments and soils. Water Qual Res J Can 36: 475–518
16. Pedersen F, Bjornestad E, Andersen H, Kjolholt, Poll C (1998) Characterization of sediments from Copenhagen harbour by use of biotests. Water Sci Technol 37:233–240
17. Amoros I, Connon R, Garelick H, Alonso JL, Carrasco JM (2000) An assessment of the toxicity of some pesticides and their metabolites using the Microtox system. Water Sci Technol 42:19–24
18. Lu Y, Allen HE (2006) A predictive model for copper partitioning to suspended particulate matter in river waters. Environ Pollut 143:60–72

Establishing a procedure to predict highway runoff quality in Portugal

AE Barbosa

National Laboratory for Civil Engineering. Hydraulics and Environment Department, Av. do Brasil, 101, 1700-066 Lisbon, Portugal

Abstract

In Portugal, there is insufficient data concerning road runoff characterization to identify the most important explanatory variables to predict pollutant concentrations. Nevertheless, there is the need for a simple methodology to predict road runoff pollutants, for use by road project designers and national services in charge of water management.

Based on literature results a set of five independent variables, proved to explain site concentrations on road runoff constituents, have been selected. Using data for six Portuguese road sites, a multiple linear equation and its coefficients, named the PREQUALE, was established to predict total suspended solid (TSS), Zn, Cu, and Pb average concentrations in road runoff. The predictions obtained were very close to the observed concentrations; therefore, the PREQUALE will significantly improve the present practice in Portugal. This tool should be improved in the near future with the acquisition of more information concerning road runoff characterization in Portugal.

The approach followed by this study may be useful for other countries that have not yet much data but need to establish a simple way of predicting road runoff pollutants.

Introduction

The importance of predicting road runoff quality

Road runoff is a linear diffuse source of pollution, with the pollutant characteristics and the mode of discharge in the environment very specific for this pollution source. It has been neglected in the past, due to the fact that road runoff transports low concentrations of pollutants in huge volumes of water. However, storm water has been a subject of concern for many years in the USA. In Europe, the Water Framework Directive (WFD) in a similar way to the USA Clean Water Act requires a combined approach for controlling point and diffuse sources, in order to attain a good ecological quality for all water bodies.

The characterization (qualitative and quantitative) of road runoff pollution must be performed at national and regional bases, for it has been proved that site-specific, climatic, and other local variables play an important role.

For water resources management purposes or decision concerning the need for treatment of storm water, there is the need for models able to predict road runoff quality. Such models are based on the establishment of relationships between pollutant concentrations and traffic flows, rainfall totals, rainfall intensity, rainfall duration, antecedent dry periods, impermeable drainage area, use of adjacent land, and among other variables.

Regression models, often used, are mathematical equations that define the average line for relationships between dependent variables and one or more independent variable. They are based on the assumption that road runoff quality variations may be scientifically explained by significant changes on the variables that control the process [1].

The regression models for storm water quality utilize concentrations or pollutant loads as variables dependent of traffic intensity, flow volume, interevent dry period, and adjacent land use [1]. Such models have been criticized for their weak prediction ability, when they are applied outside the original database or the regions for which they were developed. Nevertheless, as pointed out by Irish et al. [1] this statement is universally valid for all modelling methods.

There are a number of references in the literature concerning studies aiming at establishing prediction tools for estimation of road runoff pollutants. The USA presents a set of proposals; several are described in FHWA [2]. Driscoll et al. [3] and Schueler [2] suggested empirical or regression models; whereas Driver and Tasker [4] propose regression models.

Kayhanian et al. [5] found that there are no simple linear correlations between highway runoff pollutants, event mean concentrations (EMCs) and annual average daily traffic (AADT), including for those pollutants known to be related to transportation activities. They concluded, as well, that there is evidence that pollutants like total suspended solids (TSSs), chemical oxygen demand (COD), or turbidity have sources other than the transportation-related activities. The authors developed multiple linear regression models for several road runoff constituents, referring that the two most important applications of the study include estimating mass loads and use of the model as a tool to address runoff management issues. For more than 70% of the constituents the effects of AADT, total event rainfall, cumulative seasonal rainfall, and antecedent dry period on pollutant concentrations in highway runoff were significant when evaluated using multiple linear regression analysis. The effects of drainage area and maximum rainfall intensity were smaller and less frequently significant [5].

Driver and Tasker [4], in a work that generated several regression equations, concluded that the most significant explanatory variables in all their linear regression models were total storm rainfall, and total contributing drainage area. Models for areas that had large mean annual rainfall were the least accurate. Concerning the different types of pollutants, models for suspended solids were the least accurate ones.

The awareness of the need for having national methodologies for predicting road runoff pollution is rising. In the UK, Crabtree et al. [6] report that in 2003 commenced a study "to develop a predictive methodology for highway runoff pollution concentrations, and resulting pollution loads, discharged to the receiving water".

The French *Sétra (Service d'Études Techniques des Routes et Autoroutes)* recently published a report [7] with a methodology for calculation of road runoff pollution loads, focusing on TSSs, COD, total hydrocarbons and polycyclic aromatic hydrocarbons (PAH), and the heavy metals zinc (Zn), copper (Cu), and cadmium (Cd). The motivation for this proposal was to provide a common base of calculation, simple and updated, for use by road project designers and national services in charge of water management.

The Portuguese national situation

In Portugal, since 1990 new projects of main roads, or roads located in sensitive areas, are evaluated under the scope of the environmental impact assessment (EIA) law. In the course of the environmental impact study (EIS) an assessment of the impacts of road runoff is made. The methodologies used to predict pollutant concentrations has several drawbacks, and frequently overestimates pollutant levels [8].

Table 1 compares, for three roads, the level of pollutants estimated in the EIS and the concentrations observed during runoff monitoring studies. For all the roads the predictions overestimated the observed values, by a factor from 6 to 467 times.

In face of this scenario, the Portuguese Roads Institute commissioned Portuguese National Laboratory for Civil Engineering (LNEC) to establish a procedure for a more sound prediction of pollutants concentrations in road runoff. The requirements were that the model has to predict at least two priority pollutants; should use simple input data; should be easy to handle, not requiring complex decisions during the calculations, and should be based on a clear procedure that could be checked out later. This verification of the EIS methodologies and results is made at the administrative level, both by the Roads and the Water Institutes.

Table 1. Comparison of predicted (Predic) and observed (Obs) pollutant loads/concentrations for IP4, IP5, and IP6

Pollutant	IP4 (mg L^{-1})		IP5 (mg L^{-1})		IP6 (g km^{-1} vehicle^{-1})	
	Predic[a]	Obs[a]	Predic[b]	Obs[c]	Predic[d]	Obs[e]
TSS	–	–	–	–	0.329	0.052
Zn	4.7	0.308	2.1	0.205	6.85E-4	2.0E-5
Cu	–	–	–	–	0.001	8.5E-6
Pb	5.6	0.012	1.1	0.005	–	–

[a] [9]
[b] [10]
[c] [11]
[d] [12]
[e] [13]

The wanted prediction methodology is therefore a tool to be used by engineers, with the purpose of supporting decisions such as the construction of treatment systems for highway runoff.

Available data and methodology

National data on road runoff characteristics

The characterization of road runoff in Portugal is available for six roads, located in different parts of the Portuguese territory, presented in Figure 1A that includes as well a map of average annual rainfall regions in Portugal

(A) (B)

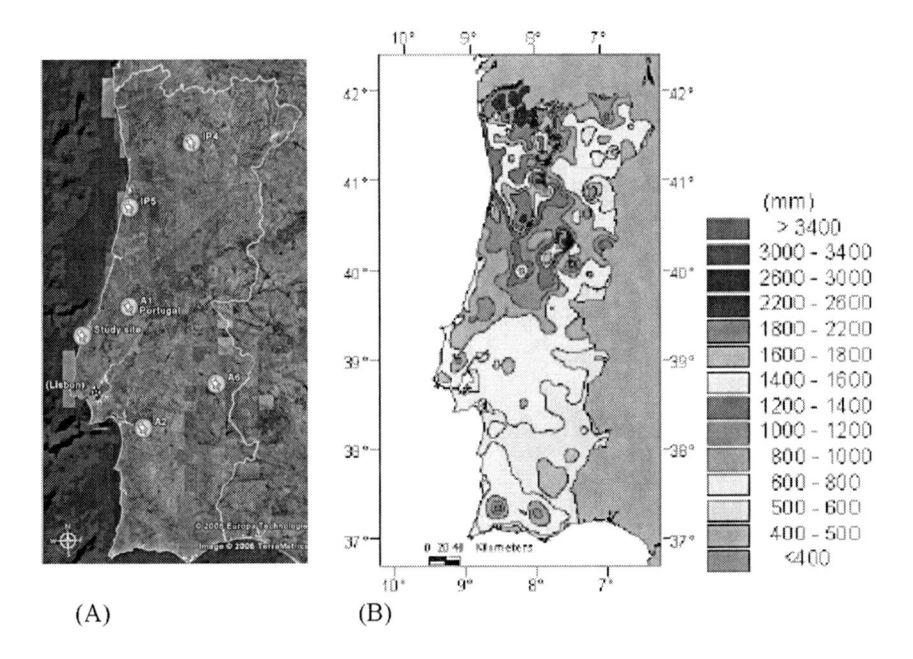

Fig. 1. (A) The six roads whose monitoring data was used in the present study. From north to south: IP4, IP5, A1, IP6, A6, and A2 (From earth.google.com.); (B) map of Portugal with regions defined based on the average annual precipitation From [14].)

(Fig. 1B). The pollutants better characterized are the heavy metals Zn, Cu, and Pb, and the TSS. Although the number of Portuguese sites could be considered significant, there are not many records of events – a total of 38 events. With such restricted data it is difficult to establish an empirical or statistical model. For instance, Kayhanian et al. [5] and Driver and Tasker [4] used, respectively, 2656 and 2813 events (for different sites) in their studies.

Table 2 presents characteristics of the six road sites and Table 3, for the same sites, summarizes the site mean concentrations (SMC) for TSS, Zn, Cu, and Pb. The determination of SMC was based on arithmetic averages of the EMCs. Each EMC is calculated from continuous flow measurements and discrete sampling during each event. The existence, for some sites, of few events is due to gaps on flow records, caused by equipment failure and/or determination of concentrations lower than the detection limit.

Table 2. Roads and monitoring information

Monitoring site characteristics	A1	A2	A6	IP4	IP5	IP6
AADT	33,299	16,344	2918	6000	27,448	6539
1st year of road operation	1990	1998	1999	1995	1998?	2004
Monitoring date	2001	2003	2004	1996/98	2003/04	2005/06
Total area (m^2)	64,600	1287	5580	5947	250	7280
Percentage imperviousness	41.2	100	100	42	100	100
Surrounding land use	Forest	Agriculture	Shallow veg.	Agriculture + forest	Water/road	Agriculture
Section type	Landfill + cut	Bridge	Landfill	Landfill	Landfill	Landfill
Average annual rainfall (mm)	1157	575	594	1064	929	902
No. of events	6	3	6	3/5	5	8
No. of samples	31	34	35	127	40	47

Table 3. Summary of SMC for six roads in Portugal

SMC ($mg\ L^{-1}$)	A1	A2	A6	IP4	IP5	IP6	Average	SD	Coef. var
Zn	0.159	0.208	0.346	0.308	0.205	0.08	0.218	0.097	2.2
Cu	0.034	0.033	0.008	0.024	0.014	0.03	0.024	0.011	2.2
Pb	0.012	0.004	0.002	0.012	0.005	0.01	0.008	0.004	1.7
TSS	84.5	7.4	19.6	8.1	44.7	224.7	64.8	83.6	0.8

Identification of the most relevant variables

Methodologies for road runoff prediction reported in the literature provide relevant information concerning the most significant variables for multiple linear regressions. Three different methods, developed in the USA and in

France, based on statistical analysis and correlation of large ranges of data, have been selected for the purposes of identifying the most relevant variables for road runoff quality prediction.

The selected three-variable models for use by engineers and in management, by Driver and Tasker [4], is a simplified version of more complex equations established by the authors. The other models that also correspond to the level of utilization required are proposals by Kayhanian et al. [5] and Hurtevent et al. [7]. Table 4 presents the input variables needed for calculations, concerning the three methods.

Multiple linear regressions based on the Portuguese data

Table 4 allows the study of independent variables already proved to explain site concentrations on road runoff constituents. It is believed that the results of the different studies do not only express regional or national patterns, but also the physical, chemical, and other processes taking place at any road environment. The fact that several explanatory variables are common among the studies supports this assumption.

Analysing Table 4 it is observed that "drainage area" (A); "% of impermeable area" (I), and "AADT" are commonly quoted by at least two of the studies and therefore should be valuable variables to be included in the regression analysis for the Portuguese data. The annual average rainfall is also included explicitly or implicitly in the studies, and therefore was selected as a possible explanatory variable and identified as P_{annual}.

Table 4. Input variables for three different methods for SMC prediction

Driver and Tasker [4]	Kayhanian et al. [5]	Hurtevent et al. [7]
Total drainage area	AADT	AADT;
Percentage impermeable area	Drainage area	Impermeable drainage area
Total rainfall volume	Event rainfall	Annual average rainfall
	Maximum rainfall	Observations: road sites
Observations: the coefficients are expressed for three different regions defined based on the mean annual rainfall value	intensity	are defined as "open" or "confined"
	Antecedent dry period	
	Cumulative seasonal rainfall	

AADT: Annual average daily traffic; *open site* when the shape of the road flanks do not prevent the atmospheric dispersion of pollutants; *confined site* corresponds to a road where both flanks limit the atmospheric dispersion of pollutants. Physically these borders should have at least 100 m of length and 1.50 m of height.

The total rainfall event or event rainfall, thought to be an important local characteristic, is usually known in Portugal after a monitoring study takes place, requiring some calculations. With the purpose of having a variable based on Portuguese rainfall data that could be easily calculated, representing local event rainfall, a search in the available information and methodologies was carried out. The result was the establishment of a P variable, representing the mean volume of a storm with the duration of the concentration time of the catchment, and a return period of 2 years. This calculation is done supported on Brandão et al. [14] study[1], that provided statistics of precipitation from several recording stations in Portugal, generating coefficients for calculation of rainfall intensity, for a chosen event duration.

Based on these five independent variables (A, I, AADT, P, and P_{annual}) and the results for the six Portuguese road sites, multiple linear regression analysis was performed (in Excel). It used the logarithmic transformation that Driver and Tasker [4] found to give the best transformation for the response variables. The objective was to find the smallest set of independent variables that could explain to a larger extent the pollutant (TSS, Zn, Cu, and Pb) concentrations.

Based on the R-square values, it was observed that Cu was the pollutant most easily predicted. This fact is noteworthy because the results from the application of the methodologies stated in Table 4 to the Portuguese sites, although not good, gave the best predictions for Cu. Possibly, the sources of Cu and the processes that this metal undergoes in the environment are less complex compared to the other pollutants.

On the other hand, the TSS was the most difficult to explain, and was very dependent on P_{annual}. Driver and Tasker [4] found out that models for TSS are the least accurate ones, agreeing with this observation. The results showed that the best approach for an equation based on three variables could be achieved with A, I, and AADT that would explain well Zn, Cu, and Pb concentrations. On the other hand, the TSS could be predicted based on A, I, and P_{annual}.

Taking into account the objectives of this study, a regression equation based on four variables – A, I, P, and P_{annual} – was finally chosen. This selection of variables fits the objectives for PREQUALE (to be based on simple input data) and is backed up by other studies presented herein. The equation coefficients and results obtained are described in the following section.

---

[1] The report can be downloaded at the Portuguese Water Institute site: www.inag.pt

The PREQUALE

The PREQUALE equation is the following:

$$C_p = a_i (A^{\beta_1} \times I^{\beta_2} \times P^{\beta_3} \times P_{annual}^{\beta_4})$$

Where:
C_p (mg L^{-1}) = Estimated concentration (SMC) of the pollutant
$a_i, \beta_1, \beta_2, \beta_3, \beta_4$ = Regression coefficients
P (mm) = Mean annual volume of the storm with the duration of the concentration time, and a return period of 2 years
A (km^2) = Drainage area
I (%) = Percentage of imperviousness of the drainage area
P_{annual} (mm) = Annual average rainfall

The regression coefficients and R-square values are presented in Table 5.

Figure 2 presents the comparison between the PREQUALE and the observed pollutant concentrations, for the six roads. The predictions produce results for TSS, Zn, Cu, and Pb concentrations very similar to the ones observed in the Portuguese roads. The average ratio observed/PREQUALE concentrations is of 0.9 for Cu and 1 for the rest of the pollutants, which is a considerable improvement in the face of the present methodologies used in EIS.

The importance of event rainfall data in runoff quality predictions is known. In the case of the PREQUALE the use of P variable was a simplification. P is calculated based on the work by Brandão et al. [14] who acknowledged themselves the low surface expression of the Portuguese rainfall records. The utilization of P based on data that for some roads was geographically distant from their location must have contributed to less accurate calculation of the coefficients.

Table 5. Regression coefficients and R-square values for the multiple linear regression (PREQUALE)

Parameter	a_i	β_1	β_2	β_3	β_4	R-square
TSS (mg L^{-1})	5.1E–28	0.675	3.656	4.194	6.972	0.9392
Zn (mg L^{-1})	2.9E+09	0.355	0.851	–3.971	–2.225	0.9648
Cu (mg L^{-1})	1.9E–04	0.299	0.975	4.718	0.400	0.9565
Pb (mg L^{-1})	5.7E–10	0.191	1.005	4.072	2.181	0.9999

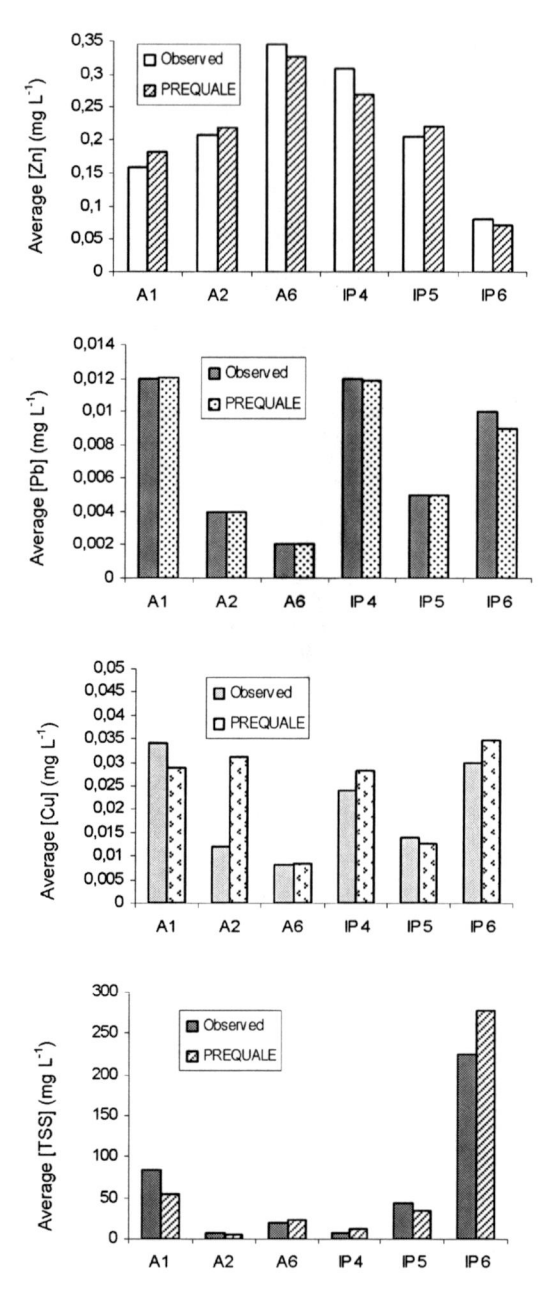

Fig. 2. Comparison between PREQUALE results and observed concentrations

Coincidently, as for the French study [7], in Portugal there is no available information concerning road runoff in regions with average annual rainfall <500 mm. Consequently the coefficients should not be able to predict so well road runoff concentrations for these regions. Figure 1B shows that not much of the Portuguese territory has such level of precipitation; nevertheless, Miranda et al. [15] refer that the latest models project reductions in the annual precipitation in Portugal within the range of 20–40% of its current values, with the greatest losses occurring in the South.

Final remarks

In Portugal, there is insufficient data concerning road runoff characterization to identify the most important variables to predict pollutant concentrations. The approach adopted in this study is considered to be useful for countries that, like Portugal do not yet possess a significant database of road runoff characteristics.

The PREQUALE, a multiple linear regression equation based on four variables (drainage area; % of impermeable area; mean volume of the storm with the duration of the concentration time, and a return period of 2 years; and annual average rainfall) provides a preliminary method that will improve the prediction of road runoff quality in Portugal. It is a tool designed for use in road runoff management and water resources protection.

The PREQUALE tool presents limitations, due to the few data that supported its construction. Moreover, any tool for road runoff quality prediction must be considered as temporary, since alterations to the use of roads, automobiles and engine construction, as well as climatic changes are expected to modify the pollutant sources and loads as known today.

Acknowledgements

The author is grateful to João Fernandes, from LNEC, for valuable suggestions.

References

1. Irish LB Jr, Barret ME, Malina JF Jr, Chabenau RJ (1998) Use of regression models for analyzing highway storm-water loads. J Environ Eng 124:10
2. FHWA (1996) Evaluation and management of highway runoff water quality. Federal Highway Administration publication no. FHWA-PD-96-032. US Department of Transportation, Washington, DC, 457 pp
3. Driscoll E, Shelley PE, Strecker EW (1990) Pollutant Loadings and Impacts from highway stormwater runoff. Federal Highway Administration publication no. FHWA/RD-88-006-9. Federal Highway Administration, Woodward-Clyde Consultants, Oakland, CA
4. Driver NE, Tasker GD (1990) Techniques for estimation of storm-runoff loads, volumes and selected constituent concentrations in urban watersheds in the United States. U.S. Geological Survey water supply paper 2363, 44 pp
5. Kayhanian M, Singh A, Suverkropp C, Borroum S (2003) Impact of annual average daily traffic on highway runoff pollutant concentrations. J Environ Eng 129:975–990
6. Crabtree B, Moy F, Whitehead M (2005) Pollutants in highway runoff, 10th international conference on urban drainage, Copenhagen/Denmark, 21–26 August, 6 pp
7. Hurtevent J, Despreaux M, Gigleux M, Caquel F, Grange D (2006) Calcul des charges de pollution chronique des eaux de ruissellement issues des plates-formes routières. Note d'information no 75 – Série Économie, Environnement, Conception. Juillet 2006, 12 pp. http://www.setra.equipement.gouv.fr/-Productions,39-.html (visited September 2006)
8. Barbosa AE (2003) Modelos para a Previsão da Qualidade das Águas de Escorrência de Estradas em Portugal, pp 93–107. In: Barbosa AE, Leitão T., Hvitved-Jacobsen T, Bank F (eds.) Curso Sobre Características de Águas de Escorrência de Estradas em Portugal e Minimização dos seus Impactes. Laboratório Nacional de Engenharia Civil, 176 pp
9. Barbosa AE (1999) Highway runoff pollution and design of infiltration ponds for pollutant retention in semi-arid climates. Ph.D. thesis, Environmental Engineering Laboratory, Aalborg University, Denmark, ISBN 87-90033-19-1
10. Mesoplano/Ecossistema/Naturibérica (1994) Estudo de Impacte Ambiental, IP5 EN109-7 Pirâmides-Barra, Projecto de Execução. A – base report; A.4 – impacts assessment; A.5 – mitigation measures, 78 pp
11. Antunes PA, Barbosa AE (2005) Highway runoff characteristics in coastal areas – a case study in Aveiro, Portugal. 10th international conference on urban drainage, Copenhagen/Denmark, 21–26 August, 6 pp
12. Barbosa AE, Fernandes, J (2005) Avaliação da Eficácia das Medidas de Minimização de Impactes Ambientais Implementadas em Portugal. Sistema de Tratamento de Águas de Escorrência. Report 424/05 – NRE, Laboratório Nacional de Engenharia Civil, 39 pp

13. Barbosa AE, Fernandes J, Henriques MJ (2006) Características poluentes duma estrada costeira e avaliação do sistema de tratamento das suas escorrências, 12 Encontro Nacional de Saneamento Básico, 24–27 de Outubro, Cascais, APESB, 15 pp
14. Brandão C, Rodrigues R, Costa JP (2001) Análise de fenómenos extremos. Precipitações intensas em Portugal Continental. Direcção dos Serviços de Recursos Hídricos, Instituto da Água, 57 pp
15. Miranda et al. (2005) O Clima de Portugal nos séculos XX e XXI, 89 pp (from http://www.cgul.ul.pt/pm/Siam2_Clima.pdf)

Websites

http://www.ha-research.gov.uk/projects/index.php?id=642 (visited May 2006)
http://earth.google.com (visited May 2006)

A field microcosm method to determine the impact of sediments and soils contaminated by road runoff on indigenous aquatic macroinvertebrates

V Pettigrove,[1] S Marshall,[2] B Ryan,[3] A Hoffmann[2]

[1]Research and Technology, Melbourne Water, PO Box 4342, Melbourne 3001, Victoria, Australia
[2]Centre for Environmental Stress and Adaptation Research, University of Melbourne, 3010, Victoria, Australia
[3]VicRoads, Private Bag 12, Camberwell, 3124, Victoria, Australia

Abstract

A field-based microcosm method was used to determine whether sediments and soils that receive road runoff are toxic to indigenous aquatic macro-invertebrates present in the Greater Melbourne Area (GMA), Australia. Sediments and soils collected from areas draining three major highways were placed in 20 L microcosms along the littoral zone of a non-polluted wetland. Aquatic insects that emerge from the wetland randomly lay eggs in the microcosms. The occurrence and abundance of key taxa in these microcosms was measured to determine toxicity of the sediments. Several taxa responded in different ways to these sediments. The abundance of *Paratanytarsus grimmii*, *Polypedilum leei*, and *Oxyethira columba* significantly increased with increased concentrations of contaminants in sediments, and appeared to be most influenced by nutrient enrichment. The occurrence of *Tanytarsus fuscithorax* significantly declined with increased concentrations of zinc in surface waters that leached from sediments. *Cricotopus albitarsis* abundance was significantly higher in nutrient-enriched sediments, but significantly declined in high zinc concentrations in surface waters.

There were significant negative correlations between the occurrence of *Larsia albiceps*, *T. fuscithorax*, and *Procladius* spp. and copper or total petroleum hydrocarbon (TPH) concentrations in sediments. This method provides unique information on the effects of contaminated soils and sediments on indigenous aquatic macroinvertebrates and can be used to determine the effectiveness of water sensitive urban designs in reducing sediment toxicity.

Introduction

Road runoff is a major source of heavy metals, mineral oils, and polycyclic aromatic hydrocarbons (PAHs) [1–3], yet it is difficult to determine what impact it has on aquatic ecosystems receiving this runoff. Biological field surveys are unable to separate the confounding effects of hydrologic disturbance created by increased catchment imperviousness and the effects of pollutants present in road runoff [4]. Furthermore, permanent, receiving waters may be some distance downstream of the roads and pollution-related effects from roads may be obscured if the drainage system receives pollutants from other land use activities (e.g., farming, residential, and industrial areas). Surveys of storm water require continuous measurements of water quality and quantity and it is expensive to install and maintain automatic sampling equipment [5]. Sediments are a sink for heavy metals, petroleum hydrocarbons, and other hydrophobic pollutants [6] and may be a comparatively cheaper alternative to directly monitoring storm waters. Water and sediment quality data can only provide an indication of potential levels of contamination, particularly as not all important parameters may be surveyed (e.g., PAHs and pesticides). Ecotoxicological studies are important in determining the toxicity of contaminants present in road runoff, but it is difficult to relate the results from a laboratory indicator species to the indigenous fauna present in receiving waters. Pettigrove and Hoffmann [7] developed an alternative field-based microcosm method to determine the toxicity of contaminated sediments on indigenous aquatic macroinvertebrates. This method involves collecting sediments from a water body and transferring them to microcosms that are then placed on the littoral zone of a non-polluted wetland. Many of the insects that inhabit the wetland randomly lay eggs in the microcosms. Toxic sediments are only able to support a pollution-tolerant fauna, whereas non-toxic sediments will support a similar fauna to that present in non-polluted reference sediments. Heavy metals and petroleum hydrocarbons tend to become associated with fine particulate matter and accumulate in soils and in quiescent areas of drainage channels and receiving waters [8].

In this study, we trial the field-based microcosm approach to determine whether it can be used to assess the toxicity of sediments in receiving waters and soils present in the drainage systems of three highways present in Melbourne, Australia. All three sections of roads investigated had asphalt surfaces and 100 km h^{-1} speed restrictions. Sediments were collected from wetlands, and soils were collected from swales and dry drainage channels downstream of these highways. Some sites were located on a newly constructed freeway and will be monitored over future years to assess changes in sediment quality and toxicity. Other sites were located on more established sections of freeway that are likely to receive higher levels of contaminants. The value of this method for assessing the toxicity of road runoff and assessing the effectiveness of water-sensitive urban designs is discussed.

Study area

Soils or fine (<65 µm) sediments were collected from the eight sites listed in Table 1. The Hallam Bypass is a new freeway that was open to traffic in 2005 and carries approximately 100,000 vehicles per day. Whole sediments were collected from a constructed South Gippsland sediment (SGS) pond and a South Gippsland east wetland (SGE) that receives runoff from approximately a 100 m stretch of freeway. The section of the Monash Freeway surveyed was constructed during the 1980s and carries approximately 170,000 vehicles per day. Soil was collected from a Heatherton Road swale (HRS) that received runoff from one three-lane carriageway and from a drain that only received road runoff from the same stretch of

Table 1. Study site locations and sediment characteristics

Site	Code	Location	Sediment	No. of microcosms
Glynns wetland	GWW	Reference site	Whole	10
East sediment pond	SGS	Hallam Bypass	Whole	5
East wetland	SGE	Hallam Bypass	Whole	5
Heatherton Road swale	HRS	Monash Freeway	Soil	5
Heatherton Road drain	HO1	Monash Freeway	Soil	5
Steele Creek wetland	SCW	Western Ring Road	Fine	5
Boundary Road North	BRN	Western Ring Road	Fine	5
West Street South	WSS	Western Ring Road	Fine	5

road (Heatherton Road drain [HO1]). The Western Ring Road was constructed between 1990 and 1995 and carries approximately 105,000 vehicles per day. Fine sediments were collected from Steele Creek wetland (SCW), and two small wetlands (Boundary Road wetland [BRN] and West Street wetland [WSS]) that were originally constructed as sediment traps during initial road construction activities. Whole sediments were collected from Glynns wetland (GWW), a non-polluted wetland where the microcosm experiments were conducted.

Methods

The soil and sediment samples were collected in September and October 2005, except SCW sediment was collected in October 2004. The method used to collect fine sediments is described by Pettigrove and Hoffmann [7]. Whole sediments were collected using a shovel. Only surface soils (<5 cm depth) were collected. Filtered water (<64 µm) from the reference GWW was added to the soils to make sediments that were then thoroughly mixed for 30 min using a paint mixer attached to an electric drill. The sediments produced were filtered through a 2 mm sieve to remove coarse particulate matter.

Sediment samples were analysed for As, Cr, Cu, Pb, Ni, Zn (detection limit = 1 mg kg^{-1}), Cd and Hg (detection limit = 0.1 mg kg^{-1}), total petroleum hydrocarbons (TPHs) (C6–C36), total phosphorus (TP), nitrogen (as total Kjeldahl nitrogen [TKN]), and oxidizable organic carbon (OOC) using the methods detailed by Pettigrove and Hoffmann [7]. At the conclusion of the microcosm experiment, water samples were collected and analysed using US Environmental Protection Agency (US-EPA) methods 3051 [9] for total and filtered Cu, Cr, Ni, Pb, Zn (detection limit = 0.001 mg L^{-1}), As, Cd and Hg (detection limit = 0.001 mg L^{-1}) and compared to the Australian and New Zealand water quality guidelines for the protection of aquatic ecosystems [10]. Water for determination of filtered metals was collected *in situ* using 45 µm cellulose acetate syringe filters.

It was anticipated that site HRS, which was a swale that had received runoff from an adjacent major highway for many decades, would have higher levels of heavy metals than what existed (see results). A pipe located about 1.5 m below the surface drained the swale. Waters were collected from this pipe about 12 h after a storm event and analysed for dissolved heavy metals to determine whether these heavy metals could pass through this drainage system.

The microcosms (20 L polypropylene tanks with 500 mL of sediment and 15 L of filtered water from GWW) were placed along the edge of GWW. A completely randomized design was used in the experiment. Briefly, insects that emerge from the wetland mate and then randomly lay eggs within the microcosms. This method provides a unique means of gathering information on the effects of sediment quality on macroinvertebrates in lentic habitats, particularly for indigenous species that cannot be easily reared in the laboratory. The experiment was terminated after 40 days when there signs of insect emergence. Microcosms were transported to the laboratory, where those with fine sediments were filtered through a 125 µm sieve to separate the macroinvertebrates from the fine (<65 µm) sediment. Whole sediment samples were washed through a 125 µm sieve and then elutriated to remove coarse particles. Processed samples were stored in 70% ethanol at 4°C. The sample was subsampled if more than 500 individuals were present. Further details regarding the method are described by Pettigrove and Hoffmann [7].

A multi-response permutation procedure (MRPP) using Sorenson distances was conducted, using the statistical package PC-ORD (version 4 for Windows, Gleneden Beach, Oregon, USA), to determine whether there was a significant difference in faunal composition. This analysis was preferred to other multivariate analyses as it is a non-parametric procedure and therefore no assumptions are required about data normality. As the MRPP found a significant difference in the faunal composition between treatments (see results), an indicator species analysis, using the statistical package PC-ORD, was then conducted to obtain a description of how well each taxon separates between treatments.

Non-parametric Spearman rank correlations were conducted between common taxa and sediment quality data to determine whether species abundance and frequency of occurrence are affected by sediment quality. Sediment quality data were summarized using non-metric multidimensional scaling (NMS) to produce factor scores (FSs) that were correlated to the abundance of common taxa or frequency of occurrence in treatments. Water quality data were correlated with the sediment quality data to determine whether any water quality parameters exhibited any patterns distinct from sediments. Filtered Ni, Cu, and Zn concentrations in water were correlated to the biological data as they did not correlate to the FS derived from the sediment quality data. Correlations were also explored between the frequency of occurrence of common taxa within treatments and Cu concentrations in sediments, as stronger correlations existed using Cu than using FS in these analyses. All correlations were conducted using the statistical package SPSS for Windows (version 14.0 SPSS, Chicago, Illinois, USA).

Results

Sediment and water Quality

Heavy metal concentrations in sediments at GWW, SGE, SGS, and HRS were below the consensus-based freshwater sediment quality guidelines [11] (Table 2). Therefore, heavy metals in these sediments were unlikely to be toxic to aquatic biota [11]. Lead concentrations in HO1 sediments and Cu, Ni, Pb, and Zn concentrations in SCW sediments exceeded the threshold effect concentration (TEC) [11]. Therefore, there is a moderate probability that these sediments may produce toxic effects on biota. Cadmium, Cr, Cu, and Ni concentrations in BRN and WSS sediments exceeded the TEC and Zn and Pb concentrations exceeded the probable effect concentration (PEC) [11]. Therefore, these sediments have a high probability of producing toxic effects on biota.

Table 2. Heavy metal concentrations (mg kg^{-1}) in sediments compared to the probable effects concentration (PEC) [11]

Site	As	Cd	Cr	Cu	Hg	Ni	Pb	Zn
GWW	<5	<0.2	16	6	<0.05	8	13	25
SGE	5	<0.2	22	8	<0.05	<5	15	18
SGS	<5	<0.2	15	13	<0.05	7	13	18
HRS	<5	<0.2	12	16	<0.05	9	32	29
HO1	<5	0.2	14	28	<0.05	12	_70_	60
SCW	<5	0.5	30	_57_	0.08	_26_	_78_	**240**
BRN	<5	_1.3_	_48_	_84_	0.14	_35_	**230**	**590**
WSS	<5	_1.5_	_54_	_110_	0.17	_38_	**290**	**640**
TEC	9.79	0.99	43	32	0.18	22.7	35.8	121
PEC	33	4.98	111	149	1.06	128	128	459

Those metals that exceed the threshold effects concentration (TEC) are underlined and those that exceed the probable effects concentration (PEC) are in bold.

Dissolved surface water concentrations of Cd, Hg, and Pb in the microcosms were below detection limits in all treatments (Table 3). Arsenic was detected at GWW, SGE, SGS, and HRS, but were below the 90% trigger value (TV) on all occasions. Nickel was only detected at BRN, but was below the 90% hardness modified TV (HMTV) of 34 µg L^{-1}. Chromium concentrations were 20 µg L^{-1} in GWW, 10 µg L^{-1} in SGS, and below the

Table 3. Filtered (<0.45 µm) heavy metal concentrations (µg L^{-1}) present in surface waters of the microcosms at the termination of the experiment and from a drain (HRS drain) that receives drainage from swale HRS

Sample	As	Cd	Cr	Cu	Hg	Ni	Pb	Zn	Hardness
GWW	2	<1	20	**10**	<1	<10	<10	10	7.6
SGE	1	<1	10	<10	<1	<10	<10	**290**	18
SGS	3	<1	20	**10**	<1	<10	<10	**470**	27
HRS	5	<1	<10	<10	<1	<10	<10	<10	84
HO1	<1	<1	<10	**10**	<1	<10	<10	<10	73
SCW	<1	<1	<10	**15**	<1	<10	<10	**65**	47
BRN	<1	<1	<10	**10**	<1	20	<10	<10	94
WSS	<1	<1	<10	**20**	<1	<10	<10	30	84
HRS Drain	2	<1	25	30	5	<10	20	<10	NA

Those numbers in bold exceed the 90% trigger values (TV) from the Australian and New Zealand water quality guidelines [10]. TVs compared to As(V) for As, Cr (VI) for Cr, and inorganic Hg for Hg

detection limit of 10 µg L^{-1} at the remaining sites. The 90% TV for Cr is 6 µg L^{-1} as Cr (VI) [10]. As the measured concentrations in this study were total Cr, it is unclear whether these concentrations exceeded the 90% TV. Copper concentrations exceeded the 90% HMTV at GWW, SGS, HO1, BRN, SCW, and WSS by factors of 6, 6, 3, 2, 6, and 2, respectively. Zinc concentrations exceeded the 90% HMTV in SGS, SGE, and SCW by factors of 31, 19, and 3, respectively. The water sample collected from the drainage piped beneath swale HRS had elevated dissolved concentrations of Cr, Cu, Hg, and Pb that were 4, 17, 3, and 4 times the 90% HMTV, respectively (assuming all Cr is Cr (VI) and all Hg is in organic form).

TPH concentrations in sediments were below the detection limit at all sites except SCW, BRN, and WSS (Table 4). TPH concentrations exceeding 860 mg kg^{-1} may impair ecosystems and increase the abundance of opportunistic species, whereas it is possible sediments may affect species presence and abundance where TPH concentrations exceed 1870 mg kg^{-1} [12]. Therefore, biota inhabiting SCW, BRN, and WSS sediments may be affected by TPH pollution. TP concentrations were comparatively low in GWW, SGE, SGS, and HRS sediments, moderate in HO1 sediments and comparatively high in SCW, BRN, and WSS sediments (Table 5). TKN concentrations were similar in most sediments with the highest concentrations occurring in BRN and WSS. OOC levels were low in SGE, SGS, and SCW, moderate in GWW, HRS, and HO1, and elevated in BRN and WSS.

Table 4. Total petroleum hydrocarbons (as mg kg^{-1})

		Total petroleum hydrocarbons			
Site	C6–C9	C10–C14	C15–C28	C29–36	Total TPH
GWW	<40	<40	<100	<100	BD
SGE	<40	<40	<100	<100	BD
SGS	<40	<40	<100	<100	BD
HRS	<40	<40	<100	<100	BD
HO1	<40	<40	<100	<100	BD
SCW	<60	<60	870	700	1570
BRN	<60	<60	2300	2700	5000
WSS	<60	<60	2100	2500	4600

BD = below detection limit

Table 5. Nitrogen (as total Kjeldahl nitrogen), phosphorus (as total phosphorus), and percent oxidizable organic carbon (OOC) concentrations in sediments

	Nutrients		
Site	Nitrogen (mg kg^{-1})	Phosphorus (mg kg^{-1})	OOC (%)
GWW	1800	160	2.0
SGE	1000	120	<0.1
SGS	890	100	<0.1
HRS	1500	180	1.9
HO1	2600	240	2.7
SCW	1300	570	<0.1
BRN	3400	790	8.4
WSS	3600	970	7.1

Macroinvertebrate results

A total of 33 taxa and an estimated 35,170 individuals were collected from the 45 microcosms used in this experiment. The fauna was dominated by chironomid larvae (*P. grimmii, T. fuscithorax, Procladius* sp., *C. albitarsis, Polypedilum verspertinus, Kiefferulus intertinctus, P. leei, Chironomus* spp., *Polypedilum watsoni, Parachironomus* sp. M1, *Ablabesmyia notabilis*, and *Cladopelma curtivalva*), other Diptera (Ceratopogonidae and Chaoboridae), Coleoptera (*Necterosoma pencillatum*), Hemiptera (*Micronecta annae* and *Mesovelia* sp.), Ephemeroptera (*Tasmanocoenis* sp.), and Trichoptera *(Oecetis* sp. and *O. columba*) (Table 6). An average of 259 animals was present in the 10 reference microcosms with GWW sediment. Comparatively fewer animals were present in SGE wetland and SGS pond that had an average of 129 and 59 animals/microcosm. The remaining sites tended to have considerably more animals than reference sediments: HRS had 394 animals/microcosm, SCW had 496 animals/microcosm, HO1 had 1460 animals/microcosm, BNN had 1922 animals/microcosm, and WSS had 2055 animals/microcosm.

An MRPP of the macroinvertebrate data demonstrated that the fauna present in treatments were significantly different to that between treatments ($p < 0.0001$). However, an indicator species analysis found that no taxon strongly reflected these faunal differences between treatments. Therefore, the faunal composition was affected by sediment quality, but no taxon strongly reflected this difference.

The first axis of a NMS conducted on sediment quality data (heavy metals, TPHs, and nutrients) summarized 92% of the variance in the data. These sediment quality factor scores (SQFS) were correlated with biological data to determine what taxa appear to be affected by sediment quality. Significant positive correlations exist between *P. grimmii, P. leei, C. albitarsis,* and *O. columba* and sediment quality (Table 7). Therefore, these taxa were more abundant in more polluted sediments.

Spearman correlations were conducted between the sediment quality and water quality data to determine whether water quality data revealed any distinct trends not evident in sediment quality data. Surface water concentrations of Cd, Hg, and Pb were not included in this analysis, as they were below detection limits. There were significant negative correlations between As (correlation = -0.73, $p = 0.02$) and Cr (correlation = -0.82, $p = 0.006$) and the SQFS, and a significant positive correlation between water hardness and the SQFS (correlation = 0.73, $p = 0.02$). Ni, Cu, and Cd surface water concentrations did not significantly vary with the SQFS ($p > 0.05$). Therefore, correlations were conducted between Ni, Cu, and Zn, and the common species listed in Table 7. A strong negative correlation occurred between *T. fuscithorax* and Zn (correlation = -0.91, $p = 0.001$) and a weak negative correlation occurred between *C. albitarsis* and Zn (correlation = -0.68, $p = 0.04$). No significant correlations occurred between remaining taxa. Therefore, *T. fuscithorax*, and possibly *C. albitarsis*, appear to be influenced by dissolved Zn concentrations in surface waters.

Correlations were conducted between the proportion of microcosms per treatment where common taxa (listed in Table 7) were present and SQFSs, Cu sediment, and Zn surface water concentrations. There were no significant correlations between the occurrence of common taxa and SQFS or Zn surface water concentrations. However, there was a significant negative correlation between Cu concentrations in sediments and the occurrence of *T. fuscithorax* (correlation = -0.83, $p = 0.005$), *L. albiceps* (correlation = -0.70, $p = 0.04$), and *Procladius* spp. (correlation = -0.68, $p = 0.04$).

Table 6. Most abundant invertebrate taxa in microcosms expressed as mean abundance per site and number of microcosms per site where taxa were present (in parentheses)

Rank Abundance	Taxon	GWW	SGE	SGS	HRS	SCW	HOI	BRN	WSS	No of Individuals	% Total Abundance
1	*Paratanytarsus grimmii*	33 (5)	44 (4)	13 (3)	206 (5)	255 (5)	695 (5)	1299 (5)	1635 (5)	21070 (38)	61.0
2	*Tanytarsus fuscithorax*	35 (10)	12 (3)	7 (2)	62 (5)	30 (1)	258 (4)	226 (4)	41 (2)	3530 (32)	9.7
3	*Procladius* spp.	64 (8)	50 (2)	11 (3)	19 (1)	108 (2)	91 (3)	132 (2)	85 (2)	3121 (23)	8.4
4	*Crictopus albitarsis*	4 (3)	0 (0)	0 (0)	18 (2)	7 (2)	123 (4)	90 (2)	94 (4)	1705 (18)	4.8
5	*Necterosoma pencillatum*	13 (4)	0 (0)	6 (2)	20 (3)	16 (3)	14 (5)	53 (4)	81 (3)	1080 (25)	3.1
6	*Larsia albiceps*	28 (7)	9 (2)	11 (2)	16 (3)	22 (1)	3 (1)	17 (2)	8 (1)	713 (20)	2.4
7	*Kiefferulus martini*	46 (2)	0 (0)	4 (2)	0 (0)	14 (1)	29 (2)	8 (2)	0 (0)	738 (9)	2.0
8	*Polypedilum vespertinus*	11 (3)	8 (2)	0 (0)	0 (0)	0 (0)	72 (1)	10 (1)	10 (1)	611 (8)	1.6
9	*Oxyethira columba*	0 (0)	0 (0)	0 (0)	2 (1)	1 (1)	106 (1)	1 (1)	1 (1)	560 (5)	1.5
10	*Kiefferulus intertinctus*	0 (1)	0 (0)	1 (2)	7 (1)	0 (0)	43 (1)	41 (2)	0 (0)	464 (7)	1.2
11	*Polypedilum leei*	1 (1)	0 (0)	0 (0)	0 (0)	0 (1)	12 (1)	4 (1)	53 (1)	356 (5)	1.0
12	*Chironomus* spp.	0 (0)	0 (1)	0 (0)	20 (1)	0 (0)	0 (0)	29 (1)	4 (1)	267 (5)	0.7
13	*Polypedilum watsoni*	5 (3)	1 (1)	1 (2)	11 (1)	5 (2)	4 (1)	5 (1)	7 (1)	224 (12)	0.6
14	*Mesovelia* sp	4 (7)	1 (2)	3 (5)	2 (5)	4 (4)	3 (2)	3 (3)	6 (3)	156 (31)	0.4
15	Ceratopogonidae	0 (0)	0 (1)	0 (0)	0 (0)	0 (0)	0 (0)	0 (0)	20 (2)	102 (3)	0.3
16	*Parachironomus* sp. M1	5 (1)	0 (0)	0 (1)	0 (0)	7 (2)	1 (1)	0 (0)	0 (0)	94 (5)	0.3
17	*Oecetis* sp.	0 (0)	0 (0)	0 (0)	0 (0)	18 (1)	0 (0)	0 (0)	0 (0)	92 (1)	0.2
18	*Micronecta annae*	0 (3)	0 (1)	0 (0)	0 (1)	2 (2)	0 (0)	1 (2)	7 (1)	54 (10)	0.1
19	*Ablabesmyia notabilis*	5 (2)	0 (0)	0 (0)	1 (1)	0 (0)	0 (0)	0 (0)	0 (0)	53 (3)	0.1
20	*Tasmanocoenis* sp.	0 (0)	0 (2)	0 (0)	8 (1)	0 (0)	0 (0)	0 (0)	0 (0)	44 (3)	0.1
21	*Cladopelma curtivalva*	0 (1)	1 (1)	0 (0)	0 (0)	0 (0)	6 (1)	0 (0)	0 (0)	39 (3)	0.1
22	Chaoboridae	2 (1)	0 (0)	0 (0)	2 (1)	0 (0)	0 (0)	0 (0)	0 (0)	29 (2)	0.1

Table 7. Spearman correlation coefficients and significance levels (two-tailed) between macroinvertebrate taxa and sediment quality, as summarized by a factor score produced from the sediment quality data

Taxon	Coefficient	
Paratanytarsus grimmii	0.93	<0.001
Procladius sp.	0.60	0.08
Polypedilum leei	0.74	0.02
Cricotopus albitarsis	0.73	0.02
Necterosoma pencillatum	0.55	0.12
Micronecta annae	0.40	0.28
Oxyethira columba	0.66	0.05
Tanytarsus fuscithorax	0.47	0.19
Polypedilum vespertinus	0.20	0.61
Larsia albiceps	–0.53	0.14
Kiefferulus intertinctus	0.29	0.44
Kiefferulus martini	–0.44	0.23

Discussion

It is difficult to interpret what effect the water and sediment quality data would have on aquatic biota without field-based microcosm data. According to freshwater sediment quality guidelines [11], the concentrations of Pb and Zn in BRN and WSS had a high probability of being toxic to biota. Previous studies that used the field-based microcosm method found that *Polypedilum grimmii, P. vespertinus, T. fuscithorax*, and *K. intertinctus* larvae are significantly less common and *L. albiceps* larvae and *Mesovelia* sp. are less abundant, in sediments with Zn concentrations of 1060 mg k^{-1}g or more, than in sediments with Zn concentrations of 320 mg kg^{-1} or less [7]. The highest Zn concentrations in this study (590 mg kg^{-1} at BRN and 640 mg kg^{-1} at WSS) are in the intermediate range and it was not known how these taxa would respond to this level of contamination. The TPH concentrations in SCW, BRN, and WSS were likely to impair aquatic ecosystems. For example, the number of adult *T. fuscithorax* that emerged from microcosms with sediments spiked with 860 mg kg^{-1} of synthetic motor oil was significantly higher than those that emerged from non-polluted sediments, and abundance significantly declined in sediments with 1860 mg kg^{-1} of synthetic motor oil [12]. Similarly, the abundance of *Procladius villosimanus* and *L. albiceps* significantly declined in TPH concentrations of 860 and 4600 mg kg^{-1} or more, respectively [12]. In this experiment, the occurrence of *T. fuscithorax*, *L. albiceps*, and *Procladius* spp. was negatively correlated with Cu concentrations in sediment. However, it is likely that TPHs rather than Cu affected these taxa, as the Cu concentrations were below the TEC [10], but the TPH concentrations in these sediments were at levels where effects had been reported in these taxa [12]. Although concentrations of heavy metals in sediments were low in SGS and SGE, these treatments had surface water concentrations of filtered Zn that were markedly higher than Australian and New Zealand water quality guidelines for the protection of aquatic ecosystems [10]. There were considerable differences in nutrient concentrations in sediments between treatments, but as there are no guidelines for nutrients in sediments, it is difficult to understand what effect these contaminants may have on biota.

The biological results helped interpret the significance of sediment and water quality data. In this experiment, the abundance of four taxa significantly correlated with increased concentrations of contaminants in sediments. This trend would appear to be due to increased nutrient enrichment of sediments, as mild levels of organic enrichment can increase species abundance and biomass [13]. Even though abundance increased, the occurrence of *T. fuscithorax*, *L. albiceps*, and *Procladius* spp. Significantly decreased with increased concentrations of contaminants in sediments. This finding suggests that the more contaminated sediments may be toxic to some invertebrates, but nutrient-enriched sediments may advantage other more pollution-tolerant taxa. Furthermore, the abundance of *T. fuscithorax* and *C. albitarsis* was significantly lower in treatments with elevated surface water concentrations of Zn. This Zn would have leached from sediments indicating that SGS and SGE sediments would have elevated porewater concentrations. In summary, HRS soils were non-toxic, SGS and SGE had potentially toxic concentrations of Zn leachate, HO1 soils were nutrient enriched, and SCW, BRN, and WSS sediments were nutrient enriched but also toxic to some taxa, possibly due to elevated TPHs.

The field-based microcosm method used in this study was effective in determining effects of soils and sediments contaminated by road runoff on indigenous aquatic macroinvertebrates. Experiments provided information on the response of indigenous species that cannot be reared or tested in laboratory conditions. Almost all macroinvertebrates in the microcosms developed from eggs, and early larval stages are more sensitive than later instars to pollutants [14,15]. If ecotoxicological tests are to be ecologically relevant, the most sensitive life stage of the test species should be used [14,15]. Therefore, this method allows a more comprehensive assessment of the effects of polluted sediments on macroinvertebrates compared to many ecotoxicological tests where only later instars are assessed.

Monitoring of sediment and soil quality and toxicity can provide valuable information to road management authorities about how effective storm water treatment measures are in intercepting and treating contaminants in road runoff. In this study, soils on a swale on the Monash Freeway contained unexpectedly low concentrations of contaminants and were non-toxic to aquatic macroinvertebrates. This finding prompted further investigations and it was found that drainage water from the swale contained elevated concentrations of heavy metals and that sediment from the HO1 outlet contained higher concentrations of heavy metals than the swale (Table 2). This suggests that the swale is ineffective in removing heavy metals from road runoff. Therefore, this swale could be retrofitted with suitable treatment media to capture contaminants and protect aquatic ecosystems in receiving waters.

Larval abundance is a less sensitive metric of sediment toxicity than adult abundance. For example, the abundance of *P. grimmii* larvae and adults was not affected by sediments spiked with up to 4630 mg kg^{-1} of synthetic motor oil, but, the abundance of *P. grimmii* adults were significantly lower in sediments with only 860 mg kg^{-1} of synthetic motor oil when compared to non-polluted sediments [12]. Future monitoring of soil and sediment toxicity should incorporate the collection of adults that emerge from microcosms, as they are more sensitive indicator of sediment toxicity than larval abundance [12].

Acknowledgements

Leanne Maas (VicRoads) for providing traffic information on the three highways studied. Graham Rooney for editorial comments on the manuscript.

References

1. Forster J (1993) The influence of atmospheric conditions and storm characteristics on roof runoff pollution with an experimental roof system. In: Proceedings of VIth international conference on urban storm drainage, Niagara Falls, Ontario, Canada, pp 411–416
2. Xanthopoulos C, Hahn HH (1990) Pollutants attached to particles from drainage areas. Sci Tot Environ 93:441–448
3. Ellis JB, Revitt DM (1991) Drainage for roads: control and treatment of highway runoff. Report NRA43804/MID.012. National Rivers Authority, Reading, UK
4. Pettigrove V (2006) Is catchment imperviousness a good indicator of ecosystem health? Proceedings of the highways and urbanization symposium

5. VicRoads (2005) Best strategy for treating stormwater at the roadside. Draft report. VicRoads design and landscape section, Kew, Victoria, Australia
6. Pitt RE (1995) Biological effects of urban runoff discharges. In: Herricks EE (ed.) Stormwater urban runoff, and receiving water systems, Lewis/CRC, Boca Raton, FL, pp 127–162
7. Pettigrove V, Hoffmann A (2005) A field-based microcosm method to assess the effects of polluted urban stream sediments on aquatic macroinvertebrates. Environ Tox Chem 24:170–180
8. Pitt R (1999) Receiving water and other impacts. In: Heany JP, Pitt R, Field R (eds.) Innovative urban wet-weather flow management systems. National risk management research laboratory, Office of Research and Development, US Environment Protection Agency, EPA/600/R-99/029. Chapter 4
9. U S Environment Protection Agency (1994) Methods for the determination of metals in environmental samples – supplement 1. Report No. EPA/600/R-94-111. USEPA, Cincinnati, OH, USA
10. Australian and New Zealand Environment and Conservation Council and Agriculture and Resource Management Council of Australia and New Zealand (2000) National water quality management strategy, Australian and New Zealand guidelines for fresh and marine water quality, vol. 1 – the guidelines
11. MacDonald DD, Ingersoll CG, Berger TA (2000) Development and evaluation of consensus-based sediment quality guidelines for freshwater ecosystems. Arch Environ Cont Tox 39:20–31
12. Pettigrove V, Hoffmann A (2005) Effects of long-chain hydrocarbon-polluted sediment on freshwater macroinvertebrates. Environ Tox Chem 24:2500–2508
13. Pearson TH, Rosenberg R (1978) Macrobenthic succession in relation to organic enrichment and pollution of the marine environment. Oceanogr Mar Biol Ann Rev 16:229–311
14. Nebeker AV, Cairns MA, Wise CM (1984) Relative sensitivity of Chironomus tentans life stages to copper. Environ Tox Chem 3:151–158
15. Gauss JD, Woods PE, Winner RW, Skillings JH (1985) Acute toxicity of copper to three life stages of Chironomus tentans as affected by water hardness-alkalinity. Environ Pollut 37:149–157

Assessment of storm water ecotoxicity using a battery of biotests

L Scholes,[1] A Baun,[2] M Seidl,[3] E Eriksson,[2] M Revitt,[1] J-M Mouchel[3]

[1]Middlesex University, UK
[2]Insitute of Environmental Resources, Technical University of Denmark, Denmark
[3]Cereve-ENPC, France

Abstract

As part of the European Union 5th Framework Programme (EU FP5) sustainable storm water management project DayWater, an international monitoring programme was established to investigate the ecotoxicity of highway and urban runoff samples. This involved the collection of samples from a total of 35 storm events from sites in Sweden (Luleå and Stockholm), Germany (Wuppertal), and France (Nantes). To enable both chronic and acute end points to be addressed, the ecotoxicity tests were performed using rotifers (*Brachionus calyciflorus*), bacteria (*Vibrio ficherii*), and algae (*Pseudokirchneriella subcapitata*). Samples collected at the Stockholm site were additionally analysed for a range of water quality parameters enabling this site to be considered from a combined physico-chemical and ecotoxicological perspective. This paper provides an overview of the results of the whole monitoring programme and demonstrates that storm water frequently exerts a toxic effect. Although the levels of ecotoxicity detected were found to vary greatly in relation to sites, storm events and test organisms, further analysis demonstrated a moderately strong correlation between the responses of algae and rotifer tests. Microtox was generally found to be the most responsive test in terms of both frequency of detection of toxicity and level at which a toxic impact could be detected.

Introduction

Current approaches for assessing the quality of aquatic ecosystems typically focus on the determination of a selection of specific physico-chemical parameters, which are then compared with water quality criteria or standards [e.g., 1]. However, chemical analysis in isolation does not reveal anything about the ecotoxicity of the sample because the bioavailability, and hence the ecotoxic impact, of many substances can vary considerably in relation to a host of environmental factors such as pH, dissolved oxygen concentration and the antagonistic and/or synergistic effects of other substances. The complexities of such interactions have been demonstrated in studies which have involved both the ecotoxicological and physico-chemical characterization of complex environmental effluents [e.g., 2]. These studies report that, despite the completion of detailed chemical analyses, complex samples frequently demonstrate a greater ecotoxic impact than that which could be explained by the presence of the chemicals identified [3–5].

Ecotoxicity is currently a priority issue in relation to the management of urban storm water for two major reasons. Firstly, because storm water runoff is widely reported to transport a wide variety of organic and inorganic pollutants, e.g., a review of the literature identified over 600 substances as being potentially present in storm water discharges [6], and secondly, because the need to control non-point source pollution is specifically identified in the EU Water Framework Directive (WFD) as a key way to enable the achievement of its ecological-based targets [7]. Evaluation of the ecotoxicity of a sample (also known as biotesting) is attracting increasing interest as a technique which can generate direct information on the toxic impact of complex effluents such as urban runoff, even if detailed knowledge of the chemical constituents are unknown and the specific mechanisms of toxicity are not fully understood [5].

Ecotoxicity testing should not be seen as an independent alternative approach as chemical analysis is essential for the identification and remediation of pollutant sources. However, biotests can generate information that cannot be obtained through chemical-only monitoring approaches. Hence, the use of a combined chemical and biological approach, as recommended by various authors [e.g., 8] is supported. But, although the physico-chemical quality of storm water has been widely reported, the ecotoxicological impact of storm water in relation to different test organisms is less well understood. It is within this context that this study sets out to investigate the use of three different types of biotests to evaluate the ecotoxicity of storm water runoff from four different sites within three different European countries. The effects and levels of the detected ecotoxicity are discussed together with the patterns of response between different tests and, where possible, the ecotoxicity results are related to water quality data.

Materials and methodology

Description of sites

The sampling site in Nantes receives runoff from an 88 ha catchment, of which 30% is impervious. The catchment is residential and the drainage network fully separate. The Stockholm site is located at a storm water treatment facility, situated beneath an elevated motorway, which consists of a sedimentation pond followed by a series of biofilters. The system receives runoff from both the motorway (traffic density of 120,000–130,000 vehicles per day) and a small local urban area (total catchment area of approximately 15,000 m^2). Storm water samples were collected from a descending well located prior to the pond. These samples were also analysed separately for a wide range of basic water quality parameters and metals. Runoff (samples 01, 02, and 03) and snowpack samples (samples 04 and 05) were collected from two locations within Luleå (North Sweden): a 660 m^2 section of road located in the city centre (traffic density; 7400 vehicles per day) and a centrally located urban park. The Wuppertal site is adjacent to an underground settlement tank and retention basin which receive runoff from an 18.5 ha catchment area, of which 11.2 ha are impermeable. The catchment is primarily residential with some minor light industry and has a separate drainage system.

Collection of samples

Samples were collected from a total of 35 separate storm events over a 10-month sampling programme between October 2004 and July 2005. All samples were collected in glass bottles and frozen within 24 h for storage and transportation to the different research centres, where the different biotests were conducted on the same sets of samples.

Selection of tests

The criteria used to identify appropriate tests included: (1) existence of validated methods or guideline protocols; (2) evaluation of ecotoxicity in relation to reproduction or metabolism (as both these end points can be directly linked to impacts on the population structure of aquatic ecosystems); (3) support for the battery approach of selecting organisms from different trophic levels; and (4) cost–time efficiency. Consideration of these objectives led to the selection of algal (*Pseudokirchneriella subcapitata*), rotifer (*Brachionus calyciflorus*), and bacterial tests (*Vibrio fischeri*; Microtox). To assist in the interpretation of data, analyses were carried out on whole effluent samples, i.e., without any preconcentration.

Sample preparation and test methodologies

Prior to analysis, samples were defrosted overnight, vigorously shaken for 1 min and then left to settle for 1 h. Water hardness, pH, and conductivity were determined using standard methods. The bacterial test employed was the Microtox acute bioluminescence inhibition test which utilizes *V. fischeri* as the test organism. All tests were completed using the whole effluent toxicity (WET) Test over a time period of 15 min. Each sample was analysed in duplicate and phenol was used as a standard reference material. All bacterial tests were undertaken at Middlesex University, London, UK. Mini-scale algal growth inhibition tests utilizing *P. subcapitata* were conducted according to the ISO standard [9] with a 48 h exposure time. Potassium dichromate was used as a reference compound. All algal tests were undertaken at the Technical University of Denmark, Lyngby, Denmark. Rotifer reproduction tests utilized *B. calyciflorus* as the test organism, and were carried out as described in the standard NF T90377 [10]. Each test involved three dilutions with eight replicates and two controls. Rotifers were exposed in the presence of 2×10^6 cells mL^{-1} *Pseudokirchneriella* and after 48 h incubation at 25°C the number of individuals were recorded and the inhibition of reproduction calculated. Potassium dichromate was used as a reference compound. All rotifer tests were undertaken at Cereve-ENPC, Paris, France.

Results and discussion

An overview of the complete ecotoxicity data set is presented in Table 1. For comparative purposes, the data is presented in terms of the inhibition caused by exposure of the test organism to undiluted storm water samples. Negative values within the table indicate stimulation of growth. The application of tests using three different species from different trophic levels to samples collected for 35 independent storm water events from four different sites within three European countries has led to the development of one of the most comprehensive storm water ecotoxicity database collected to-date.

The data in Table 1 show that the reported levels of ecotoxicity vary considerably both between sites and between the different tests carried out on samples from the same site. For example, the first sample from the Stockholm site inhibited the level of bacterial light production by 16%, algal growth by 6%, and rotifer reproduction by 100%. This difference in the impact on test organisms is thought to be associated with differences in the susceptibility of different species to different pollutants, indicating that the results of single biotests should be treated with caution. Although each storm water sample resulted in some level of inhibition using Microtox (ranging from a minimum of 7% in sample 06 from Stockholm to a maximum of 86% in both sample 05 from Stockholm and sample 02 from Nantes), 24% of events tested in relation to algae and 62% of events tested in relation to rotifers stimulated growth and reproduction, respectively. The fact that the Microtox test reported a toxic impact in every sample may also indicate that this test is the most responsive.

Table 1. Toxicity of undiluted storm water samples towards bacteria (*V. fischerii*), algae (*Pseudokirchneriella subcapitata*), and rotifers (*Brachionus calyciflorus*) (% inhibition)

Sample	Stockholm			Nantes			Luleå			Wuppertal		
	B	A	R	B	A	R	B	A	R	B	A	R
01	16	6	100	27	24	−7	36	1	−7	37	30	−13
02	9	53	59	86	19	−7	51	8	−9	37	−3	−34
03	30	11	−4	19	19	−4	NT	32	−39	19	8	−30
04	18	13	−24	82	21	33	21	3	3	35	−8	3
05	86	−13	−11	34	5	13	19	−13	−22	35	5	10
06	7	20	−14	54	14	21	–	–	–	42	−11	−23
07	32	−3	−43	63	18	−19	–	–	–	39	1	−41
08	58	100	100	18	5	3	–	–	–	18	−8	−41
09	57	99	100	NT	16	14	–	–	–	17	−17	−50
10	56	100	100	33	NT	−68	–	–	–	30	14	NT

B: bacteria; A: algae; R: rotifer; NT: not tested

Concentration–response relationships

In the algal and bacterial tests, the test design allowed for concentration–response curves to be established (it was not possible to adopt this approach using the standard rotifer methodology because the majority of undiluted samples showed a very low inhibition using this test). EC values were estimated using the Weibull model [11] to which the data was fitted using an approach developed by Andersen et al. [12]. The derived data is presented in Table 2 and again suggests that the Microtox test is the most responsive, with data permitting the calculation of EC_{10} values in the vast majority of tested samples in contrast to the algal data where it was possible to calculate EC_{10} values in less than 50% of analysed samples.

It is also interesting to note that, whilst it was not possible to calculate EC values for bacteria in samples 02 and 06 from Stockholm and sample 06 from Nantes, it was possible to calculate EC values for algae in these samples. This finding is of importance because if only bacterial tests had been used, the conclusion may have been that these samples did not exert an ecotoxicological impact. However, the additional use of algal tests indicates that this is not the case, demonstrating both the interspecific effect of storm water on different receiving organisms as well as clearly supporting the approach of using a battery of biotests to gain a more complete overview of the ecotoxicological impact of storm water on surface water quality. For sample 01 from Stockholm and samples 03, 09, and 10 from Wuppertal, the toxic responses in both algal and Microtox tests were too low to allow for the calculation of any EC values. As with the data reported on the impact of undiluted samples (Table 1), the EC value data indicates there is considerable variation between the different storm water samples both within a site and between sites, with the differences thought to represent factors such as different land use patterns, traffic densities, and antecedent weather conditions.

The presentation of an ecotoxic effect by an EC value is the standard approach to presenting toxicity data. It supports a more quantitative evaluation of the data than percentage inhibitions of undiluted samples (as presented in Table 1) because of the extrapolation to specified end points (e.g., EC_{20}) enabling the direct comparison of data. However, it should also be noted that protocols for the determination of ecotoxicity data have not yet been fully established with regard to the identification of "meaningful"

Table 2. Effect concentration (EC) values obtained for bacteria (*V. fischerii*) and algae (*Pseudokirchneriella subcapitata*) exposed to storm water samples from each site (mL sample per litre diluent)

Sample	Bacteria			Algae		
	EC_{10}	EC_{20}	EC_{50}	EC_{10}	EC_{20}	EC_{50}
STO/02	–	–	–	358	522	965
STO/03	307	617	>1000	1000	>1000	>1000
STO/04	570	775	>1000	–	–	–
STO/05	139	>1000	392	–	–	–
STO/06	–	–	–	327	1000	>1000
STO/07	393	626	>1000	–	–	–
STO/08	182	319	837	182	242	374
STO/09	229	358	770	137	196	333
STO/10	236	383	877	152	216	368
NAN/01	275	649	>1000	545	904	>1000
NAN/03	170	248	469	581	965	>1000
NAN/06	–	–	–	619	973	>1000
NAN/07	175	259	508	603	933	>1000
NAN/08	199	446	>1000	–	–	–
NAN/09	55	151	850	–	–	–
NAN/10	1	5	214	–	–	–
NAN/11	340	881	>1000	–	–	–
NAN/12	NT	NT	NT	–	–	–
NAN/13	41	220	>1000	–	–	–
LUL/01	281	523	>1000	–	–	–
LUL/02	114	254	1000	–	–	–
LUL/03	NT	NT	NT	390	579	>1000
LUL/04	418	964	>1000	–	–	–
LUL/05	397	>1000	>1000	–	–	–
WUP/01	208	397	>1000	406	749	>1000
WUP/02	272	515	>1000	–	–	–
WUP/05	380	607	>1000	–	–	–
WUP/06	214	476	>1000	–	–	–
WUP/07	246	435	>1000	–	–	–
WUP/08	172	390	>1000	–	–	–
WUP/11	42.3	64.5	>1000	–	–	–

–: Estimation of EC value not possible; NT: not tested; STO: Stockholm, NAN: Nantes; LUL: Luleå; WUP: Wuppertal

levels of toxicity, i.e., the effect concentration (EC) value at which concern should be expressed and/or action taken. There is still considerable discussion over the most appropriate EC value to use [e.g., 13] with various authors supporting the use of EC_{10} or EC_{20} values as quantitative measures of the lowest observed effect concentration (LOEC). This is an important decision, as it will influence the development of environmental quality standards under the WFD, and it is also a clear example of an issue which may be better supported and informed by the use of a combined, as opposed to an "either/or", ecotoxicity and chemical water quality approach.

Correlation between test organisms

To determine the existence of any relationships between the different tests, the responses of the test organisms to undiluted storm water samples were further analysed. The scatter plot shown in Figure 1 compares algae and rotifers and indicates a moderately strong positive correlation ($r = 0.759$, $p = 0.000$) between the responses of these different test organisms.

Samples 08, 09, and 10 from Stockholm show 100% inhibition for both tests. These samples were collected during the winter period when the

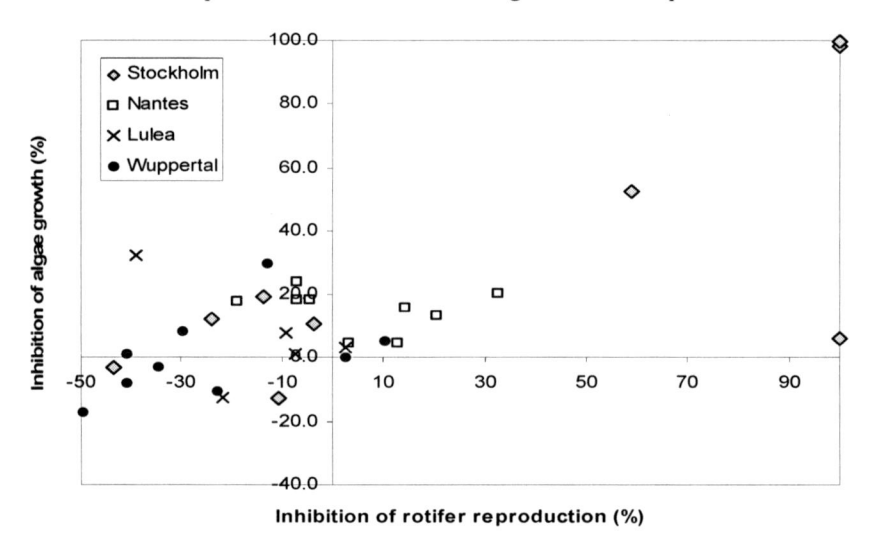

Fig. 1. Inhibition of rotifer growth versus inhibition of algal growth on exposure to undiluted storm water samples (% effect)

reported salinity was extremely high (37.6–39.3 mS cm^{-1} which is within the range of salinity reported for seawater) due to the high levels of de-icing salts used on the elevated motorway sections at this time of year. However, the bacteria utilized in Microtox is a marine bacteria and the light production in the bacteria tests was also notably reduced (56–58%) suggesting that the toxic effect of these samples is not entirely due to the high salt content.

As noted earlier, rotifer reproduction is stimulated in a large number of samples, i.e., it exceeds the rate associated with the blank sample. This also occurs with algae, but to a much more limited extent and the effect remains moderate (up to a value of 17%). This may be due to the presence of certain substances in the growth media in suboptimal concentrations which are enhanced by the presence of these substances in storm water in greater concentrations. In the case of rotifers, reproduction higher than that reported for the test blank was frequently observed. Although the reproducibility of the rotifer test did show considerable variation (8–22%; expressed as standard deviation divided by the average blank value), such a systematic shift of rotifer data to a negative response (i.e., stimulation) strongly suggests that the combination US Environmental Protection Agency (US-EPA) Moderately hard water and feed used in the tests was itself limiting rotifer reproduction. Due to these effects, the obtained blank results should be considered as a reference value only and not as the optimum growth reference value.

Scatter plots of rotifer and bacteria tests and of algal and bacteria tests (data not presented) did not indicate there were any correlations between the responses of these test organisms. Although the three biotests do not provide the same answers, confirming the need to use more than one type of test organism, the convergence of results between algae and rotifers demonstrates the existence of toxic effects in urban runoff samples.

Comparison of toxicity results with chemical analyses

The results of the chemical analyses carried out on the Stockholm samples are given in Table 3. All three test organisms showed a strong toxic response in relation to samples 08, 09, and 10 which, as described earlier, is thought to be associated with de-icing activities (as indicated by the elevated conductivity values reported in Table 3). Samples 01 and 02 contain

Table 3. Aqueous concentrations for a range of water quality parameters recorded during storm events, Stockholm site, Sweden

Sample	SS (mg L^{-1})	NO$_2$/NO$_3$ (µg L^{-1})	PO^{4-} (µg L^{-1})	Cond. (mS m^{-1})	Zn (µg L^{-1})	Cu (µg L^{-1})	Cd (µg L^{-1})	Pb (µg L^{-1})
01	13,000	110	66	145	12,000	1900	8.5	900
02	1750	540	12	23	3800	320	1.0	110
03	780	370	7	17	2400	170	1.1	120
04	210	690	15	31	500	52	0.3	44
05	150	213	6	23	450	62	0.3	26
06	17	250	12	46	66	16	0.2	4
07	210	17	66	28	3900	380	2.4	300
08	270	850	19	3910	180	62	0.5	10
09	190	240	23	3930	200	61	0.9	11
10	320	130	30	3760	410	95	0.9	14

elevated concentrations of several of the determined pollutants and both these samples were toxic towards rotifers. However, only sample 02 was noticeably toxic towards algae, with Microtox showing limited toxic response to either sample (Table 1). Sample 07 contained elevated levels of Zn, Cu, Cd, and Pb and this sample substantially inhibited bacterial light production (58%), whereas both algal growth and rotifer reproduction were stimulated (3% and 43%, respectively). Urban runoff is known to be a complex effluent with the potential for its constituents to interact in complex way which can be difficult to predict, and these data are considered to be indicative of such a scenario. The variable intraspecific response of these tests to the same samples provide further support for the need to use more than one test organism, with the use of three tests involving three different toxicological end points in this study clearly increasing the possibility of detecting a realistic toxic effect.

Conclusions

The described study has provided a substantial database relating to the ecotoxicity of storm water, with toxic effects being frequently detected. The potential ecotoxicological impact varies in relation to the test organism utilized, between sites as well as between storm events at a single site. In terms of frequency of detection, Microtox was the most responsive of the three tests utilized, with impacts being detected in all of the samples analysed. Microtox was also most responsive in terms of the quantitative detection of toxicity, providing viable EC_{10} values in 88% of samples tested as opposed to 34% of samples tested using algae. However, the level of concern which should be attached to EC values within the range reported using any of the test organisms has yet to be established.

The results of this study support the use of a "battery of biotests" as this approach resulted in a higher detection of toxicity than would have been achieved using a single test alone. The fact that the various chemical and ecotoxicological analyses highlighted potential concerns in relation to samples from different storm events at the Stockholm site provides further support for the use of a combined physico-chemical and ecotoxicological approach for evaluating the status of aquatic ecosystems.

Acknowledgements

The results presented in this publication have been obtained within the framework of the EC-funded research project DayWater "Adaptive Decision Support System for Stormwater Pollution Control", contract no. EVK1-CT-2002-00111, coordinated by Cereve at ENPC (F) and including Tauw BV (Tauw) (NL), Department of Water Environment Transport at Chalmers University of Technology (Chalmers) (SE), Environment and Resources DTU at Technical University of Denmark (DTU) (DK), Urban Pollution Research Centre at Middlesex University (MU) (UK), Department of Water Resources Hydraulic and Maritime Works at National Technical University of Athens (NTUA) (GR), DHI Hydroinform, a.s. (DHI HIF) (CZ), Ingenieurgesellschaft Professor Dr. Sieker GmbH (IPS) (D), Water Pollution Unit at Laboratoire Central des Ponts et Chaussées (LCPC) (F), and Division of Sanitary Engineering at Luleå University of Technology (LTU) (SE). This project is organized within the "Energy, Environment and Sustainable Development" Programme in the 5th Framework Programme for "Science Research and Technological Development" of the European Commission and is part of the CityNet Cluster, the network of European research projects on integrated urban water management.

References

1. EU Directive on Pollution Caused by Dangerous Substances Discharged in the Aquatic Environment on the Community-Dangerous Substances Directive. 76/464/EEC. 1976
2. Galassi S, Guezzella L, Mingazzini M, Viganó L, Capri S, Sora S (1992) Toxicological and chemical characterization of organic micropollutants in river po waters (Italy). Water Res 26:19–27
3. Hendriks AJ, Maas-Diepeveen JL, Noordsij A, Van der Gaag MA (1994) Monitoring response of XAD-concentrated water in the Rhine Delta. Water Res 28:581–598
4. Tonkes M, van de Guchte C, Botterweg J, De Zwart D, Hof M (1995) Monitoring water quality in the future. Vol. 4: monitoring strategies for complex mixtures. Ministry of Housing, Spatial Planning and the Environment, Zoetermeer, The Netherlands
5. Baun A, Ledin A, Reitzel LA, Jonsson S, Bjerg PL, Christensen TH (2004) Xenobiotic organic compounds in leachates from ten Danish MSW landfills – chemical analysis and toxicity tests. Water Res 38:3845–3858
6. Eriksson E, Baun A, Mikkelsen PS, Ledin A (2005) Chemical hazard identification and assessment tool for evaluation of stormwater priority pollutants. Water Sci Technol 51:47–55
7. EU Directive of the European Parliament and of the Council of 23 October 2000 establishing a Framework for Community Action in the Field of Water Policy. 2000/60/EC
8. Smoulders R, De Coen W, Blust R (2004) An ecologically relevant exposure assessment for a polluted river using an integrated multivariate PLS approach. Environ Pollut 132:245–263
9. International Organization for Standardization (ISO) (1989) Water quality – fresh water algal growth inhibition test with Scenedesmus subspicatus and Selenastrum capricornutum. ISO 8692. International Organization for Standardization, Geneva, Switzerland
10. AFNOR (2000) Détérmination de la toxicité chronique vis-à-vis de *Brachionus calyciflorus* en 48 h. NF T90-377. AFNOR, Paris, France
11. Christensen ER, Nyholm N (1984) Ecotoxicological assays with algae: Weibull dose-response curves. Environ Sci Technol 18:713–718
12. Andersen JS, Holst H, Spliid H, Andersen H, Baun A, Nyholm N (1998). Continuous ecotoxicological data evaluated relative to a control response. J Agric Biol Environ Stat 3:405–420
13. Chapman PM, Caldwell RS (1996) A warning: NOEC's are inappropriate for regulatory use. Environ Tox Chem 5:77–79

Is catchment imperviousness a good indicator of ecosystem health?

V Pettigrove

Melbourne Water, Research and Technology, PO Box 4342, Melbourne, Victoria 3001, Australia

Abstract

Reducing the impervious area (IA) of a watershed is becoming a common practice to mitigate the effects of urban runoff on the health of aquatic ecosystems. Yet, it is often argued that this approach is rather inefficient and it is debatable whether percentage IA targets can be applied to all urban environments, as local factors (e.g., soils, geology, terrain, and rainfall patterns) may influence the effect of imperviousness. Furthermore, different receiving waters may be affected by urban runoff in different ways: the effect of flows is likely to be strongest at the sub-watershed level, whereas quiescent waters are likely to be more affected by pollutants. A more effective approach is to develop an understanding of the effects of urban runoff on all receiving waters. This can be achieved by recognizing all the values in all receiving waters and the threat urban runoff poses to these values.

Introduction

A dominant feature of urbanization is a decrease in the perviousness of the catchment to precipitation, leading to a decrease in infiltration and an increase in surface runoff [1]. Increased imperviousness and the efficient drainage systems typically associated with urban areas result in a rapid transfer of storm waters into receiving waters. Consequently, flows in receiving waters receive a higher frequency and intensity of storm events. These altertions to stream hydrology affect stream geomorphology, leading to greater bed and bank erosion and the channel will tend to become wider and waters shallower [2].

Catchment imperviousness has commonly been portrayed as the dominant driver of degradation in urban streams, based on the widespread observation of a negative correlation between stream condition and imperviousness [e.g., 3–5]. The impervious area (IA) in a watershed is increasingly being used as a benchmark for determining the health of watersheds and has been popularized by the Centre for Watershed Protection [5]. The amount of catchment imperviousness has also been claimed to be a useful management tool for local planners and engineers to prevent cumulative impacts within sub-watersheds. In particular, it can be used to predict future stream quality, used to set realistic management goals, test whether existing zoning can support or maintain aquatic resources in the future, and assist in the development of future land use decisions and generally drive planners and engineers to reduce the creation of impervious at development sites [6]. Impervious percentages are attractive to regulators, as thresholds can be set for new developments [7].

Much of the literature on development impacts has focused almost exclusively on level of imperviousness (or directly connected imperviousness) of a watershed as the factor that indicates structural integrity of the stream or health of its aquatic species [8–10]. It has been argued that almost all storm water problems relate to increases in imperviousness from increased runoff of water and pollutants and reduced infiltration [11,12]. Therefore, some people have recommended that the primary means to protect or minimize impacts on receiving waters from urbanization is to reduce catchment imperviousness [5,6,12–18]. Some promoters of the impervious threshold theory assume that reducing imperviousness will maintain an acceptable hydrologic regime and maintain the ecological integrity of receiving waters [19]. Others assume that reducing imperviousness will also reduce the amount of pollutants being transported to receiving waters [6]. The significance of catchment imperviousness as a dominant factor influencing the health of receiving waters is reviewed to determine whether reducing IA is the most effective and efficient management tool to minimize effects of urban catchments on ecosystems in receiving waters. The significance of pollutants on urban aquatic ecosystems is considered as well as whether reducing IA is the most efficient and effective means of reducing pollutants. An alternative approach is proposed whereby a greater appreciation is required to understand the ecological values in receiving waters, the threats posed by urbanization and established urban areas on these values and what actions can be taken to reduce these potentia threats.

What is the maximum level of catchment imperviousness to protect aquatic ecosystems?

There has been considerable debate about what minimal level of catchment imperviousness leads to ecological impairment in receiving waters [10,12,17,20-22]. Paul and Meyer [23] and Beach [3] each present a table that summarizes key findings of various studies that found the rivers and streams become seriously impaired when the amount of impervious surfaces in the watershed exceeds 10%. The so-called 10% rule was popularized by the Centre for Watershed Protection [5] which stated that there was a direct link between IA and stream health. Supporters of the 10% rule believe that there is a threshold level of imperviousness that will result in a substantial deterioration in the ecosystem. If urban watersheds have an IA below 10%, receiving waters may have minimal impact from urbanization, but if this threshold is exceeded then ecosystem function usually declines because of individual and cumulative stresses. Others believe that there is no threshold of imperviousness, as any level of urban development will tend to produce additional aquatic system degradation [7].

The level of catchment imperviousness would categorize all watersheds correctly if it was the sole factor influencing stream health, but there are considerable exceptions to this rule [19], as some receiving waters appear to be more resilient to urbanized watersheds than others. For example, Pitt and Boseman [24] found biological effects at 5% urbanized catchment where there was no industry or municipal discharges. Horner and May [25] reported that 9 of 31 basins with IA >25% were ranked as good habitat, whereas it is predicted that where IA was >25%, the streams will be degraded having a low level of stream quality and an inability to support a rich aquatic community. On other occasions, geomorphic changes such as changes in channel width have been associated with IAs as low as 2–10% [1,10]. Reasons for this observed variability in effects of a certain level of catchment IA to receiving waters may be attributed to two factors. Firstly, some impervious surfaces in a watershed or attributes of the drainage system may have more impact on receiving waters than others. Therefore, it is important to understand the significance of these factors and develop better measurements of catchment imperviousness. Secondly, local factors may have a substantial influence on the effects of catchment imperviousness on receiving waters; if this is the case, no general rule of maximum percentage IA can be applied to all urban environments.

Factors influencing impervious area

The type of pervious surfaces and drainage systems and their proximity to receiving waters appear to influence the amount of runoff and impact on receiving waters. Estimating the amount of IA can be a difficult task. Generally, IAs include roofs, roads, sidewalks, and private driveways. Pervious surfaces may be affected by urbanization and contribute more to surface runoff (e.g., lawns, compacted soils), but are not accounted for in estimates of IA [26]. Not all IAs in a watershed are thought to have a similar impact on receiving waters. IAs nearest to receiving waters will have a greater influence than receiving waters further away [27]. The amount of IA directly connected to drains is a major source of runoff from small storms that may substantially impact receiving water bodies [28–30] and the hydraulic efficiency of these storm water drains is also important [17]. Categorizing of watersheds may be more effective if these factors are considered. For example, Walsh et al. [17] found that a strong threshold occurs between stream condition and effective catchment IA (the product of the proportion of impervious surfaces of a catchment directly connected to streams by storm water pipes by total imperviousness).

The utility of landscape indicators will depend on the response variable, the strength of the relationship between the indicator and the relative importance of other strong controlling variables [4]. The wide range of stream conditions observed in receiving waters with watersheds with low to moderate imperviousness may be due to local factors such as soils and surficial geology [31]; the spatial patterning of IA and other modified land cover on the landscape [e.g., 32]; developmental densities [26] and the effect of point sources of pollution and other human activities [33]. IAs are major sources of flow in small rainfall events (5–10 mm), whereas pervious areas may contribute the majority of runoff in larger storms [34]. In the south-eastern USA, pervious surfaces will dominate outfall discharges in low-density developments where large to moderate sized rains frequently occur. Conversely, impervious surfaces will dominate outfall discharges where most rains are relatively small (such as the arid USA) [34].

It is possible that local factors may have a considerable influence on the effects of watershed imperviousness within an urban area. For example, terrain, rainfall patterns, soils and geology, and natural river flows vary considerably within the Greater Melbourne Area (GMA), Australia. Streams in the western and northern areas of Melbourne are in low rainfall areas and drain basaltic plains, and upstream of the urban hinterland these streams are ephemeral. In contrast, streams in the eastern region of the GMA receive up to three times more rainfall than western streams, are permanently flowing systems and drain Silurian sedimentary soils [35].

These basaltic and sedimentary streams also have different water quality characteristics. Sedimentary streams generally have higher turbidity and suspended solids, lower conductivities, and lower pH than basaltic streams [35,36]. Basaltic stream sediments tend to be contaminated by heavy metals at much lower levels of catchment urbanization than sedimentary sediments. For example, zinc sediment concentrations were predicted to exceed the probable effects concentration (PEC), which indicates that there is a high probability of toxicity [37] at 11.4% IA in basaltic and 46.9% IA in sedimentary streams [35]. Bank erosion widely occurs in the erosive soils of sedimentary streams, whereas it appears to be comparatively less common in the rocky basaltic streams (personal observation). Further research is required to understand how watershed characteristics may influence the effects of imperviousness.

Relative significance of polluted urban runoff and modified hydrology

There has been considerable debate about whether modified flow regimes or polluted urban storm water is the first factor influencing receiving waters. The interest in isolating what factor is most important is based on the premise that efforts should be made to minimize the impact of the factor that deteriorates receiving waters at the lowest level of urbanization.

Few studies have considered the specific mechanisms leading to the observed effects of urbanization. This is a difficult task because of the multivariate nature of urban disturbance [23]. Assessment of the impact of storm water on a stream is complicated, as seldom does just one parameter produce an impact and impact is typically caused by the combination of physical alteration of habitat, chemical contamination, and the introduction of exotic organisms [38] and simple correlations with land-cover percentages can lead to incorrect interpretations [27]. There is compelling evidence indicating that the first factor associated with urbanization that impacts aquatic ecosystems is modifications to flow regimes [20,39–41]. If the primary aim of controlling development is to reduce IA, then it is assumed that modified flow regimes are the primary factor influencing receiving waters, although reducing IA will also reduce the effects of pollution [e.g., 16]. Yet, the chemical effects of urbanization are far more variable than hydrologic or geomorphological effects and depend on the extent and type of urbanization (residential versus commercial or industrial), presence of wastewater treatment plants, effluent and/or combined sewer overflows, the extent of storm water drainage [23], and the type of water body receiving the storm water.

The influence of impervious cover is strongest at the catchment or sub-watershed level (0.2–10 km^2) than in larger watersheds and basins, as flows become more attenuated in larger watersheds [6]. Receiving waters nearest the urban development (i.e., at the sub-watershed level) are usually the focus of storm water management controls as they potentially have the most visible impact from erosion and general habitat degradation. Any effects of toxicants at the sub-watershed level in a receiving stream are likely to be masked by decline in habitat quality caused by modified hydrology. However, the hydrologic effects of urbanization may abate in older streams. Sediment loads have been reported to decline in older streams [42], presumably when construction activities in the sub-watershed that create land disturbance decline, from increased paving of the sub-watershed and stabilization of earth, and when stream morphology has adjusted to the new flow regime created by increased imperviousness. Management actions, such as retarding basins and bed and bank stabilization works can reduce the time required for stream morphology to reach a new equilibrium with modified flow regimes from the urban watershed [43].

Urban runoff will affect different types of water bodies in different ways. Wetlands, lakes, estuaries, and quiescent areas of streams and rivers are prone to the effects of pollution from storm water. The growth of plants in estuarine systems is generally controlled by the amount of available nitrogen. Consequently, additional nitrogen from urban development can cause algal blooms [3]. Reduced water clarity can damage seagrass beds, coral reefs, and other critical habitats [3]. In the GMA, nitrogen pollution from catchment storm flows has been identified as the major threat to the health of Port Phillip Bay [44,45], whereas the loss of seagrass in the adjacent Western Port Bay is probably due to increased sedimentation [46]. Storm water runoff may also elevate nutrient concentrations in urban lakes and wetlands, causing algal blooms or excessive macrophyte growth [47].

Urban waters may be toxic [48]. For example, pesticides are present in all waters receiving urban runoff in the USA [e.g., 49] and are likely to be often toxic to aquatic life. Yet it appears that most biological effects associated with urban runoff are likely to be caused by polluted sediments affecting benthic organisms [34]. Many urban pollutants, such as heavy metals, petroleum hydrocarbons, and some pesticides, are hydrophobic and accumulate in sediments.

Fine (<64 μm) sediments collected from urban streams and wetlands in the GMA were toxic to indigenous aquatic macroinvertebrates [50] and would have imposed a long-term impact on the aquatic community. About 28% of urban aquatic ecosystems in the GMA may be impaired by long-chain high molecular weight petroleum hydrocarbon pollution [51]. During periods of dry weather, contaminants such as heavy metals and ammonia can be released from sediments as a result of degradation of organic materials and possibly oxygen depletion [52] and can be toxic to aquatic organisms [53–55]. The wide dispersal of contaminated sediment is difficult to remove and can cause significant detrimental effects on biological processes [34].

Wetlands and estuaries are highly prone to pollution effects [47]. In New Zealand, there is widespread exceedance of Pb, Zn, and organochlorines and sediments in the estuary of Auckland (New Zealand's largest city) appear to be toxic to aquatic life [56]. Estuaries in Sydney, Australia that are polluted with heavy metals had significantly different assemblages compared to non-polluted estuaries [57]. These contaminants may also be bioaccumulated through the food chain and pose a potential health risk to human consumers of seafood [e.g., 58]. An unpublished study conducted by Melbourne Water found that 47% of 114 wetlands and lakes in the GMA had heavy metal concentrations in sediments and/or porewaters that exceeded the PEC for sediment quality [37] or 95% trigger values for water quality [58], respectively. Sediments from five of these wetlands were examined to determine their effects on the aquatic ecosystem. Fine (<64 µm) sediments from all five wetlands were found to be toxic to indigenous aquatic macroinvertebrates [50]. Therefore, quiescent waters that receive urban runoff may be more prone to the effects of contaminants than modified urban flows.

Is reducing imperviousness the most effective way of treating toxicants?

Reducing catchment imperviousness can be achieved by various treatment technologies, including the use of dual purpose rainwater tanks on all properties, grassed swales along some sections of roads and streams, streetscape systems that infiltrate storm water into the underlying soil, and biofiltration basins near streams and wetlands [18]. Many of these treatments will be effective in removing some contaminants such as nitrogen and suspended solids [18]. Rapid small storms disperse the most highly concentrated contaminants (oil, metals, and other toxicants), particularly in the first flush [59]. Therefore, this approach would also seem to treat toxicants.

However, a disproportionate amount of pollutants are generated from various land use activities. Street runoff is often contaminated by heavy metals, mineral oils, and polycyclic aromatic hydrocarbons (PAHs) [60,61]. Parking lots and major streets cover 6% of the watershed, but have been reported to contribute about 25% metals, 64% of petroleum hydrocarbons (PAHs) produced in the watershed [62]. Highways may occupy only 5–8% of a catchment area, but can contribute 50% of suspended solids, 16% of hydrocarbons, and 35–75% of heavy metals [63].

Industrial estates have the potential for contributing very high concentrations of contaminants to storm water in the GMA [64], New Zealand [65], and in a county of Los Angeles, California [66]. In the GMA, one industrial estate on Darebin Creek substantially increased nickel and chromium concentrations in sediments downstream and a second industrial estate further downstream increased zinc and copper sediment concentrations [35]. Therefore, if heavy metal and/or petroleum hydrocarbon pollution is a threat to aquatic ecosystems, it would be effective to focus on appropriate treatment of runoff from industrial areas, parking lots, and major roads and highways.

Different treatment trains may be required to remove toxicants than if only treating flows was considered. Water quality can be improved by targeting removal of the finer particle fraction in storm water [67]. If sedimentation pretreatment was used before infiltration, then many of the pollutants (e.g., nitrates, malathion, atrazine, diazinon, chlordane, PAHs, enteroviruses, *Pseudomonas aeruginosa*, Ni, Cd, Cr, Pb, and Zn) will likely be removed before infiltration [34].

Setting management priorities, understanding catchment management goals

The main purpose of treating storm water is to reduce its adverse impacts on receiving water beneficial uses (e.g., flood prevention, protection of aquatic ecosystems, recreation, and water supply) [34]. When determining the most appropriate actions to treat urban runoff, it is important to understand what impact runoff (whether modified flows or pollution) has on all receiving waters. It is likely that urban runoff is likely to impact ecosystems in different types of receiving waters in different ways. Successful stream rehabilitation requires a shift from narrow analysis and management to integrated understanding of the links between human actions and changing river health. It requires coordinated diagnosis of the causes of degradation and integrative management to treat the range of ecological stressors with each urban area [68].

Urbanization does not affect all streams the same way and any effort to manage a specific stream must relate stream biological condition to specific human activities and their effects in that watershed [67]. An example of managing storm water to protect management goals occurs in Auckland, New Zealand. Zinc-contaminated sediments have been reported as toxic to aquatic life in estuaries receiving urban runoff from Auckland [56]. Roof runoff has been identified as a major source of Zn in New Zealand and source control options have been proposed to reduce Zn concentrations in receiving waters [65].

Several objectives exist for maintaining or enhancing aquatic ecosystems in the Melbourne region. Nitrogen pollution from catchment storm flows has been identified as the major threat to the health of Port Phillip Bay [44]. Catchment management activities in the Port Phillip Bay catchment have therefore aimed to reduce nitrogen loads. Ecological objectives have also been developed in freshwater streams for native species of fish, frogs, water rats, and riparian and aquatic vegetation in freshwater streams [45]. These objectives were developed by determining the ecological values and potential threats to these values in a particular stream [69]. Further actions are currently being prepared to strategically manage potential toxicants in watersheds.

One ecological objective is to have sustainable populations of platypus (*Ornithorhynchus anatinus*) in Melbourne's urban streams. Considerable research has been gathered to understand the ecology of the platypus and to identify the primary factors that restrict the distribution and abundance of platypus [70]. Platypus is affected by inadequate flows, the physical condition of streams, and sediment and water quality. Targeted management programmes have been successful in enabling platypus to return to some urban streams. Platypus returned to the lower reaches of the Yarra River, and Diamond and Mullum Mullum Creeks (the later site with about 44% IA), once appropriate habitat had been created by stream improvement works [70]. Sediment pollution appears to be a major factor preventing platypus from inhabiting many other urban streams. As toxicants in sediments can substantially reduce the amount of macroinvertebrates in sediments, and it is hypothesized that there is insufficient food available to support platypus [71]. Therefore, efforts to return platypus to these streams should focus on reducing levels of toxicants in sediments. In a broader sense, the management of potentially toxic contaminants in the GMA is essential to protect the ecosystems in many urban streams, wetlands, and estuaries.

Reducing catchment imperviousness will decrease the amount of heavy metals and long-chain petroleum hydrocarbons entering receiving waters, but other more cost-effective management options may be available. A more effective approach might be to identify the target pollutant, any point sources, and major land use activities generating this pollution. If management aims to reduce heavy metals, efforts should be made to reduce heavy metals from major sources (e.g., parking lots, industrial estates, and roads). Other actions to reduce heavy metals may be more cost-effective than reducing catchment imperviousness in these land use activities.

Pollution abatement at the source can be effective. Such actions include the introduction of lead-free gasoline, catalytic converters, and road furniture with low levels of zinc [72]. Therefore, indiscriminately reducing catchment imperviousness could be a wasteful approach in reducing heavy metals and long-chain petroleum hydrocarbons to achieve an ecological outcome of enabling platypus and other organisms to return to urban streams. We should know how effective each best management practice (BMP) is in mitigating various adverse impacts of urbanization on receiving waters, including their geomorphic stability and their aquatic ecology. Different BMPs are required for reducing changes in hydrology, reducing trash and debris, or reducing pollutants [73].

Conclusions

In theory, catchment imperviousness is an integrative measure of urban development. However, reducing catchment imperviousness is a rather blunt instrument that does not do justice to the complex, physical, chemical, and biological processes that must be considered to manage the ecological integrity of watersheds [19,33]. Urban stream managers should not blindly use the level of catchment imperviousness as a measure of ecosystem health. Informed management decisions to maintain or improve urban stream health can only occur through understanding the primary factors, whether physical or chemical, that impact stream health. Once this is known, a more focused and cost-effective strategy can be developed to protect these ecosystems.

Acknowledgements

Graham Rooney and Professor Ary Hoffmann for providing editorial comments to the manuscript.

References

1. Dunne T, Leopold LB (1978) Water in environmental planning. W.H. Freeman, San Francisco, CA, 818 pp
2. Booth DB (1991) Urbanization and the natural drainage system – impacts, solutions, and prognoses. Northwest Environ J 7:93–118

3. Beach D (2001) Coastal sprawl. The effects of urban design on aquatic organisms in the United States. Pews Ocean Commission, Arlington, VI, http://www.pewtrusts.org
4. Gergel SE, Turner MG, Miller JR, Melack JM, Stanley EH (2002) Landscape indicators of human impacts to riverine systems. Aquac Sci 64:188–128
5. Centre for Watershed Protection (2003) Impacts of impervious cover on aquatic ecosystems. Watershed Protection Research Monograph No. 1. Centre for Watershed Protection, Ellicott City, MD, USA
6. Scheuler T, Claytor R (1997) Impervious cover as a urban stream indicator and a watershed management tool. In: Roesner LA (ed.), Effects of watershed development and management on aquatic ecosystems: proceedings of an engineering foundation conference. Snowbird, Utah, USA. American Society for Civil Engineers, New York, pp 513–529
7. Booth DB, Hartley D, Jackson CR (2002) Forest cover, impervious-surface area, and the mitigation of stormwater impacts. J Am Water Resour Assoc 38:835–845
8. Scheuler T (1994) The importance of imperviousness. Watershed Protect Tech 1:100–111
9. May CW, Welch EB, Horner RR, Karr JR, Mar BW (1997) Quality indices for urbanization effects in Puget Sound Lowland streams. Water Resources Series Technical Report No. 154. Urban Water Resources Centre, Department of Civil Engineering, University of Washington, Seattle, Washington, DC
10. Booth DB, Jackson CR (1997) Urbanization of aquatic systems: degradation thresholds, stormwater detention, and the limits of mitigation. J Am Water Resour Assoc 33:1077–1090
11. Tourbier JT (1994) Open space through stormwater management: helping to structure growth on the urban fringe. J Soil Water Conserv 49:14–21
12. Wang L, Lyons J, Kanehl P, Bannerman R (2001) Impacts of urbanization on stream habitat and fish across multiple scales. Environ Manag 28:255–266
13. Sieker H, Klein M (1998) Best management practices for stormwater-runoff with alternative methods in a large urban catchment in Berlin, Germany. Water Sci Technol 38:91–97
14. Prince George's County Department of Environmental Resources (PGC) (2000) Low impact development design manual. Department of Environmental Resources, Prince George's County, MD, USA.
15. Zielinski J (2002) Open spaces and impervious surfaces: model development principles and benefits. In: France RL (ed.) Handbook of water sensitive planning and design. Lewis Publishers, Boca Raton, FL, pp 49–64
16. Walsh CJ (2004) Protection of in-stream biota from urban impacts: minimise catchment imperviousness or improve drainage design? Mar Freshwater Res 55:317–326
17. Walsh CJ, Fletcher TD, Ladson AR (2005) Stream restoration in urban catchments through redesigning stormwater systems: looking to the catchment to save the stream. J N Am Benthol Soc 24:690–705
18. Ladson AR, Lloyd S, Walsh C, Fletcher TD, Horton P (2006) Scenarios for redesigning an urban drainage system to reduce runoff frequency and restore

stream ecological condition. In: Delatic A, Fletcher T (eds.) Proceedings of the 7th international conference on urban drainage modelling and the 4th international conference on water sensitive urban design, Melbourne, Australia, V2.233–240
19. Coffman LS (2002) Low-impact development: an alternative stormwater management technology. In: France RL (ed.) Handbook of water sensitive planning and design. Lewis Publishers, Boca Raton, FL, pp 97–124
20. Horner RR, Booth DB, Azous A, May CW (1997) Watershed determinants of ecosystem functioning. In: Roesner LA (ed.) Effects of watershed development and management on aquatic ecosystems: proceedings of an engineering foundation conference. Snowbird, Utah, USA. American Society for Civil Engineers, New York, pp 251–274
21. Veni G (1999) A geomorphological strategy for conducting environmental impact assessments in karst areas. Geomorphology 31:151–180
22. Stepenuck KF, Crunkilton RL, Wang L (2002) Impacts of urban landuse on macroinvertebrate communities in southeastern Wisconsin streams. J Am Water Resour Assoc 38:1041–1051
23. Paul MJ, Meyer JL (2001) Streams in the urban landscape. Annu Rev Ecol Syst 32:333–365
24. Pitt R, Bozeman M (1982) Sources of urban runoff pollution and its effects on an urban creek. EPA-600/52-82-090. US Environment Protection Agency, Cincinnati, OH, USA, December
25. Horner RR, May CW (1999) Regional study supports natural land cover protection as leading best management practice for maintaining stream ecological integrity. In: Proceedings of comprehensive stormwater and aquatic ecosystem management 1st south Pacific conference, 22–26 February 1999, vol. 1, pp 233–247
26. Strecker EW (2001) Low impact development (LID): how low impact is it? Water Resour Impact 3:10–15
27. King RS, Baker ME, Whigham DF, Weller DE, Jordan TE, Kazyak PF, Hurd MK (2005) Spatial considerations for linking watershed land cover to ecological indicators in streams. Ecol Appl 15:137–153
28. Heany JP, Pitt R, Field R (1999) Summary and conclusions. In: Heany JP, Pitt R, Field R (eds.) Innovative urban wet-weather flow management systems. National risk management research laboratory, Office of Research and Development, US Environment Protection Agency, EPA/600/R-99/029. Chapter 12
29. Lee JG, Heany JP (2002) Directly connected impervious areas as major sources of urban stormwater quality problems-evidence from south Florida. 7th Biennial stormwater research and watershed management conference, 22–23 May, pp 45–54. http://www.stormwaterauthority.org
30. Sample DJ, Heaney JP, Wright LT, Fan CY, Lai FH, Field R (2003) Costs of best management practices and associated land for urban stormwater control. J Water Resour Plan Manag Jan/Feb:59–68
31. Booth DB, Haugerud RA, Troost KG (2003) Geology, watersheds and Puget Lowland Rivers. In: Montgomery DR, Bolton S, Booth DB, Wall L (eds.)

Restoration of Puget Sound Rivers. University of Washington Press, Seattle, Washington, DC, pp 14–45
32. Alberti M, Marzluff J, Shulenberger E, Bradley G, Ryan C, Zumbrunnen. (2003) Integrating humans into ecology: opportunities and challenges for studying urban ecosystems. BioScience 53:1169–1179
33. Karr JR, Chu EW (2000) Sustaining living rivers. Hydrobiologia 422:1–14
34. Pitt R (1999) Receiving water and other impacts. In: Heany JP, Pitt R, Field R (eds.), Innovative urban wet-weather flow management systems. National risk management research laboratory, Office of Research and Development, US EPA EPA/600/R-99/029. Chapter 4
35. Pettigrove V, Hoffmann A (2003) Impact of urbanization on heavy metal contamination in urban stream sediments: influence of catchment geology. Australas J Ecotoxic 9:119–128
36. Fletcher TD, Breen PF, Pettigrove VJ (1997) Influences of geology and land-use on surface water turbidity and suspended solids: implications for stormwater management. In: Proceedings of 24th hydrology and water resources symposium, Wai-Whenua, New Zealand, 24–28 November 1997, pp 385–391
37. MacDonald DD, Ingersoll CG, Berger TA (2000) Development and evaluation of consensus-based sediment quality guidelines for freshwater ecosystems. Arch Environ Con Tox 39:20–31
38. Herricks EE (2001) Through the pipe: down the creek! Water Resour Impact 3(6):24–27
39. Klein RD (1979) Urbanization and stream quality impairment. Water Res Bull 15:948–963
40. Booth DB, Reinelt LE (1993) Consequences of urbanization on aquatic systems – measured effects, degradation thresholds and corrective strategies. In: Proceedings of watersheds '93' conference, sponsored by US EPA, Alexandria, VI, 21–24 March, pp 545–550
41. Shaver E, Maxted J, Curtis G, Carter D (1995) Watershed protection using an integrated approach. In: Torno HC (ed.), Stormwater NPDES related monitoring needs. American Society of Civil Engineers, New York, pp 435–459
42. Finkenbine JK, Atwater JW, Mavinic DS (2000) Stream health after urbanization. J Am Water Resour Assoc 36:1149–1160
43. Heede BH (1986) Designing for dynamic equilibrium in streams. Water Res Bull 22:351–357
44. Commonwealth Scientific & Industrial Research Organisation (1996) Port Phillip Bay environmental study, 1992–1996. Final report. Dickson, ACT, Australia
45. Melbourne Water (2005) Melbourne Water annual report 2004–05. http://www.melbournewater.com.au
46. Environment Protection Authority (1996) The Western Port marine environment. Publication 493. April 1996.
47. Pouder N, France R (2002) Restoring and protecting a small, urban lake (Boston, Massachusetts). In: France RL (ed.), Handbook of water sensitive planning and design. Lewis Publishers, Boca Raton, FL, pp 317–339

48. Crunkilton R, Kleist J, Ramcheck J, DeVita J, Villeneueve W (1997) Assessment of the response of aquatic organisms to long-term in situ exposures of urban runoff. In: Roesner LA (ed.) Effects of watershed development and management of aquatic ecosystems. American Society of Civil Engineers, New York, pp 95–111
49. Schiff K, Sutula M (2004) Organophosphorus pesticides in storm-water runoff from southern California (USA). Environ Tox Chem 23:1815–1821
50. Pettigrove V, Hoffmann A (2005) A field-based microcosm method to assess the effects of polluted urban stream sediments on aquatic macroinvertebrates. Environ Tox Chem 24:170–180
51. Pettigrove V, Hoffmann A (2005). Effects of long-chain hydrocarbon-polluted sediment on freshwater macroinvertebrates. Environ ToxChem 24:2500–2508
52. Ellis JB, Hvitved-Jacobsen T (1996) Urban drainage impacts on receiving waters. J Hydrol Res 34:771–783.
53. Pratt JM, Coler RA, Godfrey PJ (1981) Ecological effects of urban stormwater runoff on benthic macroinvertebrates inhabiting the Green River, Massachusetts. Hydrobiologia 83:29–42
54. Mederios C, LeBlanc R, Coler RA (1983) An in situ assessment of the acute toxicity of urban runoff to benthic macroinvertebrates. Environ Tox Chem 2:119–126
55. Mayer T, Marsalek J, Delos Reyes E (1996) Nutrients and metal contaminants status of urban stormwater ponds. J Lake Reserv Manag 12:348–363.
56. Morrisey DJ, Roper DS, Williamson RB (1997) Biological effects of contaminants in sediments in urban estuaries. In: Roesner LA (ed.), Effects of watershed development and management on aquatic ecosystems: proceedings of an engineering foundation conference. Snowbird, Utah, USA. American Society for Civil Engineers, New York, pp 228–250
57. Stark JS (1998) Heavy metal pollution and macrobenthic assemblages in soft sediments in two Sydney estuaries. Aust J Mar Freshwater Res 49:533–540
58. Australian and New Zealand Environment & Conservation Council and Agriculture and Resource Management Council of Australia & New Zealand (2000) National water quality management strategy, Australian and New Zealand guidelines for fresh and marine water quality, vol. 1 – the guidelines
59. France RL, Craul P (2002) Retaining water: technical support for capturing parking lot runoff (Ithaca, New York). In: France RL (ed.) Handbook of water sensitive planning and design. Lewis Publishers, Boca Raton, FL, pp 175–202
60. Forster J (1993) The influence of atmospheric conditions and storm characteristics on roof runoff pollution with an experimental roof system. In: Proceedings of VIth international conference on urban storm drainage, Niagara Falls, Ontario, Canada, pp 411–416
61. Xanthopoulos C, Hahn HH (1990) Pollutants attached to particles from drainage areas. Sci Tot Environ 93:441–448
62. Steuer J, Slebig W, Hornewer N, Prey J (1997) Sources of contamination in an urban basin in Marquette, Michigan and an analysis of concentrations, loads, and data quality. US Geological Survey. Water Resources Investigation

Report 97-4242. Wisconsin Department of Natural Resources and US EPA, Middleton, WI
63. Ellis JB, Revitt DM (1991) Drainage for roads: control and treatment of highway runoff. Report NRA43804/MID.012. National Rivers Authority, Reading, UK
64. Pettigrove V, Hoffmann A (2003) Major sources of heavy metal pollution during base flows from sewered urban catchments in the City of Melbourne. Proceedings of third south Pacific conference on stormwater and aquatic resource protection combined with the 10th annual conference of the Australasian chapter of the International Erosion Control Association, Auckland, New Zealand, 14-16 May
65. Kingett Mitchell Ltd. (2003) A study of roof runoff quality in Auckland, New Zealand implications for stormwater management. Auckland Regional Council, Auckland, New Zealand
66. Lee H, Stenstrom MK (2005) Utility of stormwater monitoring. Water Environ Res 77:219-228
67. Tov P, Lee B, Tonto F (2006) Comparison of stormwater quality treatment guidelines and critical structural best management performance factors and parameters to consider. www.wateronline.com
68. Booth DB, Karr JR, Schauman S, Konrad CP, Morley SA, Larson MG, Burges SJ (2004) Reviving urban streams: land use, hydrology, behaviour, and human behaviour. J Am Water Resour Assoc 40:1351-1361
69. Coleman R, Pettigrove V (1999) Managing Melbourne's waterways: the Tributary Investigation Program. In: Proceedings of second Australian stream management conference, Adelaide, 1 February, 187-192
70. Pettigrove V (2000) A future for Melbourne's platypus. J Aust Water Assoc September 2000:51-54
71. Serena M, Pettigrove V (2005) Relationship of sediment toxicants and water quality to the distribution of platypus populations in urban streams. J N Am Benthol Soc 24:679-689
72. Berbee R, Rijs G, de Brouwer R, van Velzin L (1999) Characterization and treatment of runoff from highways in the Netherlands paved with impervious and pervious surfaces. Water Environ Res 71:183-190
73. Urbonas BR (2001) Our receiving waters with BMPs. Water Resour Impact 3:3-6

V. Storm Water Treatment

Evolution on pollutant removal efficiency in storm water ponds due to changes in pond morphology

TJR Pettersson,[1] D Lavieille[2]

[1] Department of Civil and Environmental Engineering, Water Environment Technology, Chalmers University of Technology, Göteborg, Sweden
[2] Department of Molecular and Macromolecular Photochemistry, Blaise Pascal University, Clermont Ferrand, France

Abstract

Ponds are frequently used to remove pollutants from urban runoff, but only a few accurate studies have been carried out to determine the long-term pollutant removal efficiency, and almost none on changes in removal over time. Removal efficiency will be affected by changes in pond morphology, vegetation growth, and sediment accumulation. This study presents the evolution of pollutant removal efficiency over a 7-year period.

The results showed that vegetation growth and increased sediment thickness affected copper, zinc, and nitrogen removal efficiency negatively. Concluding recommendations are removal of vegetation in the autumn and sediment removal after approximately 7–10 years in operation.

Introduction

Storm water contains several pollutants harmful to the environment and ecosystems. Many pollutants are attached to particles and can therefore be removed through sedimentation. The size of storm water particles may vary in the range from 0.004 to 2 mm [1]. The specific particle area is of importance for the heavy metal lead that is highly bound with fine sediments [2], which then might be easier to remove than, e.g., zinc, cadmium, and copper which have a larger dissolved fraction than lead in road runoff [3]. These heavy metals are of importance in our study because the catchment area is in part covered by highways.

During the last 25 years, there has been an increased interest in treating storm water by using ponds. Originally ponds were designed only to reduce flow peaks during heavy rain events; but since early studies showed that ponds in storm water systems improved the storm water quality, the scope of these measures has been extended and now mainly includes storm water quality improvements, through settling of particles and removal of attached pollutants. To improve the performance of the ponds, more focus has to be put on proper pond design [4].

In order to quantify the effectiveness of such measures, long-term studies need to be carried out over a large period of time to achieve accurate results and also include seasonal variations, different rain depths, dry period lengths, etc. [3]. Even though a great number of ponds have been built, very few have been evaluated to determine their pollutant removal efficiency. For the few ponds studied, there is a tendency to study ponds only right after or very close in time to when they were built. Ponds are rarely evaluated after being in use for several years and almost never studied at two different time periods.

Changes in pond morphology such as sediment accumulation can lead to resuspension of bottom sediments by physical or chemical processes [5–7], causing lower removal efficiency. Increased vegetation mass in the pond may affect the removal efficiency [8]. Vegetation can also be a nutrient source or sink, where seasonal variations may cause nutrient uptake during the growing season and nutrient leakage in the autumn and winter due to decaying plants [9].

Natural factors that may impact the removal efficiency are hydrological characteristics such as variations in dry periods and rain depths [10], seasonal variation [11], and vegetation growth in and around the pond [8].

This paper presents the results on long-term pollutant removal efficiency evolution for a storm water pond over a period of 7 years. The first study of that pond was carried out in 1997 and the current study in 2004. Suggestions for proper storm water pond management will be presented.

Experimental method

Pond and catchment area characteristics

The Järnbrott pond is a wet off-stream storm water treatment pond with a permanent pool and with no constant inflow (e.g., from creek or river) during dry periods. However, there is a small constant groundwater leaking inflow from the storm water pipe system.

The pond is situated 5 km from the city of Göteborg and was constructed in 1996 to reduce storm water pollution discharges to the nearby, and at that time heavily polluted, River Stora Ån [3].

The pond inlet consists of a 1000 mm steel pipe which is connected to an upstream overflow chamber that starts to discharge storm water directly into the river when the flow exceeds 700 L s^{-1}. Annual discharge of untreated storm water amounts to 20% of the total inflow per year [12]. Upstream of the overflow chamber, a storm water sewer system is connected to a catchment that has a total area of 480 ha, where nearly 160 ha consist of impervious surfaces such as highways and industrial, commercial, and residential areas. The highway has a traffic load of up to 6000 cars per day, which therefore is one of the largest contaminating sources. The pond outlet consists of a concrete crest, 8 m broad, discharging treated storm water to the River Stora Ån [3].

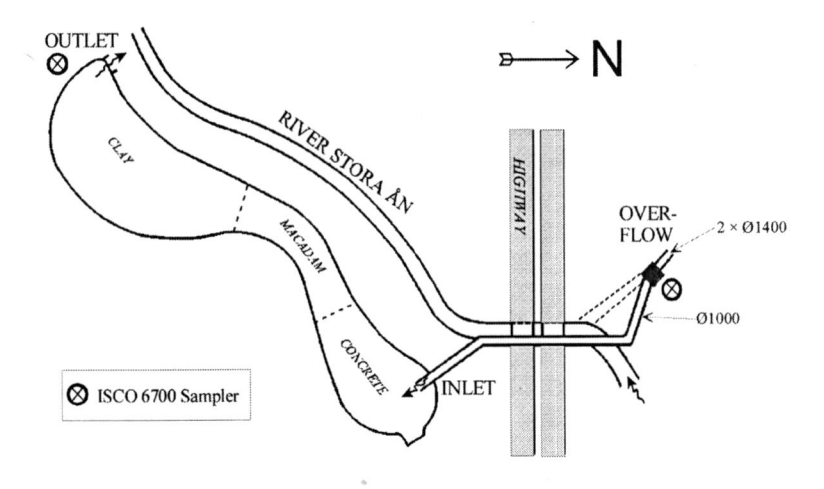

Fig. 1. Järnbrott storm water pond schematic, including sampler location

The pond is divided into three sections with different bottom materials (Fig. 1), due to local variations in soil strength. The inlet section has a dry weather depth of 1.5 m and consists of a concrete slab, to facilitate removal of sediments by using a wheel loader. In the middle section, the shallow bottom consists of penetration macadam that has a depth of only 0.6 m. Finally, the outlet section consists of clay and the depth is about 1.6 m. The pond slope is composed of clay and has a gradient of 30%.

Measuring equipment

In order to calculate the pollutant removal efficiency of a storm water pond, flow measurements and storm water sampling are necessary. Automatic samplers of model ISCO 6700 were installed at both the pond inlet and outlet samplers, the same as those used 7 years ago during the first measurement campaign carried out in this pond [3]. Each sampler has a capacity of twelve 900 mL bottles: six polyethylene and six glass bottles. The samplers were also equipped with flow meters, which monitored the flow and triggered flow-weighted sampling during the storm events. The inlet sampler had a velocity–height (V–H) probe installed (ISCO 750 area velocity module) at the bottom of the inlet pipe calculating the inflow from measured velocity and water level in the pipe. At the outlet sampler, a pressure probe (ISCO 720 submerged probe module) was installed measuring the outflow, which is calculated from the water surface level over the concrete crest. The samplers were programmed to take a subsample at every 800 m^3 of storm water passing the measuring point at the inlet and outlet, respectively. Each subsample was added to a set of two bottles (in total six sets), one plastic and one glass bottle, and to each set six subsamples were added, i.e., 4800 m^3 of storm water had passed for each filled set. All sampling and flow data were stored in the samplers' built-in data loggers.

Laboratory analyses

After each storm event the filled bottles were collected from the samplers and transported to the laboratory within 8 h after the storm. At the laboratory the bottles were mixed to a composite sample representing the event mean concentration (EMC) of the analysed pollutants for the inflow and the outflow samples, respectively. Samples from glass bottles and plastic bottles were kept separate, which made two identical composite samples per sampling location. The composite sample of the glass bottles was used for analyses of nitrogen and organic compounds, whereas the sample of the plastic bottles was used for the analyses of the remaining pollutants.

Total suspended solids (TSSs) and volatile suspended solids (VSSs) were analysed according to the Swedish standard method SS 02 81 12. Heavy metals (zinc, copper, lead, cadmium, nickel, and cobalt) were analysed using an inductively coupled plasma mass spectrometry (ICP-MS) instrument according to standard method SS-EN ISO 11885:1997. Nutrients have been analysed using Swedish standard methods. The nitrogen compounds included total nitrogen (total-N), nitrite (NO_2-N), nitrate (NO_3-N), and ammonium (NH_4-N); the phosphorus compounds included total phosphorus (total-P) and phosphate phosphorus (PO_4-P).

The polycyclic aromatic hydrocarbon (PAH) analysis of the composite water samples was carried out at a commercial laboratory, following the method NEN 6524 [13].

Data processing

A mass balance budget of total inflow and outflow storm water pollutant masses was compiled for each storm event. The pollutant masses were calculated using the recorded storm water volumes at the inlet and at the outlet by multiplying these with corresponding pollutant EMCs, as $M_i = V_i \cdot C_i$ (see principle in Fig. 2). The site mean concentration (SMC) for a specific pollutant represents the long-term mean concentration from several successive storm events at the studied site, and is calculated as the accumulated pollutant

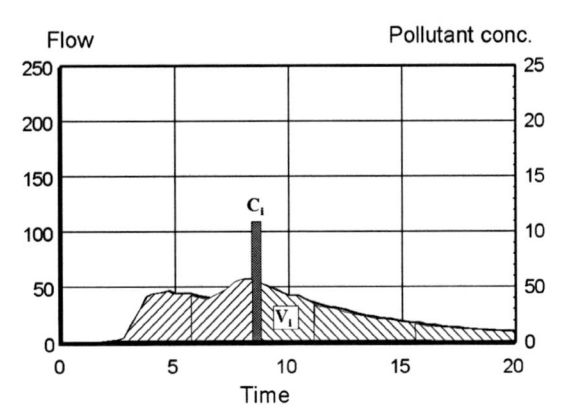

Fig. 2. Pollutant mass ($M_i = C_i \cdot V_i$) calculation principle for each storm event (where C_i is the pollutant EMC and V_i is the storm water volume of the event)

mass over accumulated storm water volume, such as SMC = $\Sigma M_i / \Sigma V_i$. Long-term removal efficiency *I* of the storm water pond for each pollutant was then calculated using the SMCs for the inflow and outflow storm water; see Eq. 1.

$$R(\%) = \frac{(SMC_{in} - SMC_{out})}{SMC_{in}} 100$$

Results and discussion

Results presented here are from the current study in 2004 [14], which are compared to the corresponding results obtained from the study carried out during 1997–1998 [3]. The latter study is divided into the two seasonal periods, autumn 1997 and summer 1998.

Storm events

In total, seven storm events have been monitored and analysed in this study during 2004 (Table 1). The measurement period lasted for 6 weeks, from 16 October to 24 November. For the last three storm events, the precipitations were a mixture of rain and snow. There was no heavy rain with large rain depth during the seven storm events with a maximum depth of 12 mm Table 1).

Table 1. Periods of long-term field studies carried out in the Järnbrott pond and corresponding rain characteristics

Study year	Observation period	No. of storm events	Rain depths (mm)	Dry periods (days)	Storm water volumes (m³)
2004	Oct–Nov 2004	7	4–12	1–8	5000–30,000
1997[a]	Aug 1997–Feb 1998	16	0.2–90	0.2–28	540–63,000
1998[a]	Apr 1998–July 1998	17	0.2–35	0.9–5.4	1200–50,000

[a]Results from the study carried out by Pettersson [3]

Pollutant removal efficiency

From the seven studied storm events in 2004, the pollutant removal efficiency was determined (Eq. 1) using the calculated SMCs for each pollutant as presented in Table 2. It is seen that the Järnbrott pond, in 2004, removes particulate matter (PM) effectively, especially TSS (62%), VSS (50%), and PAH (70%). Good removal efficiency (about 45%) was also obtained for the metals lead and cobalt, and for lead this is not surprising since lead is known as a highly particulate bound metal [2,3]. The other metals studied show low or even negative (copper and nickel) removal efficiency. For nutrient compounds it is found that nitrogen was not removed (–52%) but produced, and that phosphorus exhibited moderate removal (33%) in the 2004 study. The only nitrogen compound that has not increased is NH_4, which indicates that the nitrification process works properly in the pond.

Evolution of the long-term removal efficiency of the Järnbrott pond from 1997 to 2004

The main objective of this study was to compare changes in long-term removal efficiency over a period of 7 years. The measurements in the study 2004 were carried out in the late autumn and the comparison is made to the corresponding period in the autumn/winter in 1997.

No changes in the pond's catchment area have been identified during the period 1997–2004. The vegetation in the pond has grown substantially and the sediment thickness has increased, by 15 cm in the inlet section and by 20 cm in the outlet section, over this period. By comparing the pollutant removal efficiency of the two field measurement campaigns from 1997 and 2004, the effects of pond morphology changes will be made evident.

Comparison of removal efficiencies in 1997 and 2004

When comparing long-term removal efficiencies obtained in 2004 with the ones obtained 7 years ago, it appears that some parameters are negatively affected by the changes in pond morphology, whereas others do not seem to be affected (Table 2).

Lead and suspended matter removal efficiencies are quite similar in both studies, and the removal of cadmium is halved for 2004. The removal of copper, zinc, PO_4-P, and total-N is significantly lower in 2004 compared to 1997. Copper has changed from about 30% removal in 1997 to be negative (–12%) in 2004, and total-N has become even more negative (producing total-N), from –7% (1997) to –52% (2004).

Table 2. SMCs and removal efficiency for studied pollutants during current study 2004 and the study in 1997 and 1998

Parameters	Autumn 2004 SMC In	Autumn 2004 SMC Out	Autumn 2004 R (%)	Autumn 1997[a] SMC In	Autumn 1997[a] SMC Out	Autumn 1997[a] R (%)	Summer 1998[a] SMC In	Summer 1998[a] SMC Out	Summer 1998[a] R (%)
TSS (mg L^{-1})	38	14	62	52	20	61	58	13	77
VSS (mg L^{-1})	14	7	50	15	8	44	17	5	73
NO_2-N (mg L^{-1})	0.014	0.023	−60	n/a	n/a	n/a	n/a	n/a	n/a
NO_3-N (mg L^{-1})	0.77	1.7	−124	n/a	n/a	n/a	n/a	n/a	n/a
NH_4-N (mg L^{-1})	0.13	0.12	9	n/a	n/a	n/a	n/a	n/a	n/a
Total-N (mg L^{-1})	1.7	2.6	−52	2.3	2.5	−7	1.7	1.3	25
PO_4-P (mg L^{-1})	0.06	0.05	13	0.08	0.06	29	0.06	0.03	53
Total-P (mg L^{-1})	0.14	0.09	33	n/a	n/a	n/a	n/a	n/a	n/a
Cadmium (µg L^{-1})	0.35	0.31	10	0.51	0.41	20	0.57	0.55	4
Cobalt (µg L^{-1})	1.5	0.8	46	n/a	n/a	n/a	n/a	n/a	n/a
Chromium (µg L^{-1})	9.5	7.4	22	n/a	n/a	n/a	n/a	n/a	n/a
Copper (µg L^{-1})	190	210	−12	50	33	33	56	41	27
Lead (µg L^{-1})	8.3	4.5	46	15	8.3	45	11	5.3	53
Zinc (µg L^{-1})	200	170	15	121	80	34	116	85	27
Nickel (µg L^{-1})	12	14	−15	n/a	n/a	n/a	n/a	n/a	n/a
PAH-16 (µg L^{-1})	0.7	0.2	70	n/a	n/a	n/a	n/a	n/a	n/a

n/a: not analysed
[a]Pollutant SMCs and removal efficiencies for the Järnbrott pond obtained in the study 7 years earlier than current study [3]

Despite an increase in the sediments' thickness from 1997 to 2004, the removal of suspended matter is not affected. A hypothesis would suggest that the increasing amount of vegetation in the pond enhanced the removal of suspended particles by vegetation filtration [8] and decreased resuspension [15].

A laboratory sediment experiment [14] showed a release of nitrogen compounds from the Järnbrott pond sediment to the water phase. This phenomenon, together with increased decaying vegetation, can explain the highly negative removal efficiency for total nitrogen achieved in 2004, when sediments' thickness and vegetation have significantly increased in comparison with results in 1997. The same experiment also showed a release of copper and zinc, which may explain the significantly decreased removal efficiency for these two metals. On the contrary PO_4-P was not released from the sediments to the water phase, which indicates the origin of this pollutant, and the decreased removal efficiency from 1997 to 2004 probably can be explained by vegetation decay during the autumn and winter period.

If we compare changes in SMCs (Table 2) for the studied pollutants from 1997 levels to 2004 levels, it appears that both the inflow and outflow SMCs for TSS are slightly increased, while the SMCs for VSS are almost unchanged. Total nitrogen has a decreased inflow SMC and an increased outflow SMC causing the significant decrease in removal efficiency. The copper SMC in the inflow storm water has increased to nearly four times as high, from 50 µg L^{-1} in 1997 to 190 µg L^{-1} in 2004, but for the outflow SMC the corresponding change is to more than six times as high, from 33 to 210 µg L^{-1}, which probably can explain the dramatic decrease in removal efficiency as a considerable leakage from the bottom sediment. Zinc exhibits the same tendency as for copper. These findings indicate that the sediment layer has reached a level (thickness) when it needs to be removed from the pond to restore the designed pollutant removal efficiency.

Seasonal variations

In the study from 1997 to 1998 we can see that the seasonal variations in SMCs impacted the pollutant removal efficiency; see Table 2.

Particulate pollutants, such as TSS, VSS, and lead, exhibited increased removal efficiency during the summer period compared to the autumn/winter period, where VSS shows the greatest increase (from 44% up to

73%) and where TSS and lead show an increase, but less pronounced. For the nutrients, total nitrogen shows dramatic increase in removal efficiency, from −7% in the autumn/winter period up to 25% during the summer period. This is also evident for PO$_4$-P, but not as pronounced, which again indicates a great impact of the vegetation, with decay in the winter releasing nutrients in the water phase and uptake in the summer period where the vegetation acts as a nutrient sink [9]. For the remaining heavy metal elements there are no significant seasonal variations, but an opposite trend with a slight increase from the summer to the autumn/winter period can be identified (Table 2).

Conclusions

This study has shown that the Järnbrott pond does not remove all pollutants to the same extent. It is particularly effective in removing particulate pollutants such as TSS, VSS, lead, and PAH.

The main objective of this study was to examine the pollutant removal efficiency evolution over a period of 7 years, and the conclusions are that changes in pond morphology affected the removal negatively for copper, zinc, PO$_4$-P, and total-N, while the removal of TSS, VSS, and lead were not significantly affected. One possible explanation for the decreased removal efficiency, for those pollutants, is the increased sediment thickness in the pond, 15–20 cm during this period, and thus amplified leakage of pollutants out into the water column, which also was confirmed by elevated SMC's in the outflow storm water. This leads to the conclusion that the pond sediment after 8 years reached a level when it needs to be removed to restore the designed removal efficiency for the most negatively affected pollutants.

Finally, the study showed that seasonal variations in removal efficiency are mainly affecting the nutrients negatively. This favours harvesting the vegetation in storm water ponds before the autumn (in the northern hemisphere) to avoid the current production of nitrogen compounds and significant decrease in removal efficiency of PO$_4$-P.

References

1. Greb SR, Bannerman RT (1997) Influence of particle size on wet pond effectiveness. Water Environ Res 69:1134
2. Nascimento NO, Ellis JB, Baptista MB, Deutsch J-C (1999) Using detention basins: operational experience and lessons. Urban Water 1:113–124

3. Pettersson TJR (1999) Stormwater ponds for pollution reduction. Ph.D. thesis no. 14, Department of Sanitary Engineering, Chalmers University of Technology, Göteborg, Sweden
4. Persson, J. (2000) The hydraulic performance of ponds of various layouts. J Urban Water 2:243–250
5. German J (2001) Stormwater sediments, removal and characteristics. Licentiate thesis, Department of Water Environment Transport, Chalmers University of Technology, Göteborg, Sweden
6. Karouna-Renier NK, Sparling DW (2000) Relationships between ambient geochemistry, watershed land-use and trace metal concentrations in aquatic invertebrates living in stormwater treatment ponds. Environ Pollut 112:183–192
7. Marsalek J, Marsalek PM (1997) Characteristics of sediments from stormwater management pond. Water Sci Technol 36:117–122
8. Palmer MR, Nepf HM, Pettersson TJR, Ackerman JD (2004) Observations of particle capture on a cylindrical collector: implications for particle accumulation and removal in aquatic systems. Limnol Oceanogr 49:76–85
9. Granéli W, Weisner SEB, Sytsma MD (1992) Rhyzome dynamics and resource storage in Pragmites australis. Wetlands Ecol Manag 1:239–247
10. Pettersson TJR, Svensson G (1998) Particle removal in detention ponds modelled for one year of successive rain events. Proceedings of Novatech 1998, 3rd international conference on innovative technologies in urban storm drainage, vol. 1, pp 567–574
11. Sansalone JJ, Buchberger SG, Koechling MT (1995) Correlations between heavy metals and suspended solids in highway runoff: implications for control strategies. Department of Civil and Environmental Engineering, University of Cincinnati, Cincinnati, Ohio
12. Månsson A. (1998). Modelling of Urban Runoff with MOUSE. Master thesis 1998:1, Department of Sanitary Engineering, Chalmers University of Technology, Göteborg, Sweden
13. NEN (1984) Water – determination of the content of six polycyclic aromatic hydrocarbons (PAH) by high-pressure liquid chromatography. Document no. NEN 6524:1984 nl
14. Lavieille D (2005) Järnbrott Stormwater Pond – evolution of the pollutant removal efficiency and release from sediments. Master thesis 2005:46, Department of Civil and Environmental Engineering, Chalmers University of Technology, Göteborg, Sweden
15. Lopez F, Garzia M (1998) Open-channel flow through simulated vegetation: suspended sediment transport modelling. Water Resour Res 34:2341–2392

Characterization of road runoff and innovative treatment technologies

M Boller, S Langbein, M Steiner

Swiss Institute of Aquatic Science and Technology, Eawag, 8600 Duebendorf, Switzerland

Abstract

The road runoff of a sewered section of a road in Switzerland with a traffic intensity of 17,000 vehicles per day was intensely investigated over a period of 2 years. At the same time the road runoff was collected and pumped to a pilot plant for road runoff treatment. The treatment facilities consisted on one hand of shafts containing removable fleece filter bags and underlying GfeH-adsorber filters and on the other rotating fleece filter drums and subsequent adsorbers. It could be shown that the investigated processes for road runoff treatment can be operated as modular systems in one-, two-, or three-step flow schemes. Depending on the process scheme between 70% and 97% of the heavy metals could be removed.

Introduction

Runoff from roads is nowadays one of the most important sources of diffuse pollution of surface waters, sediments, and soils. According to a study in Germany by Hillenbrand et al. [1], diffuse emissions of heavy metals contribute to the total emissions by 84%, 80%, and 76% for lead (Pb), Zinc (Zn), and Copper (Cu), respectively. Motorized traffic may be responsible for 30%, 25%, and 36% of the diffuse loads for the respective metal. From these emissions, only a certain fraction is finally transported through runoff water. The fraction depends strongly on local conditions such as particle retaining construction elements along the roadside. In a mass balance study of a road shoulder, only 25–30% of the emissions were found in the runoff water [2]. In paved urban areas, the runoff fraction may rise to 45–70% of traffic emissions (this study). An urban catchment area study in Switzerland showed that motorized traffic may account for 45%, 35%, and 20% of the heavy metals Pb, Zn, and Cu contained in domestic sewage [3].

Nowadays, road runoff is often directly discharged to receiving water or infiltrated over the road shoulder. Direct discharge of the runoff into small receiving waters may cause ecotoxic shock loads of a mixture of substances during limited time (pulse pollution). If the runoff is infiltrated into the underground, e.g., into the road shoulder, especially heavy metals and polycyclic aromatic hydrocarbons (PAHs) are subject to rapid accumulation in the top soil. In order to gain improved control on diffuse pollution, abatement of the use of certain materials and substances (source control) is certainly the best way on a long-term basis. Although source control measures should be prioritized, there will be a need for efficient barrier systems in near future to overcome undesirable and uncontrolled spreading and accumulation of hazardous substances along roads and highways. In principle, there are two ways of applying barrier systems for pollutants in road runoff: (1) construction of new types of road shoulders which allow for the retention of the pollutants in a decentralized way and (2) treatment of sewered runoff in special installations along roads and highways. In the latter case different approaches have been adopted in the last years showing pond and retention filter systems with large detention times close to nature on one hand and smaller technical high rate facilities on the other.

Based on detailed studies concerning important characteristics of road runoff, this investigation shows new innovative possibilities of technical solutions to treat surface runoff from roads focusing on the removal of heavy metals and PAHs. The proposed technology may be applied in decentralized and centralized arrangements. It is important to notice that the major hazards in road runoff are mainly bound to particulate matter (PM) [4]. Therefore, primary issue is the retention of particles down to the colloidal size range. In cases of stringent effluent requirements, further removal of the smaller dissolved heavy-metal fractions to a very high percentage has shown to be feasible with a combination of textile filters and granulated iron hydroxide adsorbers (GfeH).

Experimental method

The road runoff of a sewered section of a road in Burgdorf BE with a traffic intensity of 17,000 vehicles per day was intensely investigated over a period of 2 years. At the same time the road runoff was collected and pumped to a pilot plant for road runoff treatment. The pilot plant served to test various technical solutions leading to innovative and sustainable measures in road construction and runoff management.

On one hand, the road runoff was continuously sampled on a long-term basis. Composite samples of 24 months were collected and analysed. The analysis of numerous water quality parameters such as TSS, DOC, P, and N and hazardous substances like various heavy metals allowed characterization of the average pollutant load contained in the runoff of this particulate road. On the other hand, detailed and extended analysis of the road runoff in the course of seven rain events allowed insight into the wash-off and first-flush behaviour of the analysed substances including also PAH and MTBE.

The road runoff from the road section with a surface area of 1500 m^2 was collected in a shaft and during rain events pumped continuously at a rate of 1.3 L s^{-1} to the nearby pilot facilities. In total, the pilot plants were operated in five experimental phases with different process schemes for runoff treatment:

1. Fleece filtration combined with granulated iron hydroxide adsorption (GfeH) filter
2. Sedimentation combined with fleece filter and GfeH adsorption filter
3. Polfabric drum filter combined with fleece filtration
4. Polymer flocculation combined with polfabric drum filter and filter fleece filtration
5. Polfabric drum filter combined with filter fleece filtration and GfeH adsorption filter.

Each of the experimental set-ups were operated over a period of at lest 4 months.

In the first experimental series the wastewater was treated in three parallel filter columns with a diameter of 1.0 m containing fleece filter bags of 0.8 m height (Feld AGEO, Germany) and in two columns combined with underlying GfeH layers of 30 and 10 cm with a grain size of 0.5–1 mm (GEH, Osnabrück, Germany), respectively. The used textile filter fleece bags consisted of a two-layer arrangement of polypropylene and polyester fibres with a total thickness of about 4 mm. In a later stage of the experiments, a sedimentation basin with a surface load of 0.65 m h^{-1} was installed as pretreatment for the filter bags. In the following experiments a rotating drum filter of 1.2 m diameter and a surface area 1.17 m² (Mecana TF 117, Polstoff) with automatic backwash was applied for solids separation. The drum filter was tested with two types of polfabric textile which consisted of hair-like Pol fibres of 20 µm and 7.5 µm fibre thickness, respectively.

Results and discussion

Long-term runoff characteristics

The average concentrations of the analysed quality parameters in the road runoff are given in Table 1 together with the standard deviations.

In the course of the 2 years investigation, the concentrations of heavy metals ranged in a relatively narrow characteristic band width which is shown in Figure 1 for the metals iron, zinc, copper, lead, nickel, and chromium. In contrary to the concentrations, the respective loads varied considerably according to runoff fluctuations (Fig. 2).

From the pollutant loads in Figure 2, the surface specific mass emissions over the whole period of 2 years were determined for the different

Table 1. Average concentrations and standard deviation in composite samples

	Units	Average	Standard deviation
TSS	mg L^{-1}	100.3	62.30
Calcium (Ca)	mg L^{-1}	26.6	10.70
Phosphorus (P)	mg L^{-1}	0.3	0.13
Nitrogen (N)	mg L^{-1}	2.3	1.00
Iron (Fe)	mg L^{-1}	2.9	1.60
Cadmium (Cd)	µg L^{-1}	0.8	0.30
Chromium (Cr)	µg L^{-1}	10.2	6.00
Copper (Cu)	µg L^{-1}	56.5	31.00
Nickel (Ni)	µg L^{-1}	7.0	3.30
Lead (Pb)	µg L^{-1}	23.1	13.00
Zinc (Zn)	µg L^{-1}	299.0	172.00

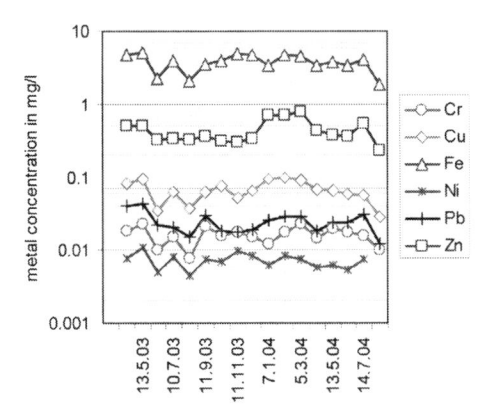

Fig. 1. Time history of heavy-metal concentrations in road runoff over a period of 2 years

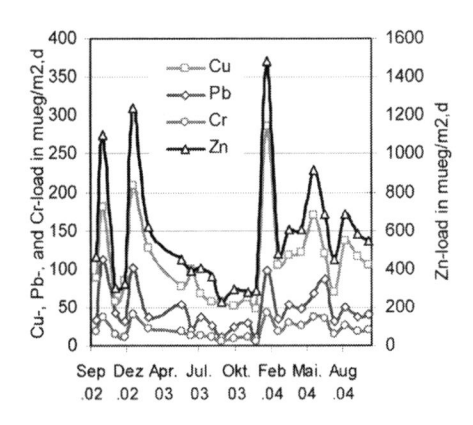

Fig. 2. Time history of heavy-metal loads in road runoff over a period of 2 years

parameters. In Table 2, these values in kilograms per hactare road surface and year are summarized and compared to results from other investigations in Switzerland (highway) and Germany (minor road) [4]. An estimate of the loads contained in road runoff compared to the total emissions revealed that in this study with a high percentage of impermeable area about 52%, 45%, 55%, and 68% of Cd, Cr, Pb, and Zn are flushed off with the runoff water.

Table 2. Surface specific pollutant emission in runoff water from different roads

Parameter	Units	Road Burgdorf 17,000 vpd 7.5 m wide	Hwy. Winterthur 26,000 vpd 25 m wide	Road Augsburg 6800 vpd 7.5 m wide
Cd	g ha^{-1} a^{-1}	4.0	30	2.7
Cr	g ha^{-1} a^{-1}	119.0	–	–
Cu	g ha^{-1} a^{-1}	471.0	1070	293.0
Pb	g ha^{-1} a^{-1}	179.0	560	122.0
Zn	g ha^{-1} a^{-1}	3231.0	1990	3400*

vpd = vehicles per day
*Zn emission by crash barrier

Dynamic runoff behaviour

Seven rain events of very different characteristics were investigated with respect to rain intensity and consequent runoff behaviour including the analysis of numerous parameters in the course of rainfall. Since the samples were fresh and immediately analysed, additional pollutants such as the 16 EPA-PAH, MTBE, hydrocarbons, TOC, DOC, and the dissolved fractions (0.45 µm filtered) of heavy metals were measured.

Considerable concentration and load variations were observed between the different events, as well as within the rain events according to rain intensity, length of preceding dry period, substance characteristics, etc. Typical first-flush behaviour could be observed for all parameters during rainfall with higher intensity, especially pronounced for PM and total metals. Figure 3 shows an example of dynamic pollutant wash-off of selected parameters during a rain event with a maximum intensity of 83 L s^{-1} ha^{-1} corresponding to a return period of approximately 0.5 years. Obviously, road wash-off during rainfall only occurs at certain minimum rain intensity. Figure 4 illustrates the correlation between maximum intensity and the total load of washed off pollutants. It can be concluded from the data that a rain intensity of at least 50 L s^{-1} ha^{-1} is necessary in order to mobilize PM and substances bound to particles on the road surface. In addition, it was shown that duration of the preceding dry period has an effect on the amount of pollutant wash-off.

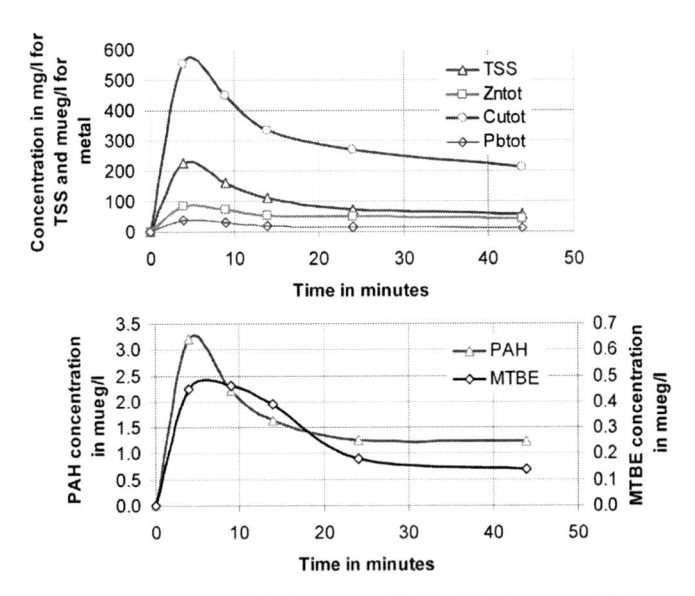

Fig. 3. Typical first-flush runoff behaviour of TSS, Pb, Cu, Zn, PAH, and MTBE during a rain event with a return period of about 0.5 years

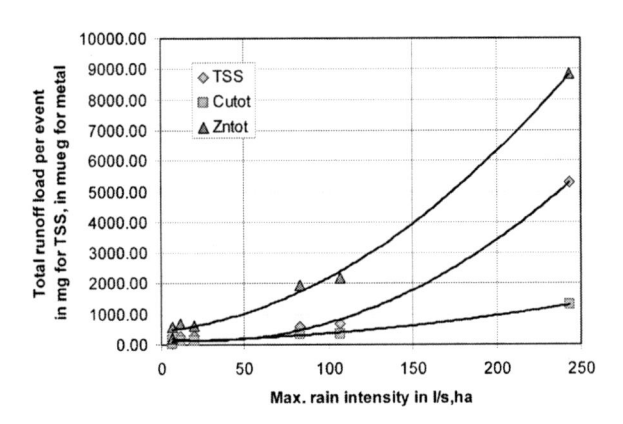

Fig. 4. Runoff load of TSS, Cu, and Zn as a function of maximum rain intensity

From TSS, total and dissolved metals, the particle bound fraction of metals, and the particles content of heavy metals could be evaluated. Table 3 shows a summary of the results. As can be seen, heavy metals are present to more than 80% in particulate form.

Table 3. Particulate fraction of calcium and heavy metals and content of metals in suspended solids

	Units	Ca	Cu	Cr	Fe	Zn
Percentage particulate	%	28.0	82.00	81.00	97.0	86.00
Content in particles	mg g^{-1}	64.2	0.43	0.07	26.2	2.07

Treatment of road runoff

Since the major pollutants of road runoff are present in high fractions of PM, emphasis was put on the application of solids separation processes as primary treatment task. Studies in Germany revealed that particles contained in road runoff may be captured in special fleece material leading to appreciable removal rates for heavy metals and PAH [5]. Although particle separation may lead to appreciable pollutant removal, smaller fractions of dissolved and fine colloidal heavy metals will remain in the effluent. In order to attain higher degree of pollutant removal, particle separation may be combined with adsorber systems enabling also the retention of dissolved substances. In several studies the application of GfeH has been shown to perform extremely well as an adsorber material especially for the removal of heavy metals [6].

The results of the pilot tests in the different phases may be summarized in three different modular systems: (1) one step treatment with solids separation; (2) two-step treatment with combinations of different solid separation processes; and (3) solids separation combined with GfeH adsorber filters.

One step solids separation was performed with fleece bags, coarse, and fine pol fibres, and sedimentation. From the results in Figure 5, it is clearly visible that the fleece bags with 80% TSS removal and 60% to more than 70% heavy-metal removal showed best performance, followed by the fine polfabric, the coarse polfabric and finally sedimentation. Particle size distribution analysis revealed that considerable fractions of road particles are in the colloidal and nanosize range and may not be settled within conventionally designed settling tanks. Therefore, ordinary sedimentation with removal rates below 30% is not considered to be an appropriate treatment for road runoff. Larger pond type systems would be required for improved settling performance. Interesting is the removal of Zn which showed poor removal for all systems except for the fleece bags. It is hypothesized that Zn is present in very fine colloidal or dissolved form and penetrates the applied fleece textiles to a certain extent. In the fleece bags, however, a build-up of sludge on the fleece surface was observed which acted as excellent adsorber

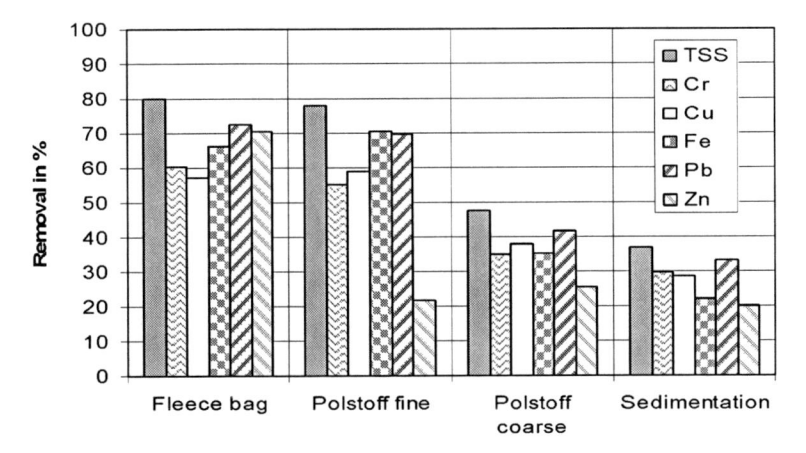

Fig. 5. Removal rates in one-step treatment applying different solids separation processes

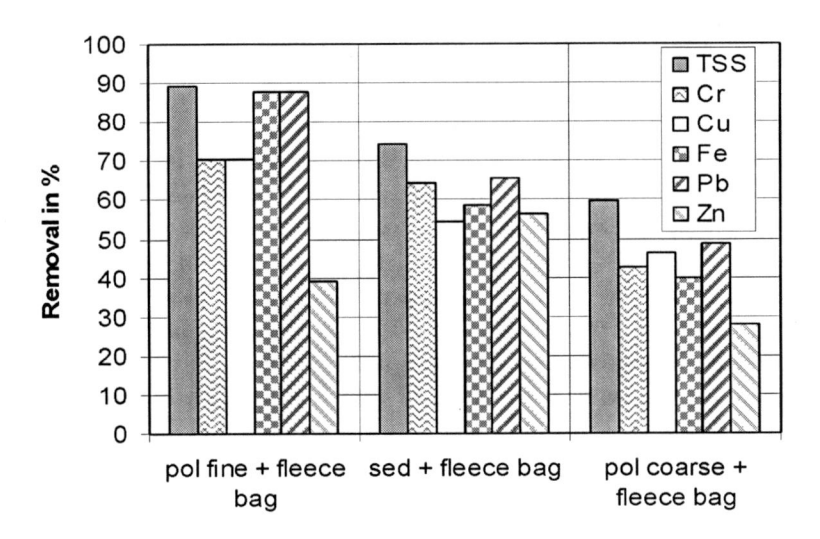

Fig. 6. Removal rates in two-step treatment applying different solids separation processes

material for Zn. In later experimental phases with new fleece bags and no sludge layer, Zn removal was also not satisfactory.

Combinations of fleece bags with sedimentation or pol fibre filtration were tested in different experiments (Figure 6). Because of pretreatment, the

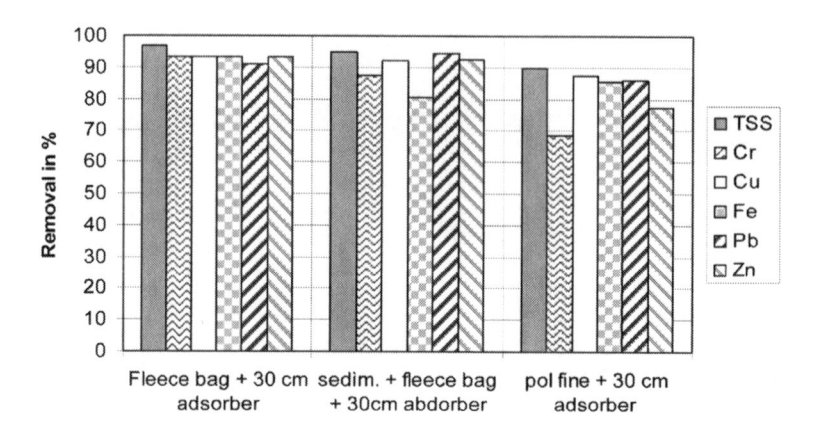

Fig. 7. Removal rates in two- or tree-step treatment applying different solids separation processes and subsequent GfeH or calcite adsorption filter

fleece bags were loaded with sludge in a lesser extent, thus, leading again to poor Zn removal. Also the removal performance of the bags alone decreased considerably because the major particle fractions were removed in the preceding treatment step. With a modest increase of only 5–20% removal, two-step treatment may economically not be justified.

The application of a mixture of GfeH or calcite sand in combination with preceding solids separation in different forms was investigated in order to test process combinations for advanced removal of heavy metals. Tests with fleece bags with underlying adsorber, sedimentation followed by fleece bags, and adsorber, as well as pol fibre drum filter followed by GfeH adsorber were performed. Layers of 10 and 30 cm adsorber material were applied. The experiments revealed that a layer of 30 cm was necessary in order to reach satisfactory removal rates. Therefore, only these results are shown here. The results in Figure 7 illustrate the excellent performance that can be reached with all combinations. Thanks to the adsorber, especially also high levels of Zn removal could be reached. Under optimized operation conditions, removal rates for all investigated heavy metals of more than 90% can be reached.

Conclusions

The long-term investigations prove that the suspended solids content with 150 mg TSS L^{-1}, the dissolved organic carbon content with 33 mg C L^{-1}, and the heavy metals copper with 65 µg Cu L^{-1} and Zinc with 440 µg Zn L^{-1}

are present in especially high concentrations in road runoff. The concentrations of environmentally hazardous substances classify the road runoff as polluted wastewater which has to be treated by law in order to be discharged into receiving waters or infiltrated into groundwater.

The results illustrate that above all pollutants bound to particles show strong dynamic effects with pronounced first flush. Peak concentrations are above average event concentrations by a factor of 2–3. The analysis of the particulate fraction of heavy metals represents especially valuable information in view of appropriate process development for runoff treatment. On average, the particulate fractions for chromium, copper, and zinc were between 81% and 86%. Focus was also put on information concerning organic hazards such as PAH and methyltertiary butyl ether (MTBE), a gasoline additive. The arithmetic mean of all runoff events was 1.4 µg L^{-1} for PAH and the mean peak concentration amounted to 3.2 µg L^{-1}, for MTBE the same values were 0.21 µg L^{-1} and 0.46 µg L^{-1}, respectively.

It could be shown that the investigated processes for road runoff treatment can be operated as modular systems in one-, two-, or three-step flow schemes. The tested filter fleece bags could typically be used as decentralized treatment facility in filter shafts along roads. The removal efficiency could further be improved by adding a GfeH or calcite adsorber layer of 30 cm according to requirements. In centralized or semi-centralized facilities for road runoff treatment, the tested polfabric drum filters are more suited. The automatic backwash of these filters requires further treatment of the backwash water.

In a next phase, the tested processes for road runoff treatment have to be transferred into practical applications. Full-scale installations will give the necessary information on operation and maintenance aspects and on economical feasibility.

References

1. Böhm E, Hillenbrand T, Marschneider F, Schempp C, Fuchs S, Scherer U (2001) Bilanzierung des Eintrags prioritärer Schwermetalle in Gewässer, Umweltbundesamt, ISSN 0722-186X
2. Steiner M, Boller M, Langbein S (2005) Bankette bestehender Strassen. Eawag (download www.regenanalyse.ch)
3. Boller M (1997) Tracking heavy metals reveals sustainability deficits of urban drainage systems. Water Sci Technol 35:77–87

4. Langbein S, Boller M, Steiner M (2006) Schadstoffe im Strassenabwasser einer stark befahrenen Strasse und deren Retention, Eawag (download www.regenanalyse.ch)
5. BLW (2001) – Bayerisches Landesamt für Wasserwirtschaft, Entwicklungsvorhaben: Versickerung des Niederschlagswassers von befestigten Verkehrsflächen, 2. Zwischenbericht
6. Steiner M, Boller M. (2002) Granulated iron-hydroxide (GEH) for the retention of copper from roof runoff. In: Chemical Water and Wastewater Treatment VII, IWA Publishing, London, pp. 233–242, ISBN: 1 84339 009 4

Development and full-scale implementation of a new treatment scheme for road runoff

M Steiner, S Langbein, M Boller

Swiss Federal Institute of Aquatic Science and Technology (Eawag), Ueberlandstrasse 133, CH-8600 Duebendorf; Institute of Environmental Engineering, ETH Zurich, 8093 Zurich, Switzerland

Abstract

Efficient heavy-metal removal from road runoff should be realized with particle separation and adsorption. As adsorbents, three zeolites and three ferric hydroxides were tested for copper and zinc removal. Sorption capacity and kinetics were evaluated with batch and specific column experiments. Results evidence that sorption capacity and kinetics of all ferric hydroxides were superior to the zeolites. Increased electrical conductivity revealed severe desorption of copper and zinc only from zeolites. For the full-scale treatment plant, zeolite 3 and Ferrosorp are used in one of the two retention filters each. The zeolite retention filter is not in service during winter to prevent desorption caused by de-icing salt.

Introduction

Nowadays, in Switzerland, runoff of motorways with a high traffic density must be treated prior to infiltration into the ground or discharge into receiving waters. As target pollutants, heavy metals such as cadmium (Cd), copper (Cu), lead (Pb), nickel (Ni), and zinc (Zn), polycyclic hydrocarbons (PAH), and dissolved organic carbon (DOC) should be considered.

For the design of new treatment schemes, the physical or chemical speciation of the pollutants must be considered. Measurement in road runoff reveal that the fraction of heavy metals passing a 0.45 μm membrane filter account for between 60% and 90% of the total heavy-metal load. Pilot studies indicate that with a textile filter fleece, 70–80% of the total copper, zinc, cadmium, and lead load in runoff can be retained [1]. However, higher removal performance may be required, if discharge occurs into for instance an ecologically sensitive river.

In such cases, an adsorption process to remove the dissolved heavy metals is necessary. Pilot plant experiments with road runoff revealed that adsorption based on granulated ferric hydroxide (GFH) mixed with calcite can reduce dissolved heavy-metal content of copper, zinc, lead, cadmium, and chrome up to 95% [1]. GFH-calcite adsorbers were already applied successfully at numerous roof runoff treatment sites for copper and zinc removal with feasible amounts of GFH [2]. The main disadvantage of GFH is its high costs, which make the application for road runoff treatment problematic. Therefore, strong need for alternative adsorbents exists, with a more advantageous ratio of cost to adsorption performance.

Among alternative adsorbents, zeolites are promising because they are up to ten times cheaper than GFH and it is known that zeolites are able to adsorb a wide range of heavy metals [e.g., 3]. In this study, three zeolites and three ferric hydroxides such as Ferrosorp, Everzit, and GFH were tested. Ferrosorp and Everzit are many times cheaper than GFH (Table 1).

Based on the result of this study, the most suitable adsorbents were chosen for the adsorber layer in a new type of road runoff treatment plant (RRTP), now built near Attinghausen, Switzerland. This RRTP combines a lamella separator for coarse particle removal with two retention filters. The retention filters consist of an adsorber layer for dissolved heavy-metal removal, which is located under a sand layer which operates as filter for fine particles. The RRTP discharges into a small river. As target pollutants, copper (Cu) and zinc (Zn) where chosen, because (1) aquatic organisms are very sensitive to these heavy metals and (2) dissolved Cu and Zn loads in road runoff account for 20–40% in runoff. Three criteria have to be considered for evaluation of adsorbents:

1. Sorption capacity under typical conditions prevailing in road runoff.
2. Sorption kinetics: Typical operating conditions for adsorbers for surface runoff treatment in general and especially for road runoff are highly variable loading patterns with respect to flow rate and

heavy metal concentrations. To assess sorption kinetics which is determined by film and/or pore diffusion, specific column experiments were performed [4].
3. Desorption of Cu and Zn as a result of increased electrical conductivity caused by salt used as deicing agent on roads during winter.

Experimental

Materials and runoff matrix composition

The adsorbents are characterized in Table 1. For a representative matrix for the batch and column experiments, data from road runoff composition were taken from a 2 years field study with 24 monthly averaged samples for Cu and Zn. The traffic density was 17,000 vehicles per day. The range of conductivity and pH is based on samples taken during seven runoff events [1]. A conductivity of 100 µS cm^{-1} was chosen because this reflects tailing after the first flush. Increased electrical conductivity of 2300 µS cm^{-1} was measured in monthly averaged samples during winter in the same field study (Table 2). For the experiments, NaCl was used for conductivity adjustment.

Table 1. Characterization of the adsorbents used in this study

	Filter bed density kg m^{-3}	Specific costs € t^{-1}	Grain diameter mm
Zeolite 1[a]	800–850	150–220	
Zeolite 2[b]			
Zeolite 3[b]			1–3
GFH[c]	1300 (650[e])	3620 (1935[e])	0.3–2
Ferrosorp[b]	650	740	0.5–2
Everzit[d]	640	1060	0.8–2

[a]Zeocem Vertriebsbuero Deutschland, 83064 Raubling, Germany
[b]Zeolith in Deutschland e.K. Waldsassen, Germany
[c]GEH Wasserchemie GmbH, Osnabrueck, Germany
[d]Everzit e.K. Hopsten, Germany. All Zeolites are Clinoptilolites
[e]If calcite (€250 t^{-1}) is added in a 1:1 weight ratio

Table 2. Road runoff matrix and concentrations chosen for the experiments

	Road runoff [1]	Batch experiments	Column experiments (sorption)	Column experiments (desorption)
pH	6.5–7.5	6.3–7.1	7–7.5	7–7.5
Electrical conductivity ($\mu S\ cm^{-1}$)	50–250, 2300*	100	100–150	2300
Zn ($\mu g\ L^{-1}$)	470 (SD 15)	50–1500	500–600	500–600
Cu ($\mu g\ L^{-1}$)	65 (SD 20)	50–1000	300–400	300–400

SD: standard deviation
*If de-icing salts are used in winter

Sorption capacity

The adsorbents were suspended into reaction tubes (PUR, 50 mL) filled with a matrix described in Table 2. The pH was adjusted using N-2-hydroxyethylpiperazine-N'-2-ethanesulfonic acid (HEPES) and NaOH. As kinetic experiments showed, equilibration was achieved after a residence time in the overhead shaker of 24 h. After equilibration, the samples were filtrated (0.45 µm, PES membrane, Oregon Scientific) acidified with 100 µL suprapure HNO_3 (69%) and analysed with inductively coupled plasma optical emission spectroscopy (ICP-OES) (Spectroflame). To assess precipitation or sorption, e.g., on tube walls, for each series, a blank sample without adsorbent was treated and analysed similar. The loading of the adsorbents was calculated based on mass balances.

Sorption kinetics

All sorption kinetics experiments were conducted with a pilot plant consisting of s even uniform transparent polyvinyl chloride (PVC) columns with an inner diameter of 5 cm and a total length of 1 m each. A sieve with a mesh size of 0.3 mm was deployed 15 cm above the column end to retain the adsorbents. To build stable and water saturated adsorber layers, the adsorbents were filled during constant up-flow. The flow direction during the experiments was always downward and the flow rate was adjusted with a needle valve. All columns were fed from an elevated tank (12 L) which was supplied by a pump from a larger storage tank (2 m³). In order to achieve a constant water head in the elevated tank, it was operated in a way that the surplus pumped into this tank flowed back to the storage tank.

Mass transport limitations due to film and pore diffusion were investigated with a variable flow rate scheme. The flow rate is highest at the start of the experiment and is reduced stepwise from 8 to 1 mh^{-1}. Afterwards, the flow rate remained constant at 1 mh^{-1} for 8 h before the same flow rate variation was applied again starting again with a filter velocity of 8 mh^{-1}. All adsorber layers have the same height of 10 cm, including GFH which was mixed with calcite at 1:1 weight ratio. The pH was adjusted with NaOH, but not buffered. In the column experiments, Cu and Zn were added simultaneously to assess competing sorption. For this purpose Cu concentration was increased to 300 µg L^{-1} instead of 65 µg L^{-1} (Table 1).

Desorption

For the desorption column experiments, the NaCl content of the inlet was increased up to a value of 2.3 mS cm^{-1} (Table 1). This value is based on measurements in road runoff taken during winter months and is therefore realistic [1]. Desorption experiments were conducted after the column experiments for assessing kinetics. After NaCl was added, a constant flow rate of 1 mh^{-1} was applied and sampling was done at regular intervals. Simultaneously, the electrical conductivity was measured.

Results and discussion

Adsorption capacity

The adsorption of Cu and Zn onto the three zeolites is shown in Figure 1. Additionally, the fitted Langmuir adsorption isotherms are displayed. In order to achieve low effluent concentrations in an adsorber layer, high affinity is most important. In general, the affinity for Cu is higher than for Zn which is advantageous because Cu concentrations in road runoff are lower compared to Zn.

The highest affinity for both metals is achieved by zeolite 3 followed by zeolite 2 and 1, which is confirmed by the decreasing parameter K_L of the

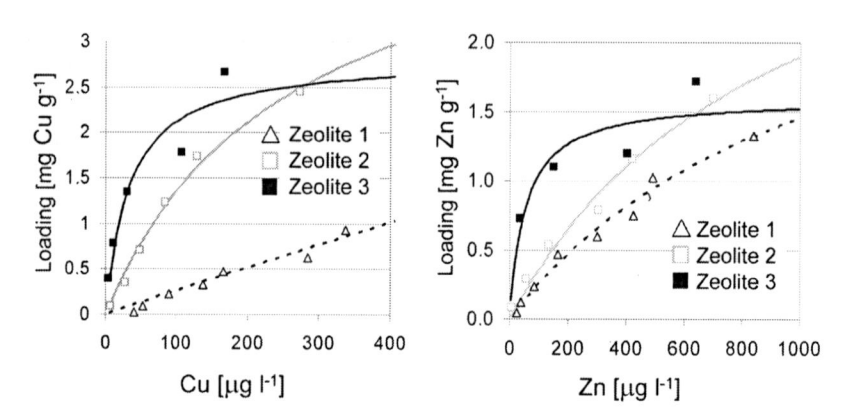

Fig. 1. Adsorption of Cu and Zn on three types of zeolite. Lines are fitted Langmuir adsorption isotherms. Langmuir parameters S_{max} (mg g^{-1}), K_L (l µg^{-1}) are (Cu/Zn): Zeolite 1 (27, 9.6E-5/3.1, 9E-4), zeolite 2 (4.9, 0.003/3.6, 0.001), zeolite 3 (2.8, 0.03/1.6, 0.018). pH was 6.9–7.2 for all experiments

Langmuir isotherm (Fig. 1). As a consequence, zeolite 3 performs best. The sorption capacity at typical Cu and Zn concentrations in road runoff (Cu: 65 µg L^{-1}, Zn: 470 µg L^{-1}, Table 1) is almost 2 mg Cu g^{-1} and 1.5 mg Zn g^{-1} zeolite 3, respectively. However, since the dissolved metal fraction is only 20–40%, the real adsorption capacity is smaller. The sorption capacity of zeolite 1 is significantly lower, especially for Cu.

The adsorption isotherms for ferric hydroxides are shown in Figure 2. As the adsorption of Cu and Zn onto ferric hydroxides is known to be strongly pH dependent, experiments were performed at pH between 6.3 and 7.1, reflecting the lower range of pH measured in road runoff (Table 1). In general, the sorption capacity for Cu is higher than for Zn for all ferric hydroxides. For Ferrosorp and Everzit, the sorption isotherm is similar for Cu and Zn, except for the sorption capacity of Everzit at pH 6.3 for Zn, which is lower compared to Ferrosorp. The sorption capacity of GFH for Zn at pH 7 can be compared to Ferrosorp and Everzit at pH 6.3. The sorption capacity of GFH for copper at pH 6.5 is comparable to the capacity of Ferrosorp and Everzit at pH 7.1.

At pH 6.3 (Ferrosorp, Everzit) and 7 (GFH), the adsorption capacity at typical Zn concentrations in road runoff (Zn: 470 µg L^{-1}, Table 1) is 9.9, 7.9, and 7.8 mg Zn g^{-1}, respectively. At pH 7.1, these values are increased to 24 and 27 mg Zn g^{-1}. Compared to zeolite 3, the sorption capacity is 5–6 and 16–18 times higher. For copper (65 µg L^{-1}), the relations are similar (9–11 times at pH 7.1 and 3–7 times at pH 6.3 and 6.5).

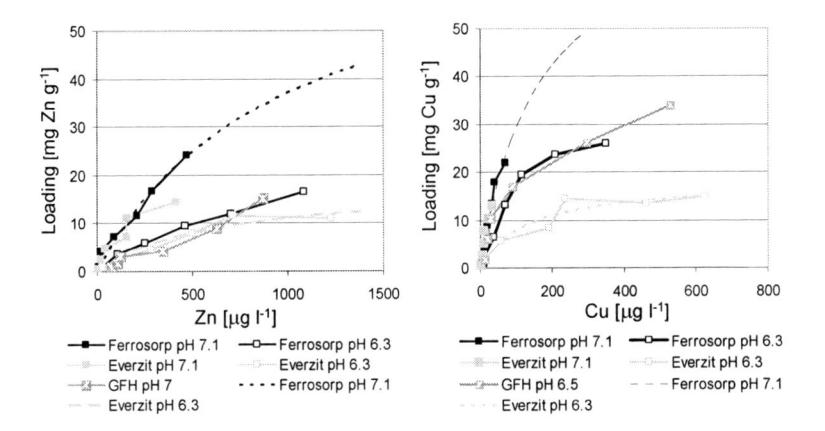

Fig. 2. Adsorption isotherms of Cu and Zn on ferric hydroxides. Calculated Langmuir isotherms reveal the suitability of this model. Langmuir parameter K_L (l µ g^{-1}) is (Cu/Zn): Ferrosorp pH 7.1 (0.006/0.001), Everzit pH 6.3 (0.007/0.001). For accurate estimation of S_{max}, more data would be required

For the design of an adsorber layer, the filter bed density, expressed in grams of metals per cubic metre must be taken into account. As shown in Table 1, the filter bed density is highest for GFH. However, based on experimental results as well as on data of full-scale application, the addition of calcite sand in a 1:1 weight ratio stabilizes a GFH layer considerably [4]. Therefore, the specific GFH layer density given in Table 1 must be divided by two, resulting in a comparable specific density of GFH in a GFH or calcite layer to Ferrosorp, Everzit, and zeolites.

Calculations of the total required amount of adsorbents based on the expected copper and zinc load in the road runoff reveal the following dimensions at the given retention filter surface area. The filter bed height of a layer of zeolite 2 or 3 should be 40 cm for an operating time of 5 years, while a layer of GFH or calcite or Ferrosorp or Everzit with a height of 20 cm can be operated theoretically for 10 years in minimum. It has to be noticed that adsorber layer heights smaller than 20 cm should not be applied in practice.

Sorption kinetics

The results of the sorption kinetics experiments with the columns for zeolites 1 and 3 and for Zn are shown in Figure 3. For all filter velocities, the effluent concentrations of copper from zeolite 1 are lower than from zeolite 3, although the sorption capacity of zeolite 1 is lower than of zeolite 3 (Fig. 1).

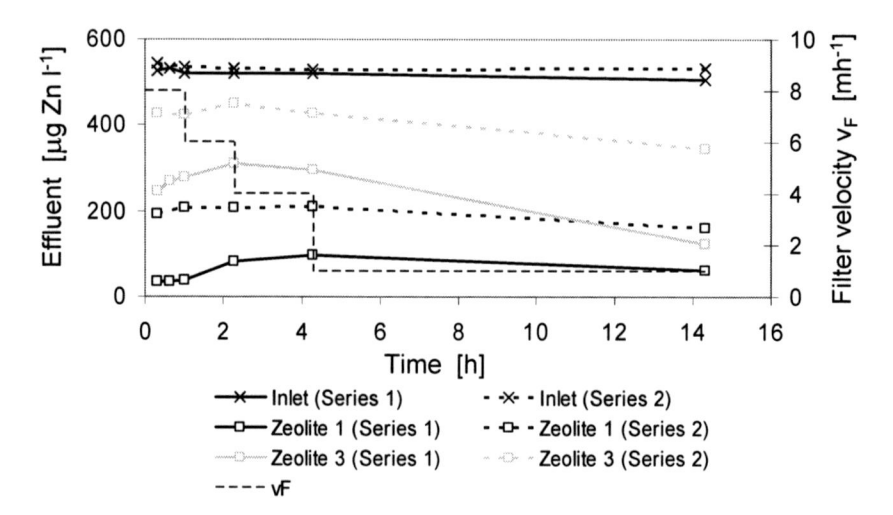

Fig. 3. Zinc concentrations in the effluent of the columns with zeolite 1 and 3 with layer heights of 10 cm each (both series). pH in inlet and effluent was 7.1

This can be explained by the smaller grain size range of zeolite 1 (Table 1), resulting in a higher specific grain surface area and therefore in a higher specific mass transfer to the grain surface (film diffusion). This experiment shows low mass transfer limitation at the beginning of the experiments due to no preloading of the surface and near surface sites. As film diffusion limitation decreases in parallel with flow rate, lowest effluent concentrations can be observed at the beginning of the first flow rate variation at highest flow rate [5].

For both zeolites, the effluent concentrations during series 2 are higher than during series 1. This is due to preloading especially at the surface of the zeolite layer resulting from series 1. At the end of series 1, only 8% and 10% of the sorption capacity of the layers were used. Similar patterns can be observed for copper.

As shown in Figure 4, the effluent concentration of Zn of all ferric hydroxides is significantly lower compared to the zeolites. The lowest effluent concentrations, especially at high flow rates, are observed for the GFH–calcite layer (series 2).

Effluent concentrations of the Ferrosorp and Everzit columns are up to three times higher at the beginning of each series compared to the ones of the GFH–calcite column. However, at decreasing flow rates, concentrations are comparable. In contrast to the zeolite experiments, differences between series 1 and 2 are small. Thus, film and pore diffusion limitations are lower

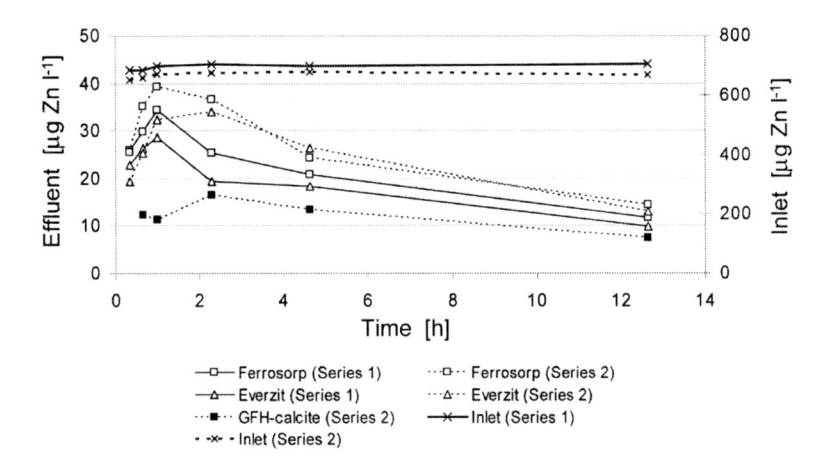

Fig. 4. Zn concentrations in the effluent of the ferric hydroxides columns. Layer heights are 10 cm each. Flow rates of series 1 and 2 are similar to Figure 3. pH in the inlet and effluent of the GFH–calcite layer was 7.4–7.6. pH in the effluent of the Everzit and Ferrosorp column pH was 8.8–9.0

compared to the zeolites. Interestingly, the pH in the effluent of the Ferrosorp and Everzit layer was increased by 1–1.3 pH units.

Due to the sand layer above the adsorber in the RRTP, the adsorber is expected to be operated at a filter velocity within a range of 0.5–1.5 mh^{-1}. Therefore, the calculated adsorber bed height of 20 and 40 cm for zeolite 3 and ferric hydroxides, respectively, based on sorption capacity and practical feasibility are sufficient, if a smaller grain size is applied for zeolite 3.

Desorption of copper and zinc

The results of the desorption experiments with zeolite 1 and 3 and the ferric hydroxides are shown in Figure 5 for Zn. Samples were taken after inlet conductivity was achieved in the effluent. The results reveal that for both zeolites, significant desorption occurs, as Zn concentrations in the effluent actually exceed many times the inlet concentrations. Effluent concentrations of the GFH-calcite are not affected by increased conductivity. In contrast, the Ferrosorp and Everzit columns show slightly increased concentrations in the first sample at 10 min. Obviously, the selectivity of the zeolites for Cu and Zn is not strong enough to prevent ionic back-exchange under high ionic strength. These results evidence that zeolites must not be used for road runoff treatment, if de-icing salts are applied in winter.

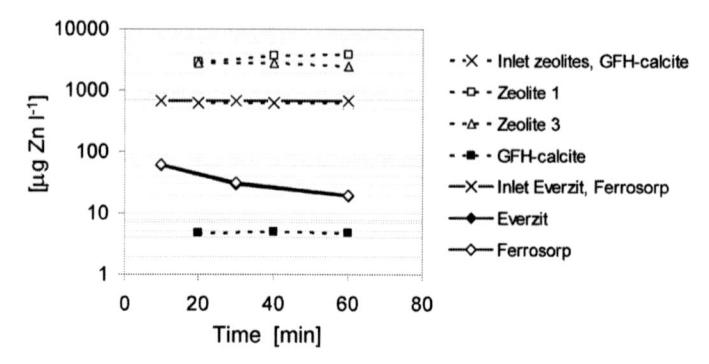

Fig. 5. Effluent concentrations of Zn during application of increased electrical conductivity of 2.3 mS cm^{-1}

Adsorber material for the full-scale treatment plant

Based on the result of this study, zeolite 3 and Ferrosorp were chosen as adsorbents for the two full-scale retention filters. The main arguments for using zeolite 3 in one filter were the lower costs compared to Ferrosorp, and the possibility to test these two adsorbent under full-scale operating conditions, as an extensive monitoring system was installed. To cope with the de-icing salt problem, the RRTP will be operated in a way that during winter, only the Ferrosorp retention filter will be in use.

Conclusions

Zeolite 2 and 3 and all tested ferric hydroxides can be used as adsorbents for road runoff treatment, as their adsorption capacity and kinetics are high and fast enough to allow feasible adsorber layer heights and satisfactory run times in practice. For the RRTP, zeolite 3 with a smaller grain diameter and Ferrosorp were chosen for the retention filters.

If zeolites are exposed to increased salt content, adsorbed copper, and zinc desorbs immediately and to a high extent. Therefore, zeolites are not applicable in RRTP, if de-icing salts are used. Ferric hydroxides were not affected by increased salt content due to specific sorption of copper and zinc. Therefore, one of the two retention filters of the RRTP equipped with zeolite 3 will not be in operation during winter time.

For zeolite 3 and the ferric hydroxides, a layer height of 40 and 20 cm is required for a 5 and 10 years operation time, respectively, as filter velocity is expected to be below 1 mh^{-1}.

Because this is a new system for road runoff treatment, it was agreed upon an extensive monitoring programme. This programme will be started in July 2006 and will give answers on the performance of each treatment component of the RRTP.

Acknowledgements

This research was founded by the civil construction department of the Kanton of Uri, Switzerland and was realized in collaboration with the consulting engineering company André Rotzetter & Partners in Baar, Switzerland.

References

1. Boller M, Langbein S, Steiner M (2006) Characterization of road runoff and innovative treatment technologies. Proceedings of the 8th highway and urban environment symposium. Springer, Amsterdam
2. Steiner M, Boller M (2006) Copper and zinc removal from roof runoff: from research to full-scale adsorber systems. Water Sci Technol 53:199–207
3. Curkovic L (1997) Metal ion exchange by natural and modified zeolites. Water Res 31:1379–1382
4. Steiner M, Pronk W, Boller M (2006) Modelling of copper sorption onto GFH and design of full-Scale GFH adsorbers. Environ Sci Technol 40:1629–1635
5. Bajracharya K, Barry D (1997) Nonequilibrium solute transport parameters and their physical significance: numerical and experimental results. J Contam Hydrol 24:185–204

Reactive filters for removal of dissolved metals in highway runoff

M Hallberg, G Renman

Royal Institute of Technology, Stockholm, Sweden

Abstract

A pilot-scale system consisting of presedimentation and a saturated down-flow reactive bed filter was used for cleaning highway runoff. Blast furnace slag (BFS) and Polonite were selected as filter materials. A total suspended solids (TSSs) removal of over 99% was achieved. High removal performance was observed for dissolved Mn, Ni, Co, and Cu. In contrast Al was released after filtration. Metals were retained in the upper layer of the bed filters while a desorption was suggested to take place in the downward layers. This was probably attributed to the elevated salt levels during winter and the intermittent operation.

Introduction

Highway runoff is being recognized as the most contaminated of the different kinds of storm water due to the loading from various sources connected to vehicles, road construction, and road management. A relationship exists between an increasing number of vehicles and a rise in the pollutant load [1,2]. In Stockholm, Sweden, an elaborate investigation was carried out to assess the impact of runoff water on receiving water bodies. Roads and motorways were identified as important sources of pollutants. Runoff from highways with an annual average daily traffic (AADT) exceeding 30,000 vehicles, as a rule needed treatment to reduce the contaminant cocentrations with regards to expected pollutant concentrations [3]. Other issues include seasonal variations, the impact of colder climates, the influence of road maintenance operations such as de-icing and snow ploughing, and the use of friction tyres. Traffic, traffic-related activities, and road maintenance have been found to be a major contributor of pollutants in snow [4]. A study showed that the pollutant load could increase dramatically during the winter season when de-icing agents are utilized [5]. Other studies have shown that Zn, Cd, and Cu were mainly dissolved and Pb, Fe, and Al to a greater extent were particulate bound [6]. Furthermore the concentration of dissolved metals in road runoff can vary significantly between summer and winter [7].

The most common treatment method for runoff water is sedimentation [e.g., 8]. Various wetland systems have been utilized for extended pollutant removal purposes from highway runoff [e.g., 9–11]. In highly urbanized areas space constraints may limit the construction of large detention ponds or wetlands. Consequently, compact removal systems are of interest, for example, a filter system using reactive materials prepared from natural minerals or from by-products of steel production such as blast furnace slag (BFS). These filters could be a possible solution for the removal of colloidal and dissolved metals in highway runoff where space is limited. However, runoff from highways displays elevated levels of TSS typically well exceeding 60 mg L^{-1} as reported by [7]. It is therefore necessary to remove solids before filtration. The reduction of solids can be achieved by sedimentation or possibly also in combination with a chemical coagulant as discussed by [12].

Several low-cost filter materials for removal of heavy metals have been investigated [13] and these could be included in systems based on filter-bed techniques [14]. Laboratory column and batch experiments have been performed with zeolites to assess their sorption of ammonium nitrogen [15, 16]. Various studies have been conducted on different filter materials for the removal of dissolved metals [e.g., 17–19]. These studies have been executed in laboratory scale and with metal concentrations far exceeding the dissolved and colloidal metal concentrations that can be found in highway runoff. No field studies have been performed to assess the removal of metals from highway runoff.

In the present study, we tested two filter media for their ability to remove selected heavy metals from highway runoff. BFS and Polonite were used in the pilot tests. The experiment was carried out between October 2004 and August 2005. The aim has been to determine the removal of the dissolved and colloidal metals during winter and summer.

Experimental

Study area

An area located at the northern exit of Stockholm City was selected for the study. A six-lane highway (E4) dominates the 6.7 ha catchment area with an AADT of 120,000 vehicles and a speed limit of 70 km h^{-1}

In order to reduce the pollutant load from the catchment area a treatment plant was constructed and commissioned in 1991. The treatment plant, named Eugenia, is located below ground and the runoff is transported by gravity to the intake chamber.

Description of filter materials

Polonite is a product manufactured from the cretaceous rock "Opoka" and is intended for use in wastewater treatment [20]. This material is known for its high sorption capacity of soluble phosphorus and usefulness for recycling of nutrients in agriculture [21,22]. The particle size of the material was 2–5.6 mm and the pore volume was 45%. The Polonite was mixed with peat, less than 1% by weight, to avoid chemical clogging [23].

BFS is an industrial by-product resulting from the process of extracting iron from iron ore at steelworks. A manufactured product from Merox AB in Oxelösund, Sweden was used in our column experiment. A further 1% by weight of CaO was added to the BFS. The particle size of the material was 0.125–2 mm with a pore volume of 47%.

Pilot trials

A pilot-scale sampling system and filter system was constructed for the study (Fig. 1). A pump of type Flygt SXM 2, located approximately 0.5 m above the intake chamber floor, pumped the runoff to the tank. The tank was filled by the use of a floating switch device to control the volume of runoff. Sampling was started either automatically or manually. All material was plastic with exception for the pump of stainless steel. After each trial the check valve system was dismantled and thoroughly washed and hosed down with potable water, as was the tank.

Fig. 1. Experimental set-up: inflow (1), tank (2), level control discharge to columns, peristaltic pump (4), sampling ports (5), column (6), discharge filter (7), level control filter (8), sampling vessel for filtrated water (9)

Discharge to the columns (F1–F4) from the tank was carried out by peristaltic pump with a constant flow. Sampling was carried out during discharge from the sample port located at the level from which the water was pumped to the filters. Columns F1 and F2 contained Polonite and columns F3 and F4 contained BFS. The four filter columns were made of PVC, each having an overall height of 90 cm and an internal diameter of 30 cm. All columns were filled with filter material to a height of 60 cm. A sintered polycarbonate plastic plate with a height of 5 mm and 100 µm pore size was placed at the bottom of the filters to retain the filter materials. The filters were operated under saturated conditions and were saturated between operations. Sampling of the treated water was carried out from a 25

L plastic vessel. The samples were analysed at an accredited laboratory (SWEDAC ISO/IEC 17025, Reg. number 1087). The metals were analysed by means of inductively coupled plasma sector field mass spectrometry (ICP-SFMS) and inductively coupled plasma atomic emission spectrometry (ICP-AES). Filtered samples (Sartorius 0.45 µm) were used for the analysis of soluble metals. Analysis of TSS, pH, and conductivity was according to Swedish Standard SS-EN 872, SS 028125, and SS-EN 27888, respectively. A HACK 2100P ISO turbidity meter was utilized for turbidity measurements. The operating wavelength of the instrument was 860 nm. The measuring range was from 0 FNU to 1000 FNU with a resolution of 0.01.

Calculation of the mass load of TSS was executed according to Eq. 1:

$$m_{TSS} = Q_F \sum_{i=1}^{j} \frac{C_{TSS_i} + C_{TSS_{i+1}}}{2} \cdot (t_{i+1} - t_i) \tag{1}$$

where "m_{TSS}" is the mass load mg, "Q_F" is the discharges to the columns, "C_{TSS}" is the concentration of TSS mg L^{-1} and "t" is the time of sampling. The mass load of metal was calculated according to Eq. 2:

$$mme = Q_F \sum_{i=1}^{j} \frac{C_i + C_{i+1}}{2} \cdot (t_{i+1} - t_i) \tag{2}$$

where "mme" is the mass load of the metal µg and "C" is the sampled concentration of the metal. The calculation of the total concentration of a metal was done according to Eq. 3:

$$C_{me} = \sum_{i=1}^{i=10} \frac{mme_i}{V_i}$$

where "C_{me}" is the metal concentration in the total volume discharged to a filter during the experiment. "V" is the volume discharged to the filter during a single pilot trial calculated according to Eq. 4:

$$V = Q_F \cdot t_{PiTr} \tag{4}$$

where, t_{PiTr}, is the total pumping time during an individual pilot trial.

Results and discussion

TSS, pH, and conductivity measurements

The pilot trials performed well without any hydraulic problems. The surface load to the filters was 0.06 m h^{-1}. No clogging was observed throughout the experiment. The TSS load varied during the trials, as did the electrical conductivity and pH in the discharge to the columns (Table 1). The average measured turbidity in the discharge.

Table 1 was 89.6 FNU and did not exceed 5.0 FNU in the treated water. The TSS concentration was below the detection limit in the treated water. The pH in the treated water was above 10 during the entire experiment.

A total of 64 grab samples were taken of the treated water. The total concentrations of Al, Cd, and Cr after passing through filters F1 and F2

Metals

For As and Hg, the analysed water samples were as a rule below the detection limit of the analysis method. The concentrations of the other studied metals are shown in Table 2.

Table 1. Parameters for the discharge to columns

PT	DBT (day)	V (L)	m_{TSS} (mg)	Cond (mS m^{-1})	pH
1	0	150	600	15	7.1
2	12	1626	2700	80	7.4
3	6	225	17,000	100	7.4
4	16	238	6500	2150	7.5
5	11	210	3300	520	7.7
6	32	250	25,500	160	–
7	31	250	5300	1940	–
8	35	260	20,900	764	–
9	72	250	34,600	45	7.8
10	90	275	7800	80	7.8

Table 2. Metal concentrations in the total discharge colume (2270 [1])

	Al	Cd	Co	Cr	Cu	Ni	Zn
C_{me} (Total, µg L^{-1})	1883	0.205	7.0	9.0	31	5.0	179
C_{me} (Dissolved, µg L^{-1})	21.9	0.149	4.7	3.2	14.2	3.0	97.9

were higher thus indicating that no reduction of dissolved matter has taken place particularly in the case of Al and Cr (Tables 2 and 3). For Co, Cu, Ni, and Zn a reduction of the dissolved part is suggested (Tables 2 and 3). The total concentration of Al after passing through filters F3 and F4 was higher indicating that no reduction of dissolved matter took place (Tables 2 and 4). For Cd, Co, Cr, Cu, Ni, and Zn it is suggested that a reduction of the dissolved part occurred (Tables 2 and 4).

After the pilot trials, F2 and F3 were dug out and the metal content in the filter materials were studied. The metal content in the original filter material are shown in Table 5. Hg and As was below the detection limit of the analysis method. Polonite showed an accumulation of Al_2O_3, Cd, Co, Cu, and Zn in the top layers (0–105 mm) of the filter material, but for the bottom layers (145–600 mm) a decrease in metal content was found. A decrease in metal contents of Cr and Ni was found throughout the whole of the filter material. BFS displayed an accumulation for Al_2O_3, Cd, Co, Cu,

Table 3. Total concentration of metals after F1 and F2 (treated water)

	ATC F1 ($\mu g\ L^{-1}$)	SD ATC	F1 Red. DM (%)	ATC F2 ($\mu g\ L^{-1}$)	SD ATC	F2 Red. DM (%)
Al	733	268		752	270	
Cd	0.185*	0.264		0.167	0.250*	
Co	0.3*	0.1	>94	0.3	0.1*	>94
Cr	13.9*	8.5		14.0	8.7*	
Cu	8.4	8.2	60	7.7	8.3	46
Ni	1.9*	1.3	>37	1.7	1.1	43
Zn	5.0	2.6	>95	5.1	2.8*	>95

ATC average total concentration, SD standard deviation, DM dissolved matter
*Analysis below detection limit

Table 4. Average total concentration of metals after F3 and F4 (treated water)

	ATC F3 ($\mu g\ L^{-1}$)	F3 SC ATC	F3 Red. DM (%)	ATC F4 ($\mu g\ L^{-1}$)	F4 SD ATC	F4 Red. DM (%)
Al	390	245		343	258	
Cd	0.066*	0.036	>53	0.071	0.048*	>52
Co	0.2*	0.1	>96	0.2	0.1*	>96
Cr	1.5*	1.3	>53	1.5	1.3*	>53
Cu	1.5*	0.8	>89	1.6	0.9*	>89
Ni	0.9	0.6	70	0.9	0.5	
Zn	4		>96	4		>96

ATC average total concentration, SD standard deviation, DM dissolved matter
*Analysis below detection limit

Table 5. Metal content in filter materials Polonite [1] and BFS [2]

	Al_2O_3 (%)	Cd (ppm)	Co (ppm)	Cr (ppm)	Cu (ppm)	Ni (ppm)	Zn (ppm)
1	5.68	0.22	4.11	88	27.1	25	35.1
2	12.9	0.02	0.315	103	6.93	2.32	2

Ni, and Zn in the top layers (0–165 mm) and decrease in the consecutive filter material. A decrease in metal content of Cr was found throughout the whole of the filter material.

The result from the water sampling suggests that the removal condition of the Polonite and BFS increases the portion of dissolved Al. Both materials showed a potential for removing the dissolved fraction of Co, Cu, Ni, and Zn. The BFS showed an overall better removal capacity for the dissolved fraction of all the studied metals. However, the analysis of the solid filter materials strongly suggests that leaching of all the studied metals had occurred during the pilot trials. The operation of the filters was intermittent with up to 90 days between operations (Table 1). During trial 4 and 7 the filters were exposed to elevated levels of road salt (NaCl) as shown by the increased conductivity [7]. The times between trial 4 and 5 and 7 and 8 were 11 and 35 days, respectively. One possibility is that the elevated levels of sodium ions by ion exchange reactions mobilized metals in the filter materials. In the study by Brogowski and Renman [20] it was shown that the major components of Polonite were SiO_2 (39.4%), CaO (42%), Al_2O_3 (4.3%), and Fe_2O_3 (2%). The contents of Cr, Cu, Ni, and Zn in the original material were 58.6, 13.0, 26.9, and 121.3 mg kg^{-1} respectively. In the analysed Polonite sample the solid concentrations of SiO_2, CaO, and Fe_2O_3 were 59.4%, 24.2%, and 4.02%, respectively. These variations can be expected since the mined Opoka is a naturally formed rock. Seventy kilograms of Polonite were used in the experiment. From this a 5 g sample was extracted for analysis of the solid material. Variations between the filter material in the columns and the sampled Polonite are a possibility. This may consequently challenge the conclusions drawn from the analysis of the solid samples. The BFS originates from a continuous and controlled industrial process thus the variations in the material are expected to be considerably less. The solid sample from BFS should be representative of the filter material used in the experiment. In accordance with this it is likely that leaching has occurred in both filter materials during the experiment based on the result from the analysis of the filter materials.

The set-up of the experiment is novel for a runoff water situation. The use of reactive filters for protecting sensitive recipients is of interest. Thus, further studies are of interest for elevated salt levels and the impact on the metal removal capability of filter materials. The surface loading during the experiment was comparable with that of slow sand filtration [24]. The results in this study showed no clogging during the experiment thus studies with raised surface loads are important. Furthermore, similar *in situ* studies should be executed utilizing continuous sampling of the filtrate or retrieving the complete filtrate volume.

Conclusions

The filter materials Polonite and BFS performed hydraulically well with removal of PM exceeding 99% regardless of the variations in TSS load after sedimentation. Similar variations can be expected in a road runoff situation. The BFS showed better removal of the studied dissolved and colloidal metals. However, a release of Al was found from the BFS and Polonite during the pilot trials. Metals were retained in the upper layer of the bed filters while desorption was suggested to take place in the downward layers. The high content of salt in runoff from highways in Nordic countries is considered to reduce the effectiveness of reactive bed filters in full-scale treatment plants.

References

1. Hvitved-Jacobson T, Yousef YA (1991) Highway runoff quality, environmental impacts and control. Highway pollution, studies in environmental science. Elsevier, New York, pp 166, 170–171
2. Barett ME, Irish L, Malina J, Charbeneau RJJ (1998) Characterisation of highway runoff in Austin Texas, Area. J Environ Eng 124:131–137
3. Stockholm Vatten (2002) Klassificering av dagvatten och recipienter samt riktlinjer för reningskrav – Del 3 Rening av dagvatten (In Swedish)
4. Glenn DW, Sansalone JJ (2002) Accretion and partitioning of heavy metals associated with snow exposed to urban traffic and winter storm maintenance activities. II. J Environ Eng 128:167–185
5. Legret M, Pagotto C (1999) Evaluation of pollutant loadings in the runoff waters from a major rural highway. Sci Tot Environ 235:143–150
6. Sansalone JJ, Buchberger SG (1997) Partitioning and first flush of metals in urban roadway stormwater. J Environ Eng 123:134–143
7. Hallberg M (2006) Suspended solids and metals in highway runoff – Implication for treatment systems. Licentiate thesis, Royal Institute of Technology, Stockholm

8. Pettersson T (1999) Stormwater ponds for pollution reduction. Ph.D. thesis, Department of Sanitary Engineering, Chalmers University of Technology, Göteborg
9. Ellis JB, Revitt DM, Shutes RBE, Langley JM (1994) The performance of vegetated biofilters for highway runoff control. Sci Tot Environ 146/147: 543–550
10. Shutes RBE, Revitt DM, Lagerberg IM, Barraud VCE (1999) The design of vegetative constructed wetlands for the treatment of highway runoff. Sci Tot Environ 235:189–197
11. Sriyaraj K, Shutes RBE (2001) An assessment of the impact of motorway runoff on a pond, wetland and stream. Environ Int 26:433–439
12. Heinzman B (1994) Chem. Water Wastewater Treatment. III. Proceedings Of the 6th Gothenburg symposium 1994, Springer, Berlin, 283–296
13. Bailey, SE, Olin, TJ, Bricka, RM, Adrian, DD (1999) A review of potentially low-cost sorbents for heavy metals. Water Res 33:2469–2479
14. Kängsepp P, Hogland W, Mathiasson L (2003) Proceedings of Sardinia 2003, 9th international waste management and Landfill symposium, Cagliari, Italy, 6–10 October, p 278
15. Papadopoulos A, Kapetanios EG, Loizidou M (1996) Application of chemical oxidation for the treatment of refractory substances in leachates. J Environ Sci Health A31:211–220
16. Demir A, Günay A, Debik E (2002) Ammonium removal from aqueous solution by ion-exchange using packed bed natural zeolite. Water SA 28:29–335
17. Gomonay VI, Golub NP, Szekeresh KY, Gomonay PV, Charmas B, Leboda R (2001) Adsorption of lead(II) ions on transcarpathian clinoptilolite. Ads Sci Technol 19:465–473
18. Al-Asheh S, Banat F (2001) Adsorption of zinc and copper ions by the solid waste of the olive oil industry. Ads Sci Technol 19:117–129
19. Ake CL, Mayura K, Huebner H, Bratton GR, Philips TD (2001) Developement of porous clay-based composites for the sorption of lead from water. J Toxicol Environ Health 63:459–475
20. Brogowski Z, Renman G (2004) Characterization of Opoka as a basis for its use in wastewater treatment. Polish J Environ Stud 13:15–20
21. Hylander LD, Simán G (2001) Plant availability of phosphorus sorbed to potential wastewater treatment materials. Biol Fertil Soils 34:42–48
22. Renman G, Kietlińska A, Cucarella Cabañas V (2003) Ecosan closing the loop. Proceedings of the 2nd international symposium, 7–11 April, Lübeck, Germany, pp 573–576
23. Kietlinska A, Renman G (2005) An evaluation of reactive filter media for treating landfill leachate. Chemosphere 61:933–940
24. American Water Works Association (AWWA) Water Quality & Treatment, A Handbook of Community Water Supply, 5th edn. McGraw-Hill, New York, p 8.79

Designing filters for copper removal for the secondary treatment of storm water

H Genç-Fuhrman, P Steen Mikkelsen, A Ledin

Institute of Environment & Resources, Technical University of Denmark, Building 115, DK-2800 Kgs. Lyngby, Denmark

Abstract

In this study alumina and granulated activated carbon (GAC) are investigated as potential storm water filtration media by testing their heavy metal (i.e., As, Cd, Cr, Cu, Ni, and Zn) removal efficiency in batch and column experiments at a starting pH of 6.5. However, only Cu removal results are presented here due to space concerns. It is found that the equilibrium time for Cu sorption is 2 h for both sorbents, and that the Freundlich adsorption isotherm moderately fits to the sorption data possibly due to the fact that the isotherm fails to take into account the competitive sorption as well as possible precipitation. The results also suggest the presence of humic acid (HA) suppresses the removal, while that of Fe colloids (FC) (FC) is insignificant.

Introduction

Storm water runoff is recognized as a major non-point source pollution that may deteriorate the quality of the receiving environment, in particular any receiving water bodies. For example, it is suggested that rainfall runoff may be a major reason for the impairment of several water bodies in the USA [1].

Many pollutants exist in the runoff (especially road runoff) and heavy metals (e.g., As, Cd, Cr, Cu, Ni, Pb, and Zn) are among the most important pollutant group [2]. Among them Cu is of special interest here, as it is a major aquatic toxic metal in storm water [2]. Furthermore, storm water runoff (especially road runoff) is an important source of Cu to the environment, and high concentration of Cu in road runoff is mainly due to tyres, brake pads, and car washing. Special interest should therefore be directed to Cu removal from storm water in order to reduce its quantities to acceptable levels for protecting the quality of receiving waters.

Storm water detention ponds are widely used to address the storm water quality and quantity problems. Although they may be effective at removing particulate form of contaminants, colloidal and dissolved pollutants including heavy metals are not effectively treated in such systems. Therefore, secondary treatment is required to assure the removal of colloidal and dissolved pollutants including heavy metals. Here, filtration is postulated as a technically attractive (as it is not very complicated) and effective secondary treatment option for metal removal from storm water runoff providing that highly efficient filtration media is used and its metal removal characteristics (e.g., effect of reaction conditions, water composition) are understood.

The first part of the study is devoted to a brief literature survey on possible storm water filtration materials, while in the experimental part alumina and granulated activated carbon (GAC) are tested for Cu removal at the presence of As, Cd, Cr, Ni, and Zn; as they often occur along with Cu [2] and they can influence the Cu removal efficiency. On the other hand, humic acid (HA) and Fe colloids (FC) are also important in terms of defining the fate and transport of heavy metals [3] due to their very large surface area, which gives them a high capacity to sorb both essential and toxic trace metals [4]. Thus, the presence of HA and FC on the heavy-metal removal is investigated along with the effect of initial heavy-metal concentration, reaction time and pH.

Consequently, the main aim of this paper was to communicate some of the important findings of the research project, which had a long-term goal of developing filters for the secondary treatment of storm water, with following specific objectives: (1) to document the sorbent groups, which can be used for storm water filtration or may have a potential to be developed as filter media; (2) to test the efficiency of the alumina and GAC in batch and column tests in terms of their Cu removal efficiency at the presence of coexisting heavy metals; and (3) to investigate the effect of reaction time, pH, and the presence of HA and FC on the Cu removal.

Experimental methods

In the batch trials the alumina and GAC are used to investigate the Cu removal at the presence of As, Cd, Cr, Ni, and Zn. Furthermore, the effect of reaction time, pH, HA, and FC on the Cu removal is tested. The sorbents particle size range was 0.6–1 mm, and the BET-N_2 surface areas were 238.9 and 784.5 $m^2\ g^{-1}$ for alumina and GAC, respectively. Initial heavy-metal concentration ranges were 1–1000, 5–2670, 7–2830, 12–1820, 160–8640, and 110–52,300 $\mu g\ L^{-1}$ for As, Cd, Cr, Cu, Ni, and Zn, respectively. The ionic strength of the water samples was controlled using 0.01 M NaCl, and all samples had 0.003 M $NaHCO_3$ to minimize the pH changes during experiments. These concentrations were selected to mimic storm water conditions.

When preparing batches with HA, a commercial HA purchased from Aldrich was used. Herein, the desired amount of HA powder is added to distilled water to prepare 1 $g\ L^{-1}$ stock solution and then kept in UV bath for 10 min before preparing batches. In the case of batches containing FC the stock FC solution is prepared using the method of [5] at 1500 $mg\ L^{-1}$, and then desired concentrations are prepared from diluting this stock solution.

Column studies are conducted over a period of 180 days in 6 PE columns with 5 cm diameter and 15 cm height, the working volume of each column was 157 mL. Herein, columns were packed with alumina or GAC and used as fixed-bed up-flow reactors; influent water was pumped through alumina and GAC-packed columns with a peristaltic pump. Samples were taken daily (for the first part) or weekly (for the second part) and analysed for the metals in the influent and effluent water. Average influent heavy-metal concentrations were 8, 10, 14, 20, 24, and 220 $\mu g\ L^{-1}$ for As, Cd, Cr, Ni, Cu, and Zn, respectively.

Metal measurements

All the samples were sent to a certified commercial laboratory (Analytica, Sweden) for quantifications of metals, where inductively coupled plasma atomic emission spectrometry (ICP-AES) or inductively coupled plasma sector field mass spectrometry (ICP-SFMS) were used. Detection limits were 1, 0.05, 0.5, 1, 0.5, and 1 $\mu g\ L^{-1}$ for As, Cd, Cr, Cu Ni, and Zn, respectively.

Freundlich isotherm

Amount of heavy metal adsorbed, q_e (mg g^{-1}), is determined for each sorbent by simply analysing the corresponding heavy-metal concentration before and after the treatment using the Freundlich adsorption isotherm given below:

$$\log q_e = \log K + \frac{1}{n} \log C_e, \tag{1}$$

where C_e is the equilibrium heavy-metal concentrations in the solution (mg L^{-1}), and K and $1/n$ are adsorption constants.

Results and discussion

Literature survey

The survey revealed that the number of studies on the sorbents or filtration materials directly used for storm water treatment are limited, thus the review is extended to include sorbents used for other water treatment purposes in view of the assumption that, at least, some of them can be potentially developed for storm water treatment purposes. The reviewed sorbents are divided into six different groups and some of the effective sorbents from the each group are presented below:

1. Sand and sorbents prepared by coating to sand (e.g., BCS, ABCS, IOCS)
2. Sorbents developed from waste materials (e.g., bauxsol, activated bauxsol [AB], FA, blast furnace sludge [BFS])
3. Sorbents reported as unconventional low-cost sorbents including biosorbents (e.g., wheat bran, mulch, activated oak shells, peat, bark, chitosan, canola meal)
4. Geological or natural materials and oxides (e.g., zeolite, aluminum oxide, goethite)
5. Sorbents developed in the laboratory (e.g., resin PAGA, polyacrylamide grafted hydrous tin (IV) oxide gel having carboxylate functional groups, nanostructured akaganeite)
6. Commercially available sorbents (e.g., activated carbon, alumina, GAC, amberlite resin, granulated iron or ferric hydroxide)

Among them IOCS, sand, granulated ferric hydroxide (GFH) and zeolites are already tested in some storm water treatment facilities. On the other hand, blast furnace sludge, mulch, zeolites, goethite, resin PAGA, amberlite, peat, coir, and GFH is highlighted as efficient Cu removal sorbents. It is noted that a quantitative comparison of the wide range of sorbents is presented elsewhere along with the detailed results of the literature survey including sorbent capacities regarding to metals, and relevant experimental conditions [6].

Cu removal using alumina and GAC in batch experiments

The results of Cu removal using alumina and GAC in batch experiments at the presence of As, Cd, Cr, Ni, and Zn are presented in Figure 1. It is found that GAC is slightly more effective sorbent than alumina for Cu removal. The heavy-metal removal using GAC is primarily enhanced by its high porosity and high surface area, and the heavy-metal ions are attached to the surface bonding sites. It should be noted that the full outcome of the batch investigation is presented elsewhere [7].

Here, the final pH after the shaking period is about 7.4 for alumina and 8.7 for GAC. Similarly it is reported that heavy-metal removal is accompanied by pH increase when using GAC [8]. The observed pH increases are attributed to the dissolution of the sorbents particles. It is suggested

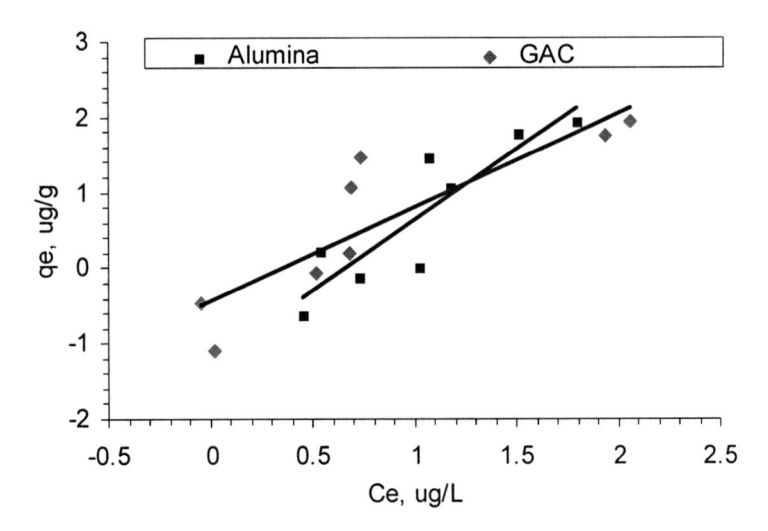

Fig. 1. Freundlich isotherms for Cu removal from storm water at the presence of As, Cr, Cd, Ni, and Zn using alumina and GAC with sorbent dosage of 20 g L^{-1}, ionic strength of 0.01 M NaCl, starting pH of 6.5, and with 3 mM NaHCO$_3$

using the outcome of the PHREEQ-C speciation calculations (data shown in [7]) that due to these pH increases some of the Cu may have precipitated during the sorption experiments (e.g., as $Cu_2CO_3(OH)_2$) especially at higher initial heavy-metal concentrations.

Freundlich isotherm is used to investigate the sorption data, but very well fits are not obtained possibly due to the fact that simple Freundlich isotherm fails to predict competitive adsorption (because of the multisolute adsorption, as competitive adsorption may result in mutual suppression of the adsorption capacity of each solute), and that precipitation may be dominant mechanism at least at the higher initial metal concentrations, and thus the isotherm might have overpredicted the adsorption.

Effect of time, pH, and the presence of HA and FC on Cu removal

The effect of reaction time, solution pH, and the presence of FC and HA on the removal of Cu at the presence of As, Cd, Cr, Ni, and Zn is investigated in batch tests using alumina, and GAC as sorbents, and the results are presented in Figures 2 and 3. Note that further results are presented elsewhere [9].

Effect of reaction time on the removal

It is found that the equilibrium time for Cu is about 2 h but a very minor sorption is continued at least up to 48 h.

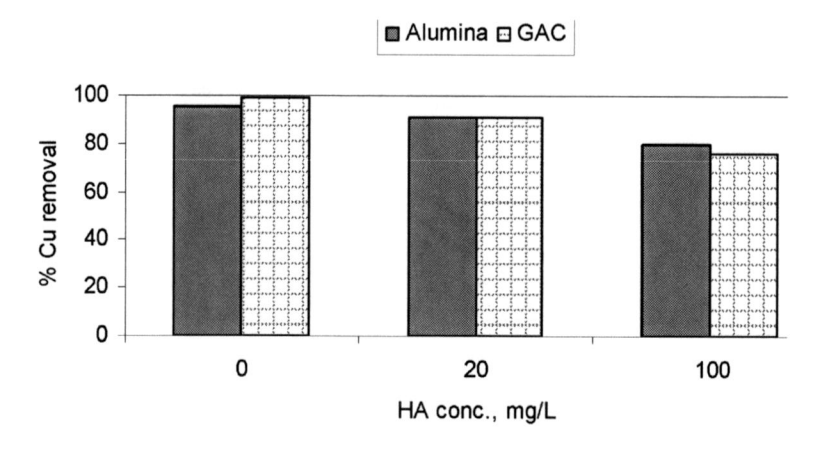

Fig. 2. Effect of HA on Cu removal from storm water at pH 8.5 with initial heavy-metal concentrations of ~1 mg L^{-1}, ionic strength of 0.01 M NaCl, and with 3 mM $NaHCO_3$

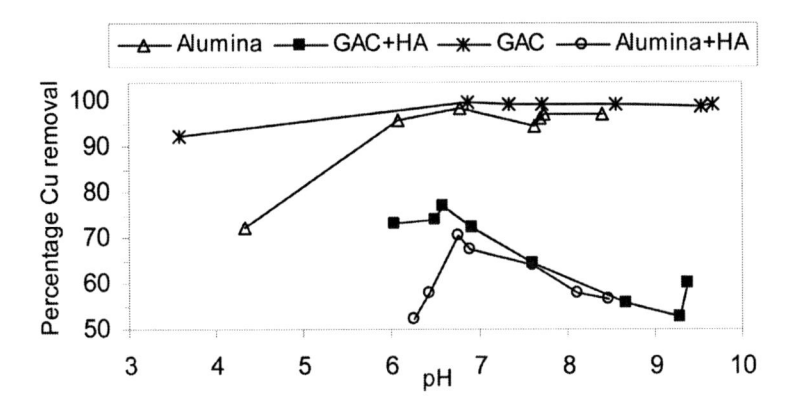

Fig. 3. Effect of pH on Cu removal from storm water at the absence and presence of 100 mg L^{-1} HA using alumina, BCS, and GAC with the initial heavy-metal concentrations of 1 mg L^{-1} As, Cd, Cr, Cu, Ni, and Zn; ionic strength of 0.01 M NaCl, and with 0.003 M NaHCO$_3$

Effect of Fe colloids on the metal removal

Effect of FC on Cu removal using the sorbents is tested and the results showed that the presence of FC has insignificant effect on the metal removal [9] both in the column experiments, where initial FC concentration was 5 mg L^{-1}; and in the batch experiments, where initial FC concentration range was 5–50 mg L^{-1}.

Effect of humic acid

The effect of 20 and 100 mg L^{-1} HA on the Cu removal is investigated, and it is found that Cu removal is suppressed due to the presence of HA (see Fig. 2), and the suppression is more pronounced at the higher HA concentration and when using GAC. It is noted that heterogeneous nature of HA is reported to make the metal sorption complicated [10]. Here, the Cu removal is reduced from 99.3% to 76.3% when 100 mg L^{-1} HA is added to the GAC system, while the corresponding numbers for alumina is 95.5% and 79.6%, respectively. Cu is known to have higher affinity towards HA, thus it may have formed un-sorbable Cu–HA complexes.

Effect of pH

The pertinent data presented in Figure 3, indicates that the metal removal follows a similar pattern at the absence and presence of HA. In general, Cu removal is enhanced when pHpzc value of a sorbent is lower than the solution pH (as the surface is negatively charged). It is noted that the pHpzc values are generally about 9 for alumina and 6 for GAC. The observed Cu sorption at pH values smaller than pHpzc of the sorbent sorption at pH values greater than pHpzc indicates that sorption is governed by additional forces and not by electrostatic attraction. For example, when the surface charge and sorbate change are same then the sorption would take place either as specific sorption or via chemical interactions with enough energy to overcome the repulsive forces between the same charges. However, it is also noted that due to lower initial metal concentrations used the pH effect may not be entirely clear, as ~100% removal is recorded for Cu at a wide pH range as can be seen from Figure 3.

Cu removal in column experiments

Sorption filters, filled with alumina and GAC, are used to remove As, Cd, Cr, Cu, Ni, and Zn when the metals coexist at the presence and the absence of HA or FC, and some of the results are presented in Figure 4. It can be seen from Figure 4 that alumina and GAC demonstrated very similar Cu removal efficiencies, but GAC was slightly more effective as also observed in the batch tests both at the absence and presence of HA and FC. The results indicated that the process is sensitive to both parameters, and columns run at the absence of HA and FC providing better column performance. It should however be noted that the magnitude of FC suppression is quite small compared to that of HA. It is emphasized that the similar lar results are observed in the batch tests.

It is postulated that the metal removal in columns takes place with several processes including diffusion into mineral lattices, sorption onto filter media, if an organic film is formed then also sorption on natural organic matter. It is suggested that possibly the competitive adsorption of six heavy metals may have decreased the capacity of the sorbents for Cu sorption.

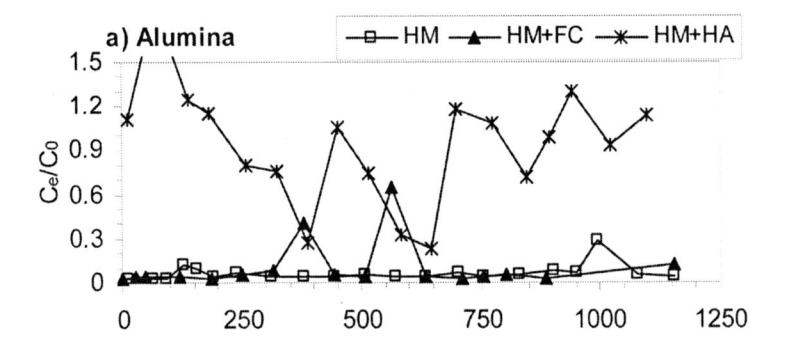

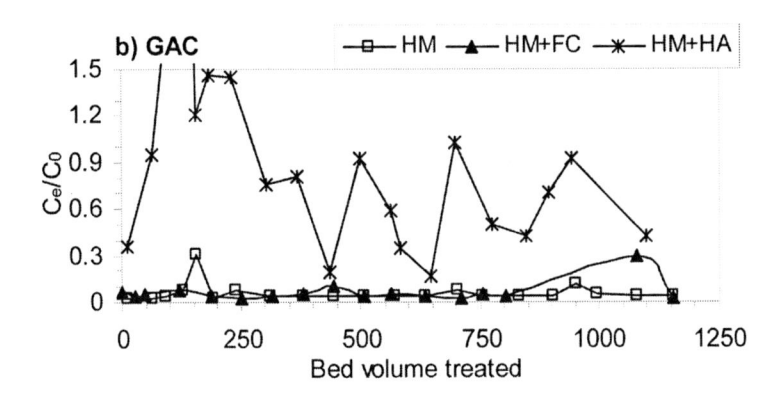

Fig. 4. Cu removal from storm water in alumina and GAC columns. All feeding waters had 0.01 M NaCl, and 3 mM NaHCO$_3$ at pH ~7. Columns those included FC or HM initial concentrations were 5 and 100 mg L^{-1}, respectively

Conclusions and suggestions

In this study batch and column experiments are carried out to investigate the effect of reaction time, solution pH, and presence of HA and FC on the combined removal of As, Cd, Cr, Cu, Ni, and Zn using alumina and GAC as sorbents, and the following conclusions are drawn:

1. When using alumina or GAC the sorption of Cu at the presence of As, Cr, Cd, Ni, and Zn reached equilibrium approximately after 2 h, but a very slow adsorption continued for 48 h.
2. The batch results also showed that the sorption is highly affected by the solution pH both at the absence and the presence of HA, in general the sorption of Cu increased with increasing pH.
3. The presence of FC had minor effect on the Cu removal both in batch and column experiments using the initial FC concentrations as high as 50 mg L^{-1}. In the case of HA a significant suppression is documented for both alumina and GAC.

Here experiments are carried out using synthetic storm water samples and effect of several parameters such as time, pH, presence of HA, and FC are checked to better understand the performance of the sorbents. It should however be noted that compared to the experimental solution used here, storm water has complex composition of pollutants, which are both bound to particulates fractions and or exist in dissolved form in the water. Furthermore, the concentration of metals in storm water changes with time, temperature, and location. Clearly, the laboratory experiments stimulating the exact storm water conditions are rather difficult, if not impossible, and thus field tests must be carried out under the light of the current laboratory findings for effective design of storm water filters for the secondary treatment.

Acknowledgements

This work is funded by Danish Research Council (Grant no.:26-03-0326). The laboratory technician Susanne Kruse from the same institute is also acknowledged for the contribution during the experiments. Thanks are also to Haldor Topsøe A/S, and Kemira Denmark for supplying the alumina and GAC, respectively.

References

1. USEPA (1996) National water quality inventory - 1996 Reports to Congress Office of Water, USEPA, Washington, DC
2. Makepeace DK, Smith DW, Stanley SJ (1995) Urban stormwater quality: summary of contaminant data. Environ Sci Technol 25:93–139
3. Grout H, Wiesner MR, Bottero J (1999) Analysis of colloidal phases in urban stormwater runoff. Environ Sci Technol 33:831–839

4. Dario M, Ledin A (1997) Sorption of Cd to colloidal ferric hydroxides-impact of pH and organic acids. Chem Spec Biol 9:3–14
5. Pedersen HD, Postma D, Jacobsen R, Larsen O (2005) Fast transformation of iron hydroxides by the catalytic action of aqueous Fe(II). Geochim Cosmochim Acta 69:3967–3977
6. Genç-Fuhrman H, Mikkelsen PS, Ledin A. Filtration materials for stormwater treatment: a review (in preparation)
7. Genç-Fuhrman H, Mikkelsen PS, Ledin A (2007) Simultaneous removal of As, Cd, Cr, Cu, Ni and Zn from stormwater: experimental comparison of 11 different sorbents. Water Res 41:591–602
8. Chen JP, Wang X (2000) Removing Copper, Zinc, and Lead Ion by granular activated carbon in pretreated fixed-bed columns. Sep Purif Technol 19:157
9. Genç-Fuhrman H, Mikkelsen PS, Ledin A. Effect of, time, pH, humic acid and iron colloids on the heavy metal removal (in preparation)
10. Alcacio TE, Hesterberg D, Chou JW, Martin JD, Beauchemin S, Sayers DE (2001) Molecular scale characteristics of Cu(II) bonding in goethite-humate complexes. Geochim Cosmochim Acta 65:1355–1366

Modelling the oxygen mass balance of wet detention ponds receiving highway runoff

HI Madsen, J Vollertsen, T Hvitved-Jacobsen

Section of Environmental Engineering, Aalborg University, 9000 Aalborg, Denmark

Abstract

The dissolved oxygen (DO) concentration is a central quality parameter for the performance of wet detention ponds used for storage and purification of storm water runoff from urban catchments and roads. A dry weather DO mass balance model was established for two ponds based on measured data and empirical relationships for the governing processes. The results of the DO model were used in a risk assessment for occurrence of low DO concentrations. An evaluation of design criteria for wet detention ponds was accomplished in terms of DO influencing parameters like water depth, temperature, and wind speed.

Introduction

A wet detention pond (wet pond) is a pond with a permanent water pool designed for storage and purification of storm water runoff from impervious surfaces. The pond behaves like a small shallow lake purifying runoff water through naturally occurring processes like sedimentation, biodegradation, adsorption, chemical precipitation, and plant uptake. The design of wet ponds is in general based on rather crude information like empirical equations for pollutant removal efficiency and runoff load [1]. In order to optimize the design of these ponds, more detailed knowledge is needed on the chemical, biological, and hydraulic processes occurring in the ponds under wet and dry weather conditions.

The dissolved oxygen (DO) concentration is a central quality parameter for the performance of a wet pond, potentially critical during long dry summer periods where the DO consumption is high, the solubility of oxygen in water is low and the calm, dry weather can result in a low reaeration rate and a low degree of mixing. Discharge of water from a wet pond with a low DO concentration can have severe impacts on the organisms and fauna present in downstream receiving waters. Furthermore, anaerobic conditions at the bottom of a pond, may lead to noxious smell and changed redox conditions resulting in, e.g., release of phosphorus and heavy metals [2]. It is therefore crucial to maintain a relatively high DO concentration (e.g., above 2 mg L^{-1}).

The objective of this study is to develop a dry weather model for prediction of DO concentrations in wet detention ponds. The model is used in an assessment of design criteria influencing the DO concentration in wet detention ponds. These design criteria are expressed in terms of particularly water depth, wind speed, and temperature.

Model set-up

A number of parameters need to be considered when setting up a reliable DO model. A wet pond is an ecosystem with characteristics similar to both a shallow lake and a wetland. The existing knowledge of modelling such systems can therefore be applied. Ecosystems consist of numerous interacting processes affecting the DO concentration. However, in case of design of wet ponds, rather limited information is available. It is therefore considered important that a DO model for risk assessment and performance of wet ponds has a low data requirement. Only simple empirical expressions that are central for the DO mass balance should be considered and included in the model, but the model should under such constraints still result in a reliable determination of design criteria. It is therefore a major task to assess which DO influencing processes should be included and to exclude those of minor importance. In this respect it is crucial that field studies have been performed as a basis for assessment of the feasibility of such a very simple DO model for design purposes.

Mixing conditions

Transport and mixing in lakes are often controlled by dispersion affected particularly by the wind at the water surface and temperature gradients [3,4]. In small lakes, the water currents are usually unevenly distributed because of shelter and boundary flow, but both horizontal and vertical gradients tend to be smoothed by wind [5,6]. Furthermore, shallow lakes respond more readily to variations in meteorological conditions compared to deep lakes [7].

Wind at the water surface mixes the lake water to a limited depth, resulting in a density difference, whereby thermal stratification may occur [6]. Stratification in shallow lakes is generally rare due to mixing by wind, however, varying with day and season stratification occurs [8].

Some of the eutrophication models for lakes include DO-influencing processes and most of these models are constructed as box models [3]. The lake is divided into a number of completely mixed boxes, where the transport in and out of each box is taken into account. This model type is typically used where full mixing does not occur. It is therefore important to determine the mixing conditions in the pond prior to modelling.

Based on the above-mentioned facts, permanent stratification is not expected in a shallow wet pond. The extent of mixing depends on the location of the pond, amount of shelter, water depth, etc. It is therefore considered essential to examine the mixing conditions prior to modelling, e.g., by addition of a conservative tracer like Rhodamine WT.

The water mass balance

The water mass balance consists of inflow, outflow, evaporation, precipitation, seepage, and infiltration. In case of a dry weather model, inflow, outflow, and precipitation can be omitted.

The evaporation rate depends on relative humidity, temperature, and wind velocity [9]. Lake evaporation was found to vary from 0.6 to 5.4 mm day^{-1} in a Swedish lake measured from June to September (surface area: 35 km^2 and average depth: 1.2 m) [10]. In a shallow lake, the presence of vegetation reduces the water loss in comparison with evaporation from an open water surface [11]. The evaporation may therefore be neglected for short simulation periods.

Seepage and infiltration can be considered negligible, however, depending on the soil porosity, the methods used for pond construction and the pond management practices [12].

The oxygen mass balance

An outline of the DO influencing parameters in a wet pond is shown in Figure 1. Chamber measurements can provide parameters for a DO model, however, the experimental procedures suffer from several problems like non-representative sampling, disruption of the enclosed benthic community, nutrient depletion, and scaling up to real systems [13,14]. Instead, *in situ* methods based on diurnal changes in DO concentrations were suggested, where it is assumed that increases in DO during the day are mainly due to photosynthesis, while decreases during the night are caused by respiration [15]. It is stated that the greatest difficulties with this method are the inaccurate corrections for reaeration and problems with precise measurements of small DO variations [16].

Reaeration

One of the most important factors that affect the air–water mass transfer rate of oxygen is the wind speed at the water surface [17]. The relation between the oxygen mass transfer rate and wind speed has been measured in three wet detention ponds using a tracer gas method with propane as the tracer gas [18]. The basic of this method is a constant ratio between the mass transfer rate for oxygen absorbed by water and the mass transfer rate for a tracer gas desorbed by the same water volume. The empirical result of this study is shown in Eq. 1.

$$K_L(20) = 0.11 \cdot w + 0.44 \tag{1}$$

where $K_L(20)$ is the oxygen air–water mass transfer rate at 20°C (m day^{-1}) and w the wind speed (m s^{-1}). The empirical equation is valid in the interval 1–8 m s^{-1} [18].

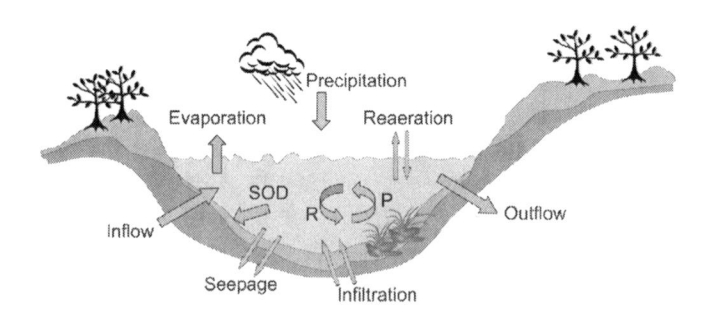

Fig. 1. The DO influencing parameters in a wet pond. P is the photosynthesis, R the respiration, and SOD the sediment oxygen demand

Application of Eq. 1 in the DO model yields a good estimate for the reaeration rate in the valid interval, solving the difficulties previously mentioned with inaccurate corrections for reaeration in the methods based on diurnal changes in DO concentrations.

Sediment oxygen demand

Sediment oxygen demand (SOD) includes both biological respiration and chemical oxidation of reduced substances [15]. The SOD rate can be considered constant over time, depending on the duration of the simulation period. The SOD rate is site specific, but is typically in the range 1–5 g O_2 m^{-2} day^{-1}.

Photosynthesis

Photosynthesis by plants and algae in the water phase depends primarily on three external parameters: Temperature, solar radiation, and the availability of nutrients [19]. At low light intensities, photosynthesis is directly proportional to the intensity of the solar radiation [20]. The variation in solar radiation during the day can be modelled as a cosine function with a maximum photosynthesis at noon [21].

Additionally, a number of models exist with nutrients as the limiting factor, but in general it can be assumed that wet ponds are not nutrient limited due to a relatively low retention time and large inflow of nutrients.

Respiration

Aerobic respiration of organisms and plants is the utilization of organic matter for energy and biomass growth, whereby, the DO concentration in the water phase is affected. Respiration is primarily determined by the type of organism, biomass size, and the temperature [9]. The respiration rate can normally be considered constant over short periods [15,19].

Materials and methods

The field sites used as basis for the development of the DO model consists of two wet detention ponds located in North Jutland, Denmark. The wet ponds receive surface water from a highway (E45) with an average daily traffic load of 9500–12,300 vehicles. Physical properties and location for each pond are shown in Figure 2.

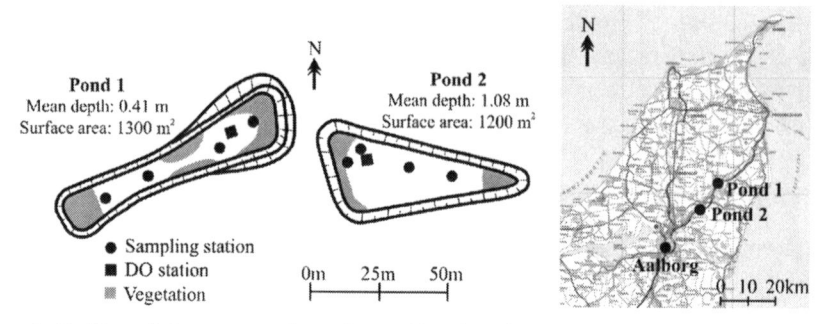

Fig. 2. Outline of the two ponds and sampling locations

The DO concentration was measured using an Evita Oxy 150 transmitter (Hach Lange, Denmark) with an accuracy of 0.5% of saturation. The DO concentration was logged in both ponds every minute during the period from 8 September to 20 October 2005. The transmitter was placed 0.18 and 0.64 m above the bottom in ponds 1 and 2, respectively.

Data for solar radiation, wind speed, and air temperature was obtained from a weather station operated by the Danish Meteorological Institute (DMI). The station is located in Aalborg 25–30 km from the two ponds. A comparison of wind speeds measured at two weather stations located in the same region and with a distance of 25 km shows similar variations [18]. It is therefore assumed, that the variations in weather conditions obtained in Aalborg is valid at the field sites. The data was recorded every 10th min.

The mixing conditions in the ponds were studied by addition of a conservative tracer, Rhodamine WT. After addition, samples for Rhodamine WT determination were drawn at four locations in each pond with time intervals of 5–10 h (Fig. 2). The concentration of Rhodamine WT was measured on a RF-1501 spectrofluorophotometer with the excitation and emission set to 558 and 582 nm, respectively. The outcome of the mixing experiments is used to determine the number of boxes applied in the model.

The DO model applied is shown in Eq. 2.

$$\frac{dC}{dt} = K_L a \cdot (C_S - C) + P(t) - R - \frac{SO}{H}$$

where C is the DO concentration (g m^{-3}), t the time (d), a the surface area per unit volume (m^{-1}), C_s the DO concentration at saturation (g m^{-3}), $P(t)$ the photosynthesis (g m^{-3} day^{-1}), R the respiration (g m^{-3} day^{-1}), SOD the sediment oxygen demand (g m^{-2} day^{-1}), and H the pond depth (m).

Two different functions for photosynthesis were used: a cosine function and a function of monitored solar radiation. The photosynthesis is inhibited below a certain depth, which is found from calculations with the calibrated parameters from pond 1 in the model of pond 2, while it is presumed that no light inhibition occurs in pond 1, because of the low depth.

The SOD rate and respiration rate were assumed constant during the modelling period. The reaeration, SOD, and respiration were temperature dependent. The time step for modelling was set to 10 min. The model was calibrated by minimization of the root mean square error (RMSE), which expresses the deviation of the modelled data from the measured data.

Results and discussion

The results of the experiments with Rhodamine WT in the two ponds showed that full mixing occurred relatively fast (within 20 h). The model applied is therefore a box model with one completely mixed box.

The two different functions for photosynthesis gave almost identical results, which reconfirms that the solar radiation is approximately a cosine function of time. The cosine function is used in the final modelling due to less input requirements.

The results from calibration and validation of the model with data from pond 1 are shown in Figure 3. Most of the time, the model fits the measured data well for both the calibration period and the validation period. The model fits the measured data poorly on 22 September and 15–17 October, which is believed due to problems with correct modelling of the DO concentration at low wind speeds (cf. Fig. 3). The wind speed influences

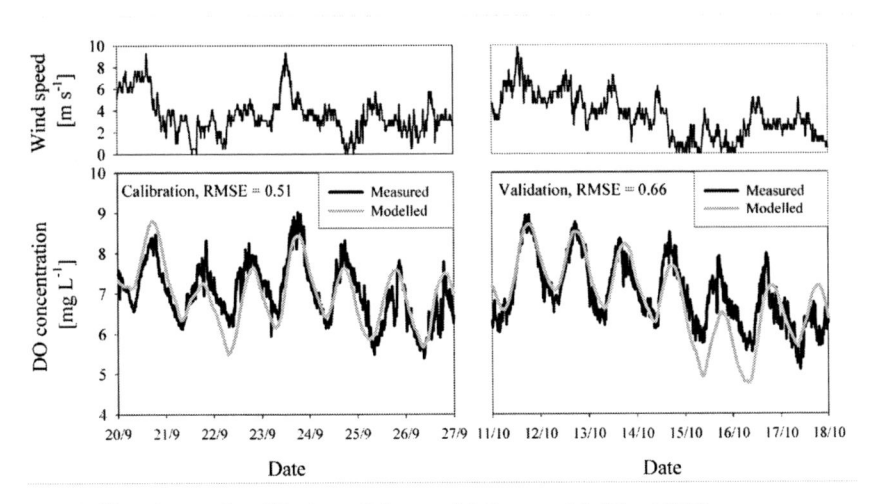

Fig. 3. Calibration and validation of the model for pond 1 ($T = 12°C$)

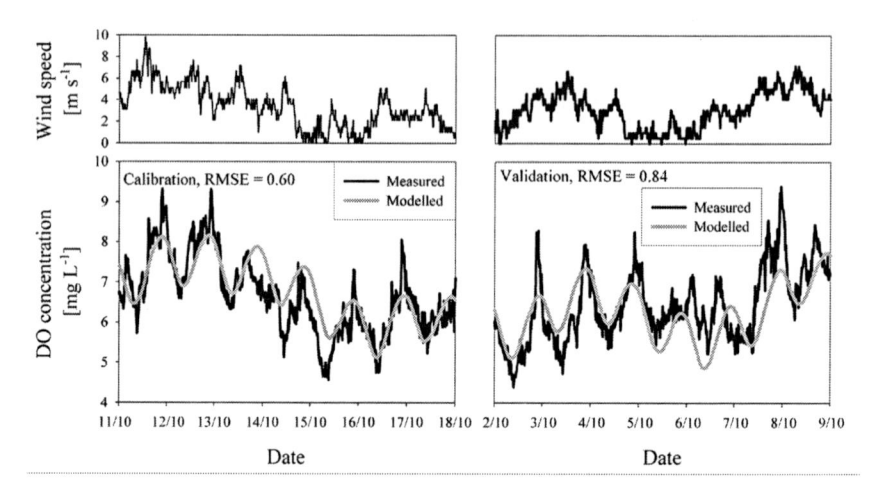

Fig. 4. Calibration and validation of the model for pond 2 ($T = 12°C$)

the reaeration rate, which was found from tracer gas experiments in the two ponds (Eq. 1). The modelling results indicate that the influence of wind on the reaeration rate is not as pronounced at low wind speeds as found in the tracer gas experiments. The model can therefore not be applied at low wind speeds (lower than 2 m s^{-1}).

A similar calibration and validation were done with the data from pond 2 (Fig. 4). The validation of the model is relatively poor in the period 5–6 October, which is believed due to the low wind speeds during the period.

Table 1 shows three calibrated parameters for the models in ponds 1 and 2. The models for the two ponds result in almost similar parameters. The SOD rates are in the upper interval of SOD values previously mentioned, which may be due to a high inflow of organic matter and nutrients.

Table 1. Values for three calibrated parameters for the model in ponds 1 and 2

	SOD (g O$_2$ m^{-2} day^{-1})	R (g O$_2$ m^{-3} day^{-1})	P_m
Pond 1	3.78	6.10	8.50
Pond 2	4.36	6.44	8.20

P_m is a dimensionless oxygen production ($P(t) = P_m \cdot \cos(\pi x \cdot t)$) [21], x is a daytime-dependent parameter

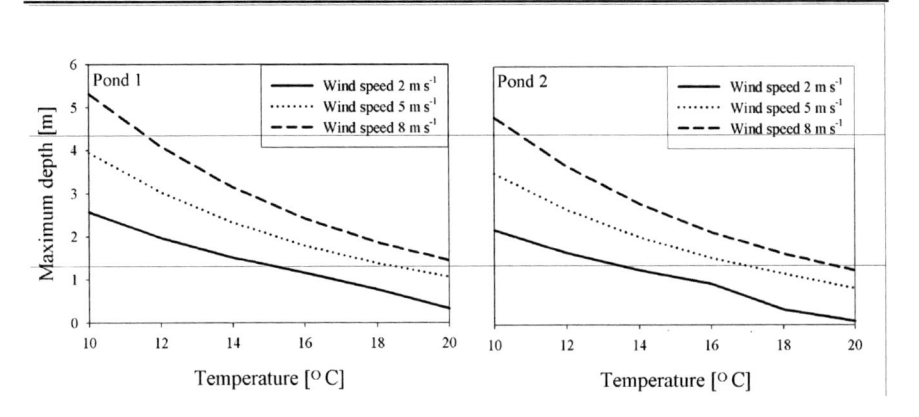

Fig. 5. The influence of temperature and wind speed on the maximum allowable pond depth applied in order to avoid average DO concentrations below 2 mg L^{-1}. The calculations are done under steady state conditions

The model is used for assessing a maximum pond depth in order to avoid DO concentrations below 2 mg L^{-1}. The influence of wind speed and temperature on the maximum pond depth under steady state conditions is shown in Figure 6. In Denmark, the summer temperature in a pond may reach 20°C during a short period, but a normal average summer temperature is 15°C. With a wind speed of 4 m s^{-1} a maximum pond depth is approximately 1.75 m (Fig. 5).

The DO concentration in pond 1 has been simulated with measured wind speed data from 7 July to 28 October 2005. In this simulation, the total time with DO concentrations below 2 mg L^{-1} and the longest period with DO concentrations below 2 mg L^{-1} were examined. As an example, a temperature of 15°C and a pond depth of 1.4 m would result in a DO concentration below 2 mg L^{-1} in 6% of the time with the longest period of 1.6 days.

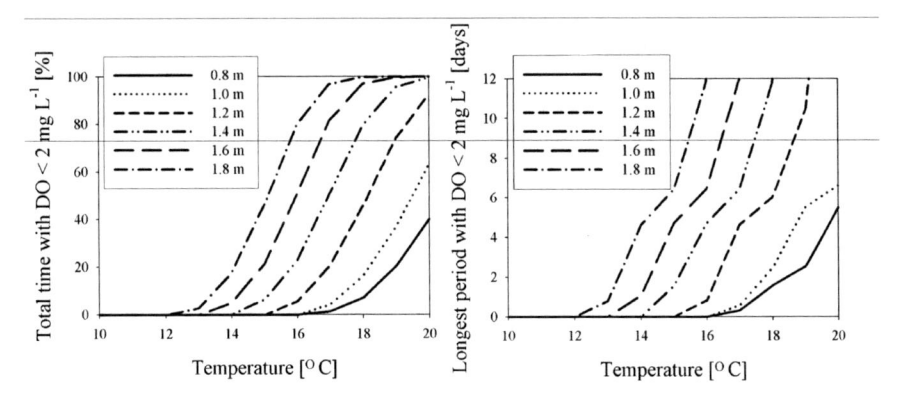

Fig. 6. Total time and longest period with DO concentrations below 2 mg L^{-1} during the period from 7 July to 28 October 2005

Different maximum depths have been suggested in the past, e.g., 1.0–1.5 m [1]. In Denmark, recommendations of a pond depth will according to the results of this study be in the order 1.25–1.5 m. Furthermore, it is important not to provide shelter around the pond.

Conclusions

The results of this study are based on measurements in real wet detention ponds and add new knowledge to an improved pond design for maintaining aerobic conditions. The present study is therefore also a step in the direction to obtain more solid design criteria for wet detention ponds receiving storm water runoff from urban catchments and roads.

A simple model can simulate the DO variability well. Under Danish climate conditions, a recommendation for a maximum pond depth is in the order 1.25–1.5 m with this model. Furthermore, it is important to restrict the amount of shelter around the pond.

References

1. Hvitved-Jacobsen T, Johansen NB, Yousef YA (1994) Treatment systems for urban and highway run-off in Denmark. Sci Tot Environ 146/147:493–498
2. Wong JWC, Yang CL (1997) The effect of pH and redox potential on the release of nutrients and heavy metals from a contaminated marine sediment. Toxicol Environ Chem 62:1–10
3. Chapra SC, Reckhow KH (1983) Engineering approaches for lake management. Vol. 2: Mechanistic modelling. Butterworth Publishers, Boston/London/Sydney/Wellington/Durban/Toronto
4. Schnoor JL (1996) Environmental modelling – fate and transport of pollutants in water, air, and soil. Wiley, New York/Chichester/Brisbane/Toronto/Singapore
5. Hansen NEO (1978) Mixing processes in lakes. Nord Hydrol 9:57–74
6. Logan BE (1999) Environmental transport processes. Wiley, New York/Chichester/Weinheim/Brisbane/Singapore/Toronto
7. Herb WR, Stefan HG (2005) Model for wind-driven vertical mixing in a shallow lake with submersed macrophytes. J Hydraul Eng 131:488–496
8. Thomas R, Meybeck M, Beim A (1996) Lakes. In: Chapman D (ed.) Water quality assessments: a guide to the use of biota, sediments and water in environmental monitoring. Chapman & Hall, Melbourne, pp 319–368
9. Straskraba M, Gnauck AH (1985) Freshwater ecosystems – modelling and simulation. Elsevier, Amsterdam/Oxford/New York/Tokyo

10. Saxena RK (1996) Estimation of lake evaporation from a shallow lake in central Sweden by oxygen-18. Hydrol Process 10:1273–1281
11. Linacre ET, Hicks BB, Sainty GR, Grauze G (1970) The evaporation from a swamp. Agric Meteorol 7:375–286
12. Boyd CE (1982) Hydrology of small experimental fish ponds at Auburn, Alabama. Trans Am Fish Soc 111:638–644
13. Bott TL, Brock JT, Cushing CE, Gregory SV, King D, Petersen RC (1978) A comparison of methods for measuring primary productivity and community respiration in streams. Hydrobiologia 60:3–12
14. Kemp WM, Boynton WR (1980) Influence of biological and physical processes on dissolved oxygen dynamics in an estuarine system: implications for measurement of community metabolism. Estuarine Costal Mar Sci II:407–431
15. Odum HT (1956) Primary production in flowing waters. Limnol Oceanogr 1:102–117
16. Mazolf ER, Mulholland PJ, Steinman AD (1994) Improvements to the diurnal upstream-downstream dissolved oxygen change technique for determining whole-stream metabolism in small streams. Can J Fish Aquat Sci 51:1591–1599
17. Gelda RK, Auer MT, Effler SW, Chapra SC, Storey ML (1996) Determination of reaeration coefficients: whole-lake approach. J Environ Eng 122:269–275
18. Madsen HI, Vollertsen J, Hvitved-Jacobsen T. Air-water mass transfer in wet detention ponds – methodology and determination. Urban Water J (submitted)
19. Jørgensen SE, Bendoricchio G (2001) Fundamentals of ecological modelling, 3rd edn. Elsevier, Oxford
20. Piedrahita RH (1991) Modelling water quality in aquaculture ecosystems. Aquaculture Water Qual 3:322–362
21. Simonsen JF, Harremoës P (1978) Oxygen and pH fluctuations in rivers. Water Res 12:477–489

Monitoring and modelling the performance of a wet pond for treatment of highway runoff in cold climates

J Vollertsen,[1] SO Åstebøl,[2] JE Coward,[2] T Fageraas,[3] HI Madsen,[1] AH Nielsen,[1] T Hvitved-Jacobsen[1]

[1] Section of Environmental Engineering, Aalborg University, Sohngaardsholmsvej 57, 9000 Aalborg, Denmark
[2] COWI AS, PO Box 6412 Etterstad, NO-0605 Oslo, Norway
[3] Norwegian Public Roads Administration, Eastern Region, PO Box 1010 Skurva, NO-2605 Lillehammer, Norway

Abstract

A wet pond in Oslo, Norway, receiving highway runoff was studied. The pond was equipped for continuous monitoring of inflow and outflow. Samples were collected over a 1-year period and analysed. The treatment performance was documented and an adverse effect of snowmelt runoff observed. The wet pond was modelled by routing the measured flow through the pond and simulating pollutant removal by first-order kinetics. The relative importance of the permanent pool of water and the design storm storage was assessed with respect to pollutant removal.

Introduction

Storm water runoff from highways and roads contain numerous pollutants in significant concentrations. The effects of the pollutants on the receiving waters are manifold and depend on the receiving water body as well as the pollutants discharged. For example, the discharge of phosphorous to lakes and estuaries causes eutrophication, whereas heavy metals and organic micropollutants accumulate in organisms, causing toxic effects. Resent investigations have clearly shown ecotoxicological effects on the receiving environment from storm water discharges [e.g., 1].

Treatment of storm water runoff is needed to ensure good ecological quality of receiving sensitive surface waters. Technologies for treatment must be robust, dependable, and easy to operate. They must furthermore be suitable for managing an intermittent and highly variable rainfall pattern. In addition hereto, the technologies must be appropriate to treat rather dilute pollutants as they occur in storm water runoff. Among the commonly used technologies, wet ponds have proven efficient performance and are in many cases the technology chosen for storm water treatment.

The Norwegian Public Roads Administration has made extensive use of wet ponds, and decided to study the performance of such a pond under Norwegian climate conditions. They constructed a wet pond in the greater Oslo area and equipped it with devices for continuous monitoring. It is the objective of this study to analyse the monitored data and draw generalized conclusions for the future design of wet ponds.

Methods

Site and pond characteristics

The wet pond investigated was constructed in 1998–1999 to protect a small river (Ljanselva) from pollution caused by runoff from the E6 highway (Fig. 1). The highway is located in the greater Oslo area and is heavily

Fig. 1. The Skullerud junction on the E6 highway

Table 1. Pond dimensions, cf. Figure 2

	Surface area (m²)	Volume (m³)	Depth (m)
Permanent pool, silt trap	68	103	1.5
Permanent pool, earthen pond	910	710	0.8
Design storm storage, silt trap	68	14	0.2
Design storm storage, earthen pond	1034	194	0.2

traffic-loaded with an ADT of 42,000 vehicles per day. The pond serves a paved area of 2.2 ha, of which 1.5 ha are the road surface of the E6, 0.5 ha are the paved surfaces of the junction ramp, and 0.2 ha are walkways and bikeways. The pond furthermore serves a vegetated area of 1.2 ha.

When designing the wet pond, no significant reduction in peak flow prior to discharge to the river was required, and the pond was consequently designed with a large permanent pool and a modest design storm storage capacity (Table 1), i.e., the pond was designed for storm water treatment but not for peak flow reduction. The hydraulic retention time was designed to 72 h with a return period of 3–4 year^{-1} [2,3].

The inlet pipe to the pond is a 400 mm PVC pipe, entering the silt trap below the water surface (Fig. 2). From the silt trap, the storm water enters an earthen pond with a length to width ratio of approximately 4:1. The outflow structure of the pond is a submerged 315 mm PVC pipe.

The pond was constructed for detailed monitoring of flow and pollutants. At the inlet and the outlet, full-flowing magnetic flow meters of type Starflow 6526-21 were placed as part of the permanent structure. The flow meters return the flow velocity as well as the water pressure; the later is interpreted as the pond water level. The accuracy of the water pressure reading

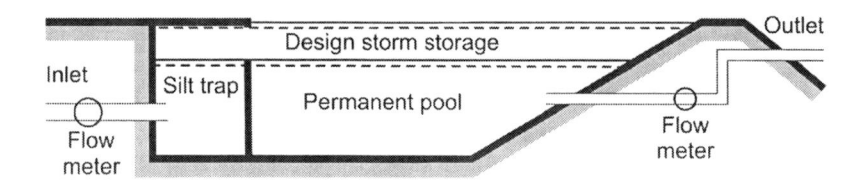

Fig. 2. Schematic layout of the Skullerud storm water pond

is 3 mm and the accuracy of the flow reading is 2% of the measured value. The measuring range for the inlet and the outlet flow meter is 2.6–565 L s^{-1} and 1.5–318 L s^{-1}, respectively. It is estimated that 5–10% of the total runoff is below the detection limits of the inlet flow meter. Precipitation is continuously measured by a tipping bucket rain gauge of type Young 52202.

Storm water sampling

One autosampler for water quality monitoring was placed at the inlet to the silt trap and another autosampler was placed at the outlet of the earthen pond. Subsamples were taken for every 8 m^3 of incoming and outgoing storm water. The samples were analysed after each runoff event.

The pond was intensively monitored during the period 1 May 2003 to 30 April 2004. During this period, a total of 28 composite samples were collected from the inlet and from the outlet. The samples covered 87% of the measured incoming storm water.

Each composite sample was analysed for water quality parameters according to Norwegian standards (in principal following internationally recognized standard methods): Total PAH, 4-PAH [benzo(b)-fluoranten, benzo(k)fluoranten, benzo(ghi)perylen and indeno(1,2,3-cd-)pyren], oil and fat, suspended solids, total nitrogen, total phosphor, bioavailable phosphor (molybdate reactive phosphor according to Blakar and Løvstad [4]), lead, cadmium, copper, zinc, pH, and conductivity.

Wet pond simulation

Pollutant removal and residence times were simulated dividing the pond into n completely mixed compartments. The compartments were coupled in series and volume equalization between the compartments occurred instantaneously. The pond water balance consisted of inflow, storage, and outflow. Evaporation and precipitation on the pond surface were omitted as these contributions were minor compared to the inflow and the outflow.

Removal of a pollutant I was simulated by first-order kinetics according to Eq. 1 as proposed by, e.g., Mitsch et al. [5], k_c being a first-order rate constant. Even though such simple kinetics does have limitations when simulating wet ponds [6], they are believed a reasonable choice where the detailed processes are not well defined or well understood [2,7].

$$\frac{dC}{dt} = -k_c C \qquad (1)$$

Pond residence times were simulated by adding a virtual tracer each minute during inflow, and following the added tracers through the pond. The residence time of the water volume was defined as the time until half the tracer added to that water volume had left the pond.

Results and discussion

Treatment performance

The accumulated precipitation during the monitoring period was 620 mm and the accumulated runoff was 16,003 m^3. Viewing the accumulated runoff together with the accumulated precipitation (Fig. 3), it was observed that the ratio between runoff and precipitation was rather constant over the summer period. However, around November the proportion of runoff to precipitation increased drastically.

Defining the effective catchment area (A_{eff}) as the accumulated runoff (ΣQ_{runoff}) divided by the accumulated precipitation ($\Sigma P_{catchment}$), the runoff coefficient Φ is defined by Eq. 2, where A_{tot} is the total catchment area.

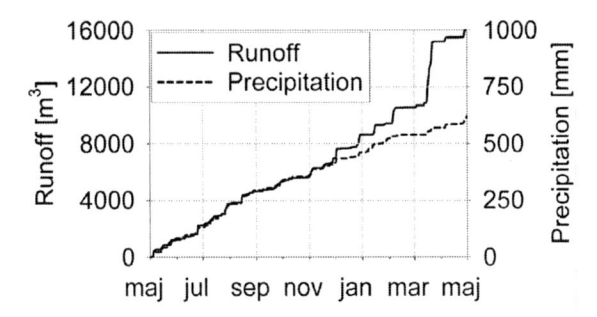

$$\Phi = \frac{\Sigma Q_{runoff}}{\Sigma P_{catchment} A_{tot}} = \frac{A_{eff}}{A_{tot}} \qquad (2)$$

Fig. 3. Accumulated runoff and accumulated precipitation

In theory, the runoff coefficient can never exceed 1, and in practice Φ is dictated by losses due to infiltration through permeable surfaces, accumulation on the catchment surface and evaporation. Figure 4 shows the runoff coefficient calculated on a monthly basis. From 1 May 2003 to mid November 2003, Φ is on average 0.47, corresponding well with the fact that 65% of the total catchment area is paved. However, as the weather gets colder, Φ increases, and in the months of December, February, and March, Φ exceeds to 1.

It is likely that saturation or freezing of the soil surface causes some decrease in the perviousness of the unpaved parts of the catchment, corresponding to an increase in Φ. An increase to values above 1, however, calls for other explanations. One possible cause is melting of snow that has accumulated over several months. However, analysing all of the period from 1 October to 30 April still yields a runoff coefficient of 1.26; i.e., the surrounding area must contribute to the runoff during snowmelt events. Looking at the Skullerud junction (Fig. 1), it is reasonable to assume that the wood-covered slopes surrounding part of the highway were the origin of this additional runoff.

In terms of the efficiencies achieved, the annual treatment performance of the pond (Table 2) is good compared to what is typically reported for wet ponds treating highway runoff [8]. However, scrutinizing the variation over the year, it becomes clear that two large snowmelt events in February and March 2004 – causing 28% of the annual runoff volume (Fig. 3) – result in disproportionate amounts of pollutants to be discharged from the pond, e.g., for total phosphorous, 70% of the annual discharge from the pond originates from these snowmelt events. The treatment efficiency for all pollutants therefore increases significantly when omitting the large snowmelt events in the calculations (Table 2).

Studying the same pond, Vollertsen et al. [7] conclude that an important factor for the high amounts of pollutants discharged during snowmelt periods is that the pond is ice-covered. Hereby the hydraulic retention time becomes severely reduced, causing diminished treatment efficiency and to some extent resuspension of pond sediments.

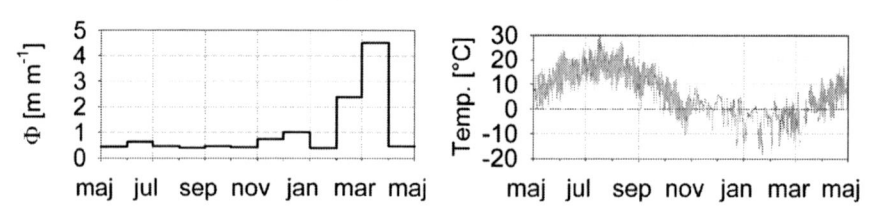

Fig. 4. The runoff coefficient calculated on a monthly basis, cf. Eq. 2 (*left graph*) and the air temperature (*right graph*)

Table 2. Annual treatment performance of the pond

	Inlet (kg year^{-1})	Outlet (kg year^{-1})	Overall treatment efficiency (%)	Treatment efficiency excl. snowmelt (%)
Suspended solids	3790	582	85	93
Total phosphorous	8.81	3.59	59	84
Bioavailable phosphorous	5.31	2.00	62	88
Total nitrogen	20.4	14.4	29	37
Oil and fat	67.9	12.0	82	89
Total PAH	0.0242	0.0035	86	92
4-PAH	0.00589	0.00067	89	96
Lead	0.234	0.056	76	83
Cadmium	0.00285	0.00115	60	70
Copper	1.17	0.50	58	66
Zinc	3.74	1.07	71	81

Residence times

Residence times were calculated with a compartment number of $n = 100$. Further increase in n did not influence the outcome of the simulations. Residence times were calculated including as well as excluding the snowmelt runoff periods of February and March 2004 (Fig. 5). A comparison of the two calculations shows that the short residence times belonged to the snowmelt runoff periods and that the residence times the rest of the year was well above 3 days.

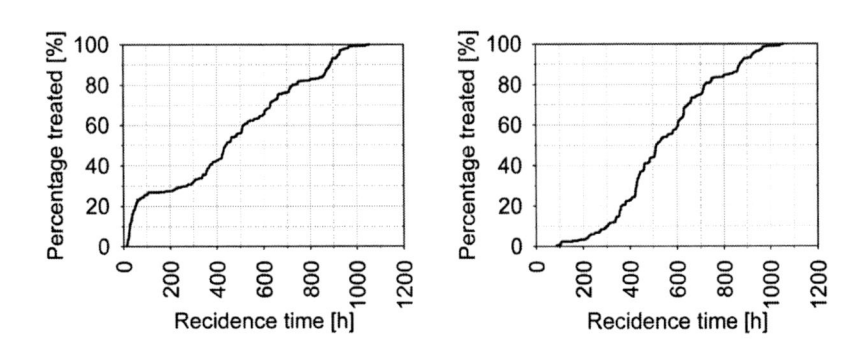

Fig. 5. Residence time distributions including (*left graph*) and excluding (*right graph*) the snowmelt runoffs in February and March 2004

Pollutant removal kinetics

The pollutant removal was simulated with a compartment number of $n = 3$, as a higher number of compartments did not result in an improved simulation of the variability in pollutant removal over the year. However, it should in this context be noted that the value of k_c (Eq. 1) depends not only on the pollutant removal to be simulated, but also on the number of compartments chosen. Table 3 gives the determined removal rate constants including and excluding the snowmelt events, and shows that some of the achieved first-order rate constants increased significantly when excluding the snowmelt runoff events from the simulations. Others, on the other hand, were unchanged or actually decreased when excluding these events.

Based hereon, it seems that process kinetics (cf. Eq. 1) are not necessarily affected by the snowmelt runoff, and that the reduction in treatment efficiency to a large extend is due to a reduced residence time, i.e., the causes of the observed reduced treatment efficiency during snowmelt runoff are mainly a reduction in treatment time and only to a lesser degree due to reduced removal rate constants. It is therefore expected that erosion of deposits plays a minor role for the performance during snowmelt periods.

Applying a similar model but only 1 completely mixed compartment, Hvitved-Jacobsen et al. [2] simulated the treatment for suspended solids and phosphorous. They simulated eight ponds in the USA and one pond in Denmark, and found specific first-order rate constants (k_c) for dissolved phosphorus, particulate phosphorus, and suspended solids to 0.1, 0.35, and 0.5 day^{-1}, respectively. When comparing with the values of Table 3, it should be kept in mind that the more compartments are applied, the lower

Table 3. First-order rate constant including and excluding snowmelt runoff

	K_c including snowmelt runoff (day^{-1})	k_c excluding snowmelt runoff (day^{-1})
Suspended solids	0.57	0.55
Total phosphorous	0.077	0.16
Bioavailable phosphorous	0.081	0.25
Total nitrogen	0.017	0.022
Oil and fat	0.53	0.29
Total PAH	0.52	0.51
4-PAH	0.86	1.25
Lead	0.22	0.15
Cadmium	0.079	0.072
Copper	0.065	0.044
Zinc	0.19	0.13

the value of k_c becomes. Vollertsen et al. (submitted) reports, e.g., for suspended solids a value of k_c of 2.0 day^{-1} for $n = 1$ and a value of 0.3 day^{-1} for $n = 100$.

Importance of the pond volume on treatment performance

Designing a wet pond, one important design decision is the volume of the permanent pool and the volume of the design storm storage (Fig. 2). Applying the present model together with the first-order rate constants obtained (Table 3), different configurations of the pond volume can be investigated.

Figure 6 illustrates the removal of suspended solids and total phosphorous, applying the values for k_c obtained including snowmelt runoff (Table 3). In the example, the pond outflow is kept at 0.0024 m^3 s^{-1} and the pond overflow is simulated to take place at the pond inlet. The discharged pollutants are the sum of the pollutants contained in the untreated overflow volume and in the treated pond outlet. The simulations show that the factor primarily determining the treatment efficiency is the volume of the permanent pool, whereas the design storm storage volume only has limited effect on the overall performance.

Simulations of the type shown in Figure 6 should, however, be interpreted with caution, as Eq. 1 is only valid as a first and rather rough description of the true processes, and probably yields over-optimistic removal efficiencies for large volumes of the permanent pool. Observations show that the treatment efficiency is not improved for permanent pool volumes above ~250 m^3 per reduced hectare [7].

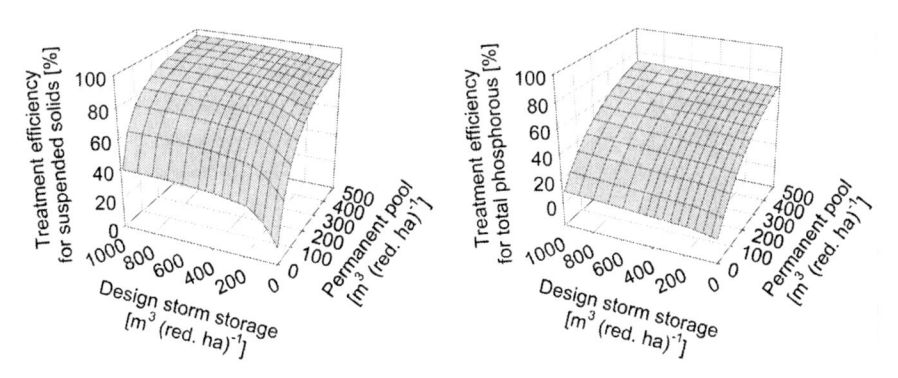

Fig. 6. Simulated removal efficiency for suspended solids and total phosphorous

Design recommendations

Evaluating the design of the Skullerud wet pond and its collection system, it would be beneficial for the overall treatment efficiency to disconnect the relatively unpolluted wood-covered slopes adjacent to the highway; e.g., by means of trenches parallel to the highway shoulders (Fig. 1). In general, it is recommended that snowmelt runoff from green areas adjacent to a highway is intercepted and not conveyed to wet ponds.

When designing a wet pond for storm water treatment, the factor limiting the pond size often is the availability of land, which therefore must be used to its uttermost to achieve the environmental objectives. Figure 6 shows that the main factor for storm water treatment is the volume of the permanent pool, whereas the design storm storage volume plays a minor role with respect to treatment.

In those cases where pollutant reduction is the main objective and hydraulic impacts in terms of, e.g., erosion and flooding of downstream receiving waters is not an issue, it is therefore as a general design rule recommended to increase the permanent pool at the expense of the design storm storage volume. Furthermore, and as a simplified rule for design, the permanent pool and the design storm storage can be sized independently: the permanent pool for storm water treatment and the design storm storage for peak flow reduction.

Conclusions

High pollutant reduction efficiency of a wet pond treating highway runoff under cold climate conditions is documented based on the results from a 1-year monitoring period. Furthermore and in general, wet ponds can be recommended as a technology for treatment of highway runoff also under such conditions. However, snowmelt can be a major challenge – partly because wet ponds become ice-covered and partly because adjacent hillsides may contribute to the runoff during winter. In general, careful consideration must be given to winter operation of wet ponds where these are placed in cold regions with significant snowfall.

The volume of the permanent pool is shown to play the by far most important role in pollutant reduction, whereas the design storm storage volume is of minor consequence. When pollution reduction is the main issue and reduction of peak flow is not required, it is recommended to increase the size of the permanent pool on the expense of the design storm

Modelling the performance of a wet pond with first-order kinetics yields a reasonable estimate of pollutant removal and allows dynamic simulation of real rainfall events. The approach is recommended as a design tool, applying long rainfall series together with empirical knowledge on removal rates to optimize the sizes of the permanent pool and the design storm storage.

References

1. Marsalek J, Rochfort Q, Brownlee B, Mayer T, Servos M (1999) Exploratory study of urban runoff toxicity. Water Sci Technol 39:33–39
2. Hvitved-Jacobsen T, Johansen NB, Yousef YA (1994) Treatment systems for urban and highway run-off in Denmark. Sci Tot Environ 146/147: 499–506
3. Åstebøl SO, Coward JE (2004) Monitoring a treatment pond for highway runoff from the E6 Skullerud junction in Oslo, 2003–2004 (in Norwegian). Report from COWI AS, Norway, to the Norwegian Public Roads Administration
4. Blakar I, Løvstad O (1990) Determination of available phosphorus for phytoplankton populations in lakes and rivers of southeastern Norway. Hydrobiologia 192:271–277
5. Mitsch WJ, Cronk JK, Wu X, Nairn RW (1995) Phosphorous retention in constructed freshwater riparian marshes. Ecol Appl 5:830–845
6. Kadlec RH (2000). The inadequacy of first-order treatment wetland models. Ecol Eng 15:105–119
7. Vollertsen J, Åstebøl SO, Coward JE, Fageraas T, Nielsen AH, Hvitved-Jacobsen T. Monitoring and modelling a wet detention pond for highway runoff. Ecol Model (submitted)
8. www.bmpdatabase.org: International stormwater best management practices (BMP) database

Can we close the long-term mass balance equation for pollutants in highway ponds?

TR Bentzen, T Larsen, MR Rasmussen

Department of Civil Engineering, Aalborg University, Sohngaardsholmsvej 57, DK-9000 Aalborg, Denmark

Abstract

The paper discusses the prospects of finding the long-term mass balance on basis of short-term simulations. A step in this process is to see to which degree the mass balance equation can be closed by measurements. Accordingly the total accumulation of heavy metals and polycyclic aromatic hydrocarbons (PAHs) in eight Danish detention ponds only receiving runoff from highways have been measured. The result shows that the incoming mass of heavy metals from short-term runoff events is accumulated. This is not observable in the same magnitude for the toxic organic compounds. The results also show that the accumulation rates significantly depend on the relative pond area (defined as the pond area divided by the catchment area). The conclusion is that the investigation indicates that a combination of short- and long-term viewpoints can close the mass balance for highway ponds with an acceptable accuracy.

Introduction

Variants of sedimentation ponds are commonly used as treatment facilities for polluted highway runoff. Many ponds have been designed only for flow control and peak reduction, but studies have shown particularly high removal efficiencies for suspended solids and thereby also for heavy metals and organic compounds due to their sorption affinity [1–3]. The removal efficiency for settleable particulate-bounded pollutants is thereby highly dependent on the pond geometry and corresponding hydraulic retention time. Optimizing of pond geometry for higher removal rates has been investigated in various studies [e.g., 4–8]. It is generally agreed that the removal efficiency varies from one facility to another [4] and from one event to another, even including negative efficiencies due to short circuiting flow, resuspension, release of pollutants due to changes, e.g., oxygen condition in the sediment [9].

In many studies, the efficiencies of pollutant removal in the detention ponds are calculated from event-based mass balances, where flow, inlet, and outlet concentrations have been measured. The question is now whether these short-term balances hold in respect to balances over many years of function. The mass balance equation runs as:

$$\text{Accumulation} = \text{influx} - \text{degradation} - \text{outflux}$$

In short-term studies only influx and outflux can be measured and in long-term studies only accumulation and rough estimates of influx can be determined. This study is based on the total accumulated masses of the pollutants in the bottom sediment in eight Danish highway wet detention ponds. The sizes of the ponds and the connected catchments areas are varying. The corresponding load (influx) to the ponds has been estimated on basis of generalized measurements from a number of locations. The advantage of dealing with the total accumulated masses in the ponds instead of concentration is that many years are taken into account and therefore event, season, and yearly variations of the pollutant loads are averaged out. The pollutants considered in this paper are chosen due to their prevalent presence in highway runoff [10] and toxicological effects onto the environment and humans [11]. The aim of the present study is to quantify the relation between the total accumulation and the total load on a long-term basis, in order to make probable that a long-term mass balance realistically seen can be calculated from a sum of short-term events. The work should also be understood as a preliminary study of an ongoing detailed description and modelling of the removal of pollutants in highway ponds.

Method

In order to state the terms in the mass balances for the pollutants in the eight wet detention ponds, following measurements and approaches have been taken.

Table 1. Site description

Pond/station number (km)	306.7	302.9	205.4	195.9
Nearby city	Hjallerup	Vodskov	Randers S	Hadsten N
Pond area (m^2)	1500	2299	2300	3480
Catchment area (ha)	1.7	2.7	3.7	6.0
Opening year for traffic	1999	1999	1994	1994
Annual day traffic in 2004	15400	14800	32100	30200
Annual precipitation (mm)	820	820	690	690

Pond/station number (km)	187.5	95.3	95.1	92.4
Nearby city	Grundfoer	Fredericia	Fredericia	Fredericia
Pond area (m^2)	2300	200	380	600
Catchment area (ha)	4.1	0.8	2.2	1.6
Opening year for traffic	1994	1994	1994	1994
Annual day traffic in 2004	33800	24100	24100	24100
Annual precipitation (mm)	690	770	770	770

The eight Danish ponds investigated in this study have been selected under four following criteria: the ponds should only receive water from highway runoff. The drainage systems should be closed so no infiltration or sedimentation in ditches occurs. The highway is established with curbs so that all runoff water is collected in gullies. The catchment area to the pond should differ from site to site. Site details can be seen in Table 1.

Accumulation

Ten samples (Fig. 1) were taken with a 56 mm cylinder in each pond representing one-tenths of the bottom area. Each sample within the ponds was taken out in the entire sediment depth, so that a mix of all 10 samples was representing the entire pond.

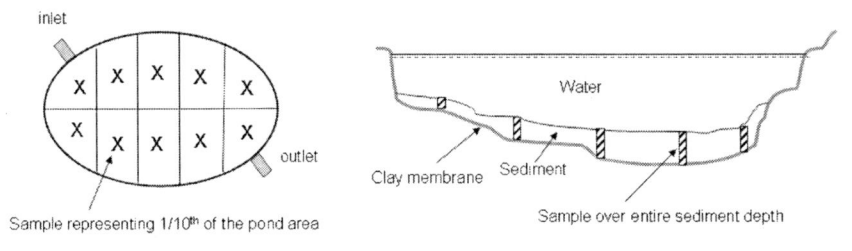

Fig. 1. Sampling method

The total mass of the ten wet sediment samples from each pond was measured and mixed heavily with a whisk on a drill machine, packed in 2 × 250 mL glass jars and kept cool until shipment to two independent accredited laboratories. Furthermore, information on sediment depths and general background information was also taken. Based on the dry matter fractions of the sediment the total masses of dry sediment in the ponds were calculated. Based on the total masses of dry sediment and the measured concentrations the total masses of accumulated pollutants were calculated and based on the total accumulations, the age of the ponds and connected catchment area the annual accumulation rate per hectare impervious catchment area were calculated.

Influx

In absence of inlet pollutant flux measurements under each rain event during the past 6 or 11 years, two opportunities are available to predict the flux of pollutants to the ponds: either a pollution build-up or wash-off model for the catchments or a mean highway runoff concentration model. It is not possible to state which model is the most suitable for this purpose but since the basis for getting concentration data are fairly good, the influx to the ponds are based on literature values for pollutant concentrations in highway rain runoff, local annual rain fall, and annual initial rain loss. The use of literature values may be highly questionable for short-term event-based balances due to the temporal and spatial variability in runoff concentrations. By dealing with long-term balances the temporal variability can be ignored. The spatial variability can of cause not be ignored due to the long time frame. But it must be remembered that this paper is not about whether one term in the mass balance equation is completely correct but about the prospects to close the mass balances for highway ponds based on short-term flux measurements and long-term accumulation measurements. The estimated annual pollutant influxes are based on following data:

Concentration and flow data

- An average of concentrations of 24 runoff samples from two highway location in Denmark, where all runoff water was collected each month over 1 year and analysed for pollutants [12] and 60 EMC from highway runoff in the UK [13]. The concentrations applied are seen in Table 2.
- Annual rainfall measured within a maximum distance of 20 km to the catchments and averaged over the years of pond function (c.f. Table 1)

[14] and a mean annual initial loss. The initial loss during one rain event was assumed to be 0.6 mm for all highway catchments. The loss has in preceding studies [15] been studied for the catchment to the Vodskov 302.9 detention pond. Based on the average number on rain events over a 20-year period and a initial loss of 0.6 mm for rain events over 0.6 mm (215 events) and a initial loss of 0.3 mm for rain events under 0.6 mm (50 events) the annual initial loss has been estimated to 140 mm year^{-1}.

Table 2. Applied average runoff concentrations in µg L^{-1}

Pollutant	Concentration	Pollutant	Concentration
ΣC6-C35	1623	Lead (Pb)	20.0
Flouranthene	0.19	Cadmium (Cd)	0.4
Benzo(b+j+k)flouranthene	0.19	Copper (Cu)	50.2
Benzo(a)pyrene	0.10	Chromium (Cr)	5.4
Dibenzo(a,h)anthracene	0.08	Nickel (Ni)	5.3
Indeno(1,2,3-cd)pyrene	0.07	Zinc (Zn)	156.7
ΣPAH	0.63		

Degradation and outflux

The degradation term in the mass balances for the heavy metals are not considered due to their state of elements. The annual outflux from the ponds can be calculated as the difference between the influx and accumulation. It has to be stated that the organic compounds including the PAHs are biodegradable either as carbon or energy source or in a co-metabolic process. The biological half-life period for the PAHs varies approximately between 6 and 12 years [16]. The organic outflux from the ponds can due to that not be calculated as for the metals. The local degradation rates are a product of many parameters such as, presence of easily biodegradable substances, oxygen, pH, and temperature conditions. A determination of annual degradation rates is subject to further investigations that cannot be done within the frames of this paper. The deficit between the annual influx and accumulation in the mass balances is owing to that a sum of annual degradation and annual outflux.

Results

The average annual increase in sediment depth in the ponds with an age of 6 years was calculated to 1.0 cm year^{-1} and for ponds with an age of 11 years to 0.6 cm year^{-1}. In previous studies [15], the annual load of suspended solids from the catchment area to the detention pond Vodskov st. 302.9 was approximately 200 kg year^{-1} ha^{-2}. Based on the measurements in this study the annual SS load is 25 times higher – showing that the contributor to the accumulated solids may not be the road runoff itself, but also solids from nearby surroundings. The mean pollutant concentrations in the pond sediments and ranges are presented in Figure 2. The concentrations in the ponds are within the range of what can be found in literature [e.g., 17–19]. The variation between the ponds is to be expected, due to very different locations with a variance in surroundings, vegetation, pH, redox potentials, microbiology, etc. These parameters are not to be considered any further in this paper. The annual accumulation rate in each pond and a catchment area-weighted mean accumulation rate for the organics and metals are presented in Table 3. It must be remembered that the calculated accumulation rates are based on ponds only receiving runoff from highways and in that case not to compared with other urban detention ponds receiving water from various areas.

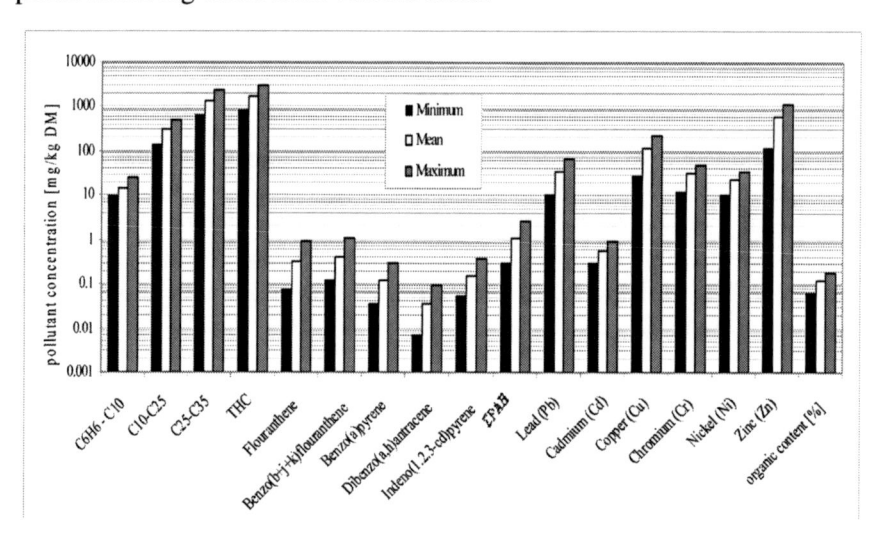

Fig. 2. Mean, minimum, and maximum pollutant concentration in the pond sediment

Table 3. Annual accumulation rates per hectare of impervious catchment in [g year^{-1} ha^{-2}]. The values in the mean column are catchment area-weighted mean accumulation rates

Pollutant pond no.	306.7	302.9	205.4	195.9	187.5	95.3	95.1	92.4	Mean
C6H6-C10	43	45	22	25	22	11	6	11	24
C10-C25	723	683	507	346	402	216	172	372	430
C25-C35	3033	3195	2131	1396	1900	1035	809	1579	1881
THC	3835	3927	2673	1765	2307	1264	986	1980	2337
Fluoranthene	0.47	0.35	0.56	0.24	0.31	0.18	0.16	0.84	0.37
Benzo(b + j + k) fluoranthene	0.77	0.57	0.74	0.32	0.46	0.20	0.21	1.01	0.51
Benzo(a)pyrene	0.21	0.17	0.21	0.08	0.12	0.06	0.06	0.27	0.14
Dibenzo(a,h) antrachene	0.03	0.05	0.06	0.02	0.04	0.02	0.02	0.09	0.04
Indeno(1.2.3-cd) pyrene	0.29	0.25	0.29	0.14	0.12	0.08	0.08	0.34	0.19
Sum PAH	1.77	1.38	1.86	0.79	1.06	0.54	0.53	2.56	1.24
Lead (Pb)	67	51	65	49	59	19	17	45	51
Cadmium (Cd)	1.8	1.5	0.8	1.3	0.8	0.3	0.2	0.7	1.0
Copper (Cu)	182	129	218	146	192	70	56	153	156
Chromium (Cr)	79	59	64	48	40	17	14	43	48
Nickel (Ni)	69	50	38	40	29	16	10	33	37
Zinc (Zn)	807	561	734	726	913	615	392	682	709

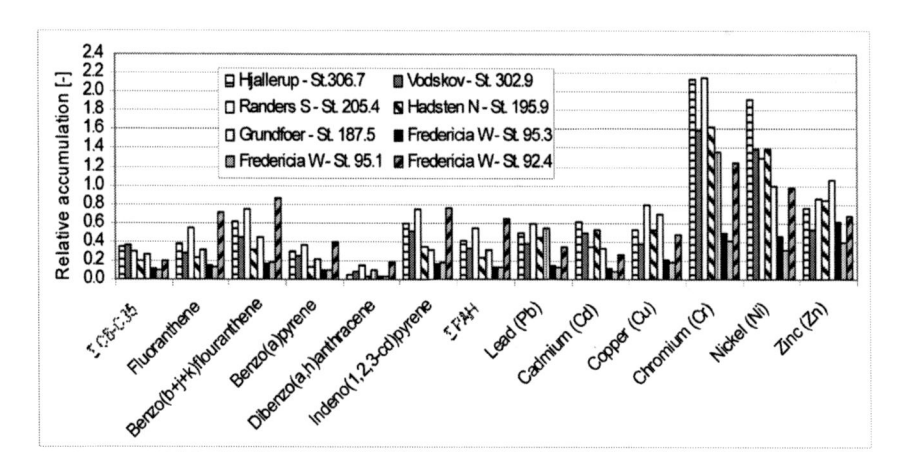

Fig. 3. Relative accumulations (annual accumulation/annual influx)

The calculated relative accumulations (Fig. 3) compared to efficiency studies based on inlet and effluent concentrations shows similarities for the metals, but with a slight tendency to show lower relative accumulation than the inlet and effluent based efficiencies does [2,3,13,20]. For primarily chromium and nickel in some of the ponds the relative accumulations are calculated to value higher than 1. Apparently this seems unrealistic but it reflects the uncertainty especially on the estimated loads from the runoff. However, this may give an indication of a high retention. For the organic compounds the relative accumulation are in general much lower (~50%) than the efficiencies, likely explained by degradation within the pond sediment. The high relative accumulation for some of the metals indicates that resuspension of sediments may have an insignificant role for the pollutant transport.

The results also show that the accumulation rates for the heavy metals significantly depend on the relative pond area (defined as the pond area divided by the catchment area) (Fig. 4). Similar dependencies are shown in [2], but as removal efficiencies as functions of the relative pond area instead. The accumulation rates in this study do not have the same flattening out tendency at a relative pond area of 250 m^2 ha^{-1} as in [2]. For direct comparison the relative accumulation could have been plotted instead. But, since the uncertainty in the calculated relative accumulations is high due to the estimated influxes this is not done. The accumulation dependency for the PAHs is not as clear as for the metals, probably due to different degradation possibilities in the very varying ecosystems.

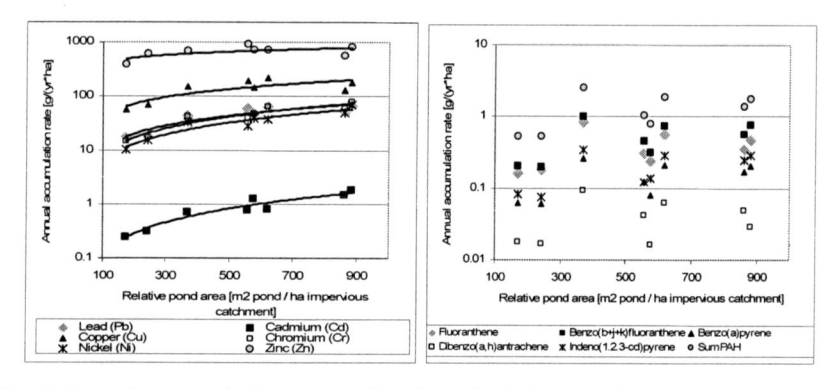

Fig. 4. Annual accumulation rate as function of relative pond area

Conclusions

As expected hydrocarbons, PAHs and heavy metals accumulates in the pond sediment. The comparison of the accumulation in relation to the load shows that the bulk of the incoming heavy metals can be found in the sediments whereas the organic compounds can only partly be found in the ponds. Although the results have a significant uncertainty the study indicates that a mass balance approach for the long-term removal of pollutants can be coupled to the short-term mass balances of the individual runoff event. The results can also be taken as indication that the resuspension from the ponds can only be of minor importance and that the relation between the pond area and the connected catchment area plays a significant role for the accumulation rates.

Acknowledgements

The Danish Road Directorate is acknowledged for their financial support and for helping gathering highway data and other practical help.

References

1. Van Buren MA, Watt WE, Marsalek J (1997) Removal of selected urban stormwater constituents by an on-stream pond. J Environ Plann Manag 40:5–18
2. Petterson TJR, German J, Svensson G (1999) Pollutant removal efficiency in two stormwater ponds in Sweden. Proceedings of the 8th international conference on urban storm drainage. Sydney/Australia, 30 August–3 September
3. Comings KJ, Booth DB, Horner RR (2000) Stormwater pollutant removal by two wet ponds in Bellevue, Washington. Comings, Department of Civil and Environmental Engineering, University of Washington, Seattle, Washington, DC
4. Van Buren MA, Watt WE, Marsalek J (1996) Enhancing the removal of pollutants by an on-stream pond. Water Sci Technol 33:325–332
5. Matthews RR, Watt WE, Marsalek J, Crowder AA, Anderson BC (1997) Extending retention times in a stormwater pond with retrofitted baffles. Water Qual Res J Canada 32:73–87
6. Walker DJ (1998) Modelling residence time in stormwater ponds. Ecol Eng 10:247–262
7. German J, Jansons K, Svensson G, Karlson D, Gustafsson L-G (2004) Modelling of different measures for improving removal in a stormwater pond. International conference on urban drainage modelling, Dresden, pp 477–484
8. Jansons K, German J, Howes T (2005) Evaluating hydrodynamic behaviour and pollutant removal in various stormwater treatment pond configurations. 10th International conference on urban drainage, Copenhagen/Denmark 21–26 August 2005

9. Lawrence AI, Marsalek JJ, Ellis B, Urbonas B (1996) Stormwater detention and BMPs. J Hydraul Res 34:799–813
10. Sansalone JJ, Buchberger SG (1997) Partitioning and first flush of metals in urban roadway stormwater, J Environ Eng 123:134–143
11. Makepeace DK, Smith DW, Stanley SJ (1995) Urban stormwater quality: summary of contaminant data. Environ Sci Technol 25:93–139
12. POLMIT (2002) Pollution from roads and vehicles and dispersal to the local environment, Final Report. Project co-coordinator: Transport Research Laboratory (TRL), UK
13. Crabtree B, Moy F, Whitehead M (2005) Pollutants in highway runoff, 10th International conference on urban drainage, Copenhagen/Denmark 21–26 August 2005
14. Danish Meteorologisk Institute (1995–2005) (Various authors) Technical Report, Operation of the The Water Pollution Committee of The Society of Danish Engineers' rain gauge system (Original in Danish: Drift af Spildevandskomitéens Regnmaalersystem)
15. Bentzen TR, Larsen T, Thorndal S, Rasmussen M R (2005) Removal of heavy metals and PAH in highway detention ponds. 10th International conference on urban drainage, Copenhagen/Denmark 21–26 August 2005
16. Environmental Protection Agency (in Danish: Miljøstyrelsen) (1996) Chemical substance behaviour in soil and groundwater (in Danish: Projekt om jord og grundvand nr. 20, 1996: Kemiske stoffers opførsel i jord og grundvand) No. 20
17. German J, Svensson G (2005) Stormwater pond sediment and water – characterisation and assessment. Urban Water J 2:39–50
18. Durand C, Ruban V, Amblès A, Clozel B, Achard L (2003) Characterisation of road sediments near Bordeaux with emphasis on phosphorus. J Environ Monit 5:463–467
19. Marsalek J, Marsalek PM (1997) Characteristics of sediments from a stormwater management pond. Water Sci Technol 36:117–122
20. Statens Vegvesen (Norwegian Road Directorate) (2005) Monitoring of runoff from Highway E6, Skullerudkrydset, Oslo 2003–2004 (Original in Norwegian: Overvåkning av rensebaseng for overvann fra E6 Skullerudkrysset i Oslo, 2003–2004)

VI. Environmental Assessment and Effects

Cardiovascular and respiratory variability related to air pollution and meteorological variables in Oporto, Portugal – preliminary study

JM Azevedo,[1] F Gonçalves,[1] AR Leal[2]

[1] Institute of Astronomy, Geophysics and Atmospheric Sciences, University of São Paulo, São Paulo, Brazil
[2] Institute of Biomedical Sciences Abel Salazar, University of Oporto, Portugal

Abstract

Urbanization is rendering cities vulnerable to the impact of air pollution on human health. In Portugal, few studies have been conducted on this subject, making this type of approach all the more important. Oporto, our study object/case study, is the second most important city in the country. The aim of this preliminary paper is to assess the effects of the meteorological conditions and atmospheric pollutants on respiratory diseases (RD) and cardiovascular diseases (CVD) in Oporto throughout 2003. Air quality variables such as nitrogen dioxide (NO_2), nitrogen monoxide (NO), carbon monoxide (CO), ozone (O_3), and particulate matter (PM_{10}); and meteorological data such as temperature (T), humidity (Hr), precipitation, and solar radiation (I) were employed. A preliminary case selection of RD and CVD morbidity of individuals over 65 years of age was carried out, employing Pearson coefficient to correlate them with meteorological variables and air quality. The results showed a close relationship between the increase in RDs, such as pneumoconiosis, and the increase in PM_{10}, O_3, and NO_2 in the summer months. The results showed a positive correlation between the cases of bronchitis and asthma, and the concentration of CO, NO_2; and a negative correlation with the variation in T and I as it was expected. In a thorough analysis it was realized that the meteorological conditions of lower T and higher values of Hr favoured the increase of RD. CVD conduction disturbances and dysrhythmia showed a positive correlation with

NO_2 (0.56, p = 0.05) and NO (0.68, p = 0.01). Hypertension (high blood pressure) and ischemic CVD were found to have a negative correlation with I, –0.64 and –0.55 (p = 0.05), respectively, without being significantly related with pollutants. As a preliminary study, this paper allows us to identify an association of RD and CVD conduction disturbance and dysrhythmia with the concentration increase of certain pollutants and extreme meteorological conditions.

Introduction

The environment is the dynamic set of all the conditions that affect daily life. The environment affects growth, development, and survival of all living species [1]. According to Gomes [2], air is the most indispensable of all the resources we use. A person of average height needs, approximately, 15 kg of air per day, to cover an alveolar surface of about 70 m^2. The importance of the environmental impact and the dimension of the risk of exposure to environmental aggressions are being explored more intensely since the beginning of the 19th century [3]. The increase in industrialization and of the anthropic action has, in the beginning, led to the belief that an age of development without deleterious consequences had begun since the atmosphere was enormous and it was thought that the pollution was restricted to the areas closer to the pollution source. According to Tromp [4] the term "meteorotropic" refers to the effect caused by one or more environmental factors on an individual or group of individuals. The climatic changes caused by cities and observed in the atmosphere can be explained by several factors: excess of pollution, the variation of radiation indexes, nebulosity, precipitation, temperature (T), humidity (Hr), and wind velocity [5]. Therefore, climatic changes and air pollution are a true problem around the whole world.

In Portugal, studies show that the atmospheric pollution affects the public health [6]. According to the Portuguese Health Ministry [7] "the absolute increase in the number of deaths (in Continental Portugal) has been higher in the group 'Diseases of the circulatory system' (more than 758 deaths) and within these, in 'Cerebrovascular diseases' (more than 370.2 deaths), 'Isquemic coronary disease' (more than 144.5 deaths) and 'Cardiac failure' (more than 118.0 deaths)". The group "Diseases of the respiratory system" (more than 255 deaths) and the set of "All the malignant neoplasias" (more than 131.2 deaths) occupy the following positions. The percentage of deaths that occurred in the various places was similar in the heat wave period and in the comparison period (Hospitals – 2003: 52.6%;

2 years 2000–2001: 56.0%; domicile – 2003: 32.2%; 2000–2001: 32.6%; other places – 2003: 15.2%; 2000–2001: 11.3%). According to the same source, the increase in deaths occurred mainly in people aged over 60. The effects of the heat wave manifested themselves in all the mainland districts, although with different intensities. Not only is the heat a factor of morbidity or mortality but, together with the other meteorological variables and high concentrations of atmospheric pollutants, contributes to the increase of the populations' vulnerability. The papers on air quality applied to public health in Portugal, particularly in Oporto, are not very numerous. Given the rapid global climatic and air quality change, knowing the local environmental conditions becomes of real importance in order to safeguard the population's quality of life.

This study will allow us to evaluate the environmental situation in 2003 and how it is affecting the health of the population residing in the city of Oporto. Variables that are usually studied separately will be studied together, which will contribute to an integrated approach to different factors that contribute to the same risk. The intention is to understand the influence of the meteorological and air quality conditions on respiratory diseases (RD) like asthma, bronchitis (B/A), and pneumoconiosis (Pn), as well as on cardiovascular diseases (CVD) such as ischemic cardiovascular diseases (DCI), conduction disturbances (AC), and hypertension (HTA), considering the age groups of people under 14 and people over 64 years.

Experimental – Method

Air quality data

Air quality variables like nitrogen dioxide (NO_2), nitrogen monoxide (NO), carbon monoxide (CO), ozone (O_3), and particulate matter (PM), were used. These were gathered from the three stations belonging to the Environment Institute (IA) situated in central places of the city (Figure 1). The pollutant concentrations were measured every hour on a daily basis, in micrograms per cubic metre ($\mu g\ m^{-3}$). This data was subjected to a statistical treatment before being made available. O_3 and CO were the results of the eight-hourly[1] average. The NO_2 and SO_2 concentrations were obtained from the average concentration for a given hour. For PM_{10}, the obtained concentrations correspond to the average (mobile average) composed by

---

[1] The eight-hourly average is composed by the current value and the seven previous hourly measurements.

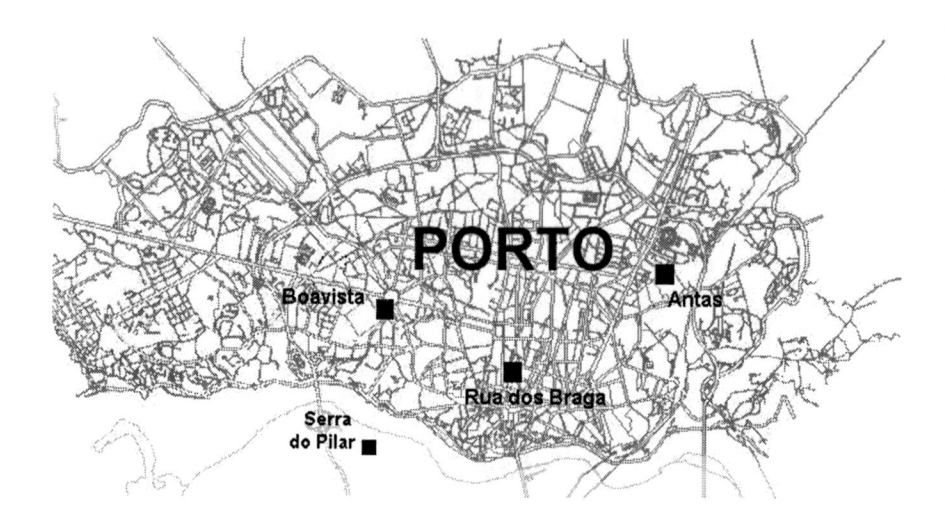

Fig. 1. Oporto city map with the location of the IA air quality stations (marked by ■)

the values received since zero hours (UTC time[2]) of the current day. The measuring stations are located at different altitudes (Boavista – 87 m; Rua dos Bragas – 98 m; Antas – 146 m) and measure pollutants of the urban type with traffic influence.

Meteorological data

The meteorological data of the daily temperature average (T), relative humidity (Hr), precipitation (P), wind velocity (Vv), atmospheric pressure (p), and solar radiation (I), belong to the Meteorology Institute weather station, situated in the Pilar Mountain Range.

Health data

The information on morbidity belongs to a central hospital in Oporto. A selection of cases of respiratory and cardiovascular morbidity from individuals over 64 years was done through their medical diagnostic code for hospital inpatients during the entire year of 2003. Only the cases admitted as emergencies were considered. The DC classes selected were 401–405, which correspond to diseases related to HTA; 410–414, which refers to

---

[2] UTC time: legal winter time = UTC time; legal summer time = UTC time + 1

DCI; and 426–428, which are related to AC. As for the RD, the diseases 490–496 corresponding to B/A and 500–508 corresponding to Pn were selected.

Analysis of main components (PCA) was also made. The data series on air quality averages (PM$_{10}$, SO$_2$, NO$_2$, CO, and O$_3$ maximum), meteorological parameters averages (T, Hr, P, p, and I), and normalized RD and CVD were used. The resulting matrix explains the variance according to their respective self values.

Results and discussion

Respiratory diseases

The link between RD and pollution was verified for the pollutants CO and NO$_2$ in the case of B/A, taking also into account the contribution from the climatic conditions I, T, and P (Table 1). The Pn shows a significant correlation with CO, as can be observed in Table 1.

Through the PCA (Table 2) we can verify the strong relationship between T and the SO$_2$ and O$_3$ levels, which in turn have an inverse association with B/A. All with significance and the first component explains 51.02% of the total variance. There still is a strong positive association between this type of disease and the CO, Hr, and P levels, with the first component having high significance levels.

The meteorological conditions could increase air pollutants and they in turn increase CO levels in the city, therefore also contributing to the asthma and bronchitis rise. These diseases show a significant correlation with this pollutant (0.65, $p < 0.05$, according to Table 1). This is supported

Table 1. Correlation values between the atmospheric variables and the respiratory diseases

Variables	B/A	Pn
CO (μg m^{-3})	0.65	0.56
NO$_2$ (μg m^{-3})	0.56	0.40
Temperature (°C)	–0.64	No sig.
Insulation (W m^{-2})	–0.64	No sig.
Precipitation (mm)	0.67	No sig.

For $p < 0.05$, No sig.: without significance

Table 2. PCA for B/A, without matrix rotation. From January to December 2003

Variables	Factor 1	Factor 2	Communalities
Temperature	**0.82**	−0.07	0.68
Humidity	**−0.79**	0.13	0.64
Precipitation	**−0.77**	0.23	0.65
Insulation	**0.86**	0.33	0.85
Pressure	−0.03	0.<u>78</u>	0.61
PM_{10}	0.55	0.<u>68</u>	0.77
O_3	**0.84**	−0.23	0.75
NO_2	−0.27	0.<u>52</u>	0.35
CO	**−0.87**	0.<u>45</u>	0.96
SO_2	**0.71**	0.<u>46</u>	0.71
B/A	**−0.81**	−0.12	0.67
Expl. variance	51.02	18.25	–
Prp. total	5.62	2.01	–

Prp. Total: Total variance explanation and accumulated; <u>underlined numbers</u> correspond to a positive association between pollutants and meteorological variables; bold numbers suggest the association with B/A and other parameters

by the PCA, which presents a strong positive association (above 0.75) of this pollutant, of P and of Hr with these diseases. NO_2 presents a low, but significant, correlation with B/A of 0.56, $p < 0.05$. That result seems to be related with the occasional increase in the NO_2 levels throughout the whole year. The meteorological variables favour the pollutant concentrations and, as a consequence, the increase of illness cases. As a general analysis, it can be seen that the low (high) T and high (low) Hr, increase (decrease) B/A respiratory disease cases.

In the autumn and winter months (October–February mainly), there is an increase in patients with B/A, which diminishes with the beginning of spring until the end of summer (Figure 2). The pneumoconiosis (Pn) presents several fluctuations throughout the year. However, during August, there are a higher number of cases when compared to the other summer months. Also, 70% of these cases were admitted in the hospital before 15 August.

According to the analysis of the main components (Table 3), there is a positive and significant association at the 2nd component. This explains 22% of the total variance between pressure, PM_{10}, SO_2, and pneumoconiosis. There is a very weak negative connection between T and Pn at the first component, but positive with precipitation.

Temperature has a negative correlation with respiratory morbidity during summer months. However, during these months, there is an increase in

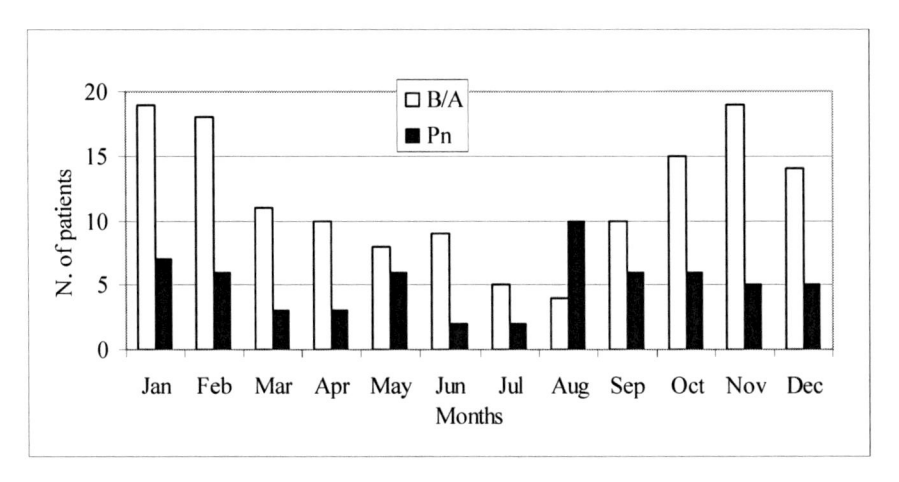

Fig. 2. Respiratory disease variability throughout 2003

solar radiation and, as a consequence, an increase in ozone concentration, which in turn induces respiratory morbidity.

A connection was verified between the increase of Pn (500–508) and the increase of PM_{10}, SO_2, and NO_2 during summer months. Throughout the PCA, a strong and positive association was found between the variables on the 2nd component. There is a significant correlation between Pn and CO (0.56, $p < 0.05$), which is supported by PCA. This suggests that Pn is associated with the increase in atmospheric pollution verified during the summer. In August, the atmospheric CO and PM_{10} concentrations increased several times, rising above the air quality index value considered "very bad" (>120 µg m^{-3}).

The Air Quality Institute (IQA) considered that from the 1–13 August the air quality was "low". PM_{10} was the pollutant with the highest concentration during this period. The level of increase and durability of this elevated pollutant concentration, which lasted for several days, must be related to the increase in temperature and decrease in P, Hr, and Vv registered from 29–30 July. The maximum temperatures during the month of August in Oporto were registered on days 6 (38.1°C) and 7 (38.0°C). The pollutants concentration variability is dependent on meteorological conditions. PM_{10}, SO_2, and O_3 concentrations have an inverse association with humidity and precipitation, and direct association with temperature and insulation. CO and NO_2 have positive associations with precipitation, and negative associations with temperature and insulation. It should be noted the decrease of B/A diseases during summer, and, on the contrary, the increase in Pn that occurred mainly in August. The maximum T values have a 1-day lag from the pollutant level peaks. During the month of August, 70% of morbidity

Table 3. Rotated PCA matrix with three iterations for Pn. Rotation method: Varimax with Kaiser normalization. From January to December 2003

Variables	Factor 1	Factor 2	Communalities
Temperature	0.81	–0.11	0.66
Humidity	–0.82	0.19	0.71
Precipitation	–0.72	0.35	0.65
Insulation	0.90	0.26	0.88
Pressure	0.06	**0.65**	0.43
SO_2	0.73	**0.44**	0.73
PM_{10}	0.62	**0.59**	0.74
O_3	0.80	–0.34	0.76
NO_2	–0.21	**0.55**	0.35
CO	–0.82	**0.54**	0.97
Pn	–0.19	**0.78**	0.64
Expl. variance	46.31	21.93	–
Prp. total	5.09	2.41	–

Prp. total: Total variance explanation and accumulated; underlined numbers correspond to a positive association between pollutants and meteorological variables; bold numbers suggest the association with Pn and other parameters

with a diagnosis referring to Pn was seen between the 8th and the 15th. This suggests the existence of a strong association between the elevated levels of PM_{10} and CO, and the increase in Pn cases (see Table 3).

Cardiovascular diseases

The AC and CVD presented positive and significant correlation with the pollutants NO_2 and NO (Table 4). However, no correlation was found with the meteorological variables. For the DCI and HTA, only with insulation was a significant correlation discovered. There is no significant correlation with air pollutants.

Table 4. Correlation values between the atmospheric variables and the DVCs for $p <$ or $= 0.05$

Variables	AC	DCI	HTA
NO_2	0.56	–0.71	–0.65
NO	0.68		
Insulation		–0.55	–0.64
CO		0.79	–0.44

Through the analysis of the main components matrix (PCA) at the 2nd component, which explains 21.74% of the variance, the CVD by AC is partially explained at the 2nd component by the presence of PM_{10}, NO_2, SO_2, and atmospheric pressure (Table 5). At the 1st component there is a weak and negative association between T, O_3, and SO_2, and the AC.

Table 5. PCA for AC, without matrix rotation, during 2003

Variables	Factor 1	Factor 2	Communalities
Temperature	0.83	−0.04	0.68
Humidity	−0.83	−0.04	0.69
Precipitation	−0.78	0.27	0.68
Insulation	0.84	0.41	0.88
Pressure	−0.07	**0.66**	0.44
PM_{10}	0.50	**0.71**	0.76
O_3	0.84	−0.08	0.71
NO_2	−0.32	**0.59**	0.46
CO	−0.89	0.33	0.92
SO_2	0.66	0.44	0.63
AC	−0.25	**0.74**	0.60
Expl. variance	45.94	21.74	–
Prp. total	5.05	2.39	–

Prp. total: Total variance explanation and accumulated; underlined numbers correspond to a positive association between pollutants and meteorological variables; bold num-numbers suggest the association with AC and others parameters.

Table 6. PCA for HTA, without matrix rotation, during 2003

Variables	Factor 1	Factor 2	Communalities
Temperature	**0.81**	−0.12	0.67
Humidity	**−0.82**	0.18	0.70
Precipitation	**−0.74**	0.28	0.63
Insulation	0.88	0.25	0.84
Pressure	0.001	0.68	0.46
PM_{10}	**0.61**	0.65	0.80
O_3	0.83	−0.26	0.76
NO_2	−0.18	0.68	0.49
CO	**−0.83**	0.52	0.96
SO_2	**0.69**	0.32	0.59
HTA	**−0.67**	−0.56	0.75
Expl. variance	48.95	20.67	–
Prp. total	5.38	2.27	–

Prp. total: Total variance explanation and accumulated; underlined numbers correspond to a positive association between pollutants and meteorological variables; bold numbers suggest the association with HTA and others parameters

It can be observed by the PCA that the CVDs by HTA are (Table 6) in the first component explained, with 48.95% of total variance, by the meteorological variables (positive with the Hr (0.82) and P (0.74), and negative with the T (0.81) and I (0.88)), and also explained by the atmospheric pollutants (negatively by O_3, PM_{10}, SO_2, and positively by CO) (Table 6).

According to the PCA, the DCI has positive association (–0.55) with CO (–0.87), P (–0.74), and Hr (–0.84), and negative association with T (0.81), I (0.88), PM_{10} (0.55), SO_2 (0.67), and O_3 (0.85) (Table 7). The variance is explained in the first component in 47.79%.

In June, the number of DCI cases increased substantially, which suggests an association between these CVDs and the higher temperatures, which occurred from the 11th to the 23rd. Additionally, an increase in temperature, no rainfall, and a decrease in Hr were registered. June 19 yielded the minimum relative humidity (30%) (average) and a maximum of 30.3°C. There was a morbidity increase before June 15 as well, which suggests that it could be caused by other variables outside of this study.

The increase of DCI cases did not repeat in August, despite a significant temperature increase. On the other hand, the relative humidity did not decrease to the values registered in June, remaining always above 50 g m^{-3}. Another possible explanation to this result is based on the fact that it is the holiday period and therefore a variable outside of the studied area.

Table 7. PCA for DCI, without matrix rotation during 2003

Variables	Factor 1	Factor 2	Communalities
Temperature	**0.81**	–0.08	0.66
Humidity	**–0.84**	0.09	0.72
Precipitation	**–0.74**	0.27	0.62
Insulation	**0.88**	0.34	0.89
Pressure	–0.05	<u>0.71</u>	0.51
PM_{10}	**0.55**	<u>0.66</u>	0.74
O_3	**0.85**	–0.21	0.77
NO_2	–0.21	<u>0.63</u>	0.44
CO	**–0.87**	<u>0.46</u>	0.96
SO_2	**0.67**	<u>0.40</u>	0.62
DCI	**–0.55**	–0.29	0.38
Expl. variance	47.79	18.56	–
Prp. total	5.26	2.04	–

Prp. total: Total variance explanation and accumulated; underlined numbers correspond to a positive association between pollutants and meteorological variables; bold numbers suggest the association with DCI and others parameters

A strong and positive association was found between DCI (0.79) and the pollutants CO, AC, NO$_2$ (0.56), and NO (0.68). The correlations between the pollutants and HTA are negative due to the fact that summer presents higher concentration of pollutants and lower CVD, existing only statistical association and not a cause–effect association. There are many other explanations other than environmental that could play a significant role on hospital admissions, which is implicated in the differences found between both of them.

Conclusions

This study, although restricted from a temporal viewpoint, suggests that the quality of the air as well as meteorological conditions affect public health in the Oporto region, especially where it concerns all the diseases herein studied (Pn, B/A, DCI, AC, HTA). NO$_2$ and CO are the pollutants whose concentration levels throughout the year present a greater importance because of their positive correlation with B/A and Pn. In the autumn and winter months (October–February mainly), there is an increase in patients with B/A. During summer, O$_3$, PM$_{10}$, and T and I have high levels, which seems to contribute to an increase of Pn. The CVD can be partially explained by the meteorological influence but also by the contribution of pollutants like PM$_{10}$, CO, and NO$_2$.

Acknowledgements

This paper was supported by CNPq (Desenvolvimento Cientifico e Tecnológico). Thanks go to Miguel Lopes for his important colaboration.

References

1. Dicionário de poluição (1996) 'Ecologica's', Ecológica Tecnologia e Controle Ambiental Ltda, Salvador BA, Brazil
2. Gomes L (1998) Apostila Didática, p 3–4
3. Becker S, Dailey LA, Soukup JM, Grambow SC, Devlin RB, Huang, Y-CT (2005) Respiratory disease – seasonal variation in air pollution particle-induced inflammatory mediator release and oxidative stress. Environ Health Persp 113:1032–1038
4. Smith RA (1872) Air and rain: the beginnings of the chemical climatology. Longman-Green, London

5. Tromp SW (1980) Biometeorology, Heyden, Inglaterra
6. Monteiro A, Fernandes A, Silva A (1998) Exemplos de agravamento de algumas patologias do foro respiratório, relacionáveis com as modificações introduzidas pela urbanização portuense na conjuntura climática e na composição química da atmosfera. Tese de Doutoramento do projecto Clias, F.C.T., PRAXIS XXI, PCSH /GEO/198/96
7. Barros N (2005) Impacte na qualidade do ar e na saúde das grandes linhas de tráfego urbano: o caso da VCI. IMPACTAir

Component analysis on respiratory disease variability at São Paulo

FLT Gonçalves,[1] MS Coelho,[1] MRD Latorre[2]

[1]IAG/Dept. of Atmospheric Sciences, [2]Faculty of Public Health, University of São Paulo; São Paulo; Brazil

Abstract

This study presents the analysis of weather and air pollutant impacts on respiratory morbidity in São Paulo metropolitan area. The daily respiratory morbidity, caused by respiratory diseases (upper respiratory tract, AVAS, and lower tract, AVAI), was used in children under 13, using principal component analysis (CP). The preliminary results show a negative association between AVAS and AVAI, and temperatures, higher for AVAS. Ozone and AVAI were associated positively. AVAS and AVAI variances are both under meteorological influences and influenced by air pollutants where SO_2 and PM_{10} play a role on AVAS variance, while ozone plays a role on AVAI variance.

Introduction

Air pollution and its impact on the human health have been considered a serious problem in the urban areas. Since the beginning of the century, many events of pollution have been associated to increases in mortality, as seen in Meuse Valley in 1930 [1], Donora in 1948 [2], and the most remarkable in London in 1952 [3].

Recently, increases in mortality have been associated with non-episodic events of air pollution, even for levels below the air quality standards [4–6].

In terms of morbidity, air pollution has been associated with decreases in pulmonary function, according to [7], increases in respiratory symptoms, attacks of asthma and chronic bronchitis, and school absences [8, 9].

São Paulo is the largest city in South America, which presents a light and heavy fleet about 6,000,000 vehicles, using diesel, alcohol, and gasohol (88% of gas and 22% of alcohol) as fuel. There are a large number of old cars and trucks running in the metropolitan area of São Paulo and no effective programme for emissions control has ever been established. In addition, during winter, frequent thermal inversion creates non-adequate climatic conditions to pollution dispersion. According to the estimates of the environmental state agency (CETESB, [10]) automotive emissions represent by large the most important source of pollutants in São Paulo. Therefore, significant associations between air pollution and mortality were observed in São Paulo [11–15]. Specific age groups, such as elderly people and children, may represent the main target affected by pollution effects [16,17].

Adding to the pollution impact, the authors [7,18] showed strong relations between respiratory diseases and weather conditions, well known since the beginning of the century. Winter in São Paulo present weather conditions characterized by strong cold air mass penetrations, though there are no adequate heating systems in most homes. Therefore, the impact of the cold and wet weather conditions on children respiratory diseases might be very impressive, summing up to the air pollution.

Experimental – Method

Daily records of hospital admissions for children under 13 years of age from 1998 to 2002 obtained from the Health State Secretary show upper respiratory tract, AVAS, and lower respiratory tract, AVAI. These records are from about 80 hospitals spread over the city of São Paulo, and that receive support from the public health system. Thus, our sample is probably representative of the poorest segment of the population that does not have private medical care insurance. Daily records of PM_{10}, SO_2, the lowest temperature, and relative humidity at noon were obtained from the state air pollution controlling agency (CETESB, [10]) and Meteorological Station from Parque Estadual das Fontes do Ipiranga (São Paulo). The 24-hours average (starting at 4 p.m. of the preceding day) of SO_2 (eight stations), and PM_{10} (eight stations) were recorded.

In a first approach, we herein explored the association between daily hospital admissions and air pollution by means of factor analysis (FA).

This multivariate method was applied to the data set in order to evaluate the relationship between the meteorological and respiratory diseases variables. The procedure adopted consisted of grouping in the same factor parameters that have the same temporal variability, finding a relation of causality. FA is mainly a statistical rather than a physical model. FA was applied to the data set of gases and particles measured in the air quality stations, data set of surface meteorological parameters measured at the climatological station of São Paulo, and the data set from the respiratory morbidity described above. The FA started with the calculation of the eigenvalues and eigenvectors, the matrix of the correlation coefficients among the variables.

PCA is the adopted multivariate technique in which a number of related variables are transformed in a smaller set of uncorrelated variables. The technique rewrites the original data matrix into a new set of principal components (hereafter PC, [19]) that are linearly independent and ordered by the amount of the variance of the original data explained by them [e.g., 20]. The number of variables p could be equal to 7 (e.g., AVAI or AVAS, PRESS, TMEAN, RHMIN, SO_2, PM_{10}, and O_3) and the number of events m is equal to 365 (i.e., the number of days in each year). Each eigenvalue is related to a corresponding eigenvector with p elements. They represent a new base in which each PC explains the variance according to the respective eigenvalue. VARIMAX rotation was used. Formally VARIMAX searches for a rotation (i.e., a linear combination) of the original factors in which the variance of the loadings is maximized.

Results and discussion

The FA is shown in Tables 1–4, where it is presented without and with moving averages.

Table 1 shows the AVAS analysis without moving average where Factors 1 and 2 explain 65% of the data variance. In Factor 1, explaining 37% of the variance, AVAS is negatively associated (–0.48) with air temperatures (TMEAN, TMIN, and TMAX, all around 0.90), and positively associated with air pressure (–0.74). That means there are cold days with polar anticyclone systems, which present high pressure, associated with high AVAS hospital admissions and vice-versa with low pressure systems. Relative humidity (RH) is very weakly and positively associated (–0.25) with AVAS. In Factor 2, explaining 28% of the variance, AVAS is weakly but positively (0.28) associated with air pollutants (SO_2 with 0.80 and PM_{10} with 0.87). Both pollutants are strongly and negatively associated

Table 1. Principal component analysis for AVAS, without moving average (Varimax row rotation)

Variable	Factor 1	Factor 2	Communality
AVAS	–0.48	0.28	0.85
TMEAN	**0.96**	0.11	0.94
TMAX	**0.84**	0.46	0.82
TMIN	**0.87**	–0.29	0.91
RH MIN	–0.25	**–0.84**	0.91
PRESS	**–0.74**	0.09	0.83
PM_{10}	–0.02	**0.87**	0.49
SO_2	–0.15	**0.80**	0.12
O_3	0.31	–0.02	0.89
Expl. variance	3.35	2.50	–
Prp. total	0.37	0.28	–

Bold form means with significance

with RH (–0.84). There were no more significant factors in this analysis. The communalities of all variables present high values, except the pollutants, despite the fact that they are strongly associated with Factor 2.

Table 2 shows the AVAI analysis without moving average, with VARIMAX rotation, where Factors 1, 2, and 3 explain 75% of the data variance. In Factor 1, explaining 36% of the variance, AVAS is weakly and negatively associated (0.25) with air temperatures (TMEAN, TMIN, and TMAX, all around –0.85 and –0.97), and positively associated with air pressure (0.74). RH is very weakly and positively associated (0.26) with AVAI, as well.

That means there are cold days with polar anticyclone systems with high AVAI morbidity and vice-versa with low pressure systems, presenting a result quite similar to AVAS, but weaker. In Factor 3, explaining 12% of the variance, AVAI is significantly and positively (–0.69) associated with ozone (–0.76). There is no significant association either positive or negative with pollutants SO_2 and PM_{10}, and both present again, a strong positive association between them (0.82 and 0.88, respectively), and a negative association with RH (–0.82), in Factor 2, which explains 27% of the variance. Communalities present high values in all variables.

That is an interesting overall result where SO_2 and PM_{10} play a role on AVAS variability, while ozone plays a role on AVAI variability.

Table 2. Principal component analysis for AVAI without moving average (Varimax row rotation)

Variable	Factor 1	Factor 2	Factor 3	Communality
TMEAN	−0.97	0.09	0.00	0.94
TMAX	−0.85	0.44	−0.08	0.93
TMIN	−0.88	−0.30	0.05	0.86
RH MIN	0.26	−0.82	0.18	0.77
PRESS	0.74	0.10	−0.05	0.57
PM_{10}	0.01	0.88	0.05	0.79
SO_2	0.15	0.82	0.03	0.69
O_3	−0.29	−0.09	−0.69	0.58
AVAI	0.25	0.13	−0.76	0.65
Expl. variance	3.23	2.45	1.10	−
Prp. total	0.36	0.27	0.12	−

Table 3 and 4 shows AVAS and AVAI FA with 3 days moving average, respectively. The overall results are quite similar to the previous tables. Table 3 presents Factors 1 and 2 playing the opposite role compared to those on Table 1. Factor 2, with 29% of the variance, presents AVAS with positive association (0.58) to the air pressure (0.83) and negative to the TMEAN (−0.86). In Factor 1, with 32% of the explained variance, SO_2, PM_{10}, and RH present quite similar results described previously, with only smaller AVAS association (0.26). Communalities show high values except for ozone.

Table 3. Principal component analysis for AVAS with 3 days moving average (Varimax row rotation)

Variables	Factor 1	Factor 2	Communality
RH MIN	−0.81	0.23	0.72
PRESS	0.06	0.83	0.69
PM_{10}	**0.88**	0.04	0.78
SO_2	**0.83**	0.14	0.70
O_3	−0.01	−0.41	0.17
AVAS	**0.26**	**0.58**	0.40
TMEAN	0.13	−0.86	0.76
Expl. variance	2.21	2.01	−
Prp. total	0.32	0.29	−

Table 4. Principal component analysis for AVAI with 3 days moving average (Varimax row rotation)

Variables	Factor 1	Factor 2	Factor 3	Communality
RH MIN	–0.79	0.26	–0.18	0.73
PRESS	0.08	0.85	0.05	0.74
PM_{10}	0.90	0.02	–0.04	0.81
SO_2	0.84	0.14	–0.02	0.72
O_3	–0.09	–0.39	**0.69**	0.63
AVAI	0.12	0.33	**0.77**	0.71
TMEAN	0.11	–0.88	0.00	0.78
Expl.variance	2.17	1.84	1.10	–
Prp.total	0.31	0.26	0.16	–

Table 4 present similar results compared to those on Table 2, with very low values at Factor 1 and 2. Factor 3 presents a high and positive association with ozone (0.77 and 0.69, respectively), with quite similar results to those on Table 2, but with high significances; Factor 3 explains 16% of the variance (against 12% on Table 2) and the communalities are higher: 0.63 to ozone and 0.71 to AVAI on Table 4 (against 0.58 and 0.65 on Table 2).

Conclusions

As a general conclusion, AVAS and AVAI variances are both under meteorological influences, such as air temperatures and air pressure systems. Cold air masses with high pressure seem to affect the AVAS and AVAI variability. RH plays a secondary role on respiratory disease variability. SO_2 and PM_{10} play a positive role on AVAS variability, while ozone plays a positive role on AVAI variability. It seems that the upper respiratory tract suffers with the first two air pollutants, while the lower respiratory tract suffers with ozone.

References

1. Firket J (1931) Sur les causes des accidents survenus dans la vallée de la Meuse. Lors des brouillards de décembre 1930. Bull Acad Roy Med Belg 11:683–741
2. Logan WPD (1953) Mortality in London fog incident. Lancet 1:336–338
3. Ciocco A, Thompson DJ (1961) A follow-up on Donora ten years after: methodology and findings. Am J Public Health 51:155–164

4. Pope CA III, Schwartz J, Ranson MR (1992) Daily mortality and PM_{10} pollution in Utah Valley. Arch Environ Health 47:211–217
5. Schwartz J, Dockery DW (1992) Increased mortality in Philadelphia associated with daily air pollution concentration. Am Rev Respir Dis 45:600–604
6. Bascom R, Bromberg PA, Costa DA, Devlin R, Dockery DW, Frampton MW, Lambert W, Samet JM, Speizer FE, Utell M (1996) Health effects of outdoor pollution. Am J Respir Crit Care Med 153:3–50
7. Moseholm L, Taudorf E, Frosig A (1993) Pulmonary function changes in asthmatics associated with low-level SO_2 and NO_2 air pollution, weather, and medicine intake. Allergy 48: 334–344
8. Brabin B, Smith M, Milligan P (1994) Respiratory morbidity in Mersey-side schoolchildren exposed to coal dust and air pollution. Arch Dis Child 70: 305–312
9. White MC, Etzel RA, Wilcox WD, Lloyd C (1994). Exacerbation of childhood asthma and ozone pollution in Atlanta. Environ Res 65:56–68
10. CETESB (1994) "Relatório de qualidade do ar na região metropolitana de São Paulo e Cubatão – 1993."[In portuguese]
11. Böhm GM, Saldiva PHN, Pasqualucci CA, Massad E, Martins MA, Zin WA, Cardoso WV, Criado PMP, Komatsuzaki M, Sakai RS, Nigri EM, Lemos M, Capelozzi VD, Crestana C, Silvia R (1989) Biological effects of air pollution in Sao Paulo and Cubatao. Environ Res 49:208–216
12. Saldiva PHN, King M, Delmonte VLC, Macchione M, Parada MAC, Daliberto ML, Sakai RS, Criado PMP, Silveira PLP, Zin WA, Böhm GM (1992) Respiratory alterations due to urban air pollution: an experimental study in rats. Environ Res 57:19–33
13. Lemos M, Lichtenfels AJFC, Amaro E Jr, Macchione M, Martins MA, King M, Böhm GM, Saldiva PHN (1994) Quantitative pathology of nasal passages in rats exposed to urban levels of air pollution. Environ Res 66:87–95
14. Braga ALF, Conceição GMS, Pereira LAA, Kishi HS, Pereira JCR, Gonçalves FLT, Andrade MF, Singer J, Böhm G, Saldiva PHN. Air pollution is significantly associated with pediatric respiratory hospital admissions in São Paulo, Brazil. J Environ Med (accepted)
15. Gonçalves FLT, Carvalho LMV, Conde FC, Latorre MRDO, Saldiva PHN, Braga ALF (2005) The effects of air pollution and meteorological parameters on respiratory morbidity during the summer in São Paulo City. Environ Int 31:343–349
16. Saldiva PHN, Lichtenfels AJFC, Paiva PSO, Barone IA, Martins MA, Massad E, Pereira JCR, Xavier VP, Singer JM, Böhm GM (1994) Association between air pollution and mortality due to respiratory diseases in children in Sao Paulo. Brazil: a preliminary report. Environ Res 65:218–25
17. Saldiva PHN, Pope CA III, Schwartz J, Dockery DW, Lichtenfels AJ, Salge JM, Barone I, Böhm GM (1995) Air pollution and mortality in elderly people: a time-series study in Sao Paulo, Brazil. Arch Environ Health 50:159–63

18. Tromp SW (1980) Biometeorology: the impact of the weather and climate on humans and their environment. Heyden, London. 346 pp
19. Hopke P (1991) Receptor modelling for air quality management. Elsevier, New York.
20. Jackson JE (1991) A user's guide to principal components. Wiley, New York.

Management and optimization of environmental information using integrated technology: a case study in the city of São Paulo, Brazil

AS Nedel,[1] P Oyola,[2] RK Fujii,[3] RC Cacavallo[3]

[1]Atmospheric Sciences, Geophysics and Astronomy Institute; University of São Paulo IAG-USP
[2]Mario Molina Centre, Santiago, Chile
[3]School of Public Health; University of São Paulo FSP-USP

Abstract

In the São Paulo metropolitan region, like other world's big urban centres, environmental monitoring plays a very important role, in view of the degradation of air quality. This work illustrates a case study of continuous monitoring of air pollutants using a differential optical absorption spectrometer (DOAS) associated with an Integrated Environmental Management System (AIRVIRO) in São Paulo, Brazil. The DOAS registered concentrations of benzene, toluene, m-p xilenes, O_3, SO_2, NO_2, NO_3, and HNO, and also of a meteorological station continuous data collection. Air quality databases from CETESB – State Environmental Authority were used for reference. Results of days with high concentration averages in winter are related to specific meteorological conditions and local traffic.

Introduction

In São Paulo metropolitan region (RMSP) the environmental monitoring subject has a very relevant role, considering the high degradation level of the air quality in this entire region. This reflects a common situation that occurs in the majority of the world's big urban centres. The pollutant load is mainly connected to the massive heavy and light vehicles' emissions, and secondarily to industrial processes [1]. The atmospheric emissions originate from approximately 2000 industries with high pollutant potential and from a fleet of 7.8 million vehicles, equivalent to one-fifth of the entire country [1]. The region's air quality degradation level requires a monitoring system that takes into account both the long period pollution level follow up, and the acute air pollution episodes [1]. Studies with a similar approach and technological resources have been carried out in a great number of cities around the world, such as Rome, Athens, Stockholm, etc [2, 3]. This work illustrates a case study of continuous air pollutants monitoring using a differential optical absorption spectrometer (DOAS) associated with an Integrated Environmental Management System (AIRVIRO) in the city of São Paulo, Brazil.

Method

An air quality and meteorological monitoring site was created in May 2005 in Cerqueira César (south/centre of RMSP), which is one of the highest altitude areas in the city of São Paulo (817 m). The participant institutions are: School of Public Health – University of São Paulo (FSP-USP), Medicine Faculty – University of São Paulo (FM-USP), and the Physics Institute – University of São Paulo (IF-USP).

The equipment is installed in the School of Public Health and in the Faculty of Medicine buildings, approximately 30 m above street level. A DOAS–OPSIS AB (model AR500) – the technique was fully described earlier [4] – registered benzene, toluene, m-p xilenes, ozone, SO_2, NO_2, NO_3, and HNO_2 concentrations with a 1 min temporal resolution. The emitter was located at the Medicine Faculty building and, 361 m across the street, in the School of Public Health building, was located the receiver and a meteorological station model MetOne, which registered data with a 5 min time resolution. The management and analyses of the information collected was performed by the software ENVIMAN (OPSIS AB) and the AIRVIRO (Swedish Meteorological and Hydrological Institute – SMHI/-APERTUM AB). The latter being an environmental information and air quality management web application that enables analysis like simulations and real-time data delivery/assessment.

For the assessment of the regions emission sources pattern an emission micro-inventory for the years 2003/2004 created by the State Environmental Authority – CETESB was used [5]. There are no point sources with

Table 1. Daily traffic volume (vehicles count) according to lane of interest and vehicle class, São Paulo, 2003–2004 (From CETESB, 2005)

Lane	Emission source[a] (vehicle class)	Daily traffic volume[a]
Dr. Arnaldo avenue	Light vehicles	83,941
	Diesel vehicles	9936
	Motorcycles	6321
Teodoro Sampaio street	Light vehicles	17,690
	Diesel vehicles	3054
	Motorcycles	808
Cardeal Arcoverde street	Light vehicles	25,590
	Diesel vehicles	5299
	Motorcycles	808

[a]Methodology described in the monitoring station characterization report – CETESB [2].

representative emission potential in the 400 m radius around the study site. Due to restrictions imposed by the municipal zoning regulations, only small stationary sources can be found in this area like gas stations, restaurants, pizzerias, and bakeries. The majority of the 400 m radius area is occupied by two cemeteries, FSP-USP, FM-USP, and the Clínicas hospital complex [5]. The most relevant emission sources in the surrounding area are a small number of high traffic lanes, such as: Dr. Arnaldo Avenue, Teodoro Sampaio Street, and Cardeal Arcoverde Street. These have both light and heavy vehicles intense traffic [5]. Daily traffic volume for each of the lanes, according to vehicle class, can be seen in Table 1.

Table 2 illustrates the regional emission estimates in ton year^{-1} for the year 2003, taking into account the traffic lanes length in the study site surroundings.

Table 2. Emission estimates according to lane of interest (road sources) and substance in ton year^{-1}, São Paulo, 2003 (From CETESB, 2005)

Source	PM	SO_2	NO_x	CO	HC
Dr. Arnaldo avenue (800 m)	3.8	3.6	47.0	332.0	37.4
Teodoro Sampaio street (450 m)	4.6	2.7	9.0	54.0	6.9
Cardeal Arcoverde street (350 m)	0.4	12.9	5.7	11.0	3.6

Emissions (t year^{-1})[a]

[a]Methodology described in the monitoring station characterization report – CETESB [5]

Results and discussion

Considering the analysed period (1 June–31 August/2005), among the seven SO_2 peak levels observed, 12 June was selected for a more comprehensive investigation since it presented the highest concentration peak level (145.0 µg m^{-3}). This one is higher than the secondary standard (100 µg m^{-3}) according to CONAMA 1990. The highlighted concentration peak levels (Fig. 1) correspond to Mondays, a day characterized by intense vehicle traffic during the first morning hours.

Figure 2 shows the meteorological parameters behaviour registered by the study's meteorological station at the FSP-USP.

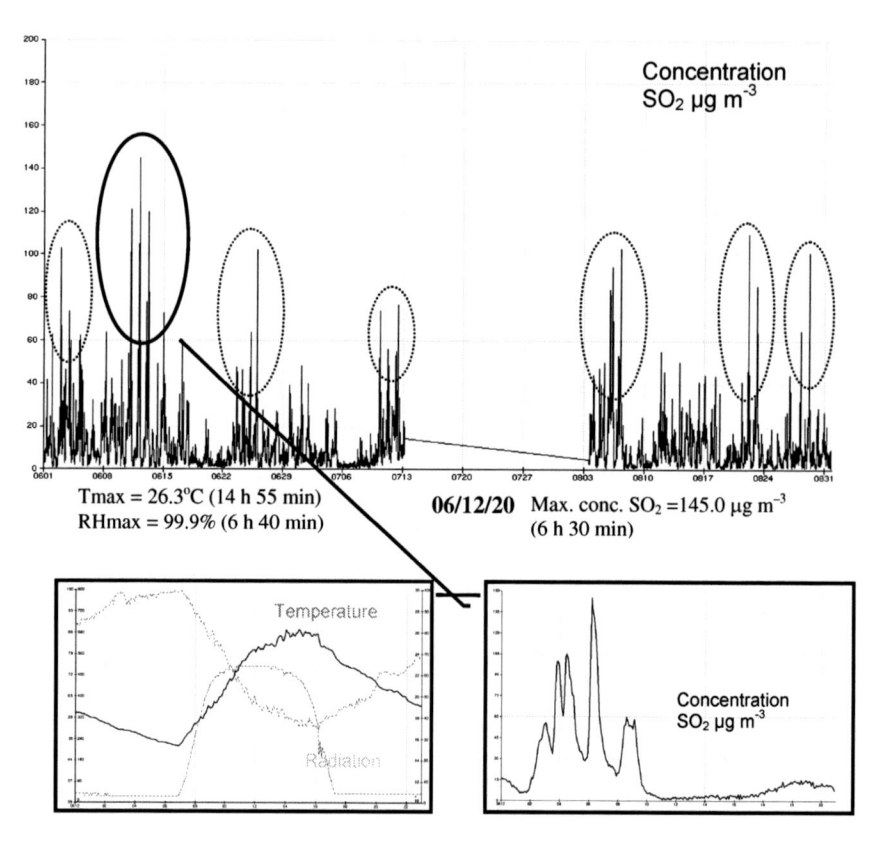

Fig. 1. SO_2 concentration distribution (µg m^{-3}) from June 1–August 31/2005. (A) Meteorological parameters: relative humidity (%), temperature (°C), global radiation (W m^{-2}) for 12 June 2005. (B) SO_2 concentration distribution during the 24 h of 12 June 2005

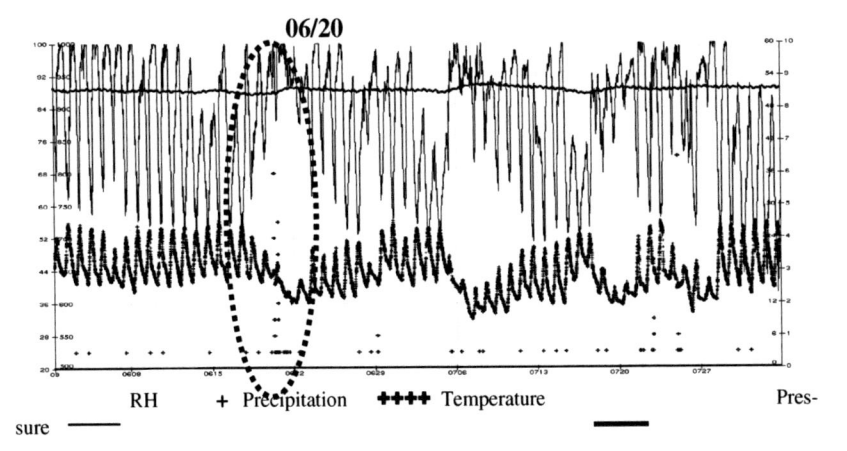

Fig. 2. Meteorological parameters distribution in the period from 1 June–31 August, São Paulo, 2005

In the period that preceded the event of interest (12 June), the São Paulo metropolitan region was under a high pressure system domain condition, generating a situation of high atmospheric stability and low pollutants dispersion. This situation formed in the first days of June and lasted several days until a cold front approached on 20 June, which increased air movement thus facilitating the pollutants dispersion.

With the intent of evaluating the feasibility of the collected data, a comparison with the State Environmental Authority records was performed. The State Environmental Authority – CETESB has 27 air quality monitoring stations spread across the São Paulo metropolitan area and is in charge of the regular monitoring of pollutants. Table 3 and Figure 3 show those comparisons.

As showed in Table 3, the data obtained in this study was very similar to the official governmental data (CETESB), with average SO_2 concentrations of 10 µg m^{-3} and 11 µg m^{-3}, respectively.

Table 3. Comparison of average and maximum SO_2 concentration values, days of data collection and period analysed between the State Environmental Authority (CETESB) winter report and the present study, São Paulo, 2005

SO_2	Average µg m^{-3}	Maximum µg m^{-3}	Days of data collection	Period
CETESB	10	22	127	1 May–30 Sept/2005
Case study	11	145	92	1 June–31 Aug/2005

June 2005

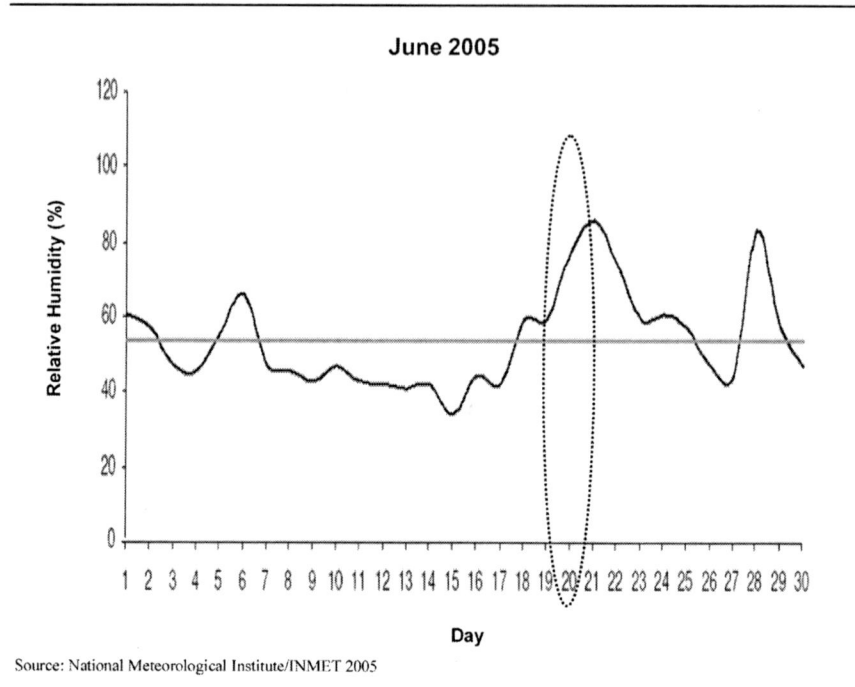

Fig. 3. Relative humidity values distribution for the 30 days of June 2005, São Paulo

Figure 3 shows relative humidity elevation in the period from 19–21 June, which confirms the data observed by the study meteorological station.

Conclusions

This study was the first opportunity of using such technology (DOAS + environmental data management software) in the city of São Paulo.

From the current results, the location of the data collection instruments can be considered suitable for the development of studies concerning the environmental situation associated with pollution and transportation. These will be mainly studies for the development of models and urban environmental condition studies in the city of São Paulo.

The presented technology, with its high efficiency performance and real time data delivery, associated with CETESB's monitoring network, could enable a better air quality situation comprehension. This will in turn facilitate feasible decisions and policies, which can gradually provide a better air quality to the city's population.

Acknowledgements

The financial support for this paper from the Conselho Nacional de Desenvolvimento Científico e Tecnológico (CNPq) is gratefully acknowledged.

References

1. CETESB – Companhia de Tecnologia de Saneamento Ambiental (2005) Relatório de qualidade do ar no estado de São Paulo, 2004. (série relatórios/ Secretaria de Estado do meio Ambiente). Also at http://www.cetesb.sp.gov.br/ AR/relatorios/relatorios.asp
2. Brocco D, Fratacangeli R, Lepore L, Petricca M, Ventrone I (1997) Determination of aromatic hydrocarbons in urban air of Rome. Atmos Environ 31:557–566
3. Kourtidisa KA, Ziomasa I, Zerefosa C, Kosmidisa E, Symeonidisa P, Christophilopoulosa E, Karathanassisa S, Mploutsos A (2002) Benzene, toluene, ozone, NO_2 and SO_2 measurements in an urban street canyon in Thessaloniki, Greece. Atmos Environ 36:5355–5364
4. Edner H, Ragnarson P, Spaennare S, Svanberg S (1993) Differential optical absorption spectroscopy (DOAS) system for urban atmospheric pollution monitoring. Applied Optics 32:327–333
5. CETESB – Companhia de Tecnologia de Saneamento Ambiental (2005) Caracterização das estações de rede automática de monitoramento da qualidade do ar na RMSP – estação Cerqueira César: http://www.cetesb.sp.gov.br/ ar/relatorios/relatorios.asp

Using Bayesian inference to manage uncertainty in probabilistic risk assessments in urban environments

E Chacón, E De Miguel, I Iribarren

Grupo de Geoquímica Ambiental, Madrid School of Mines, Alenza 4, 28003 Madrid, Spain

Abstract

A Bayesian risk assessment to assess the potential adverse health effects of the exposure of children up to 6 years of age to urban trace elements in municipal playgrounds in Madrid was carried out. Bayesian statistical methods were used to adapt the distributions of some of the exposure variables taken from the literature to the specific exposure conditions found in the playgrounds of Madrid, Spain. The exposure variables borrowed from the scientific literature were revised with the population-specific data acquired through two limited surveys of 75 and 56 parents, respectively. The predictive distributions of two exposure variables, i.e., body weight and exposure frequency, were subsequently used to better define the distribution of risk estimates.

Introduction

The relevance of exposure to trace elements in urban environments can be evaluated by means of a risk assessment. In Madrid, exposure of children – the most sensitive segment of the population – to trace elements in urban particulate materials occurs mainly during their games in municipal playgrounds after school.

Standard environmental risk assessments very often borrow exposure values from the literature and apply them to estimate the "maximum reasonable exposure" for the population under study. The uncertainty associated with these point-estimate risk assessments cannot be statistically evaluated.

This uncertainty, however, can be estimated if what is borrowed from the literature and introduced in the model is the density function of the exposure variable (i.e., probabilistic risk assessments). Bayesian methods go one step further and allow updation of the prior, generic information supplied to the model as new site – and population-specific data come along. These theoretical considerations have been applied in a Bayesian risk assessment of the exposure of children to urban trace elements in the sandy substrate of municipal playgrounds in Madrid, Spain.

Materials and methods

In November 2002 and November 2003, two samples of approximately 500 g of surficial (approx. top 2 cm) substrate were collected in 20 playgrounds in parks and urban green areas of Madrid, Spain. The samples were oven dried at 45°C for 48 h, sieved below 100 µm, aqua regia-digested in a hot water bath (>95°C), and analysed by inductively coupled plasma mass spectrometry (ICP-MS). For the purpose of this discussion, only the results for mercury and arsenic – as a carcinogen – have been used as examples.

A small survey (Survey A) of 75 parents was carried out in November 2002 in order to obtain specific values for children body weight and exposure frequency. A second survey (Survey B) was undertaken in May 2006 and the data collected from the questionnaires, handed out to 56 parents, were used to update the information gathered in the previous campaign.

Results and discussion

The model used to estimate the risk for children from the exposure to trace elements during their games at municipal playgrounds in Madrid considers three complete exposure pathways: (1) ingestion of substrate particles; (2) inhalation of re-suspended substrate particles; and (3) dermal absorption of trace elements in substrate particles adhered to the exposed body parts. For mercury, the only element with a significant vapour pressure at ambient temperature, inhalation of vapours was also taken into account. The doses

Table 1. Distributions of the exposure variables

Exposure variable	Symbol	Distribution function	Reference
Body weight (kg)	BW	TNORMAL (15.6, 3.7, 6, 30)	[7]
Exposure duration (years)	ED	UNIFORM (1, 6)[a]	[7]
Exposure freq. (hour year^{-1})	EF	BETA (2,2)[b]	judgement
Skin exposed surface (cm^2)	SA	UNIFORM (1400, 2800)[c]	[9]

[a]Concawe, 1997 assigns for this variable a distribution UNIFORM [1, 5]. Since the sampled playgrounds have been designed for children up to 6 years of age, this distribution has been modified to a UNIFORM [1, 6]
[b]Concawe, 1997 assigns to this variable a CONSTANT (350) days year^{-1}. Due to the lack of information on hourly exposure frequencies to playground substrate a beta distribution has been used for this variable
[c]USEPA, 2002 assumes a fourth of the total body surface is exposed. A UNIFORM (1400, 2197) results from division by four of the p5 and p95 of the total body surface distribution for children between 3 and 6 years old. The upper limit of the resulting distribution has been raised to 2800, value recommended by USEPA (2002)

received by children have been calculated using hourly exposure variables in the standard equations proposed by USEPA [1], which are routinely used for deriving risk-based guideline values [2, 3]. Risk equations have been modelled with the BRugs package, linked with OpenBugs, in R language [4].

Reference doses and slope factors have been taken from USEPA Integrated Risk Information System [5]. ABS, VF, and PEF constants have been taken from [6].

For all the other exposure variables hourly distributions were searched in the literature [7–9]. When an hourly distribution was not found (soil ingestion rate, air inhalation rate, and skin adherence factor) point estimates were taken from [8, 9]. Table 1 shows the original distributions representing these variables.

The information collected in surveys A and B was used to modify the prior distributions of the parameters of the density functions for body weight and exposure frequency. The specific exposure frequency in hours per year was estimated from the exposure in hours per week reported in the surveys by considering that the number of weeks per year that a child plays in the playground is equivalent to the number of days without rain in Madrid [10], assuming that this child spends 5 weeks per year away from his usual residence.

Based on this model three different situations were tested. Firstly, only distributions found in the literature were used to make predictions about the exposure variables (i.e., no specific data were taken into account). Secondly data from the survey carried out in November 2002 were introduced in the model to modify the prior distributions of the parameters of body

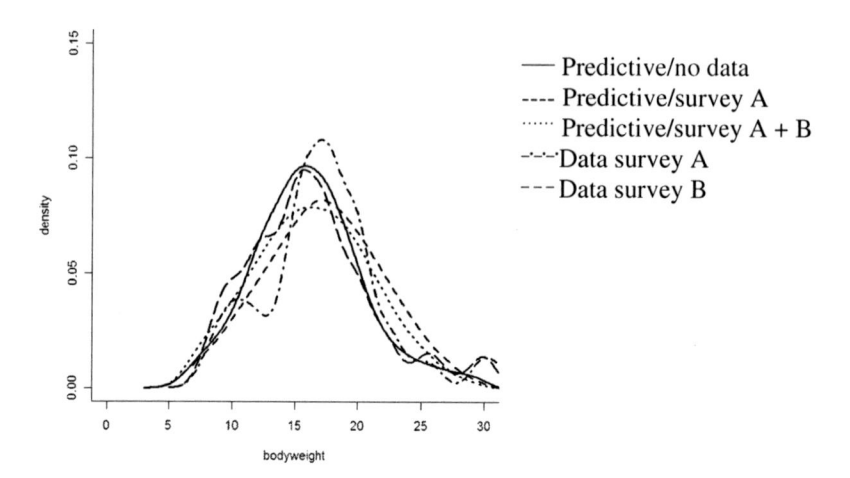

Fig. 1. Body weight predictive distributions

weight and exposure frequency with the sampling distributions of these variables. Finally, the data gathered in May 2006 was included in the model in order to check how the prior distributions and the resulting risk distribution were modified. The model was run 10,000 iterations.

Figure 1 shows the distribution of body weight data from surveys A and B, the body weight predictive distribution from the prior and the body weight predictive distribution from the posteriors once the data from surveys A and A + B have been taken into account. It can be seen how the original body weight distribution found in the literature is progressively modified as new data are incorporated in the model.

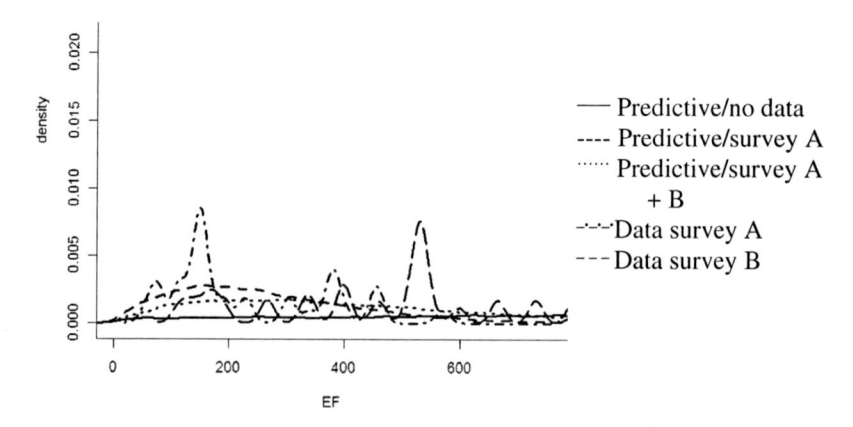

Fig. 2. Exposure frequency predictive distributions

A different situation appears when a convenient distribution for a selected variable cannot be found in the literature, as in the case of exposure frequency (Fig. 2). For this variable, a wide beta distribution has been adopted. Again, as more specific data for the playgrounds in Madrid is introduced, the distribution for the studied variable becomes better defined.

In both situations it can be seen how the importance of the prior distribution diminishes when more site-specific data is included in the model.

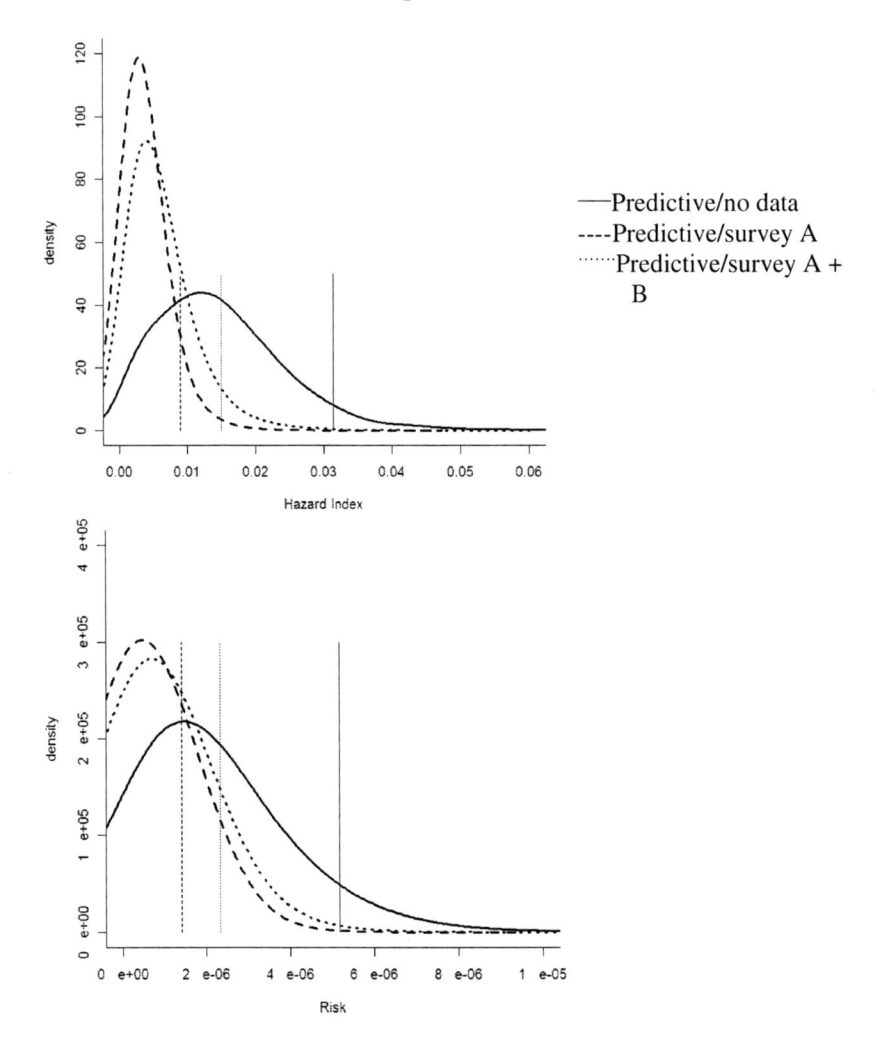

Fig. 3. Resulting hazard quotient and risk distributions for non-carcinogenic Hg (above) and carcinogenic As (below)

Lastly, Figure 3 shows the resulting hazard quotient and risk distributions, for non-carcinogenic mercury and carcinogenic arsenic, respectively, with no *in situ* data; with data only from survey A; and with data from surveys A + B. The vertical lines represent the *p*95 of each distribution.

The discrepancies between the data gathered in surveys A and B explain the non-monotonic behaviour of the *p*95. If more homogeneous data had been used, the *p*95 of the predictive distributions when data from surveys A and A + B are used, would converge to the specific risk/hazard index value.

The shorter tail of these distributions means a higher confidence in that there is no risk from the exposure to playground substrate, particularly for As for which risk levels are close to the legal unacceptable limit of 10^{-5} set by most environmental legislations.

Conclusions

Given the large differences in climate, urban landscape, social habits, etc. between cities across the world, the use of exposure variables borrowed from the scientific literature, which have been defined for a particular population and urban setting, may highly underestimate or overestimate exposure somewhere else. On the other hand, site- and population-specific surveys are costly, a fact that will normally lead to reduced sample sizes and, as a consequence, to large degrees of uncertainty in the estimates of exposure. Bayesian risk assessments, as demonstrated in Madrid for the exposure variables, body weight, and exposure frequency, integrate – and benefit from – both sources of information: they modify the prior distributions defined from density functions borrowed from the literature with limited information gathered *in situ*. This approach allows to better define the predictive distributions of risk resulting from the model, to better manage the uncertainty associated with these estimates, and therefore to make more confident decisions as to whether there is risk or not in a particular situation of exposure to urban pollutants.

References

1. USEPA (1996) Soil screening guidance: technical background document. US Environmental Protection Agency, Office of Solid Waste and Emergency Response, EPA/540/R-95/128 online
2. CCME (1996) A protocol for the derivation of environmental and human health soil quality guidelines. Report no. CCME-EPC 101E. Canadian Council of Ministers of the Environment. Winnipeg, Manitota, Canada

3. DEFRA (2002) The contaminated land exposure assessment (CLEA) model: technical basis and algorithms. R&D publication CLR 10. Department for Environment, Food and Rural Affairs. UK Environment Agency
4. R Development Core Team (2004). R: A language and environment for statistical computing. R Foundation for Statistical Computing, Viena. Also at http://www.R-project.org
5. USEPA. Integrated risk information system website: http://www.epa.gov/iris
6. USEPA (2001) Supplemental guidance for developping soil screening levels for superfind sites. OSWER 9355.4-24. Office of Solid Waste and Emergency Response. US Environmental Protection Agency. Washington, DC
7. Concawe (1997) European oil industry guideline for risk-based assessment of contaminated sites. The Oil Companies European Organisation for Environment, Health and Safety, Report No. 2/97, Brussels

An assessment framework for urban water systems – a new approach combining environmental systems with service supply and consumer perspectives

C Lundéhn, GM Morrison

Water Environment Technology, Department of Civil and Environmental Engineering, Chalmers University of Technology, SE-412 96 Göteborg, Sweden

Abstract

There is increasing awareness that improvements to urban water systems (UWS) should be based on an assessment of sustainability that safeguards the well-being of both humans and ecosystems. There is a need for coherent and participatory frameworks, which integrate existing assessment tools.

A combined framework for UWS assessment is presented here and tested by application in a case study in Accra (Ghana). The approach combines the results of interviews with, and questionnaires to consumers and stakeholders with data collection of environmental sustainability indicators (ESIs) for the UWS.

Safe water quality, safe access to water, and affordable water were identified as key criteria in the Accra dialogues, while healthy ecosystems and reducing environmental effects were considered least important. The ESI results reveal the need for an increased awareness and monitoring if the requirements of and challenges for a satisfactory UWS are to be met.

The approach expands from existing methodologies towards a greater involvement of the general public and thereby provides a link between environmental sustainability and the service supply, and consumer perspectives.

Introduction

There is increasing awareness that improvements to urban water systems (UWS) should be based on an assessment of sustainability [1]. This implies both safeguarding the well-being of people (meeting present and future demands) and the vitality of ecosystems (protection of water resources, efficient use of resources and energy, waste and emission minimization).

Assessment frameworks have been proposed as a basis for making sustainable decisions on the management of UWS [1–4]. The driving force behind existing frameworks has been defined as either the people or the environment [4]. Assessments have been applied fully or on parts of UWS and differ in comprehensiveness in terms of stakeholder/consumer involvement and inclusion of sustainability aspects.

Indicators are recognized tools for evaluating, setting targets, communicating, and monitoring performance over time [5]. They have been applied to UWS in several studies, although the choice of indicator frameworks vary [6–10]. International organizations concerned with sustainability issues stress the importance of harmonizing indicators and data collection, and linking local perspectives to global targets [5].

Performance indicators, which to date have been mainly applied to UWS operation and maintenance at the local level, have proven useful for UWS management [11]. An analytical life-cycle assessment-based procedure has allowed the selection of UWS environmental sustainability indicators (ESI) and permitted extension of system boundaries from a local level to a regional/global level [6]. Current frameworks that combines environmental and socio-economic considerations for the UWS are based on system analysis such as Multi-Criteria Analysis [1,9,12,13]. It is emphasized that successful frameworks for sustainability assessment need to be coherent, feasible, and transparent, and that more attention must be paid to ensure adequate participation of special interest groups and the general public [14–16]. It has also been proposed that there is a need for a framework that integrates existing assessment tools rather than developing new tools [17].

This article presents a combined approach, which aims to bring together commonly employed techniques for stakeholder and consumer dialogue (interviews and questionnaires), with life-cycle-based ESIs under defined system boundaries. The approach is applied to an African setting, Accra, where water management is currently being shifted, supported by multilateral organizations, towards private sector participation (PSP).

Assessment framework for urban water systems 561

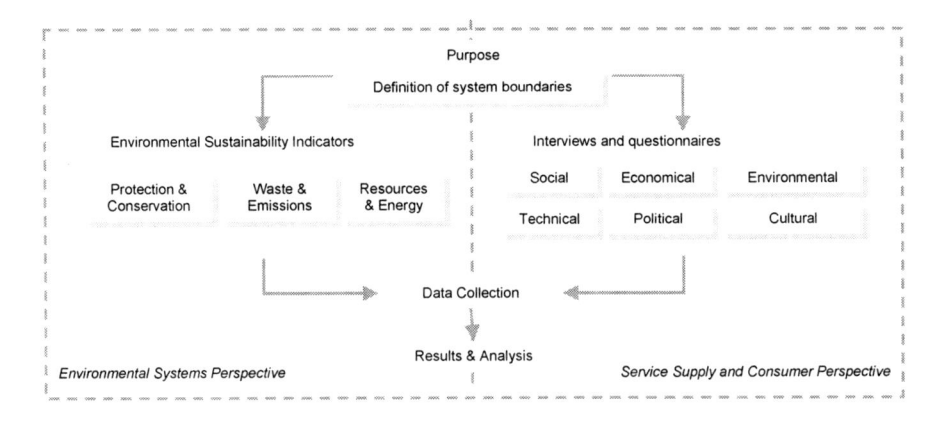

Fig. 1. Conceptual model of the combined approach

Method

The combined stakeholder–indicator approach is based on two parallel perspectives illustrated in Figure 1: (1) the environmental systems perspective; and (2) the service supply and consumer perspective.

System boundaries

The chosen system boundaries follow three previously defined levels (Fig. 2) [6]:

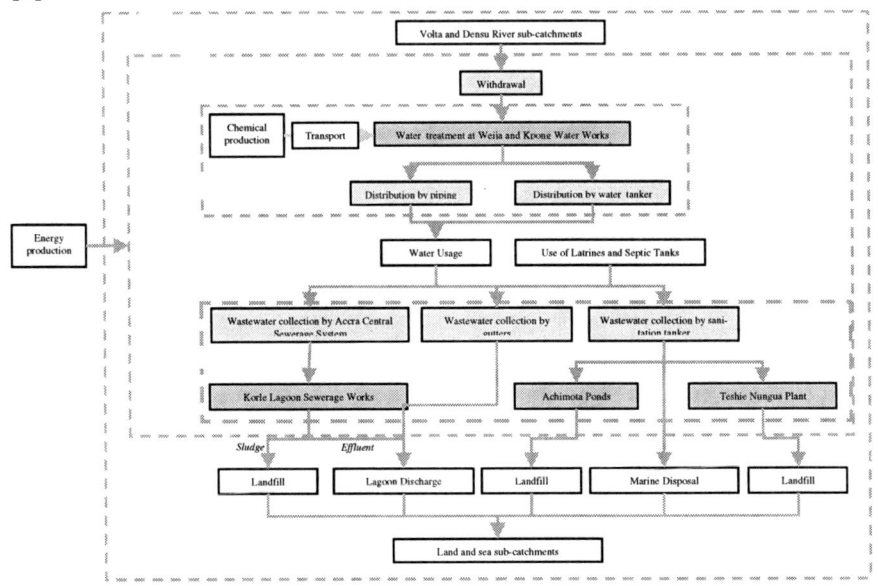

Fig. 2. Accra UMS and system boundaries

- Level 1 – which is restricted to the technical system such as treatment plants and the physical infrastructure for water and wastewater.
- Level 2 – which includes water usage, local management, recovery and transport of water, wastewater and by-products.
- Level 3 – which extends from a local to a regional catchment and resource management level. As the improvement of local conditions alone might result in the depletion of resources elsewhere, level 3 also includes a life-cycle extension, which is relevant in terms of material and energy flows [6,18].

The study boundaries are expanded to include a catchment perspective, in line with the EU Water Framework Directive (WFD). In WFD, system boundaries are based on water resources and hydrologically based catchment boundaries, rather than administrative boundaries [14].

Environmental systems perspective

The environmental systems perspective was assessed using ESI following the urban water cycle (freshwater resources, water production and distribution, usage, wastewater collection, and treatment and handling of by-products). The ESIs were selected based on collated life-cycle assessment (LCA) studies as reported elsewhere [6] (Table 1). Data collection was through field visits to water works, wastewater treatment plants, relevant

Table 1. Selected ESI for UWS assessment

Urban water cycle	Study boundary	Environmental sustainability indicator
Freshwater resources	3	Freshwater availability
		Raw water protection
		Raw water quality
Drinking water production and distribution	1, 2	Drinking water production
		Chemical and energy consumption
		Leakage
Usage	2	Drinking water consumption
Wastewater collection and treatment	1, 2	Wastewater collection
		Treatment performance
		Loads to receiving water
		Chemical and energy consumption
Handling of by-products	2, 3	Sludge disposal/reuse
		Nutrient recycling
		Loads to receiving soils

authorities, and institutions as well as published national and international reports. Data availability and accuracy was problematic and required great effort; extended time-series for indicators were not available for Accra.

Service supply and consumer perspective

The service supply and consumer perspective was assessed through interviews and questionnaires. Structured interviews were arranged with key stakeholders involved in the UWS including local authorities and multilateral bank representatives, employees within the water and sanitation sector, academic expertise, as well as national and international non-governmental organization (NGO) representatives operating in Accra. Interviews focused equally on the environmental, economic, and social dimensions of sustainability. The stated aim of the interviews was to highlight local sustainability issues for the UWS. Stakeholders selected for interviews were at a senior operational level. Interviews involved a brief introduction to the research aim and methodology followed by a set of open questions. All interviews were documented and had the same outline and questions. The interviewee was given a brief introduction to the research followed by a set of broad questions in order to capture stakeholder concerns related to the three UWS boundary levels (Fig. 2) and the sustainability concept illustrated in Figure 1.

Certain issues arose in all interviews. These were identified as local sustainability issues and formulated as statements in a questionnaire. The questionnaire survey was used to map correlations and differences in perception of the local sustainability issues within and between stakeholder groups, as well as the general public. A number of sustainability criteria for the UWS were also extracted from the interviews and included in the questionnaire for a broader evaluation. Interviews thus formed the basis for local data collection through a three-tiered questionnaire:

- Part I: Household facts to provide data on household type, size, usage, access to water, and experienced water deficiencies
- Part II: Sustainability criteria where the participant was asked to rank a given set of criteria that derived from the interviews
- Part III: Statements of local sustainability issues where the participant was asked to agree/disagree (on a five grade scale) to statements on the site-specific critical issues derived from the interviews

The same questionnaire was distributed to the participants regardless of level of water sector involvement (Table 2). All questionnaire survey participants were permitted to provide personal comments in the questionnaire.

Table 2. Questionnaire distribution

Stakeholder group	Handed in	Distributed	Response rate
Politicians/government	3	5	0.6
Professional employees	15	18	0.8
Academics	4	4	1.0
NGOs	5	5	1.0
General public	22	25	0.9
Total	49	57	

Industrial and agricultural consumers were not part of the questionnaire survey. The distribution of questionnaires to the general public was based on face-to-face questioning without an even geographical distribution.

Case study area

Accra UWS (Fig. 2) can be described as a segmented system at a stage of having to safeguard adequate water and sanitation supply for all. The UWS is currently enforcing PSP management supported by the World Bank, with sustainable development as a stated goal.

It is estimated that 2,101,400 people live in Accra (and that 3.6 million people live in Accra Tema Metropolitan Area; ATMA) with a population growth rate of 4.2–6.6% [19]. ATMA, is supplied with water from two sources; the Densu River (Weija Water Works) and the Volta River (Kpong Water Works). The urban water supply in Ghana is managed by the Ghana Water Company Limited (GWC).

To a large extent, the population in Accra and its surroundings lack water pipe network connections or suffer from dry pipes and thus depend on water being delivered by water tanker service (Table 3).

Pipe-borne water is paid by meter or by a fixed price according to tariffs. Shared water is from public taps but to a large extent these taps are run by someone that stores the tap water to secure supply when the pipes are dry, and thus charges the consumer additional costs when collecting their

Table 3. Accra water supply sources as of 2000 (From GSS, 2002)

Water source	% of population
Private water access	35.9
Access to shared water source	44.9
Access through water tanker service	7.3
Boreholes, rivers etc.	11.8

water. Water tankers and trucks loaded with sachet water are a common sight. The sachet water is mainly sold to retail vendors that cool it and sell it in the streets.

Only the very central parts of Accra are served by a water-borne sewage network (that transports the wastewater to the Korle Lagoon Sewerage Works), while the rest is served by on-site sanitation systems from which sanitation tankers, private or public, collect the sewerage and transport it to the nearest of three plants (Achimota Ponds, Teshie Nungua Plant, and the Marine Disposal Site). The Marine Disposal Site is a section of the beach on the outskirts of Accra where sewage is discharged untreated into the sea.

GWC is responsible for the Accra Central Sewerage System (Korle Lagoon Sewerage Works), while the Waste Management Department of Accra Metropolitan Assembly manages the Marine Disposal Site, Achimota Ponds, and Teshie Nungua Plant.

The Korle Lagoon receives final effluent from the Korle Lagoon Sewerage Works and virtually all the runoff from the city of Accra. This includes upstream solid waste, as all wastewater from places and facilities that are not connected to the Accra Central Sewerage System end up in gutters that run into the lagoon, and finally to the sea. The wastewater is often mixed with municipal waste that is washed into the lagoon during storm events.

Results and discussion

Participants in the questionnaire survey were asked to rank what they find most important for the future based on the identified criteria. Table 4 shows

Table 4. Results of the sustainability criteria ranking where low value represents most important and high value represents least important

Rank	Criterion
1	Safe water quality
2	Safe access to water
3	Affordable water
4	Water treatment plants and delivery that function well
5	Development of the drinking water production
6	Development of the sanitation system
7	Public awareness on water and sanitation issues
8	Integrated water and sanitation planning
9	Safe access to flush toilet/latrine
10	Healthy ecosystems and reduced environmental effects

the results of the ranking based on all participants of the questionnaire survey; the stakeholder groups were unevenly represented in terms of numbers. The results rank safe water quality, safe access to water, and affordable water as the top three criteria. Healthy ecosystems and reduced environmental effects were ranked least important.

Freshwater resources

Water availability is high in Ghana. With the Volta Lake and other river basins surface water assets of 3000 m^3 capita^{-1} and groundwater availability of 2100 m^3 capita^{-1} are provided [20]. Ghana is not considered as having water stress conditions. Of the total land area 15.4% is under water protected area status (both land and water).

Water quality monitoring made by the Water Research Institute [21] has identified very high colour, turbidity, and nutrient levels in the Densu Basin due to uncontrolled human activities that generate waste, untreated sewage, fertilizer, and pesticide runoff. Weija water works reports raw water BOD to 10 mg L^{-1} and COD to 49 mg L^{-1}.

The quality of raw water from the Volta is better than Densu due to two large dams that serve as sedimentation basins for the raw water. However, as the population in the surrounding villages continue to grow, the situation is likely to worsen both at Weija (Densu) and Kpong (Volta) in the near future. Estimates of nutrient input reveal a 25% increase between 1997–2010 [22,23]. The increased water stress calls for improved raw water protection and regulation, and currently a Ghana Water Policy is being formulated to provide a framework for the development of Ghana's water resources for all stakeholders.

Drinking water production and distribution

The Weija and Kpong water works have a total water production of 121 Mm^3 year^{-1} (92 L capita^{-1}day^{-1} equivalent, based on the ATMA population estimate). The total chemical consumption for water production in 2004 was 7968 t (Table 5). The consumption of chemicals is higher at Weija where water quality is poor.

Stakeholder interviews demonstrate that high chemical cost is a barrier to the production of high quality drinking water, as all chemicals are imported (total cost for chemicals in 2004 was US$20,400 Mm^{-3} produced drinking water).

Table 5. Chemical consumption for drinking water production at Weija and Kpong Water Plants 2004

Chemical	Weija Water Works		Kpong Water Works	
	t	mg L^{-1}	t	Mg L^{-1}
Lime	850	14.2	276	4.5
Chlorine gas	180	3.0	96	1.6
Bleaching powder	10	0.2	56	0.9
Aluminium sulphate	6500	108.3	0	
Salt	0		4.8	0.08
Total	7540		428	

The total energy consumption for water production and distribution in 2004 was 101,900 M Wh (842 Wh m^{-3}). Consumption is high at Kpong due to the long pumping distance to Accra, whereas consumption is low at Weija as distribution is based on gravity feed. Consumption at both plants has slightly increased during the past 5 years when comparing collected data with previous data [22]. Data collection reveals that energy consumption for water distribution (and wastewater collection) by truck is the main critical issue. The results in Figure 3 are based on data collection and estimations. Results are thus uncertain but indicate the need for further investigation. Understanding the share of water that is delivered through tap, tanker, and sachets is essential in order to interpret the impact of these results.

The level of unaccounted water is as high as 50–65% including leakage, non-revenue water, and commercial loss, and has remained the same since data became available in 1997 [24]. Much of the water that is not lost is

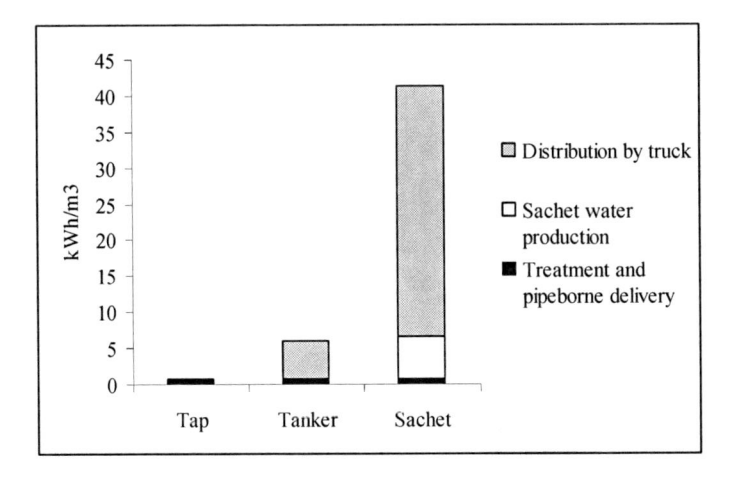

Fig. 3. Energy consumption for water supply

still not distributed directly to the consumers, as the piping network does not reach all potential consumers. Even for those with a pipe connection, water delivery cut-offs are frequent. Water Aid [25] reports that only 25% of the population in Accra have 24 h water service, while Ghana Statistical Service [26] states that 76% of the urban households have water supply all the time (Table 6). The questionnaire evaluation revealed that only 9.4% of the participants have access to water all the time and that participants experience frequent cut-offs in water delivery (Fig. 4A).

It is understood that water delivery cut-offs can last for days, weeks, or even months. In terms of water cut-off tolerance 34.5% of the participants find that it is not acceptable with cut-offs at all, while 27.6% find short weekly cut-offs acceptable and 20.7% find short monthly cut-offs acceptable. The acceptability curve lags the experience curve, indicating that acceptability is related to personal experience (Fig. 4). Of the survey participants

Table 6. A: Review of Ghana water supply coverage data; B: review of Accra water supply coverage data

A. National data	Year	%	Reference
Access to drinking water	1995	86[a]	HCGR (2005)[c]
	2003	76[a,b]	GSS (2003)[c]
	2005	73	
Access to improved water	1990	54[a]	WHO/UNICEF
	2000	91[a]	UNSD (2004)
	2002	79[a]	WHO/UNICEF (2004)
	2004	61[a]	WB PAD (2004)
Access to safe drinking water	1997	93[a]	GSS (2001)[c]
	2000	70	

B. Accra data	Year	%	Reference
Access to 24 hr day^{-1} water supply	2005	25	WaterAid (2005)
Access to pipe-borne water	1958	24	Acquah (1958)
	1984	89	GSS
Access to drinking water	2002	33.3[d]	GSS (2003)
	1992	99.8[e]	Benneh et al. (1993)
	2000	88.1[f]	GSS (2002)

[a] Urban population
[b] Service 24 hr day^{-1}
[c] Reported Ghana official data
[d] Private water access
[e] Private or shared water access
[f] Private, shared or tanker water access

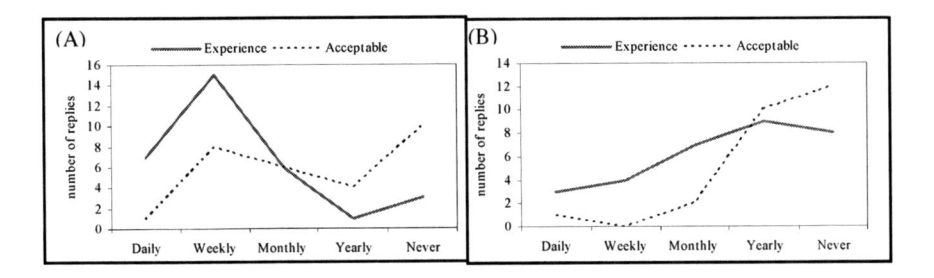

Fig. 4. (A) Water delivery cut-off frequency (consumer experience) and tolerance (acceptable cut-off frequency); (B) water quality deficiency frequency (consumer experience) and tolerance (acceptable deficiency frequency)

there are those that want an improved situation but can accept cut-offs and those that do not accept cut-offs under any circumstances.

Usage

Access to water is a critical issue and a concern (Table 4). The total water production of 121 Mm^3 year^{-1} provides water for households, industry, and institutions. This figure equals a water consumption of 92 L capita^{-1} day^{-1} (based on population estimates). However, based on domestic and commercial (tanker service, public standpipe, etc.) water sales records, consumer sources of water supply and taking leakage into consideration, the consumer pipe-borne water consumption is in the range 31–46 L capita^{-1} day^{-1}. This is below the World Health Organization (WHO) level of intermediate water access; 50 L capita^{-1} day^{-1} provided on-plot through at least one tap per yard [27].

Benneh et al. [28] state that access to pipe-borne water in Accra is closely related to the degree of urban planning and to the degree of household wealth. However, the case study interviews indicate that regardless of economic level consumers commonly suffer from dry taps and depend on water tanker service and sachet water supply.

Table 6 presents data for water supply coverage from various sources. Data differ greatly between sources and national water supply coverage range between 54% and 93% for the past 20 years. The difficulty to monitor access to water and the problems related to comparability, difference in definitions, and data collection have been highlighted elsewhere [29]. This case study demonstrates the problems of comparability and data availability. A number of definitions for water supply data are in use (Table 6) and definitions differ in terms of water source and quality (e.g., tap, tanker, well improved/unimproved), type of connection (e.g., household connection, public standpipe, borehole), and availability (e.g., 24 h service, distance to source).

Monitoring of drinking water quality is limited. Accra has a high frequency of water-borne disease, and drinking water quality deficiency is a problem. About 22.4% of the questionnaire participants report having suffered from water-borne diseases, and 74.2% experience daily, weekly, monthly, or yearly deficiency in water quality (Fig. 5). Participants find water quality deficiencies unacceptable (48%) or acceptable only yearly (40%).

The experience–acceptability curves for water quality (Fig. 4B) do not show the lag found for water delivery cut-offs, indicating a low tolerance for water quality deficiency. Of the survey participants there are those that want an improved situation but can accept cut-offs and those that do not accept cut-offs under any circumstances.

There are strong indications of a limited consumer trust in water quality. Participants are concerned when drinking tap water (only 43% feel safe) with a much higher trust in bottled/sachet water (90% feel safe). There is a clear divide between the general public and stakeholders involved in water production in terms of trust in water quality. The general public representatives strongly believe that bottled water is healthier than tap water (82%) with the reverse expressed by government and employee representatives (16.7%).

Affordable water is a concern (Table 4) and survey results show consumer water expenditures varying between 5% and 45% of the consumer food expenditures. Private access to tap water is by far the cheapest for the consumer. Dependence on a shared stand-pipe increases prices almost four times for the consumer involved in this study. Private water delivery through tanker or sachet/bottled water is the most expensive with tanker water delivery costing 13 times the tap water price.

The questionnaire survey results reveal that all stakeholder group participants, apart from the government representatives, expect an increase in consumer water costs with the current shift to PSP for Accra water supply management (Fig. 5A). Government and employee representatives find this acceptable in order to reach sustainability objectives but both academic and NGO representatives object to an increase in consumer water costs (Fig. 5B). NGO representatives also strongly oppose that consumers should pay for investments in improvements towards a more sustainable UWS. Representatives from the general public, however, seem willing to pay for water supply improvements. In a previous consumer survey of Accra it was shown that for those with an existing water supply, 66% are willing to pay more in principle for supply improvements [30].

(A) Private sector participation in Ghana Water Company Ltd. promotes a sustainable development of the urban water system.

(B) Water costs should pay for investments in improvements towards a more sustainable water system.

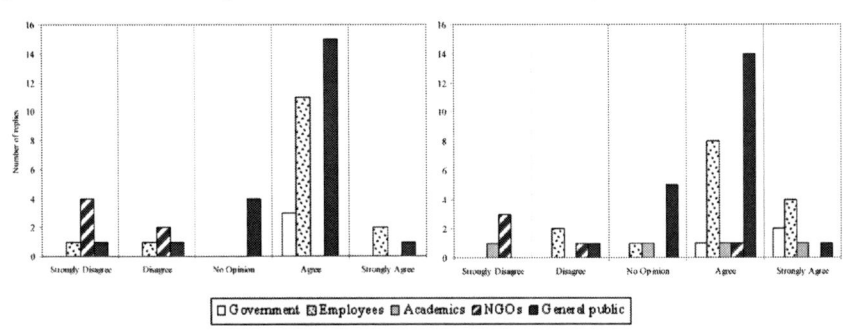

Fig. 5. (A) Water costs and sustainability investments; (B) expected consumer costs with PSP

Wastewater collection and treatment

It has been reported that as of 2000 [24] 34% of the urban population in Ghana had access to "safe sanitation", while other reports [31] state that 74% have access to "improved sanitation". The same insecurity in data availability, comparability, and accuracy as for water supply coverage data, was found for sanitation coverage.

The total wastewater collection volume (by sanitation tanker and the pipe-borne system) in Accra is only 6.2 Mm^3 year^{-1} and the treatment of collected wastewater is limited. The Korle Lagoon Sewerage Treatment Plant uses USAB reactors for purifying the water, which is claimed to reach 95% biological oxygen demand (BOD) removal [22]. However, the plant was not fully functioning at the time of this case study and the plant has no facility for N or P treatment.

Of the total collection of sludge about 0.2 Mm^3 year^{-1} is collected by sanitation tankers and transported to the Marine Disposal Site (42%), the plant in Teshie-Nungua (11%) or the ponds in Achimota (47%).

The Marine Disposal Site is a section of the beach where sanitation tankers discharge untreated sewage sludge into the sea. It is reported that the Korle Lagoon is one of the most polluted in Ghana with extreme nutrient levels due to human activities including the treatment plant load to receiving waters [21]. There is no chemical use for the wastewater treatment and energy consumption is not reported.

(A) The recycling of nutrients from sewage sludge either through agricultural application or extraction should be prioritized to reduce the burdens on natural resources.

(B) Potential human health and environmental risks associated with the use of sewage sludge on agricultural land are acceptable trade-offs to meet nutrient recycling goals.

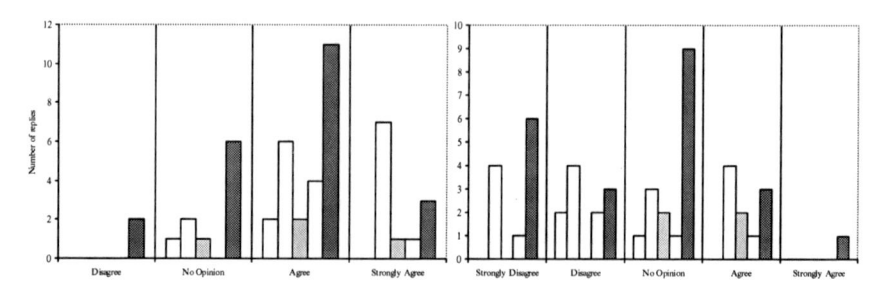

Fig. 6. (A) Nutrient recycling and reduced environmental burden; (B) sludge re-use and potential risks

Handling of by-products

Sludge is placed in drying beds and the former facility for sludge composting is currently not in use. From the drying beds sludge is collected by truck and reused as soil amendment. The sludge quantity and quality is not monitored and levels of nutrient recycling and loads to receiving soils are uncertain.

Participants in the questionnaire survey agree that nutrient recycling from sewage sludge should be prioritized to reduce environmental burdens. However, when explicitly stating potential risks associated with such practices, only participants representing academics seem to find risks of sewage sludge application on land acceptable to meet nutrient recycling goals (Fig. 6).

UWS management

Water supply management

Representatives from NGOs, government, and employees agree that the current institutional organization structure in Accra does not promote a sustainable development of the UWS. The debated water management reform of Ghana Water Company has been a long political process, finally

(A) Private sector participation of Ghana Water Company Ltd. increases management efficiency.

(B) Private sector participation in Ghana Water Company Ltd. promotes a sustainable development of the urban water system.

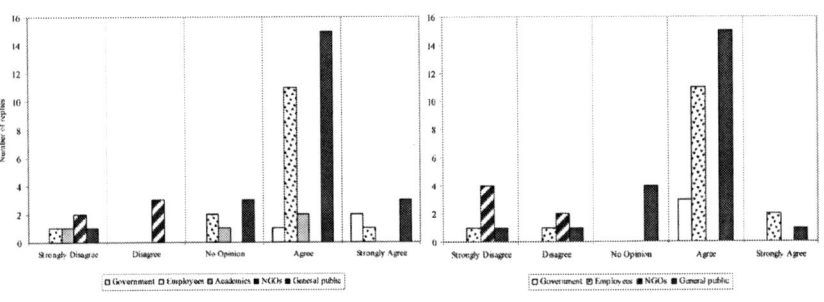

Fig. 7. (A) PSP and management efficiency; (B) PSP and sustainability

leading to PSP in water supply management. In agreement with the representatives of the general public, representatives of the government and employees clearly believe that the reform will increase management efficiency and promote sustainability (Fig. 7). NGOs have objected to the water management reform since its start, which is also seen in this study, and believe that PSP neither promotes a sustainable development nor improves management and efficiency of the water supply sector.

Wastewater management

Participants agree that water and sanitation management should be viewed as one system (Fig. 8), although this is not necessarily the case in the water management reform [32]. These results together with the criteria ranking (Table 4) indicate a lack in perceiving sanitation as part of the UWS, even

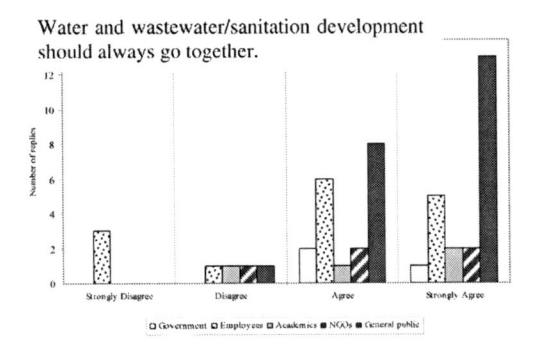

Fig. 8. Water and sanitation management

though this is explicitly emphasized in the Millennium Development Goals as critical for a sustainable UWS development [33].

Conclusions

Findings for the developed approach

- The presented framework where a stakeholder–consumer approach (with emphasis on a local perspective) is combined with an environmental approach (with a regional/global perspective) provides both qualitative and quantitative data.
- The approach covers the whole urban water cycle and expands from current stakeholder targeted methodologies towards a greater involvement of the general public.
- The approach provides UWS information by integrating ESI data collection and questionnaire survey results. The combination of indicators and questionnaires in this approach has proven successful as both give indicative results, providing a balanced level of information between the two.
- The results demonstrate that the approach captures not only water system data but also consumer and stakeholder perceptions. The stakeholder–indicator approach thus serves to combine environmental sustainability facts for the system and human perception.
- As the framework is based on straightforward assessment techniques (indicators, interviews, and questionnaires) it is not viewed as complex or limited in feasibility by practitioners.

Environmental sustainability indicators

- The ESI results show that leakage, drinking water quality, and energy for water distribution by truck are critical issues. Significant limitations in wastewater collection and treatment result in high loads to receiving water and soil.
- A major constraint for the ESI data collection in Accra was data availability, uncertainty, and comparability, which limited the accuracy and the possibility to view trends over time. For the local stakeholders greater effort needs to be placed on obtaining reliable data, valuation of identified issues, and applying statistical methods to safeguard adequate results.

Critical issues from interviews

- For a system like Accra where the infrastructure for water supply is not yet fully established, the identified critical issues are mainly related to the technical system (study boundary level 1) and its management (study boundary level 2).
- Safe water quality, safe access to water, and affordable water are recognized as the most important issues for the future and it is clear that a well functioning pipe-borne water supply should be strived for to keep the system both cost and energy effective.

Questionnaire findings

- Although some groups are critical (e.g., NGOs), individual consumers believe that PSP will bring improvements to the UWS. Most believe that PSP will mean increased consumer water costs but the questionnaire reveals a willingness-to-pay.
- The questionnaire participants rank the criterion healthy ecosystems and reduced environmental burden, as least important for the future, while, on the contrary, the need for improved environmental status is confirmed through the ESI results.
- There seems to be a lack of awareness and limited recognition of environmental aspects as crucial for a sustainable UWS. This indicates the need for a clearly stated definition of what sustainability implies. However, such definition and acceptance processes amongst the different stakeholders as well as the general public will be difficult and complex.
- The results of the questionnaire survey stems from the questionnaire formulation, which is based on stakeholder interviews. Thus the selection of interviewees is crucial and it is unknown whether the results would differ by expanding the number of interviews made and/or expanding the selection of stakeholder interviewees to include the private and agricultural sector.

References

1. Sahely HR, Kennedy CA (2005) Developing sustainability criteria for urban infrastructure systems. Can J Civ Eng 32:72
2. Vari A, Kisgyorgy S (1998) Public participation in developing water quality legislation and regulation in Hungary. Water Polic 1:223–238

3. Starkl M, Brunner N (2004) Feasibility versus sustainability in urban water management. J Environ Manage 71:245
4. Rijsberman MA, van de Ven FHM (2000) Different approaches to assessment of design and management of sustainable urban water systems. Environ Impact Assess 20:333–345
5. Albertini M (1996) Measuring urban sustainability. Environ Impact Assess 16:381–424
6. Lundin M, Morrison GM (2002) A life cycle assessment based procedure for development of environmental sustainability indicators for urban water systems. Urban Wat 4:145
7. Zinn E (2000) Sustainable development indicators for urban water systems: A case study of King William's Town South Africa, in Water Environment Transport. Chalmers University of Technology, Göteborg, Sweden
8. Borgström S (2004) Urban wastewater systems – assessment with environmental sustainability indicators: a case study in Santiago de Chile, in Water Environment Technology. Chalmers University of Technology, Göteborg, Sweden
9. Palme U et al. (2005) Sustainable development indicators for wastewater systems – researchers and indicator users in a co-operative case study. Resour Conserv Recy 43:293
10. Verocchi M, Codner GP (2003) Sustainability of potable water supplies in developing countries – Baguio case study. Australian J Multi-diciplinary Eng 1:17–24
11. Janssens JG (1996) Development of a framework for the assessment of operation and maintenance performance of urban water supply and sanitation. Water Supply 14:21–23
12. Hellstrom D, Jeppsson U, Karrman E (2006) A framework for systems analysis of sustainable urban water management. Environ Impact Assess 20:311
13. Brunner N, Starkl M (2004) Decision aid systems for evaluating sustainability: a critical survey. Environ Impact Assess 24:441
14. European Commission (2000), Water Framework Directive 2000/60/EC. EC, Brussels
15. Doelle M, Sincleir AJ (2006) Time for a new approach to public participation in EA: promoting cooperation and consensus for sustainability. Environ Impact Assess 26:185–205
16. Bertrand-Krajewski JL, Barraud S (2000) Need for improvement methodologies for sustainable management of urban water systems. Environ Impact Assess 20:323–331
17. Walton JS et al. (2005) Integrated assessment of urban sustainability. In: Proceedings of the institute of civil engineers: Engineering Sustainability, London
18. Malmqvist PA, Palmquist H (2005) Decision support tools for urban water and wastewater systems – focussing on hazardous flows assessment. Water Sci Technol 51:41
19. Ridder D (2002) Umweltmanagement im urbanen Ghana. University of Dortmund, Dortmund, pp 1–259

20. Esty DC et al. (2005) Environmental sustainability index: benchmarking national environmental stewardship. Yale Center for Environmental Law and Policy, New Haven, CT
21. WRI (2003), Water Research Institute annual report 2003 Accra Ghana. Council for Scientific and Industrial Research:,Accra, Ghana
22. Uusitalo K (2002) An evaluation of urban water systems using environmental sustainability indicators: a case study in Adenta Ghana, in Water Environment Transport. Chalmers University of Technology, Göteborg, Sweden
23. Salifu LY, Mumuni F (2000) Sanitation and health. World Health Organisation Water Series 389
24. DANIDA (2003) Water and sanitation sector programme support. Phase II, G.o.G. Ministry of Foreign Affairs, p 11–20
25. WaterAid (2005) Ghana national water sector assessment. Water Aid, Accra, Ghana
26. GDHS (2003) Ghana demographic and health survey 2002. G.S. Service (ed.). Ghana Statistical Service, Ghana
27. WHO/UNICEF (2005) Water for life: making it happen. Joint Monitoring Programme for Water Supply and Sanitation, WHO/UNICEF, Geneva
28. Benneh G et al. (1993) Enironmental problems and the urban household in the Greater Accra Metropolitan Area (GAMA)–Ghana. Stockholm Environment Institute, Sweden
29. Bommelaer O, Labre J (2003) Evaluating and monitoring the access to water supply and sanitation. In: 3rd World Water Forum, Water and Poverty Theme. Kyoto, Japan
30. WATP (1999) Willingness and ability to pay: demand assessment and tariff structure study Ghana Urban Water Sector. London Economics John Young and Associates Ministry of Works and Housing, Accra, Ghana/London
31. UNSD (2000) Indicators on water supply and sanitation. United Nations Statistics, Division Demographic and Social Statistics, Social Indicators
32. World Bank (2004) Project appraisal document on a proposed credit in the amount of 103USD million to the republic of Ghana for an urban water project. World Bank, Water and Urban 2, Country Department 10, Africa region
33. UNTFWS (2005) Health dignity and development: what will it take? Achieving the millennium development goals. Stockholm international water house and UN millennium project (Task Force on Water and Sanitation), New York

Large area noise evaluation

E Chung,[1] A Bhaskar,[2] M Kuwahara[3]

[1]Institute of Industrial Science, University of Tokyo, Tokyo 153-8505, Japan
[2]Ecole Polytechnique Fédérale de Lausanne, Station 18, CH-1015 Lausanne, Switzerland
[3]Institute of Industrial Science, University of Tokyo, Tokyo 153-8505, Japan

Abstract

A software tool (DRONE) has been developed to evaluate road traffic noise in a large area with the consideration of network dynamic traffic flow and the buildings. For more precise estimation of noise in urban network where vehicles are mainly in stop and go running conditions, vehicle sound power level (for acceleration/deceleration cruising and ideal vehicle) is incorporated in DRONE. The calculation performance of DRONE is increased by evaluating the noise in two steps of first estimating the unit noise database and then integrating it with traffic simulation. Details of the process from traffic simulation to contour maps are discussed in the paper and the implementation of DRONE on Tsukuba city is presented.

Introduction

Generally traffic noise is not evaluated in a comprehensive manner, i.e., it is limited to road side area with static traffic flow consideration. The effect of buildings especially in built-up area is also not taken into account. For the noise evaluation in a large area with the consideration of network dynamic traffic flow and the buildings (forming the built-up area) a computerized model, DRONE (area-wide Dynamic Road traffic Noise) [1–3], has been developed. This paper elaborates the methodology and structure of DRONE to estimate noise levels on a large urban area where the vehicles are generally in stop and go running conditions. For computationally efficient noise estimation on a large network, DRONE generates unit noise database for the study area, which is then efficiently integrated with traffic simulation.

The structure of paper is as follows: first the general methodology and framework of DRONE is discussed, followed by the discussion on the explicit consideration of acceleration/deceleration in DRONE. Then a two-step procedure of first generating unit noise database and then area-wide noise estimation is discussed. Finally the results from the implementation of DRONE at Tsukuba city, Japan are presented.

General methodology and framework of DRONE

Methodology and framework of DRONE

The methodology of DRONE is to integrate the time-dependent traffic characteristic simulated from the traffic simulation model with the noise estimation model to estimate noise on an area-wide region by considering the time-dependent variations in traffic flow on the whole network. It is then further linked with Geographic Information System (GIS) to provide visual representation to the estimated noise in the form of time-dependent noise contour maps.

In the present research, for traffic simulation, a mesoscopic traffic simulation model AVENUE [4] is used. Standard Japanese road traffic noise estimation model (ASJ Model [5–7]) developed by the Acoustic Society of Japan is used for noise estimation. In the model, the building attenuation in the built-up area is estimated by considering the statistical model by Useaka [8]. For detailed discussion about how the buildings and built-up area attenuation is considered for noise estimation in DRONE, refer to [2].

For area-wide noise estimation, the study area is divided into a number of receptor points. The road network is also divided into a number of small sections (source sections). For each receptor point, the source sections, which contribute noise level at the receptor point are searched. Then for a particular source and receptor point, noise calculations, with appropriate consideration of attenuations such as building, distance etc., are performed

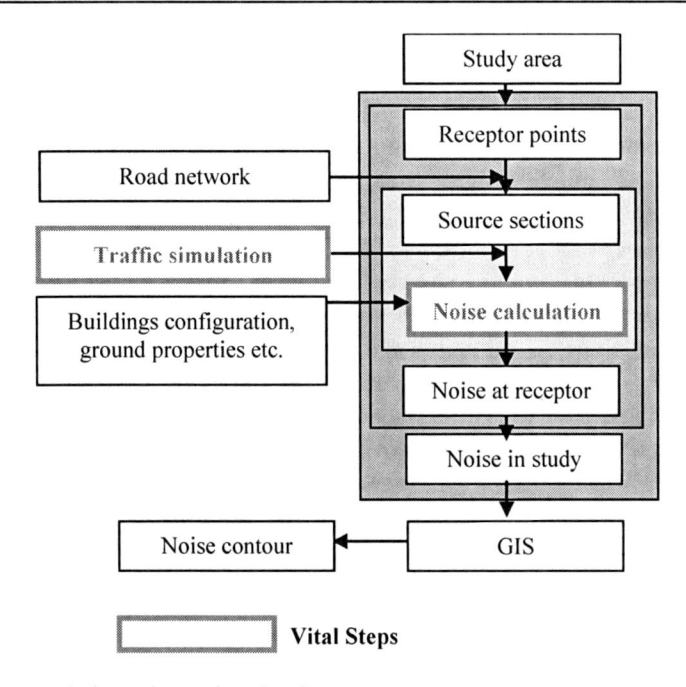

Fig. 1. Framework for noise estimation in DRONE

based on the time-dependent traffic simulation on the network. The process is repeated for all the sources and for all the receptor points to estimate time-dependent noise level in the area-wide region (Fig. 1).

A brief outline of the noise calculation procedure is as follows: The noise generated from a source section is defined in terms of sound power level (L_{WA},) for a vehicle based on the vehicle type and its running condition (transient or steady). The corresponding sound pressure level (L_p, Eq. 1) at the receptor point is calculated by appropriate consideration of the sound propagation from the vehicle position to the receptor point. The sound propagation from the source to the receptor point is dependent on the number of factors, such as ground properties, buildings, etc., between source and receptor point. These factors contribute to the correction terms to the sound power level (L_{WA}) generated by the source.

$$L_p = L_{WA} + \text{Correction terms} \qquad (1)$$

Then, the sound pressure exposure (L_{AE}) at the receptor point is estimated based on the time during which the vehicle is in the corresponding section. Finally, equivalent continuous sound pressure level (A weighted) L_{AE} is estimated by considering the traffic flow at the corresponding section.

Explicit consideration of acceleration in noise estimation

The last application of DRONE [1] is based on the ASJ Model 1998, which differentiates the vehicle motion in terms of transient and steady running conditions. Vehicle power level for transient running condition is higher than those from steady running condition, which implicitly accounts for the increase in the noise level for accelerating vehicle. However, in an urban environment, the vehicles are in stop and go running condition. For example, at control intersection they decelerate, stop, and then accelerate. It has been found that the noise level for an accelerating vehicle can be significantly higher than those from cruising vehicle at the same speed [9]. Hence, for more precise and accurate estimation of noise in urban environment, it is necessary to consider the more detail driving condition.

In ASJ Model 1998, the sound power level from the source does not explicitly consider the acceleration of the vehicle. So, in order to capture the effect, a detailed relationship between vehicle sound power level and running conditions defined in terms of speed, and rate of change of speed (acceleration/deceleration) is used in DRONE. The required relationship is obtained from the latest research from Japan Automobile Research Institute [10]. The template for the relationship is shown in Table 1. The revised power levels are for acceleration (m s^{-2}) ranging from –1.5 to 1.5 with step of 0.25. The speed (km h^{-1}) range is from 0 to 120 km h^{-1}. with zero speed corresponding to an ideal vehicle. In DRONE, Table 1 is a lookup table for the relationship between vehicle type, its running condition, and corresponding sound power level. Note, in this paper we use the word "running condition categories"; these categories are related to column 1 of Table 1.

Table 1. Template for relationship between vehicle running conditions and A-weighted sound power levels

Category	Acceleration (m sec^{-2})	Speed (km h^{-1})	Passenger car	Small size	Medium size	Large size
			\multicolumn{4}{c}{Sound power level dB (A)}			
1	0	0–5	68.6	80.6	87.0	85.0
2	–1.5	5–10	75.3	80.8	82.5	85.1
3	–1.25	5–10	75.3	80.8	82.5	85.1
…	…	…	…	…	…	…
155	1	115–125	109.3	–	–	–

Unit noise database

The sound pressure level (Eq. 1) at the receptor point depends on the vehicle sound power level (source model) and the correction terms for the sound propagation (propagation model) from source to receptor point. The sound power level depends on the vehicle running condition obtained from the traffic simulation. However, the correction terms are independent of the traffic simulation and depend on the geographical conditions. In a dense urban network, the calculation of correction terms (for building attenuation, etc.) for a grid of large number of receptor points requires considerable computation time. For noise abatement policy evaluation we need to run the noise simulation for different transportation policies to evaluate and implement the cost-effective and efficient transportation policy [1]. During different simulation runs for the same area, only traffic conditions in the network changes. Hence, for a large study area, we propose to compute the correction terms for all the sources to respective receptor point only once. And then integrate this with different traffic simulations to generate noise maps in more computationally efficient manner.

The unit noise database is defined in terms of correction terms for each receptor point from different source sections and is visualized in terms of unit noise map. Unit noise is obtained from a hypothetical situation of unit flow on the whole network and for each vehicle running category. The corresponding contour map is unit noise contour map. The physical significance of the unit noise map is that it visualizes the correction terms for sound propagation (effect of buildings, ground, and distance) on the whole network.

Traffic simulation provides the flow for each vehicle category type on the network, which is then integrated with unit noise database to obtain the area-wide noise map as discussed in the following section.

Area-wide noise estimation

Traffic simulation output

The standard traffic simulation output from AVENUE is the flow and speed (mesoscopic characteristics) of vehicles with minimum resolution of 1 s. For considering the time dependent accelerating and deceleration of the vehicle (microscopic characteristics) individual vehicle trajectory is tracked during traffic simulation from AVENUE (Fig. 2) represents an individual vehicle trajectory from AVENUE and its corresponding time-dependent information for acceleration, deceleration, cruising, and ideal conditions.

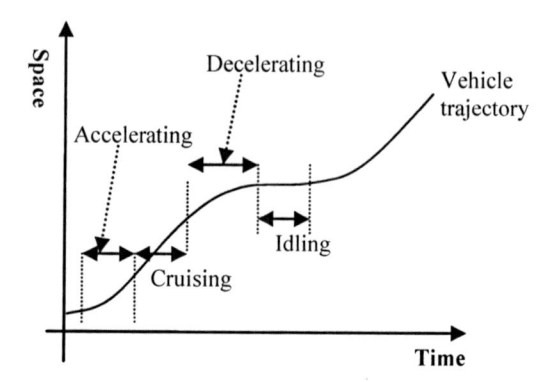

Fig. 2. Individual vehicle trajectory from traffic simulation

More precise data for noise estimation is obtained by transforming individual vehicle trajectory into its corresponding running conditions.

Individual vehicle trajectory is time-dependent space coordinates of the vehicle. These coordinates are related to the respective sections in the road network. For each section, vehicles are aggregated according to its type and running condition category (defined in column 1 of Table 1). Finally, area-wide noise is obtained by integrating the above flow with the pre-calculated unit noise database

The following section discusses about the implementation of the above mentioned procedure on Tsukuba city, Japan.

Implementation

Study area

The study area is 1.8 km × 2.3 km in Tsukuba city, Japan. Figure 3 shows the site of the study area where buildings are represented by blocks. Different colours represent different height of the buildings, which varies from 4 m to 58 m.

To generate noise map for the above mentioned area, we consider a larger area for traffic simulation, which is 2.8 km × 3.3 km as shown in Figure 3. Noise for a receptor point is contributed by all the sources in the network. Farther the source from the receptor point, lesser is the noise contribution. It has been observed that sources beyond 300 m have very little contribution in the noise at the receptor point and beyond 500 m the contribution is negligible 2. For the present application we consider 500 m as a buffer distance for noise estimation at a receptor point, i.e., all the sources within 500 m from the receptor point are considered for noise estimation. For estimating

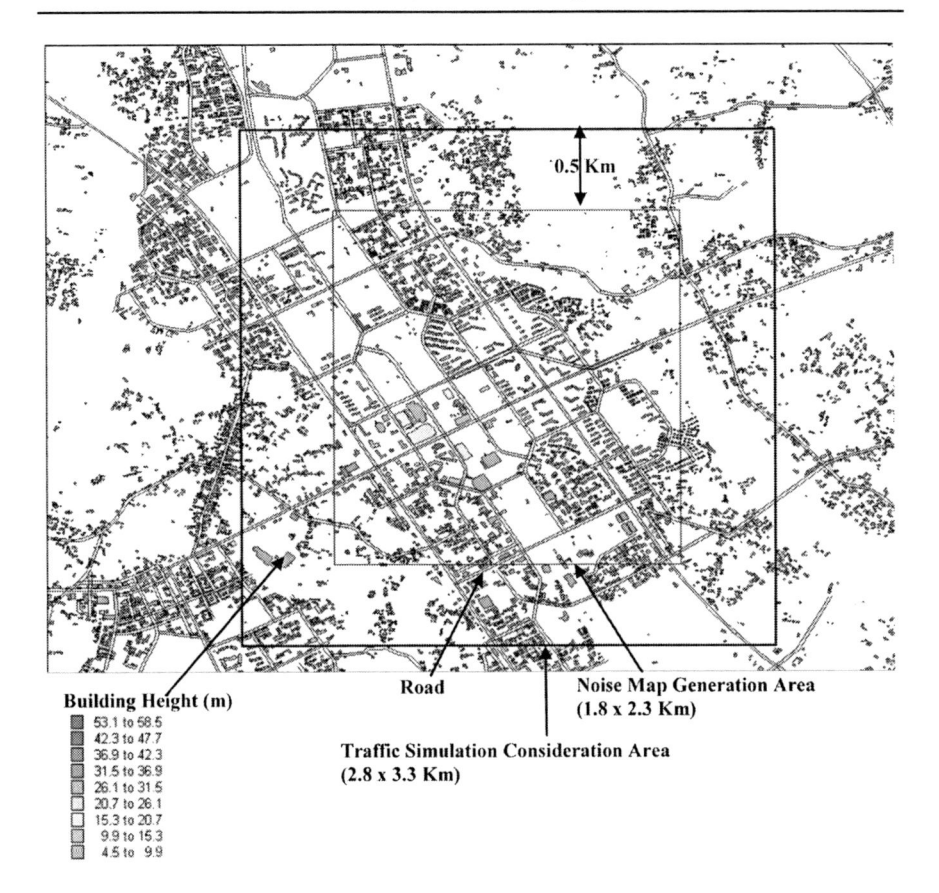

Building Height (m)
- 53.1 to 58.5
- 42.3 to 47.7
- 36.9 to 42.3
- 31.5 to 36.9
- 26.1 to 31.5
- 20.7 to 26.1
- 15.3 to 20.7
- 9.9 to 15.3
- 4.5 to 9.9

Road

Noise Map Generation Area
(1.8 x 2.3 Km)

Traffic Simulation Consideration Area
(2.8 x 3.3 Km)

Fig. 3. Study area

noise around boundary points of the study area, we need to consider sources beyond the study area; hence for traffic simulation we consider the area more than 500 m beyond the noise estimation area, which is 2.8 km × 3.3 km area as shown in Figure 3. The entire building infrastructure in the area is considered for noise simulation. Traffic simulation at the morning peak between 7 a.m. and 9 a.m. was executed to estimate area-wide traffic noise.

In the present application receptor points are spaced at 10 m grid, and the road network is divided into 10 m source sections.

Results

Figures 4–6 represent the results of the application of DRONE in the above mentioned area. First, unit noise map database is generated for the area, which contains the correction terms for each receptor point and source section. This is visualized by considering the unit flow on the network for each vehicle running category and corresponding unit noise map is presented in Figure 4. In the contour maps different colour represents different noise level. Blue, green, yellow, and red represents the increasing intensity of the noise level.

For unit noise map, we assume unit flow (for each vehicle running category) on the whole network, irrespective of the road type. So, each section of the network has a source corresponding to each vehicle running category. This can be seen as intense noise level along the road network in Figure 4. However, during traffic simulation we only have very low flow on the residential roads and high flow on the major roads. The integration of which provides the noise map for the area.

In Figure 5, traffic simulation result is also visualized in terms of hourly flow on the network. The flow on the residential roads is negligible. The diagonal running highway has the major flow, of the order of 1000 vehicles per hour. Noise contour map (Fig. 6) for the study area is finally obtained from the integration of the traffic simulation with unit noise database. The traffic simulation output indicates that the diagonal running highway has significant amount of flow, which is reflected in the noise map with more intense noise level along the same road. Similarly, residential roads have negligible flow and corresponding low noise levels in the noise contour map.

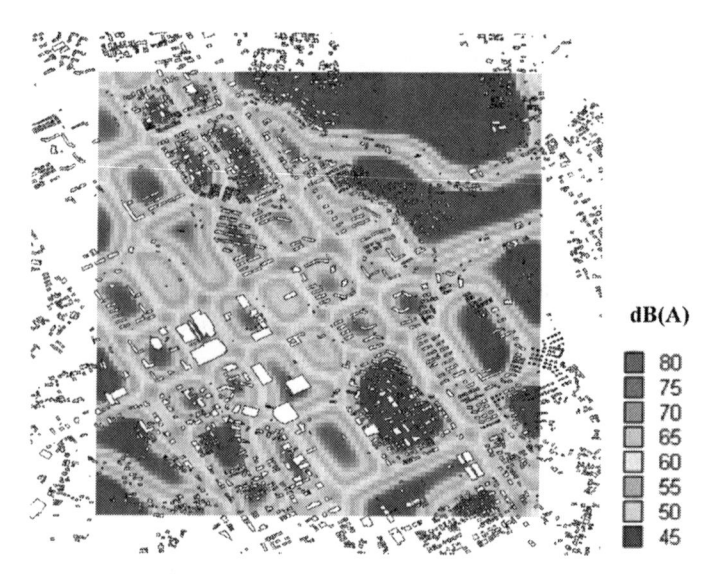

Fig. 4. Unit noise contour database

Large area noise evaluation   587

Fig. 5. Traffic simulation database

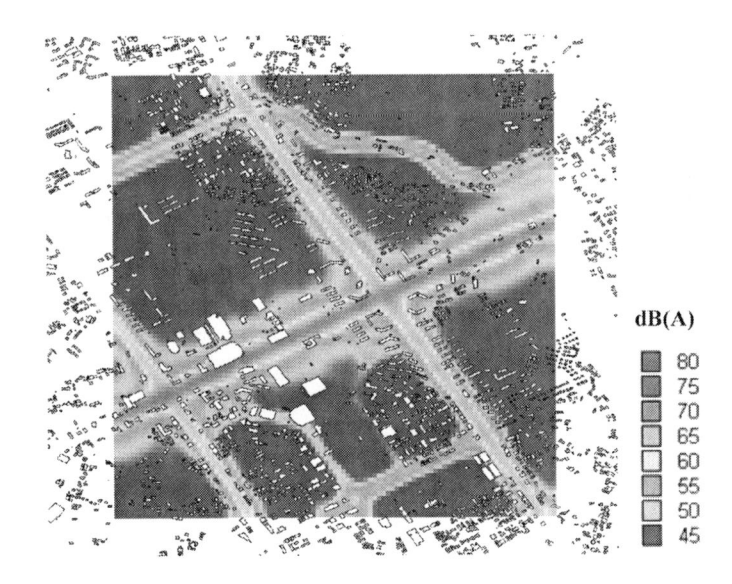

Fig. 6. Noise contour map

The unit noise database for the area is fixed and for further research it can be used for evaluating different transportation policies in computationally efficient manner.

Conclusions

A software tool DRONE has been developed by the integration of the dynamic output from a traffic simulation model with the noise estimation model. It can estimate comprehensive area-wide noise, considering network time-dependent traffic flow and buildings forming built-up area. DRONE considers the state-of-the-art relationship between vehicle sound power level and running conditions along with detailed vehicle trajectory for estimating noise in urban environment.

The calculation performance of DRONE is improved by evaluating the noise in two steps. The two-step procedure is applied on Tsukuba city, Japan and unit noise database is created, which can be used for testing different traffic management scenarios on the study area.

References

1. Bhaskar A, Chung E, Kuwahara M (2005) DRONE – a tool for urban traffic noise abatement policy evaluation. Proceedings of 5th Swiss transport research conference, Monte Verità/Ascona, Switzerland
2. Bhaskar A (2004) Areawide dynamic road traffic noise simulator. Master thesis, University of Tokyo, Japan
3. Bhaskar A, Chung E, Kuwahara M (2004) Integration of road traffic noise simulator (ASJ) and traffic simulation (AVENUE) for built-up area. Proceedings of 10th international conference on urban transport and the environment in the 21st century, Urban Transport X, pp 783–794
4. Horiguchi R et al. (1996) A network simulation model for impact studies of traffic management "Avenue Ver. 2". Proceedings of the 3rd annual world congress on intelligent transport systems, Orland, Florida
5. Oshino Y et al. (2000) Road traffic noise prediction model "ASJ MODEL – 1998" proposed by the acoustical society of Japan – Part 2: calculation model of sound power levels of road vehicle, Internoise 2000. Nice, France
6. Tachibana H (2000) Road traffic noise prediction model "ASJ MODEL – 1998" proposed by the acoustical society of Japan – Part 1: its structure and the flow of calculation, Internoise 2000. Nice, France
7. Yamamoto K et al. (2000) Road traffic noise prediction model "ASJ MODEL – 1998" proposed by the acoustical society of Japan – Part 3: Calculation model of sound propagation, Internoise 2000. Nice, France
8. Uesaka K et al. (2000) Prediction and evaluation methods for road traffic noise in built-up areas, inter-noise. Internoise 2000. Nice, France

9. Bhaskar A, Chung E, Dumont A-G (2006) Study of vehicle noise under different operating conditions. Proceedings of the 6th Swiss transport research conference, Monte Verità/Ascona, Switzerland
10. Oshino Y, Tachibana H (1998) Prediction of road traffic noise taking account of transient running conditions of vehicles, Internoise 1993.Leuven, Belgium, pp 629–632

Alliance for Global Sustainability Series

1. F. Moavenzadeh, K. Hanaki and P. Baccini (eds.): *Future Cities: Dynamics and Sustainability.* 2002 ISBN 1-4020-0540-7
2. L. Molina (ed.): *Air Quality in the Mexico Megacity: An Integrated Assessment.* 2002 ISBN 1-4020-0452-4
3. W. Wimmer and R. Züst: *ECODESIGN Pilot. Product-Investigation-, Learning- and Optimization-Tool for Sustainable Product Development with CD-ROM.* 2003 ISBN 1-4020-0965-8
4. B. Eliasson and Y. Lee (eds.): *Integrated Assessment of Sustainable Energy Systems in China. The China Technology Program. A Framework for Decision Support in the Electric Sector of Shandong Province.* 2003
  ISBN 1-4020-1198-9
5. M. Keiner, C. Zegras, W.A. Schmid and D. Salmerón (eds.): *From Understanding to Action. Sustainable Urban Development in Medium-Sized Cities in Africa and Latin America.* 2004 ISBN 1-4020-2879-2
6. W. Wimmer, R. Züst and K.M. Lee: *ECODESIGN Implementation. A Systematic Guidance on Integrating Environmental Considerations into Product Development.* 2004 ISBN 1-4020-3070-3
7. D.L. Goldblatt: *Sustainable Energy Consumption and Society.* Personal, Technological, or Social Change? 2005 ISBN 1-4020-3086-X
8. K.R. Polenske (ed.): *The Technology-Energy-Environment-Health (TEEH) Chain in China.* A Case Study of Cokemaking. 2006 ISBN 1-4020-3433-4
9. L. Glicksman and L. Leon (eds.): *Sustainable Urban Housing in China.* Principles and Case Studies for Low-Energy Design. 2007 ISBN 1-4020-5412-9
10. C. Pharino (ed.): *Sustainable Water Quality Management Policy.* The Role of Trading: The U.S. Experience. 2007 ISBN 1-4020-5862-2
11. N. Choucri, D. Mistree, F. Haghseta, T. Mehzer, W.R Baker and C.I Ortiz (eds.): *Mapping Sustainability.* Knowledge e-Networking and the value Chain. 2007
  ISBN 1-4020-6070-0
12. G.M Morrison and S. Rauch (eds.): *Highway and Urban Environment.* Proceedings of the 8th Highway and Urban Environment Symposium. 2007
  ISBN 978-1-4020-6009-0

springeronline.com

Printed in the United States
151604LV00004B/13/P